Basic Electrical Measurements

PRENTICE-HALL ELECTRICAL ENGINEERING SERIES
W. L. EVERITT, Ph.D., *Editor*

Melville B. Stout

PROFESSOR OF ELECTRICAL ENGINEERING
UNIVERSITY OF MICHIGAN

Basic Electrical Measurements

SECOND EDITION

Prentice-Hall, Inc.

Englewood Cliffs, N. J.

Current printing (last digit):

16   15   14   13   12   11   10   9   8

LIBRARY OF CONGRESS CATALOG CARD NUMBER 60-15089

PRINTED IN THE UNITED STATES OF AMERICA

05980-C

Preface

Measurements are becoming ever more important with the increase of technology. The complexity of devices has increased, placing greater dependence on the correct functioning of parts, frequently with the further requirement of interchangeability of sub-assemblies. Moreover, the accuracy required of the devices has increased most drastically. It is inevitable under these circumstances that many problems of accurate measurement should arise. In addition, electrical methods are being applied to measurements in many fields other than in electrical technology.

For all these reasons, and many more, a student of electrical science should have a thorough grounding in the science and art of measurements, beginning with basic processes. He should know how the standards are derived and propagated, and the accuracy that may be expected from them. An understanding of such matters is important both from the great need of good measurements and from the necessity for evaluating the extravagant claims of accuracy that are sometimes met.

In spite of easily demonstrable importance, the study of measurements is frequently not appreciated. Technical education must introduce many new developments, and in the process it is under great pressure to add various specialties that are represented as necessities. The specialties are new and attractive, and a persuasive argument can always be made for them. Consequently, measurements are in danger of being crowded out of an overfull program, along with other older subjects. This would be unfortunate, for a truly forward-looking program should include more, rather than less, in the field of electrical measurements.

A study of measuring processes is directly related to electrical field and circuit theory and serves as a review and integration of many branches. The electrical units originate in paper definitions, and are not available for use until they have been realized by a measuring process. The field of measurements thus includes activities ranging from the absolute determination of the units to the simplest kind of everyday measurement. Students must be thoroughly prepared for this broad field, but in a way not too exclusively pointed toward any one portion of it. Some understanding should be given of basic high-precision methods but not to the exclusion of less precise but more commonly used methods.

Selection of material for a text in this field is difficult. A book intended as an undergraduate text cannot present all equipment and arrangements, for to do so would make the book unwieldy and cause it to partake of the nature of a catalog or encyclopedia, rather than a text. An attempt has been made in this book to select the methods that are most important and most frequently used in present-day practice, and to omit special procedures and those now mainly of historical interest. It is realized that no two persons will agree completely on the choice of material, but it is felt essential to clearness and usability to limit the scope of the treatment.

Problems were introduced in the first edition, and receive continued emphasis in the revision. Experience has confirmed their value in the development of the subject, so many new examples have been added. A few problems are of a qualitative, or letter-equation nature; however, most are of a quantitative sort. There are several reasons for the emphasis on quantitative relationships. Basically, the whole subject of measurements is quantitative, so it is natural and desirable that the problems illustrate the relationships of the components of a measuring system. A numerical example gives a sense of the relative importance of the terms, which a general formula does not. A letter equation made up of various terms in R_1, R_2, L, and C, and so on, gives little or no idea of the importance of the factors in influencing the result. For example, a small capacitance shunting a resistor in alternating-current work may have negligible effect upon the magnitude of its impedance, but may introduce a phase-angle error of serious amount; such effects may be truly appreciated only by a numerical study. In many cases a measurement may be made in several ways, some of which may be more subject to disturbing effects than others, and only by quantitative study can these differences be evaluated. Quantitative analysis, as illustrated by problems, permits the designer to proportion a circuit to give the most favorable operating conditions.

A good problem is one that requires the maximum of thought and teaches the most in comparison with the labor involved, while avoiding a formula pattern. Problems of unbalanced bridges have been shunned in the past probably because they tended to degenerate into the substitution of numbers into a lengthy formula. This is not necessarily the case, however, as the use of Thevenin's theorem removes the need for a formula, simplifies computations, and teaches more about the operation of the circuit. Thevenin's method may be used in both d-c and a-c bridges, in potentiometers, and in other measuring circuits. If completely understood, it yields all needed information about a circuit and has advantages over other circuit theorems in simplicity of use and insight into the operations.

The student is urged to examine the importance of various terms that are met in the solution of problems. It is generally true in the solution of a scientific undertaking of appreciable size that the worker cannot keep all terms in the "equation," but must make judicious approximations to reduce the problem to manageable proportions. Similarly, in the computations met in measurements problems it is helpful to know which terms are important and which may be dropped in order to simplify the solution without the loss of essential accuracy. The development of "number sense" is desirable in any technical work, but is particularly valuable in measurements.

The second edition of *Basic Electrical Measurements* incorporates several new elements in addition to a general modernization. One item of great interest is the recent development of standards of capacitance based upon Lampard's theorem (chap. 13), and bridges to extend the measurements to larger values. Attention is given to microvolt potentiometers and to new voltage standards. The statistical study is broadened to apply not only to random errors met in measuring processes, but also to the planning of experimental programs, to obtain more reliable results in the presence of scatter of data. Hall elements are considered for magnetic measurements and as multiplying devices. Attention is given to resistance strain gauges, both as measuring devices and as examples of measuring circuits involving unbalanced bridges.

The placement of the measurements course in the curriculum is a matter of considerable interest. Knowledge of some simple measurements is needed in the early courses in circuits and machinery, but may well be taught as needed in these courses. A course in the science of measurements should come after the student has acquired some background and maturity in electrical matters and has considerable facility in handling circuits. Instruction in electrical measurements can be made more valuable and significant if given under these conditions.

The study of electrical indicating instruments is included in various places in the curriculum in different schools. Accordingly, a chapter on this subject is included at the end of the book and may be used if desired; it will not interfere with the program for those who have covered the material elsewhere.

The author is indebted to many persons for help in preparation of the book and the illustrations. While it is difficult both to express appreciation adequately and to avoid omissions, he must acknowledge some particular contributions. The author considers himself fortunate to have known and to have had the gracious help of Dr. Francis B. Silsbee of the National Bureau of Standards before his recent retirement. Other members of this fine organization who have helped on many occasions are Drs. J. L. Thomas, F. K. Harris, F. M. Defandorf and F. L. Hermach. Information and illustration material has been generously supplied by Mr. L. J. Lunas of Westinghouse, H. R. Brownell of Sensitive Research, Kenneth Brierley of Minneapolis-Honeywell, Rubicon Instruments, and J. B. Kelley of Physics Research Laboratories. Helpful suggestions were made by Professor R. F. Knox of the University of Washington and Dr. R. E. Frese.

M. B. S.

Contents

8 APPLICATIONS OF POTENTIOMETERS   241

9 ALTERNATING-CURRENT BRIDGES: 253
Principles; Basic Circuits

10 ADDITIONAL MEASURING CIRCUITS   290

11 MEASUREMENT OF MUTUAL INDUCTANCE   313

17 ELECTRICAL INDICATING INSTRUMENTS   466

APPENDICES

INDEX   **561**

CHAPTER 1

Development of the Electrical Units

1-1. Early History of Electrical Measurements

INTRODUCTI

Accurate electrical standards are as essential to the electrical engineer and the physicist as an accurate tape line to the surveyor or an accurate time source to the navigator. As in any other field, technical progress depends on units of measurement that are reliable and accurate and available to workers everywhere. Electrical scientists have gradually built up a system of such units to a refinement far removed from the crude beginnings.

The early stages of electrical science were marked, as would be expected, by qualitative observations of phenomena. Only later, as the nature of the actions became better understood, could quantitative relationships be derived. The first observations were made on electrification by rubbing, leading to the study of the effects of charges at rest, usually referred to as "static" electricity. From this came the first quantitative result, expressing the force between two charges in the form generally known as Coulomb's law.

Production of steady flow of electric current became possible as a result of the discovery of the voltaic pile by Volta in 1800. Succeeding years saw great activity in the use of electric currents in many ways, including electrochemical experiments and the production of electric arcs and of magnetic effects. The quantitative study of electrical circuits originated in 1827, when Ohm made known the relationship or "law" that now carries his name. Ohm expressed the fact that the magnitude of the current flowing in a circuit depends directly on the electrical force or pressure and inversely on a property of the circuit known as the resistance. Obviously, however, he did not have units of the size of our present ampere, volt, and ohm to measure these quantities. Relative current values could be determined by a compass needle and a coil (a tangent galvanometer, in other words), but the proportionality constant of such an instrument depends on its construction; hence no standard could be expected immediately as between different laboratories. A "unit" of resistance in the early days usually consisted of an arbitrary length of iron or copper wire of the size that the worker happened to have available to him.

It soon became evident that a universal system of units was needed to permit the interchange of information between different experimenters. Also, it was evident that the electrical units should not be a law unto themselves, but should be related to the mechanical units of length, energy, and so on. Gauss took the first step in this direction in 1832 when he measured the horizontal component of the earth's magnetic field in terms of length, mass, and time. Kohlraush, in 1849, measured resistance in terms of the mechanical units. Weber made an important contribution in 1851 in his proposal of a complete system of electrical units based on the mechanical units. His principles form the basis of our present system.

In 1861 the British Association for the Advancement of Science appointed a committee to establish resistance standards. This committee is notable not only for the pioneer work it performed but also for the famous men included on it—Maxwell, Joule, Lord Kelvin, and Wheatstone, to name a few. These men made a thorough study of the many possible systems of units and decided to use Weber's electromagnetic ideas based on the centimeter-gram-second system of mechanical units. They decided to make the practical unit of resistance equal to 10^9 absolute electromagnetic units and fostered research to establish the working standards in accordance with this decision. The British Association unit, announced in 1864 and widely circulated thereafter, consisted of a coil of platinum-silver alloy wire sealed in a container filled with paraffin, a construction that was expected to protect the wire and give a high degree of permanence. This unit was rather generally used as a standard for nearly two decades.

The practical unit of resistance, the ohm, has been equal, in theory, to 10^9 abs. e.m.u. ever since that time. The magnitude of the unit in actual use, however, has changed from time to time as continued improvement of equipment and measuring technique has made possible the closer realization of the theoretical value. The development of the electrical units is an interesting story in itself,[*] and important in the part it played in the development of electrical science as a whole. We may be inclined

---

[*]See references 1 and 8 for a more complete account than can be given here.

to think casually of the units as fixed and immutable objects which of necessity have always been as they are, but we find that they have unavoidably undergone an evolutionary process in company with other branches of science. [1]

1-2. Absolute Units and Concrete Standards

Before going further in the history of the electrical units, it seems advisable to point out the difference between a unit as defined theoretically and a standard used for actual measurement purposes. The defined units begin with the quantity that seems most convenient in establishing a complete and logical system. Measurement, on the other hand, is concerned in the main with setting up concrete standards that can be used in the laboratory to determine the quantities commonly used in practical work —resistance, current, voltage, etc. It is natural that a different sequence be followed in the two cases.

As an example, it has been customary in the absolute CGS electric system to begin by defining the unit charge (statcoulomb) as the charge that exerts a force of 1 dyne on an identical charge placed at a distance of 1 cm from it in vacuum. With this start, unit current is defined by the change of unit charge per second, and potential difference between points by the work done in carrying unit charge from one point to the other. Resistance is defined in terms of potential difference and current. It is obvious that this sequence is not a practical laboratory procedure for ordinary measurements of current and resistance. In the early days an analogous course was followed in defining the absolute electromagnetic units, beginning with the idea of the unit pole.

For measurement purposes it is necessary to have concrete working standards of resistance for determining the values of resistors which are brought to the laboratory for calibration. The duty of maintaining such standards is usually assigned to the national laboratories. It is desirable, of course, that standards set up for this purpose be constant and permanent to a high degree, and also that they possess values consistent with the units set up on the basis of the theoretical definitions. Bridging the gap between definitions and concrete standards is more difficult than is generally realized.

The term "concrete standard" may be used to cover either a reproducible standard or a semipermanent working standard. A reproducible standard is one that can be constructed from specified materials and dimensions. A working standard is a fixed unit that will have, as nearly as possible, a permanent value, although no standard is absolutely permanent and must be checked at intervals. A reproducible standard may be used as a working standard, but in most cases is not convenient for such use.

One purpose of a reproducible standard is to make possible periodic checks of working standards. Another purpose is to enable laboratories in different parts of the world to have the same standard. However, it has been found in some cases that the production of a reproducible standard to the required accuracy is as difficult, or more difficult, than the basic absolute measurement.

1-3. Development of the Resistance Standard

The British Association ohm was believed at the time of its introduction in 1864 to be an accurate realization of the intended 10^9 abs. e.m.u. In 1878, however, Rowland announced that the unit was in error by nearly 1.5 per cent. This touched off renewed activity in the determination of the units. At an international congress in Paris in 1884 it was decided to base the ohm on a reproducible standard consisting of a mercury column 106 cm long and of 1-mm^2 cross section, measured at 0°C. Mercury was selected because it is possible to produce it in very pure form by repeated distillation. The unit thus defined was called the "legal ohm."

The Chicago Congress in 1893 changed the length to 106.3 cm and specified the column by weight, 14.4521 grams, rather than by cross-sectional area. This unit was generally adopted, and is called the "international ohm." The London Conference in 1908 specified the length as 106.300 cm in order to make it definite, although realizing that the value was not known surely to that degree of exactness.

The "mercury ohm" was the legal basis for resistance measurements in this and other countries for many years. It was adopted because it seemed at that time to offer easy reproducibility for laboratories all over the world. Mercury has the advantages of freedom from internal stress and of being purifiable to a high degree. It is in both respects superior to iron or copper wire, the resistance of which is affected considerably by small amounts of impurities and by internal stresses. In actual use, however, the mercury ohm proved a disappointment, due to the great difficulty in obtaining accurate results. There were minor troubles due to the large temperature coefficient of resistance of mercury, to contamination of the mercury by the electrodes, and to the end-terminal arrangements. The greatest difficulty lay in the manufacture of glass tubing that was straight and of uniform bore to the required degree of accuracy. The internal diameter of the tubing was measured at short intervals, and after selection of the better pieces, a correction was computed to take into account the departure from uniformity of bore. In addition, it was found that an undulation in the axis of the tube of an amount too small to be detected by eye could produce an increase in the resistance of the mercury column of 30 or more parts per million. On account of the difficulties and the great amount of work involved, only a few truly reliable mercury-ohm determinations have ever been made.

It had been expected that resistance values would be carried along by resistance-wire standards, with periodic checks by mercury-ohm determinations. Actually, little use has been made of the mercury ohm, although it was the ultimate standard of resistance by decision of the London Conference in 1908. The International Technical Committee held a meeting in Washington in 1910, following which the members carried back to their respective countries standards calibrated in terms of the mean of the British and German units of that date. (These countries had just completed extensive mercury-ohm determinations.) The various countries for some years used standards based on these supposedly identical 1910 units, but subsequent comparisons indicated that there had been a drift in values between laboratories. In fact, the Germans decreased their ohm by 33 ppm in 1927 as comparisons with other laboratories indicated a discrepancy. They also made a mercury-ohm determination at this time as a further check.

During this period increasing study was given by several workers to the absolute determination of the ohm. "Absolute" refers to measurement in terms of the basic units, length, mass, and time. Several methods have been devised for this purpose. All of them make use of a rotating conductor or commutator whose speed must be known, thus bringing in the time element. Also, they all use either a self or a mutual inductance whose value is assigned by computations involving the dimensions.* The absolute-ohm determinations were so improved over a period of years that they were more reliable than the mercury-ohm measurements and no more difficult to perform. Accordingly, there was an increasing tendency to depend on absolute methods and to discard the mercury ohm.

Absolute measurements, beginning about 1912, indicated that the value selected for the international ohm was greater than the intended 10^9 abs. e.m.u. Stated in terms of the mercury ohm, it appeared that the length of the column should have been about 106.245 cm, instead of 106.300 cm, and hence the international ohm was greater than the absolute ohm by about 0.05 per cent. This difference became of more and more importance as measurements in general were made with greater accuracy. Various relationships and formulas were derived on a theoretical basis (absolute units) and accordingly required a correction factor when the international units were used in experimental work. By 1930 there was considerable sentiment in favor of changing the measuring units to bring them in agreement with the absolute units. It was felt that a change of standards, though troublesome, would become increasingly difficult if long postponed.

---

*See chap. 12 for an alternative method based on a capacitive standard.

The decade 1930–40 saw a great deal of activity in the way of international committee meetings and in new absolute measurements at the National Bureau of Standards and the other national laboratories. The International Committee on Weights and Measures agreed to change the value of the ohm, along with the other units, the new values being scheduled to go into use on January 1, 1940. The outbreak of the war prevented final comparisons between the various national laboratories and, consequently, the date for the new standards was postponed. They finally went into effect January 1, 1948. Table 1 at the end of the chapter shows the new values in comparison with the old.

The use of the mercury ohm is now definitely abandoned. In the future, resistances will be certified at the national laboratories in terms of wire-resistance standards. The standards, in turn, will be checked periodically by absolute measurements as guard against the possibility of gradual drift in value.

1-4. Standards for the Determination of Current

Early measurements of current were made by the tangent galva-nometer in terms of the earth's magnetic field. This proved to be un-satisfactory for precision purposes as a result of the variability of the earth's field and because of the disturbances set up by neighboring circuits as electricity came into wider use for power purposes. Absolute measure-ments of current were made by the force between conductors by a current balance, or electrodynamometer, but it was desired to adopt a reproducible standard that could be set up independently at any place. It was decided to define the ampere for measurement purposes by the electrolytic depo-sition of a metal from a solution, and selection was made of silver as deposited from a standard silver nitrate solution. Before 1880 results were reliable to about 1 per cent, but by 1900 this had been improved to 0.1 per cent. The accuracy was further improved by refinements in the form and mounting of the electrodes, method of weighing, and so forth. The international ampere was defined by the London Conference as the current that deposits silver at the rate of 0.00111800 grams/sec from a standard silver nitrate solution.

The electrolytic cell, called a silver voltameter, consists of a silver anode, silver nitrate solution, and a platinum cup as the cathode within which the silver deposit is made. Great care must be taken to prevent stray particles and the "slime" that forms at the anode from reaching the cathode, where it would cause a false indication of weight. In one form of voltameter the anode is wrapped with filter paper to prevent the slime from reaching the cathode; in another the anode is surrounded by an unglazed porcelain pot; and in still another a shallow glass cup is placed below the anode. It was found that any one form of voltameter could be made to yield regular results, but that consistent differences existed between one form and another. For example, the form using filter paper gave deposits that were heavier by about 150 ppm. Some of the sources of error were later tracked down, and operating procedures were formulated but never officially adopted. The silver voltameter is not used today, due to increased accuracy in the absolute measurement of current and to the proved reliability of standard cells.

Definition of the ampere by electrolytic action is strictly a calibration procedure to be performed at one of the national standards laboratories and is far from being a working method for ordinary purposes. It requires careful control in all respects, including the solution, weighing of the electrode, and holding current constant for a considerable period to minimize the error of timing. The practical working standard of current is achieved in a different manner. It is interesting to note in this connection that there was a great deal of debate at the London Conference regarding which two of the three units — volts, amperes, and ohms — should be defined by separate calibration procedures. It is necessary (and desirable) to establish only two by separate means as the three quantities are related by Ohm's law. There was general agreement on the ohm as one of the quantities to be defined basically, but there was great debate on the choice of the other unit. Some members of the Conference wanted to define the volt in terms of the emf of the standard cell and then derive the ampere from the ohm and volt. However, the other group argued the (apparent) simplicity of the silver voltameter as the standard of current, and finally won out. Actually, the standard cell is now used both as a working standard and as the basis for the volt in the national standardizing laboratories.

Absolute measurements of the ampere are made by a current-balance arrangement in which the force between a movable coil and a set of coaxial fixed coils is measured by a weighing balance. A formula for the force between the coils can be derived from theoretical considerations in terms of the current and the dimensions of the coils. The experimental work thus involves the three fundamental units mass, length, and time, as the balancing mass must be translated into force units by the acceleration of gravity, which involves time units. At the time the balancing process is carried out, the voltage is measured across a known resistor in series with the coils of the current balance. The voltage drop is determined with reference to a standard cell, thus establishing the emf of the standard cell in terms of the absolute ampere and the absolute ohm. The standard cell is used to carry forward the value of the volt between the times of absolute-ampere determinations.

1-5. Standards of Electromotive Force

The London Conference, as related above, defined resistance and current in terms of concrete standards. They then defined the unit of emf in the following way:

An international volt is the potential difference produced by an international ampere flowing through an international ohm.

The idea of a voltaic cell of some sort as a concrete standard of emf, however, had occurred to various people at a much earlier time. The Daniell cell was first used for this purpose but was soon abandoned because of its short life. Latimer Clark in 1873 introduced a cell having zinc and zinc amalgam as the electrodes and zinc sulphate as its solution. It showed a much improved constancy of emf but frequently gave trouble by cracking of the glass around the lead-in wires. In 1893 Weston produced a new cell, similar to Clark's, except for replacement of the zinc by cadmium. This change eliminated many of the previous troubles and produced a cell that has made a remarkable record for constancy of emf over long periods of time.

After some preliminary values for the new cell and upon recommendation of the London Conference, an International Technical Committee met in Washington during 1910 and made extensive measurements on the cadmium cell. As a result of their work the value 1.01830 v was assigned to it, and this value stayed in force until the change to the absolute values of the units on January 1, 1948. The National Bureau of Standards and the other national laboratories keep large banks of these cells, each laboratory making frequent comparisons among the cells of its own bank and occasionally comparisons between laboratories. Any cell that shows a serious drift of emf in comparison with the average is removed from the group. The old cells are also compared with new cells that are made up from time to time.

An idea of the consistency of these cells may be obtained from the following quotations from reference 8 (p. 9):

As an indication of the mutual consistency of the neutral, saturated cadmium cells, data taken between 1919 and 1933 on the primary reference group showed that 18 of these cells had changed relative to their mean by less than 15 μv (microvolts). Nine cells showed a net increase, while the other nine had decreased, the average net change during the 14 years, taken without regard to sign, being ± 5.0 μv. This corresponds to an average drift rate of ± 0.25 μv per year, the greatest mean drift rate being 1 μv per year.

Of the 18 neutral cells in the group in 1947, 13 had been in the primary group since 1914 and 5 since 1908.

Theoretically, the procedure of maintaining a unit of measurement solely by the use of a large group of standards is subject to the fundamental limitation that a simultaneous progressive drift of all the standards cannot be detected by the intercomparisons and will result in an equally rapid drift of the unit. It is therefore important to seek corroborative evidence of the constancy of the unit.

One such line of evidence derives from the construction of new groups of cells from freshly prepared batches of pure materials. One example of this is a batch of six cells constructed in 1925, the mean value of which was found to be greater in terms of the 1925 unit by only 12 μv than the mean of the large number of cells set up 15 years earlier by the International Technical Committee.

1-6. MKS Units and the Practical Units

International agreement for the change from the CGS to the MKS system of units occurred in the same period (1930–40) that witnessed a great deal of the discussion of the change in the size of the volt, ampere, and ohm units. Also, the basic electrical units of the MKS system are the same volt, ampere, and ohm. For these reasons the change from CGS to MKS and the change of the size of the working units have tended to be connected and confused. Actually, these are two separate and distinct problems. The size of the ohm (and the other units) needed to be changed in the interests of increased accuracy, regardless of whether the ohm is regarded as the basic unit of resistance, as in the MKS system, or whether it is considered a decimal multiple of the basic unit, as in the CGS system. It is a convenient feature of the MKS system that its basic units are also the usual working units, but this is apart from the exact size of the unit.

1-7. Accuracy of the Present Standards

The continued improvement of the electrical standards, as outlined briefly in this chapter, presents a picture in miniature of the development of electrical science as a whole during the past century. However, it is more than a mirror reflecting that development, for it has contributed greatly to continued progress in all branches. Accurate standards are essential to accurate measurements in the first place, and, in addition, many methods and techniques devised in the determination of units have proved useful in other measurement applications.

The National Bureau of Standards is entrusted with the duty of maintaining and improving the basic standards in use in this country and has done work that makes it outstanding nationally and internationally. The workers there have developed methods of making both absolute measurements and precise comparison measurements and have devised highly reliable and permanent standards of resistance, inductance, and capacitance, as well as standard cells. The Bureau has set for itself the goal of establishing the basic electrical standards accurate to one part in a million and has made impressive progress in that direction. Time is maintained to better than one part in ten million by carefully regulated quartz-crystal oscillators, and time signals and standard frequencies are broadcast by radio. Standards of resistance, as well as of emf, can be compared to better than a part in a million, and absolute measurements of resistance and current can now be made to a few parts in a million. This may seem an extreme degree of precision; however, high precision is needed in some cases, and the general requirements over the years have been more exacting, so the Bureau endeavors to stay well in advance of possible needs.

1-8. Summary

The electrical units as applied in concrete form for measuring purposes have evolved from crude beginnings to the present high degree of refinement. The intended size of the units as adopted in 1864 was the same as our present-day units. However, only when measuring methods were improved was it possible to realize the units in practice to a good degree of accuracy. Further progress will no doubt be made, but future changes will necessarily be of small amount. The "International" units were formulated in the years 1893 to 1908 and were the best that could be produced at that time. They continued in use until superseded by the new units in 1948.

One very practical reason why electrical experimenters should know the history of the units is the fact that electrical measuring components, such as resistors, decade boxes, bridges, inductors, and capacitors, built before 1948, were calibrated in terms of the international units. Similar equipment since 1948 is calibrated in the new absolute units. It is probable that many laboratories will have a mixture of the two sorts for many years to come, so it is essential that workers understand the difference. In the case of ohms, henries, and farads the change amounts to about 0.05 per cent, in the volt 0.033 per cent, so the difference is appreciable in precision work. New equipment being made at present has the letters "abs" inscribed on the name plate to permit identification.

TABLE 1

NEW VALUES ADOPTED JANUARY 1, 1948, STATED IN TERMS
OF THE PREVIOUS INTERNATIONAL VALUES

1 international ohm	= 1.000495 abs. ohm
1 " volt	= 1.000330 abs. volt
1 " ampere	= 0.999835 abs. ampere
1 " coulomb	= 0.999835 abs. coulomb
1 " henry	= 1.000495 abs. henry
1 " farad	= 0.999505 abs. farad
1 " watt	= 1.000165 abs. watt
1 " joule	= 1.000165 abs. joule

REFERENCES

1. Curtis, Harvey L., *Electrical Measurements*, New York: McGraw-Hill Book Co., Inc., 1937.

2. *British Association Reports on Electric Standards*, Cambridge University Press, 1913.

3. Brooks, H. B., "The Unit of Electrical Resistance; Past History and Impending Change," *Trans. A.I.E.E.*, **50**, 4, 1318 (1931).

4. Curtis, H. L., C. Moon, and C. M. Sparks, "An Absolute Determination of the Ohm," *J. Research N.B.S.*, **16,** 1 (1936).

5. Curtis, H. L., C. Moon, and C. M. Sparks, "A Determination of the Absolute Ohm, Using an Improved Self-Inductor," *J. Research N.B.S.*, **21,** 375 (1938). RP1137.

6. Curtis, H. L. and R. W. Curtis, "An Absolute Determination of the Ampere," *J. Research N.B.S.*, **12,** 665 (1934).

7. Curtis, H. L., R. W. Curtis, and C. L. Critchfield, "An Absolute Determination of the Ampere, Using Improved Coils," *J. Research N.B.S.*, **22,** 485 (1939). RP1200.

8. Silsbee, F. B., "Establishment and Maintenance of the Electrical Units," *Circular N.B.S.* 475, 1949.

CHAPTER 2

Experiments and Statistical Analysis

2-1. Performance of Experiments

The term "experiment" is quite broad, in that it may cover anything from an afternoon's set of measurements to an extended research project, so our remarks at this stage must necessarily be of very general nature. All experiments, however, have many points in common regardless of the size of the undertaking. An essential component in all good measurements is a careful and inquiring attitude of mind in approaching a problem. A thorough understanding of basic principles is necessary to visualize an approach, or preferably several approaches, to the task in hand. Imagination, ingenuity, and creativity lead to solutions beyond the range of ordinary routine processes. The ability to visualize possible sources of error and disturbing factors that may interfere with the attainment of satisfactory results is also needed. Otherwise the outcome may range from second-class, to misleading, to the actually erroneous. Measurements combine both art and science; a knowledge of scientific principles is not enough, for skill in perfecting the techniques is essential for producing results of high quality.

Measurements involve the use of instruments of one sort or another. An instrument, in a general way, is a physical means of obtaining measurements of greater refinement than are possible to the unaided human faculties, or measurements which human faculties would be unable to sense and measure — the latter is particularly true in the field of electrical science. Many wonderful instruments have been devised over the years; our work in electrical measurements involves understanding and using these instruments, making new and better applications of them, and visualizing and perfecting new methods and instruments for a constantly widening technology.

The first step in a measurement problem is the choice of the method to be used. This choice, made in consideration of available equipment, must take into account the required accuracy and also conditions such as speed and convenience. Speed is of minor concern in precise measurements, but is essential if large numbers of routine tests are to be made; in the latter case a quite different procedure may be indicated, and a considerable expenditure of money for new equipment may be justified. In this book we shall study basic methods and apply them mainly to laboratory measurements. The application to special conditions and the use of electrical methods to measure nonelectrical quantities are outgrowths of the basic ideas and often provide the measurements engineer with new and fascinating problems.

Not only must the general method be considered, but also the details of procedure must be carefully planned. The same method and similar equipment may be used by two workers, yet one may obtain much better results, the difference being in the care with which small disturbing factors are eliminated. It is frequently possible to plan the procedure so that shortcomings in the equipment or stray effects will balance out in the final result. This sort of skill is something over and above direct manipulative accuracy, which is of course one essential to accurate results.

Frequently it is possible to make the required measurements by more than one method, and then it is necessary to consider the relative merits of these methods for the problem at hand. One way may be desirable in some respects, but less desirable in others, so that a choice must be made in view of the conditions of the test. If the experiment carries very important consequences, it is advisable to use two methods, if available, for verification; if the results are in reasonably close agreement, additional force is given to the conclusions. If the results of the two methods do not agree closely (as often happens), further study must be made to find the reason. This necessity may appear a nuisance at the time, but it frequently leads to considerable enlightenment about methods and equipment, in addition to settling the immediate problem regarding a particular magnitude.

Experiments involving an integrated series of tests require particularly careful planning, so that the outcome will not be disappointing after a large expenditure of time and effort. The large program is likely to involve a study of many factors, and the effect each factor has on the whole. Unless a wise program is mapped out in advance, it may be found at the conclusion that the data do not give the desired information. Self-checking features can frequently be built into the program, requiring possibly the help of a trained statistician. This is a very important aspect of an extensive investigation, and we shall return to it after we have acquired some background in statistical terms.

2-2. The Record of Experimental Work

The record made in the course of experimental work is extremely important. It must be written and it must be complete, since human memory is notoriously unreliable. Furthermore, the record should be made with the attitude that it can be understood by someone other than the one who made the measurements. The beginning should include a clear statement of what is being investigated, and the purpose of the test. Clearness and completeness of the record are, of course, essential, so that all needed material is included, and in such form that it will not be misunderstood.

Inexperienced experimenters too often make the mistake of jotting down a few measured values, possibly on a scrap of paper, trusting to memory to fill in details of equipment and circuit at a later time. This practice is certain to lead to mistakes and to introduce uncertainties that may even make the whole experiment worthless. Memory is a poor reliance because details that seem perfectly clear and obvious today have a way of being completely lost in a week or a month. The only safe way is to make a complete record of circuit, equipment, and data; better to record an occasional superfluous item than to discover at a later date that some important matter has been omitted.

Rules and conditions differ so much from laboratory to laboratory that our remarks must necessarily be very general. In commercial laboratories, the original data must generally be entered in formal data books. Correction of a faulty bit of information must usually be made by overruling, and not by erasure, since it is impossible at a later date to determine when and from what a change was made. The rules reflect not only good data-taking practices, but also the need for using the original record in patent applications or legal actions. We may, however, list a few general principles that are worthy of attention. The data record should include, in addition to the title and purpose of the experiment, the following items:

(1) A clear and complete circuit diagram, identifying all equipment and showing the connections between parts exactly as in the actual circuit. If a certain resistor is referred to as R_1 in the data, be sure it is marked R_1 on the diagram.

(2) A list of all calibrated equipment by type and serial number (or laboratory number). This information makes it possible to check back at a later time if discrepancies appear when the data are worked up or if a question arises as to the accuracy of some element. It is not necessary to list serial numbers for equipment whose calibration does not affect

results, as for example, a rheostat that is used merely to regulate current and does not enter into calculations.

(3) A neat tabulation of data. The columns should be headed with the name or symbol of the quantity being measured and the units used — as volts or millivolts, for example. Sometimes there are special conditions that should be entered in a column headed "Remarks." If one or more columns in the data table are used to enter computed results, these columns should be marked "computed," so that there shall be no confusion as to which figures represent readings and which computed values.

(4) The method of making computations should be indicated on the data sheet or on an accompanying sheet. This should give the formulas used and, where desirable, a sample computation. Such material is a help to a stranger, or to the observer himself at a later time.

2-3. The Report

The purpose of the report is to tell somebody what was done and why, what information was obtained, and the significance of the results. To this end, it should give a formal statement of the objective and the method used in obtaining the desired information. The form of the report will depend on the length and scope of the investigation, and on the uses to which the report will be put, so we can speak only in very general terms. Diagrams and descriptions of the equipment will usually be needed, plus a tabulation of at least the principal data and calculations made from the data. Graphs are useful in many cases, sometimes for purposes of calculation, and generally for illustrating the trend or relationships disclosed by the measurements and calculations. One important aspect of the data is the reliance that can be placed upon it — that is, its accuracy.

Finally, the most important part of the report is the discussion of the results and the presentation of the conclusions drawn as a result of the experiment. The test program was instituted to answer certain questions, so the value of the experiment rests (usually) on the extent to which the objectives have been attained. Sometimes the results of an experiment are negative, but that can be important information. Occasionally the experiment may not succeed in the desired direction, but may point to a new and different line of attack. In any event, the discussion should point out the things that have been learned in the investigation. Sometimes it is appropriate to point out desirable extensions to the test program.

The preparation of graphs might well be discussed at considerable length, for it is an important subject in which students are frequently weak and in which they do not receive adequate training. A poorly drawn curve may be anything from a total loss to an actual misrepresentation. Some of the high spots are touched in the following paragraphs:

(1) A graph should be given a title showing what it represents and as much as possible of the conditions.

(2) Mark the axes to show what they represent and what the units are, and mark the numbers on the axes.

(3) Select a convenient and readable scale, so that interpolation may be made easily. Do not use an inconvenient number of units, such as 3 or 7, etc., for a division. Preferably, use 1, 2, or 5.

(4) Select a scale so that the graph is spread to a convenient and readable extent, but not overspread beyond the accuracy of the data. This matter requires judgment.

(5) The position of experimental points should be shown, in general, by some type of symbol. In case several curves are drawn on the same set of axes, different symbols should be used for each curve and explained by a key.

(6) Curves are drawn through experimental points, in most cases, to "smooth" the data and show the general law of variation, free of experimental errors. The extent to which the curve should smooth, or average, the data depends upon the nature of the relationship being shown (i.e., whether a smooth law of variation is to be expected). General rules cannot be given, and judgment must be used, based on a study of the case. (Exception — instrument correction curves are usually drawn as straight lines, point-to-point.)

(7) If several curves are shown on the same graph, they should be identified in an easily understandable manner. If experimental points are being compared with a theoretical curve, the curve should be so marked.

(8) It is customary in the calibration of instruments to estimate readings to one-tenth of the smallest division if the divisions are of a size to permit such estimation, and if the reliability is such that this degree of refinement has some meaning. (This is sometimes called the "least count," or "L.C.")

2-4. Accuracy and Precision

The terms "accuracy" and "precision" are met in everyday usage, frequently with little distinction as to their meaning. In fact, a casual inspection of a dictionary might lend support to the idea that the terms are synonymous, although this is not strictly true even in their common use. In measurements work the terms denote quite different ideas.

The following definitions were found in one dictionary:

"Precise — Sharply or clearly defined; strictly accurate; exact."

"Accurate — Conforming exactly to truth or to a standard."

Note, although "precise" is linked to "accurate" in one part of the definition, that there is another meaning given, carrying a considerably different idea. It is in this other sense that the term is used in measurement work. So, let us extract the following short definitions that will distinguish between the terms as used in measurements:

Precise — Sharply or clearly defined

Accurate — Conforming to truth

As an example of the difference in meaning of the two terms, suppose that we have a machinist's micrometer, normal in every respect except that the stationary member, or anvil, has been displaced from its correct position. This micrometer is as *precise* as ever, as we can take readings to a thousandth of an inch, and the readings are consistent and "clearly defined." The readings of dimensions taken with this micrometer, however, are not accurate, since they do not conform to truth.

We may compare two decade resistor boxes, each with four dials, giving resistance increments of 1, 10, 100, and 1000 ohms per step. Suppose one is a cheap, low-quality box, guaranteed to be within one per cent, while the other is a high-grade box guaranteed to be within 0.1 per cent. Both boxes may be set to the *same precision*, that is, to any value up to 10,000 ohms by 1-ohm steps. The *accuracies*, however, are very different.

Or, we may compare two voltmeters of the same make and model. They have carefully ruled scales, knife-edge pointers, and probably mirror-backed scales to help avoid parallax errors. The two instruments may be read to the same *precision*. However, if the series resistor in one meter becomes defective, its readings may be considerably in error. The *accuracies* of the two instruments may be quite different, and whether this is the case or not can be determined only by comparison with a standard. A beginner in the use of measuring devices is inclined all too often to accept the indication at face value, without considering that the accuracy may not match the precision of reading. Skepticism as to accuracy, and promised accuracy, is a very necessary ingredient in good measurement work.

If a magnitude is to be determined with accuracy to a required number of digits, it is necessary that the measuring equipment have precision of this order. That is, *precision is a necessary prerequisite to accuracy*, but as we have discussed above, *precision does not guarantee accuracy*. Accuracy is a matter of careful measurement in terms of an accurately known standard. Precision is essential in the detection of possible inaccuracy, as in comparative measurements of a quantity by two methods, but does not insure accuracy.

We say that a set of readings shows precision if the results agree among themselves. Agreement is, however, not a guarantee of accuracy, as there may be some systematic disturbing effect that causes all the values to be in error. For this reason it is highly desirable in important measurements to make an independent set of observations by a second method that will probably not be subject to the same systematic errors. Where this is not possible, we should at any rate make a careful study of our method to try to discover and eliminate any systematic disturbing factor.

2-5. Indications of Precision — Significant Figures

Significant figures are those that convey actual information regarding the magnitude of a quantity. Thus, if we specify a resistance as 105 ohms, we express our belief that the value is closer to 105 ohms than it is to 104 or 106, and we have three significant figures. If we write 105.0 ohms, we have four significant figures and we give more information, for we now claim that the resistance is closer to this value than to 104.9 or to 105.1 ohms, and we should not add the additional figure unless we intend to so limit the result. A resistance of 105 ohms might also be written as 0.000105 megohm. This is also three significant figures, as the zeros between the decimal point and the 105 are present on account of the size of the unit used, but give no greater precision to the result.

One situation that may cause some uncertainty is encountered when we have a large number with some places between the last significant figure and the decimal point. Thus, strictly, 52,300 means that the true value lies between 52,299 and 52,301, which is five significant figures. If the information is good to only three figures we could use a different style of type, as 52,3*00*, but this is not general practice. A technically correct notation makes use of powers of ten, as 5.23×10^4, but this may appear awkward and unduly technical if the number is used in such a way as to indicate rounded values. Thus, reference to the population of a city as 2,000,000 would be interpreted automatically as an approximate number, or reference to the velocity of light as 186,000 miles per second would cause no misunderstanding to anyone with a technical background. Uncertainty caused by zeros to the left of the decimal point is usually resolved by the context; however, this matter should be watched more carefully in technical writing than in ordinary usage.

It is customary in measurement work to report a result with all the digits of which we are sure, and a final digit which is believed nearest to the true value (usually implying that the result is good to ±1 in the last place, unless a qualifying statement is made of greater uncertainty). It is obviously useless to give more than one doubtful figure; that is, to put down a particular value for the *tenths* of a unit would be useless if we do not know the number of *units* definitely. Sometimes, to express doubt regarding the last place, we use small type, depressed. Thus, "5.1₇ amp" indicates that we know the current to better than one tenth of an ampere, that we are not sure of the hundredths but believe the value is close to seven. A similar but more definite way of expressing doubt is, "5.17 amp within ±0.02 amp." or "5.17 amp within ±0.01 amp."

The number of significant figures in a quantity is one measure of precision, though not as definite as a percentage statement. Thus 101 ohms covers values between 100.5 and 101.5, which is ±0.5 per cent, or a total range of doubt of 1 per cent. The value of 999 ohms also has three significant figures and a range of doubt of one ohm, but as a percentage the range is now 0.1 per cent. Thus three significant figures may cover uncertainties from 1 to 0.1 per cent, and so is indefinite as a measure of precision. It is, however, a convenient approximate index, and is used to a considerable extent.

Superfluous figures are sometimes allowed to accumulate in computational processes involving addition, subtraction, multiplication, or division. For example, we may have two resistances added in series, with values

$$R_1 = 24.3 \text{ ohms},$$
$$R_2 = 0.516 \text{ "}$$

$$\text{Sum} = 24.8 \text{ "}$$

(This is not the usual notation — the italic figures are used here to help us keep in mind which figures may be in doubt.) Note that both resistances have three significant figures and the result is known to the same number. There would be no value whatever in expressing the sum as 24.816 ohms, as we cannot be sure even of the tenths of an ohm.

Suppose in multiplication that we have $I = 2.56$ amp and $R = 45.73$ ohms, and find $E = IR$ by longhand multiplication:

$$
\begin{array}{r}
45.73 \\
2.56 \\
\hline
2\ 7438 \\
22\ 865 \\
91\ 46 \\
\hline
117.0688
\end{array}
$$

Here the original doubtful figures are written in italics, and also figures that result from their use in multiplication and are thereby placed in doubt. The answer should be written as 117 v, as 7 is the first doubtful figure. It would obviously be absurd to write as answer the entire product shown above.

It is interesting to note in this illustration that I has three significant figures and R four, and the answer is known to three. This is the general nature of the result, as the answer cannot be known to greater accuracy than the more poorly defined of the factors. However, the number of doubtful figures in a product such as we have above may be changed, depending on whether one of the original doubtful digits is low or high, 1 or 9, for example. This is another illustration of the fact, discussed above, that significant figures are not a completely satisfactory index of precision. If the problem above is stated in per cent of error, I is known to ±1 part in 256, or ±0.39 per cent, and R to ±0.022 per cent. The error in the product is ±0.39 per cent ±0.022 per cent, which is not far different from the error in I alone.

If extra digits accumulate in computations of a magnitude, beyond the inherent limits of precision, they should be discarded, or the number is "rounded out." In the usual practice, if the digit in the first place to be discarded is less than 5, it and any succeeding numbers are dropped; if the digit is 5 or greater, the previous number is increased by one. Thus, for three-digit precision, 67.238 would be rounded to 67.2, and 67.267 to 67.3.

Special attention should be given to the loss of precision and accuracy that occurs in taking the *difference* of two quantities, particularly when they become of nearly the same magnitude. Suppose that we write the two numbers, N_1 and N_2, with a $\pm$ notation to denote the range of doubt in the last digit:

$$N_1 = 637 \pm 4 \quad (= \pm0.63\%)$$
$$N_2 = 426 \pm 2 \quad (= \pm0.47\%)$$
$$\text{Difference} = \overline{211 \pm 6} \quad (= \pm\ 2.8\%)$$

Note that the doubtful parts are considered additive, since the presence of a $\pm$ doubt means that one quantity may possibly be high and the other low. The precision of the difference suffers in two respects; the range of doubt is greater, and it is judged in comparison with a smaller resultant amount. The percentage doubt is seen to be increased very greatly. Another case with worse consequences for precision is given by:

$$N_1 = 427 \pm 4 \quad (= \pm\ 1\%)$$
$$N_2 = 392 \pm 4 \quad (= \pm\ 1\%)$$
$$\text{Difference} = \overline{35 \pm 8} \quad (= \pm23\%)$$

These illustrations hold a lesson for us in the planning of experiments. We should avoid a method that requires taking the difference of experimental quantities that are of anything approaching equality of magnitude. As one example, we may determine the efficiency of a machine by reading output power and input power, and taking the ratio. This is a poor method if the efficiency is high, for small errors in the power measurements become very serious, as shown by the last illustration above, which would apply for a machine of about 92 per cent efficiency. Greater difficulties would be encountered in determining the losses of a transformer, since the efficiency may be expected to fall in the range 96–98 per cent; in such case errors of

0.5 per cent in output and input power quantities could cause an error of 50 per cent in losses as computed by the difference of input and output. A much better method for the transformer is to *measure* the *losses*, as may be done by open-circuit and short-circuit tests. (Note that the *sum* of two quantities is not subject to this difficulty; in the last illustration the sum is 819 ± 8, which maintains the approximate ± range of N_1 and N_2 in percentage terms.)

The discussion of significant figures deals very largely with precision, in that it is largely concerned with matters of reading and recording the results of experiments. Sometimes accuracy is involved; in reporting the results of an investigation we should generally give only information that we believe to be true — i.e., accurate. If the accuracy limit is considerably lower than the precision limit — as it well may be — we should limit the reported figure to the assured value (with the last digit possibly in doubt). There are exceptions to this last statement. If the data of the experiment are to be subjected to statistical analysis, we should report the figures to the precision limit, as the added information may be of value in establishing the regularity and reliability of the data, and in analyzing the contribution of component factors. In such cases, the accuracy limit should be stated apart from the precision limit.

2-6. Classification of Errors

Since no measurement is made with complete accuracy, a study of errors is necessary in the study of measuring processes. We wish to separate disturbing effects from the main element which we are investigating. The fact that we are studying the subject of errors does not mean that we expect to make all measurements with an extreme degree of accuracy. Measurements cannot be called good or bad merely on the basis of the degree of accuracy, but rather on their adequacy under the given conditions. In general, closer results can be attained by time, care, and expense; whether the improvement is justified in a particular case is, of course, another question. In any event, we must know *what* the accuracy of our results really is, whether high or low, otherwise we are in the dark, and have figures which we do not know how to evaluate. We cannot study the matter of accuracy without knowing something of the different types of error that may enter, and what we can do about them.

A study of errors is important as a step in finding ways of reducing them, and also as a means of estimating the reliability of the final result. A quantity is measured in terms of a standard, which in itself cannot be perfect. In addition, errors can and do occur in the process of comparison. Errors may originate in a variety of ways, but we may group them under three main headings, as shown below.

I. Gross errors.
II. Systematic errors.
    A. Instrumental errors.
        (1) Due to shortcomings of the instrument.
        (2) Due to misuse, or loading effects, of the instrument.
    B. Environmental errors ("errors due to external conditions").
    C. Observational errors.
III. Random errors (also called residual errors).

We shall discuss these classes briefly, to define them, and to furnish a preview for further study. Some items can be disposed of completely, while others will enter frequently in the analysis of measuring methods.

I. GROSS ERRORS

This classification covers, in the main, mistakes in the reading and recording of data. The responsibility usually lies with the worker for slips such as the gross misreading of a scale, or the transposition of figures in recording the result. For example, he may read 28.3 and record 23.8. We all make mistakes of this sort at times, and only by care can they be prevented from appearing in the data record. Errors of this type may be of any amount, and are not subject to mathematical treatment.

Two things can be done to avoid such difficulties; the first, mentioned above, is to exercise great care in reading and recording the data. The other countermeasure against the possibility of gross errors is the making of two, three, or more determinations of the desired quantity, preferably at entirely different reading points to avoid re-reading with the same error. Then, if the readings show disagreement by an unreasonably large amount, the situation can be investigated, and the bad reading eliminated. Complete dependence should *never* be placed upon a single determination. Indeed, the advantage of taking at least three readings lies not so much in the use of an average as in the confidence given by agreement of values that no gross error has been committed.

II-A. SYSTEMATIC ERRORS — INSTRUMENTAL

(1) *Due to shortcomings of the instrument.*

All instruments and standards possess inaccuracies of some amount. As supplied by the maker, there is always a tolerance allowance in the calibration, and additional inaccuracies may of course develop with use and age. As a simple example, suppose that measurements of length are made with a yard stick after a small amount has been cut from the zero end; all measurements made with this yard stick will be systematically in error by a *constant* amount. Or, for an electrical example, the ratio arms of a Wheatstone bridge may have an actual ratio different from the marked value. This causes a systematic instrumental error (of *proportional* amount) for all measurements using these arms. An indicating instrument such as a voltmeter has scale errors; these errors are generally different at different parts of the scale, and so do not partake of either the constant or proportional type, and must be expressed by a correction curve. (An incorrect series resistor in a voltmeter would give it a systematic error proportional to the scale reading.)

It is important to recognize the possibility of such errors when making precision measurements, for it is often possible to eliminate them, or at least to reduce them greatly by methods such as the following:

(a) Careful planning of procedure. This may be a substitution method, measuring against a nearly equal standard; or by interchange of bridge arms, where possible, and so on.

(b) By determination of instrumental errors and application of correction factors.

(c) By careful recalibration of the instrument.

(2) *Due to misuse, or loading effects, of the instruments.*

There is an old saw that instruments are better than the people who use them. Or, stated more moderately, shortcomings in measurements may be traced all too often to the operator, rather than to the equipment. A good instrument, used in an unintelligent way, may give poor results. This may come from such things as failure to make a needed zero adjustment in a bridge or meter (set voltmeters and ammeters to zero before applying power!), poor initial adjustment, the use of connecting leads of too high resistance for the measurement being made, and many, many other possibilities. Note that the malpractices just mentioned are of a kind that result in poor results at the time, but do no permanent harm. We should emphasize at every opportunity that careless or uninformed use of an instrument *may do permanent damage* as a result of overloading and overheating an instrument; in this case the value of the instrument, and of future as well as present readings, is depreciated until the trouble is detected and repairs (frequently expensive) are made. One mark of a person with true "measurement sense" is a constant regard for allowable operating conditions of the equipment.

Another source of erroneous results which should be blamed upon the user, rather than upon the instruments, lies in the loading effects of the instruments. A well-calibrated voltmeter, for example, may give a valueless and misleading reading if it is connected across two points in a high-resistance circuit. Indicating instruments always change conditions to some extent when connected into a complete circuit; sometimes the effect is negligibly small, as when a voltmeter is connected across a large generator. Sometimes the effect, though not negligible, can be corrected by computation (see voltmeter-ammeter measurement of resistance, par. 4-8). At other times, the presence of the meter produces so great a change in circuit conditions that operating conditions in the circuit are altered radically; this happens if a low-resistance voltmeter is connected in plate or grid circuits of a vacuum-tube amplifier. An answer to this condition is to use a high-resistance voltmeter, or a vacuum-tube voltmeter, as conditions demand. The loading effect of instruments should be considered in setting up measuring systems so that corrections may be made if needed, or more suitable instruments may be selected, or a completely different method may be used. As another example of instrument loading, in obtaining a resonance curve for a series R-L-C circuit, an ammeter of one range was replaced by another of a different range; the curves for the two portions of the test did not join; the reason was found in the different resistance and inductance constants of the two ammeters. The *user* must take into account the effect that the measuring equipment has on the circuit, and plan the measurements accordingly.

II-B. Systematic errors — Environmental

This classification is also called "errors due to external conditions," i.e., conditions external to the measuring device. This includes any condition in the region surrounding the test area that has an effect on the measurements. One common source of variation comes from temperature changes of the equipment, since temperature affects the properties of materials in many ways, including dimensions, resistivity, spring effect, and others. Some instruments may be affected by humidity, barometric pressure, the earth's magnetic field, gravity, stray electric and magnetic fields, and others. There are several kinds of action that can be taken to eliminate, or at least to reduce, the undesirable disturbances:

(1) Arrangements to keep conditions as nearly constant as possible — for example, enclosing the equipment in a temperature-controlled cabinet.

(2) By use of equipment largely immune to such effects — for example, some resistance materials have a very small temperature effect over a moderate range.

(3) By computed corrections. We try to avoid the need for such corrections when possible, but computed corrections must be made in some cases.

It should be noted that any of these methods can be hoped, at best, to neutralize the *major* part, *but not all*, of the error. Correction is accomplished to a "first-order" of approximation, but leaves "second-order," or "residual" errors. Temperature affects the time-keeping of a watch by changing the size of the balance wheel, and also by changing the constant of the hair spring. Compensation may be effected by use of a bimetal rim to the balance wheel. The designer attempts the greatest possible cancellation of temperature error, but it is too much to expect that he will be completely successful, or, in fact, that the physical laws of error and compensation are such that they can cancel exactly over a large temperature range.

II-C. SYSTEMATIC ERRORS — OBSERVATIONAL

"Errors of the observer" is a name that recognizes that there exists a "personal equation" for the observer, so that several people using the same equipment for duplicate sets of measurements do not necessarily produce duplicate results. One observer may tend characteristically to read a meter higher (or lower) than the correct value, possibly because of his reading angle and failure to eliminate parallax. Tests have been conducted in which a number of persons read fractional divisions on a voltmeter scale under carefully controlled conditions and consistent individual differences were found. In measurements involving the timing of an event, one observer may tend to anticipate the signal and read too soon. Very considerable differences are likely to appear in determinations of light intensities or sound levels. Important readings which may be subject to this type of error should be shared by two or more observers to minimize the possibility of a constant bias.

III. RANDOM (OR RESIDUAL) ERRORS

It has been found repeatedly that the data from experimental tests show variations from reading to reading, even after attention to all known sources of error discussed above. There is no doubt a reason, or rather a set of reasons, for these variations, but we are not able to account for them. The physical event that we are measuring is affected by many happenings throughout the universe, and we are aware of only the more obvious; the remainder are lumped together and termed "random," or "residual." (They are sometimes called "accidental," but in a different sense than the accidental *gross* errors discussed above. We shall avoid use of the term as applied to random effects.)

The errors considered in this classification may be regarded as the residue of error when all known systematic effects have been taken into account — hence the term "residual." Test conditions are subject to variations due to a multiplicity of small causes which cannot be traced out separately. Also, when corrections are made for known effects, the correction is likely to be approximate, leaving a small residue of error.

The unknown errors are probably caused by a large number of small effects, each one variable, so that they are additive in some cases and subtractive in others in their effect on the measured quantity. In many observations the positive and negative effects are nearly equal, so the resultant error is small. In a few cases the positive (or negative) errors prevail, giving a comparatively large error. If we assume the presence of a large number of small causes, each of which may give either a plus or a minus effect in a completely random manner, we obtain the condition of *scatter* around a central value. This condition is frequently met in experimental data, and so we are justified in using the concept as a basis for our studies of those discrepancies to which we cannot assign a known cause. We see also why "random error" is an appropriate description of the situation. The supposition of randomness is useful in permitting correlation with the mathematical laws of probability, and thus leading to an analytical study of this type of error.

2-7. The Place of Statistics in Experimental Work

No measurement is made with complete accuracy. There is always some error, which changes from one determination to another, superposed upon the value which we are seeking. It is the function of statistics to separate, as far as possible, the truth from error by narrowing and defining the region of doubt. Note, however, that statistical study is concerned primarily with the *precision* of measurement. It cannot reveal anything that is not implicit in the data, and cannot remove systematic errors from a set of data. *Accuracy* is a matter of *experimental* work. The added precision afforded by statistical study may permit *detection* of a *discrepancy*, which is a necessary step in the elimination of faulty results. That is, systematic errors are usually revealed when two or more methods are used to measure the same quantity; each method may yield a closely grouped set of data, but the averages may differ by more than can be accounted for, considering the precision of the measurements. A study of the methods may bring to light a bias in one procedure. The measurement of precision is thus an indispensable step in the detection of inaccuracy.

The simplest application of statistics is in finding the average and a measure of precision of a simple set of measurements. Dependence should never be placed upon a single determination because of the possibility of gross error in measurement, reading, or recording. In addition, the average is generally expected to be a better measure of the unknown than is given by a single reading. We should mention in this connection the policy of discarding out-of-line readings. It is general practice to discard readings which show a large divergence from the average of the set. It is possible, statistically, for such an event to happen, but the odds against its happening by pure chance are so great that the presumption of gross error is more reasonable — in any event, it should not be averaged into a small set of readings. Sometimes the "best-two-out-of-three" is used — i.e., discard the one reading in a set of three that differs most from the others. This practice should be treated with caution, however, depending on the amount of divergence. For example, when readings of resistance are taken by a procedure that should be reliable within 1 per cent, if the readings are 95.2, 95.4, and 91.6, the presumption of error in the 91.6 reading is very strong, and it should be discarded. For a smaller difference the answer may not be clear. For 95.1, 95.2, and 95.6, exclusion of the 95.6 *may* actually give a *less* accurate average. A statistical investigation was conducted at the National Bureau of Standards in which sets of triads were drawn at random from normally distributed data (par. 2-14), and were examined for their spacing. It showed that a decidedly unbalanced spacing occurs rather frequently; in fact, in an average of one out of twelve sets, one of the measurements is at least 19 times further away from its neighbor than the difference separating the two closest. Since no gross reading errors were included in this test, the automatic and habitual discarding of one reading in three is indicated to be fallacious — any discarding should be done after study of the particular circumstances.

However, statistical analysis applies in a very much larger way than in the mere adjustment of a set of data. It is an important component in the planning and analysis of data in a research program, and not just an adjunct to such a program. It is so important that it should be considered at the beginning, in the planning stage of the experiment, rather than after the experimental work is completed. We may list some of the broad classifications of its use in physical science:

(1) Determination of a physical constant, involving a long and inter-related set of measurements, with a requirement of the highest attainable accuracy in the result.

(2) Investigation of a product or process, involving several factors, to determine the influence of each on the quality or functioning of the whole.

(3) Inspection sampling, and quality control of production processes.

We shall attempt to lay the groundwork of statistical ideas and terminology. Statistical methods are coming into use in so many ways that an introduction to the subject has several possibilities of usefulness, but the subject is so large that we can make only a beginning. The field of inspection sampling has such specialized techniques and literature that we shall not go into it at all.

2-8. The Characteristics of Experimental Data

We noted before that *scatter* is a noticeable feature in measurements data. Variations can be minimized by using care in all manipulations, and by holding conditions as steady as possible during the period of test. Complete steadiness is never attained for all conditions of test, and the difficulty increases with the length of the period. Even with maximum care an unavoidable uncertainty remains, so that we cannot make a measurement with absolute definiteness. It might be thought that as the measuring methods become more refined the uncertainty would disappear. This is not the case. It is as the measurements are made to closer limits that the presence of small disturbances becomes more evident. The disturbances may become smaller in absolute size through care, but they become relatively more important as maximum precision is attempted.

The first step in working up the results of an experiment is to correct the data for all known systematic effects, calibration errors, and so on, since we seek to approach as closely as possible the "true" value of the quantity being measured. Statistical treatment applies to the remaining variations. We must remove all known errors before applying statistical methods, for they are based on laws of chance, and not on consistent factors. Analysis allows us to determine the best value possible *from the given data,* and to set the limits of uncertainty inherent in the scatter of the data.

The distribution of data in a set of readings may be presented in several ways, one of which is a block diagram, or "histogram." A set of assumed readings is shown in the table; it may be regarded for our present purpose either as the measurements on a group of nominal 100-ohm resistors drawn at random from a production line, or as a set of measurements on one particular resistor (if we should take as many as fifty readings on one resistor!). The data are taken and recorded to the nearest tenth of an ohm. The diagram, as drawn in Fig. 2-1, represents the data of the table,

TABLE OF ASSUMED RESISTANCE READINGS

Resistance Value Ohms	Number of Readings	Fraction of Total Readings
99.7	1	0.02
99.8	3	0.06
99.9	12	0.24
100.0	18	0.36
100.1	11	0.22
100.2	4	0.08
100.3	1	0.02
Total	50	1.00

with the ordinates indicating the number of times a particular value was found. There is a large group at the central value, 100.0 ohms, with other

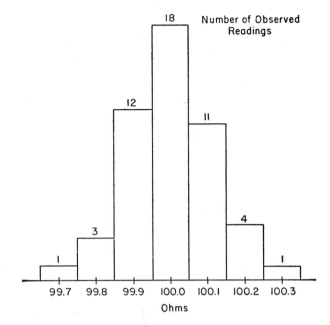

Fig. 2-1. Diagram showing frequency of occurrence of various resistance values in an assumed set of 50 readings.

values arrayed almost symmetrically on the high and low sides. Note that the figures in the right-hand column of the table, showing the *relative* occurrence of each value, might just as well have been used as the ordinates. The shape of the diagram would have been the same, with merely a change of scale. The relative-occurrence values are frequently more convenient and significant than the particular number of cases. The relative-occurrence curve is of particular interest to us, as it ties in closely with the mathematical probability curve that we shall discuss in a short time. Note, for future reference, that the relative-occurrence column adds to unity.

If we were to take more readings and group them by smaller increments, say 200 readings by 0.05-ohm intervals, we could expect a diagram of the same general form, except that the steps would be smaller and the shape somewhat smoother. With more and more data at smaller and smaller intervals, the stepped diagram may be regarded as merging finally into a smooth curve — or, the smooth curve may be obtained by drawing a curve through the steps in such way as to average them (i.e., to keep the same area enclosed). The stepped form of curve is used, (a) for small amounts of data, (b) for situations with natural group divisions, and (c) for situations in which no regular law of variation is to be expected. The smooth curve is the form we shall use in most of our analytical studies.

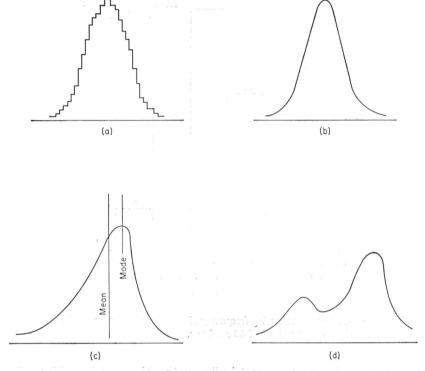

Fig. 2-2. Different forms of frequency-distribution diagrams.

    (a) Histogram, small steps
    (b) Limiting case, smooth curve
    (c) Unsymmetrical (skewed) distribution curve
    (d) Multi-modal curve

Figure 2-2(a) is a nearly symmetrical histogram with many small steps. Diagram (b) is the smooth curve that might be used to represent the general trend of (a). Many physical cases have been found, by experimental results, to yield data agreeing quite well with the symmetrical curve. There are situations, however, in which the relationships produce a "bias" more to one side of the peak point; a "skewed" curve of this type is shown in (c) of the figure. Finally, it is possible to meet a "multi-modal" distribution with two or more peaks, as illustrated in (d). This form, rare in occurrence, indicates a combination of two distinct distributions, and should be specially investigated.

2-9. Description of Dispersed Data

It is necessary to have ways of characterizing a set of data, other than presenting the entire table or drawing a graph. We need simple numerical attributes to describe and summarize the results of the measurements, both to permit easy comprehension of the nature of the findings, and to use in combinations or comparisons of different sets.

Our discussions have indicated three attributes of interest in describing dispersed data, of which the first two are of particular concern:

(1) Measure of central or typical value
(2) Measure of dispersion from central value
(3) Measure of skewness

(1) *Measure of central or typical value.*

Mean. The most commonly used measure of central value in statistical work is the arithmetic mean, or average. It is found by taking the sum of the readings ("variates") and dividing by the number of readings. In symbols, if $x_1, x_2, \ldots, x_n$ represent n readings, the arithmetic mean is,

$$\text{A.M.} = \bar{x} = \frac{x_1 + x_2 + \ldots + x_n}{n} = \frac{1}{n}\sum_{i=1}^{n} x_i \qquad (1)$$

There are other forms of mean that have engineering uses in some places. The *geometric mean* is defined as the nth root of the product of n terms,

$$\text{Geometric mean} = \sqrt[n]{x_1 \cdot x_2 \cdot \ldots \cdot x_n} \qquad (1a)$$

This is equivalent to carrying out an averaging process on the logarithms of the terms. The geometric mean is used in calculating inductances, and in some other applications.

The *harmonic mean* is found by summing the reciprocals of the variates, as

$$\frac{1}{H} = \frac{1}{n}\left[\frac{1}{x_1} + \frac{1}{x_2} + \ldots + \frac{1}{x_n}\right] \qquad (1b)$$

The harmonic mean of a set of resistance values, for example, equals the reciprocal of the arithmetic mean of the corresponding conductances.

Median. The median is defined as the middle value of the set, if the variates are arranged in numerical order. The definition is clear if there is an odd number of readings in the set. If the number is even, the definition is ambiguous; the general practice in this case is to use the average of the two central values.

Mode. The mode is the value of the variate for which the frequency of occurrence is maximum — that is, the value at which the peak of the distribution curve occurs. For a symmetrical distribution curve, these three measures of central value (average, median, mode) are equal. For a skewed curve, as in Fig. 2-2(c), the quantities are different; the mean and mode are marked on the diagram, and the median lies between them.

(2) *Measures of dispersion from the mean.*

It is important to have a measure of the dispersion from the central value, for it gives an index to the degree of consistency and regularity of the data. A large dispersion indicates that some elements of the measuring process were not under close control, and consequently we are unable to define the measured quantity with the desired definiteness. If we compare two sets of data and find that one has much less dispersion than the other, we place more reliance on it and regard it as the better set, at least with respect to the control of *random* disturbances.

Before giving a measure for dispersion, we must define the term *deviation*. The *deviation of a reading* is the amount by which it differs from the mean. If we have a set of readings, x_1, x_2, ..., with mean $\bar{x}$, the deviations of the individual readings are,

$$\text{Deviation of } x_1 = w_1 = x_1 - \bar{x}$$
$$\text{Deviation of } x_2 = w_2 = x_2 - \bar{x} \qquad (2)$$

$$\cdot \ \cdot \ \cdot \ \cdot \ \cdot \ \cdot \ \cdot \ \cdot \ \cdot \ \cdot$$

The dispersion as defined for a *set* of readings must be based in some way upon the deviations of all readings in the set. The most commonly used value is the *root-mean-square deviation* (sometimes called the *root-mean-square error*), or *standard deviation* = S.D. As the term r.m.s. deviation implies, the process of finding this quantity involves taking the *squares* of the deviations, averaging them, and taking the square root. There is one rather important detail to consider before we commit ourselves to an equation to represent the S.D., so this will be postponed for the moment. The standard deviation is an extremely important quantity in all statistical study.

Another expression for essentially the same quantity is the *Variance*. The variance is the *mean-square deviation*, which is the same as the S.D., except that the square root is not extracted. That is,

$$\text{Variance} = \text{Mean-square deviation} = (\text{Standard Deviation})^2$$

Variance is a convenient quantity in many computations, for variances of different factors in a result have an additive property, as we shall see later. The S.D., on the other hand, has the advantage of being of the same units as the variate (not squared), so comparison of magnitude of the variate and its S.D. is easier. Both terms will be useful to us.

Note that the finding of the variance or standard deviation involves the summation of the squares of the deviations from the mean, regardless of the form of distribution of the points. For discrete points, evaluation may be made by a simple summation. If the probability distribution is defined by a curve, the summation may be made by integration. The fractional number of cases in a narrow band is $y \, dw$, which is multiplied by w^2 and integrated over limits that include all cases. The *mean* square is obtained by dividing this quantity by the area under the distribution curve (unity, for a curve stated in probabilities). That is,

$$\text{Variance} = (\text{S.D.})^2 = \frac{\displaystyle\int_{-\infty}^{+\infty} w^2 \, y \, dw}{\text{Area under curve}} \tag{3}$$

The concept of mean-square and root-mean-square should come easily to electrical engineers, who are accustomed to similar computations, which are used, however, under much different circumstances.

The average deviation is sometimes used. It is computed by averaging the deviations *without regard to sign;* i.e., it is the average of the absolute values of the deviations (the average of the deviations *with* sign is zero, if no error of computation has been made). Average deviation is a less desirable measure of dispersion than the standard deviation, as it is less convenient mathematically. It is less regular for comparison of small and large samples, and in combinations of terms. We shall employ standard deviation entirely in our studies.

(3) *Measures of skewness.*

Skewness is a term for a lack of symmetry in the distribution. A measure of skewness is useful in expressing the degree of departure from the symmetrical condition. One simple and somewhat rough measure is (Mean − Mode)/S.D. A more precise measure, useful on a statistical basis, is computed from the moments of the deviations as $u_3/(u_2)^{3/2}$, where the subscript denotes the order of the moment (u_3 is found by summation of the cubes of the deviations). The matter of skewness is of less concern in industrial work than in some other fields, as it is encountered only occasionally.

2-10. Terminology

Statistical theory is based upon the idea of chance happenings in an infinitely large number of determinations. In a practical case, when we make measurements of a quantity, the determinations are always finite in number, and frequently, from the nature and cost of experimental work, rather limited in number. *Random* effects cancel completely in an infinite set of measurements, but this is not true, in general, in a small set. We must recognize that the mean of our sample is not necessarily the mean of the universe, and the S.D. as obtained from our readings may not be the S.D. of a larger sample, or of the universe. Precision increases, in general, with the size of the sample, but the means and S.D.'s of various samples have the property of scatter around the universe values.

In a limited set of data, we consider that we are obtaining a small sample drawn from a possible infinity of such observations. We must have a notation that distinguishes between the limited set and the universe of values. The difference in numerical values may seem rather trifling in a large set; however, the distinction between limited set and universe is a basic concept in statistical study. It was considered formerly that reliable results could be obtained only from a large sample. Advances have been made in recent years in the field of small-sample theory, so that more precise conclusions can be drawn from a small sample, with tests to determine the degree of reliability that can be attached to them.

The generality of all possible determinations is referred to frequently as the "Universe" of values, and sometimes as the "Population" (the latter term probably arising from sociological applications of statistical analysis). Efforts are made to distinguish between symbols by using Greek letters for Universe quantities, and ordinary letters for sample quantities, where corresponding letters are available. There are several exceptions, however, so the parallels are not complete. We may tabulate some of the quantities that are of immediate interest to us:

	Universe, or Population	*Finite Sample*
Mean..........................	μ	$\bar{x}$
Deviation = departure from mean........................	$x - \mu$	$x - \bar{x}$
Standard deviation..............	σ	s
Variance......................	$V_u = \sigma^2 = \dfrac{\Sigma (x - \mu)^2}{N}$	$V = s^2$

$$(4)$$

A purist may object to the idea of finding the mean or standard deviation of an "infinite" number of readings by summing them and dividing by the number. However, we can consider for our purpose that we have "a very large number," so that additional readings would not change the result appreciably. (The variance of a finite sample will be defined later, in par. 2-22.)

2-11. Types of Probability Distribution

A probability distribution, or frequency-of-occurrence distribution, expresses the likelihood of a particular event, or of a deviation of a particular amount. A type of variation may be characterized by a curve or equation which gives the relative probability of a particular result, or of a deviation of a given amount. There are several types of variation that occur in statistical work for different purposes. We shall list the following three forms, and give particular attention to the first:

(1) "Normal" law of error, or Gaussian law
(2) Binomial distribution
(3) Poisson distribution

We shall dispose of the second and third items with quite brief treatment, so that their general nature may be understood, and then we shall return to the Normal law for a fuller study, since it is of greatest importance to us.

2-12. Binomial Distribution

The binomial distribution deals with distributions composed of *discrete* chance possibilities, such as "good" or "bad" articles in a production run, or, "heads-or-tails" in tossing a coin, and so forth. The probability P of exactly a "good" units (or whatever event is being considered) in a set of n events is given by

$$P = \frac{p^a q^b}{a! b!} n!$$ (5)

where

p = chance of occurrence of the event in a single trial

q = chance of nonoccurrence $(p + q = 1)$

a = number of times event occurs in set of n, $(a = 1, 2, \ldots n)$

b = number of times event does not occur in set of n, $(a + b = n)$

Note that "chance" and "probability" mean the average we would expect if we repeat the test a great number of times.

As an example, suppose that we toss a coin 5 times (or 5 coins at once) with care to avoid bias. The probability of 5 heads is $(\frac{1}{2})^5 = \frac{1}{32}$, since in this case $p = q = \frac{1}{2}$. The probability of 4 heads and 1 tail is somewhat greater. The probability of 4 heads and 1 "not head" is $(\frac{1}{2})^4(\frac{1}{2})^1$, but the "not head" may occur in any one of 5 positions, so the total chance of 4-heads, 1-tail is $\frac{5}{32}$. This agrees with equation (5) for $a = 4, b = 1, n = 5$. It is of interest to note that the chance of occurrence of any particular number of heads from (5) may also be obtained from the correspondingly numbered term in the *binomial expansion* of $(p + q)^n$. To illustrate, for the case of the 5 coins, we take the expansion,

$$\left(\frac{1}{2}+\frac{1}{2}\right)^5 = \left(\frac{1}{2}\right)^5 + 5\left(\frac{1}{2}\right)^4\left(\frac{1}{2}\right) + \frac{5\cdot4}{2!}\left(\frac{1}{2}\right)^3\left(\frac{1}{2}\right)^2 + \frac{5\cdot4\cdot3}{3!}\left(\frac{1}{2}\right)^2\left(\frac{1}{2}\right)^3$$

$$+ \frac{5\cdot4\cdot3\cdot2}{4!}\left(\frac{1}{2}\right)\left(\frac{1}{2}\right)^4 \cdot \cdot \cdot + \frac{5\cdot4\cdot3\cdot2\cdot1}{5!}\left(\frac{1}{2}\right)^5$$

Probability →
$$= \frac{1}{32} + \frac{5}{32} + \frac{10}{32} + \frac{10}{32} + \frac{5}{32} + \frac{1}{32} = \frac{32}{32}$$

Number → of heads

↑	↑	↑	↑	↑	↑
5	4	3	2	1	0

The case above is special to the extent that $p = q$. For greater generality we may take the case of chance in the throwing of dice. For a throw of a single die the chance of an "ace" is 1 in 6, that is, $p = \frac{1}{6}$, $q = \frac{5}{6}$. If we throw 5 dice simultaneously, the chance of 5 aces is $(\frac{1}{6})^5$. For the chance of exactly 4 aces showing on the 5 dice, the chance of having 4 aces in the first 4 we look at is $(\frac{1}{6})^4$, and the chance is $\frac{5}{6}$ that the next one will be a "not ace." However, the one that is "not ace" might equally well occur on any of the 5 dice insofar as the total is concerned. Therefore, the chance of exactly 4 aces is $5(\frac{1}{6})^4(\frac{5}{6})$. Note, again, that this result might have been obtained as the second term of the binomial expansion, which becomes for this case,

$$P = (\tfrac{1}{6} + \tfrac{5}{6})^5 \tag{6}$$

We may tabulate the results of this expansion:

No. of Aces in 5	Probability			
5	$(1/6)^5 =$	1 ÷ 7776	= 0.00013
4	$5(1/6)^4(5/6) =$	25 ÷ "	= 0.00321
3	$10(1/6)^3(5/6)^2 =$	250 ÷ "	= 0.03215
2	$10(1/6)^2(5/6)^3 = 1250 ÷$	"	= 0.16075
1	$5(1/6)(5/6)^4 = 3125 ÷$	"	= 0.40188
0	$(5/6)^5 = 3125 ÷$	"	= 0.40188
	Total:	= 7776 ÷ 7776	= 1.00000

From this table we can determine also the probability of finding 4-or-more aces $(26/7776)$, 3-or-more $(276/7776)$, and so on.

The *mean* expectation of the event being considered by the binomial theorem is given,

Mean occurrence of event $a = np$. (7)

This may be checked from the coin or dice examples. If we toss 5 coins ($n = 5$, $p = \frac{1}{2}$), the number of heads showing per trial would be expected to average to 2.5. (The number of heads on a single trial must be an integer, but the average may be fractional.) In the dice example, the probability of an ace on the throw of a single die is $\frac{1}{6}$, so the mean expectation in a group of 5 is $\frac{5}{6}$, as the average of many trials.

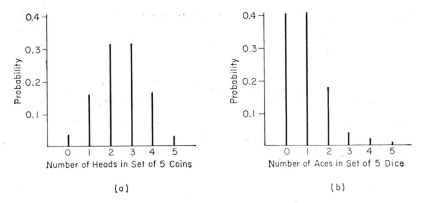

Fig. 2-3. Graphs of probability values in two binomial distribution cases.
(a) Probabilities of various numbers of "heads" in throw of 5 coins
(b) Probabilities of various number of aces in throw of 5 dice

The probability figures for both the coin and the dice illustrations are shown in Fig. 2-3. The points are shown, but are not connected by curves, as intermediate points would have no significance in these cases. The binomial distribution is distinguished from the Normal law in applying only at discrete values. For $p = q$ the graph is symmetrical as in (a), but for $p \neq q$ the distribution of points is unsymmetrical as in (b).

The variance of a, distributed binomially, may be found as the mean of the squares of the deviations from the mean. It can be shown to be,

$$V = \sigma^2 = npq \qquad (8)$$

The binomial distribution, though quite different from the Normal law in nature, does give a distribution of points for a large value of n that follows a form resembling the Normal law. That is, if a graph is drawn similar to Fig. 2-3(a), but for large n (and $p = q = \frac{1}{2}$), the points, becoming closely contiguous, give the impression of Normal law contour. In fact, Normal law functions are sometimes used for this case, with fairly good approximation, if both a and b are large. (For small inequality of p and q the binomial distribution is not markedly unsymmetrical about the maximum, but the maximum point is shifted from the center of the series.)

2-13. Poisson Distribution

The Poisson distribution is another discontinuous distribution, with only discrete values of occurrence. It is related to the binomial, and is in fact the form to which the last few terms of the binomial tend if n is very large and p very small. It applies to cases in which the probability p of occurrence of the event within a definite interval is small, yet the sampling interval is large enough so that there is a finite total probability of observing a few occurrences. The form of the Poisson distribution of the probabilities follows the series:

Number of

occurrences	0	1	2	k
Probabilities	$\epsilon^{-\mu}$	$\epsilon^{-\mu} \cdot \mu$	$\epsilon^{-\mu} \cdot \mu^2/2!$	$\epsilon^{-\mu} \cdot \mu^k/k!$

The best estimate of the quantity μ is the mean of the number of occurrences. The expected number of occasions on which a particular number of occurrences would be observed is found by multiplying the total number of occurrences by the probabilities given above. Note that the sum of the probabilities is,

$$\text{Sum} = \epsilon^{-\mu}\left[1 + \mu + \frac{\mu^2}{2!} + \ . \ \ . \ \ . + \frac{\mu^k}{k!} + \ . \ \ . \right]$$

and that the quantity in brackets is the series expansion for $\epsilon^{+\mu}$, so the sum approaches the value, $S = \epsilon^{-\mu} \cdot \epsilon^{+\mu} = 1$, which is one of the requirements, at least, for a probability distribution function.

For the Poisson distribution,

$$\text{Mean} = \mu$$

$$\text{Variance} = \mu$$

Applications of the Poisson distribution might be met in the study of wrong numbers in a telephone exchange, or of the number of dust particles observed in a very small volume of dusty gas.

2-14. The Normal Law of Error

The Normal, or Gaussian, law of error is the basis for the major part of the study of random effects. This type of variation, or a good approximation to it, is met frequently in experimental work (after *known* effects are removed). Assumptions leading to the Normal law are rather simple:

(1) All observations include a large number of small random disturbing effects, or "random errors."

(2) The small random effects may be both positive and negative.

(3) There is equal probability of positive and negative disturbing effects.

We expect, as a consequence of these assumptions, that many observations will include plus and minus errors in more or less equal amounts, giving a small total error. In a good many cases there will be a small preponderance of plus errors, in other cases a small minus total. Only in rare cases will the errors be mostly plus, or mostly minus. This is something like tossing a large group of coins; only rarely will all be heads, or all tails; we expect generally a mixture of heads and tails. As a consequence, we can state the probabilities as to the form of the error-distribution curve:

(1) There will be a strong central tendency — i.e., small total errors are more probable than large errors.

(2) Large errors, plus or minus, are very improbable.

(3) The curve of probability of a given error plotted against the magnitude of the error will be symmetrical about the zero value (because of the equal probability of plus and minus component errors).

The Normal curve may be regarded as the limiting form of a histogram such as Fig. 2-1 when the amount of data becomes increasingly large, gathered in smaller and smaller bands. A derivation of the Normal law can be given, based upon the assumptions stated above; however, we shall for brevity accept the results without formal proof. In fact, we may consider that the law is actually proved by experiment, and that the assumptions rather serve to explain the law.

One form of the equation for the Normal law may be written in terms of a simple constant. (Another form, normalized in terms of a derived constant, the standard deviation, will be given later.) The equation is,

$$y = \frac{h}{\sqrt{\pi}} \epsilon^{-h^2 w^2} \tag{9}$$

where

h = a constant

w = magnitude of deviation = $x - \mu$ (μ = universe mean)

y = probability of occurrence of deviation w.

The equation leads to a curve of the form shown in Fig. 2-4, which may be seen in a general way to match the requirements of the probability behavior as outlined in the assumptions given above. The abscissa represents the magnitude of the deviation from the central value, and the ordinate is the probability of the occurrence in an infinite set of an observation with a deviation of this magnitude. The presence of the square, w^2, in the exponent gives symmetry for plus and minus deviations, and gives flatness of slope at the origin. The use of "probability" should be explained. It means that in a large group of n observations, the expected number with deviations between w and $w + \Delta w$ is given by $ny \cdot \Delta w$, or as a fraction of the number of cases,

$$\text{fraction of total cases in band } \Delta w \text{ wide} = \frac{ny \cdot \Delta w}{n} = y \cdot \Delta w \tag{10}$$

where y = the ordinate of the curve at abscissa w.

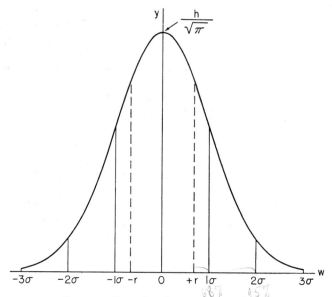

Fig. 2-4. Curve for the normal law of error.

The constant $h/\sqrt{\pi}$ stems from the necessity that the summation for the entire curve must account for the entire number of cases; that is, the summation $\sum ny\,\Delta w$ for all Δw bands must equal n, or stated as an integral,

$$\int_{-\infty}^{+\infty} ny\,dw = n \tag{11}$$

Or, after division by n,

$$\int_{-\infty}^{+\infty} y\,dw = \frac{h}{\sqrt{\pi}} \int_{-\infty}^{+\infty} \epsilon^{-h^2w^2}\,dw = 1 \tag{12}$$

which states simply that the summation of all possibilities must be certainty, which is unit probability. The infinite integral may be found in a table of integrals; the quantity under the integral sign in (12) has a value $\sqrt{\pi}/h$, so the multiplier $h/\sqrt{\pi}$ shown in (9) and (12) gives unity for the entire result of the summation. (The integral of (12) cannot be expressed in closed form for finite limits. However, it may be evaluated by series expansion. Also, the function has been used so much that tables are available in many places. See Appendix IV for a short table.)

The quantity h is one possible measure of the degree of dispersion of a set of data represented by a curve such as Fig. 2-4. For $x = 0$, the ordinate is $h/\sqrt{\pi}$. If we compare two distribution curves, as in Fig. 2-5, the curve with the greater value of h has a greater central probability, and drops off more rapidly with increase of w. Curve B has a lower h, a lower central

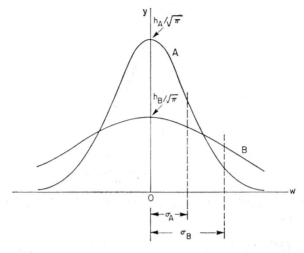

Fig. 2-5. Comparison of two frequency distribution curves with different degrees of dispersion.

value, and a slower drop-off at the sides. (Both curves have the same unit area under them.) In summary, a large value of h means a more closely grouped set of observations, with smaller dispersion, and should therefore be regarded as a better set, with respect to the control of random effects. Another criterion, the standard deviation, which will be determined in the next section, is more convenient, and is used in preference to h as a measure of dispersion.

2-15. Variance and Standard Deviation for the Normal Law

Variance is defined for unlimited data as the mean value of the squares of the deviations. Standard Deviation is the square root of Variance.

Variance for a population distributed according to the Normal law may be determined by summation in the form of an integral. Each band of "cases" (see Fig. 2-6) is multiplied by the *square* of the deviation, and the result is summed over the entire range of values. That is,

$$\text{Variance} = V = \frac{\int_{-\infty}^{+\infty} w^2 (y \, dw)}{1} = \frac{\frac{h}{\sqrt{\pi}} \int_{-\infty}^{+\infty} w^2 \epsilon^{-h^2 w^2} \, dw}{1} \qquad (13)$$

$$= \frac{2h}{\sqrt{\pi}} \int_{0}^{\infty} w^2 \epsilon^{-h^2 w^2} \, dw \qquad (14)$$

Division by unity is indicated in (13) to denote that the summation is averaged over all cases, which is unity in terms of the probability curve.

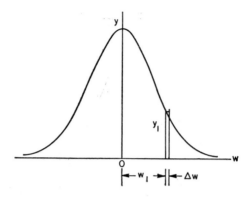

Fig. 2-6. Determination of variance and standard deviation for the normal curve.

Equation (14) may be evaluated by an integration by parts, one part being the same as the integral in (12)*. From this,

$$V = \sigma^2 = \frac{1}{2h^2} \tag{15}$$

$$\text{Standard deviation} = \text{S.D.} = \sigma = \frac{1}{\sqrt{2}\,h} \tag{16}$$

If we substitute h from (16) into (9), we have an equation for the Normal law expressed in terms of the standard deviation:

$$y = \frac{1}{\sigma\sqrt{2\pi}}\,\epsilon^{-w^2/2\sigma^2} \tag{17}$$

This form of equation is particularly useful, as σ is the quantity we ordinarily know and are interested in. σ is a quantity of the same units as the observed quantity and its mean. This makes it easy to visualize the dispersion in comparison with the mean.

Figure 2-4 has the deviations divided in terms of σ units — that is, for $w = 1\sigma, 2\sigma, \ldots$. This is a useful form of curve, for it can represent the distribution of observations for any Normal law situation we wish, once we have evaluated σ for that case.

---

*Separation of (14) into parts for integration:

$$V = \frac{-1}{h\sqrt{\pi}} \int_0^\infty w(-\epsilon^{-h^2w^2} \cdot 2h^2w\, dw)$$

$$= \frac{-1}{h\sqrt{\pi}} \left[w \cdot \epsilon^{-h^2w^2} \right]_0^\infty + \frac{1}{h\sqrt{\pi}} \int_0^\infty \epsilon^{-h^2w^2}\, dw$$

$$= 0 + \frac{1}{h\sqrt{\pi}} \frac{\sqrt{\pi}}{2h} = \frac{1}{2h^2}$$

2-16. Area under the Probability Curve

As we discussed in par. 14, the area under the probability curve within certain limits represents the number of cases among the observations within those deviation limits, expressed as a fraction of the total cases. The area between $-\infty$ and $+\infty$ is unity, representing the entire number of cases. If we find the area between $-\sigma$ and $+\sigma$, we know the relative number of cases that differ from the mean by no more than the S.D. This area can be found by integration in series, or more simply by reference to tables. The value is found to be 0.68, so 68 per cent of the cases for Normally dispersed data lie between the limits of $\pm\sigma$ from the mean. Corresponding values for other deviations are given in the table.*

Deviation ($\pm$)	Fraction of Area Included
0.6745σ	0.5000
1σ	0.6828
2σ	0.9546
3σ	0.9972
$(1.96\sigma$	0.9500)

If we find from measurements on a large number of resistors that the mean is 100.0 ohms, with a S.D. of 0.2 ohm, with Normal distribution, we know that on the average 68 per cent (roughly $\frac{2}{3}$) of the resistors lie between limits of ±0.2 ohm of the mean. There is thus approximately a 2-to-1 change that any resistor we select from the lot by random will lie within these limits. If we wish longer odds, we may take $\pm2\sigma = \pm0.4$ ohm, which includes about 95 per cent of the cases, giving somewhat over 19-to-1 odds that any resistor selected at random lies within ±0.4 ohm of the mean. (1.96σ gives the exact 19-to-1 proportion.) For even greater assurance, $\pm3\sigma$ includes 99.7 per cent of the cases, which is close to certainty.

The first entry in the table shows that limits of $\pm0.6745\sigma$ include *half* of the cases. These are the 50 per cent chance limits, or 1-to-1 odds, that carry the title of "probable error." This will be discussed in the next section.

2-17. Probable Error

If ordinates are erected at deviations of $\pm r = 0.6745\sigma$, as shown in Fig. 2-4, half the area under the curve is enclosed between these limits. The quantity r is called the "Probable Error" (P.E.). This value is "probable," as shown, in the sense that there is an *even chance* that any one observation will be greater or less than r. If we take a set of measurements and determine r, and then make one additional reading, the chances are

---

*For a more complete table, see Appendix IV.

50-50 that the new value will lie between $-r$ and $+r$. Stated in another way, the chances are even that any one reading will have a random error no greater than $\pm r$.

The situation with respect to errors of measurement may be compared with that of a person shooting at a target. Gravity pulls the bullets down, a side wind deflects them from a straight line, and the "observer" may characteristically shoot high (or low) and to the right (or left). These are systematic errors, which can be corrected by proper setting of the sights. With correct adjustment there are still erratic effects (such as irregularities of wind, variations of powder charge, and personal inaccuracies) that cause the hits to be scattered on the target, with closest grouping near the center and decreased density in all directions from the center. After a series of shots a circle can be drawn such that half of the hits are inside and half outside the circle. The chances are even that an additional shot will fall inside the circle, which in this case represents the "probable error" of a single shot.

Probable Error has been used to a considerable extent in reporting experimental results in the past. Standard Deviation is more convenient in statistical work and is now generally given preference. However, when a quantity is encountered in the literature (particularly in older reports) with a $\pm$ term, such as 106.245 ± 0.004, it is likely that the P.E. is intended. Of course, a statement should accompany the data to avoid uncertainty as to the meaning of the term, and in the absence of such information the value should be treated with caution until the uncertainty can be cleared up. A better form for the above quantity is "106.245 with a S.D. of 0.006."

The objections to the use of P.E., in addition to its being less convenient than σ, come from the feeling that while even odds are a good sporting proposition, the results of scientific investigations should have greater certainty. There is some support for the idea of using $\pm 2\sigma$ limits, which gives about 95 per cent assurance. However, if σ is specified, the user understands the magnitudes, and can set limits appropriate to the situation. Probable Error was discussed here, as it will most likely be encountered at some time in the literature, and so its meaning should be understood.

2-18. Combination of Variances

The fact was mentioned earlier that variances are additive, and that this constitutes one of the conveniences of this term. We can now show why this is the case. We shall study the effect on a combined function of the variances of its components.

If X is a function of several component variables, each of which is subject to random effects, which may be expressed as follows,

$$X = f(x_1, x_2, \ldots, x_n)$$

and if x_1, x_2, ... are independent variates, then for small variations in x_1, x_2, ... from their mean values, denoted by Δx_1, Δx_2 ... , the resulting variation of X from its mean value for any one determination is given by,

$$\Delta X = \frac{\partial X}{\partial x_1} \Delta x_1 + \frac{\partial X}{\partial x_2} \Delta x_2 + \ldots \tag{18}$$

ignoring differentials of higher order, where Δx_1 ... are the variations occurring in that particular determination. By squaring this equation we obtain,

$$(\Delta X)^2 = \left(\frac{\partial X}{\partial x_1}\right)^2 (\Delta x_1)^2 + \left(\frac{\partial X}{\partial x_2}\right)^2 (\Delta x_2)^2 + \ldots$$
$$+ 2\left(\frac{\partial X}{\partial x_1}\right)\left(\frac{\partial X}{\partial x_2}\right)(\Delta x_1 \cdot \Delta x_2) + \ldots \tag{19}$$

Now, if the variations of x_1, x_2, etc., are independent, as assumed, positive values of one increment are equally likely to be associated with positive or negative values of other increments, so the sum of the cross-product terms tends to zero in repeated observations. By definition of V as mean-square error, the mean of $(\Delta X)^2$ for repeated observations becomes the variance of X, denoted $V(X)$, so we may write,

$$V(X) = \left(\frac{\partial X}{\partial x_1}\right)^2 V(x_1) + \left(\frac{\partial X}{\partial x_2}\right)^2 V(x_2) + \ldots \tag{20}$$

since in repeated measurements $(\Delta x_1)^2$ tends to the mean value $V(x_1)$, etc. This shows that the component variances are additive, *with* the weighting factors $(\partial X / \partial x_1)^2$, etc., which express the *relative influence of the various components* on the combined function. The S.D. of X may be found from this equation, and may be expressed either in variances, or in the S.D.'s, of the components.
That is,

$$\sigma(X) = \sqrt{V(X)}, \quad \text{with } V(X) \text{ from (20)}, \tag{21}$$

or,

$$\sigma(X) = \sqrt{\left(\frac{\partial X}{\partial x_1}\right)^2 \sigma_{x_1}^2 + \left(\frac{\partial X}{\partial x_2}\right)^2 \sigma_{x_2}^2 + \ldots} \tag{22}$$

The important qualifications are that x_1, x_2, ... be *independent* of each other, and that the increments be *small*, so that terms of order higher than the first may be neglected. The latter qualification is usually not a serious limitation in engineering applications, in which it is generally possible to keep random effects under fairly close control. Workers in some other branches — agricultural experiments, for example — are not so fortunate.

2-19. Significance of the Mean

The arithmetic mean is used so generally as the measure of central value of a set of observations that we should investigate its properties. For this purpose, let x_1, x_2, ... x_n represent the n observed values of a quantity x. Call the central value X, *determined in a way not yet specified*. Then the deviations from this central value are,

$$(x_1 - X), (x_2 - X), . \quad . \quad . (x_n - X)$$

The sum of the squares of the deviations is given by,

$$S_{sq} = (x_1 - X)^2 + (x_2 - X)^2 + . \quad . \quad . + (x_n - X)^2 \tag{23}$$
$$= (x_1^2 + x_2^2 + . \quad . \quad . + x_n^2) - 2X(x_1 + x_2 + \\ . \quad . \quad . + x_n) + nX^2 \tag{24}$$

We wish to make this quantity as small as possible by selection of the best value of X. For this purpose, differentiate (24) with respect to X, and equate to zero,

$$\frac{dS_{sq}}{dX} = -2(x_1 + x_2 + . \quad . \quad . + x_n) + 2nX = 0 \tag{25}$$

From this we find for X,

$$X = \frac{1}{n} (x_1 + x_2 + . \quad . \quad . + x_n) = \bar{x} \tag{26}$$

That is, the *sum* of *squares* of the deviations is *smallest* if the central value is taken as the *arithmetic mean* of the observed values. The statement is often made,

THE MOST PROBABLE VALUE OF A SET OF DISPERSED DATA IS THE
ARITHMETIC MEAN

As we have proved, the arithmetic mean is the most probable value in the sense that it makes the sum-of-squares of the deviations a minimum.

We cannot declare that the arithmetic mean as determined from a small set of data is the true value of the quantity we seek, but only that the mean is the most probable value *on the basis of the data available to us*. If we should take another set of readings, the mean may well be different — and in general, would be expected to be different. For several sets of readings, the means would show some scatter, though to a lesser extent than would the readings within a single set.

The reliability of the mean increases as the amount of data on which it is based increases. If we can imagine obtaining the mean of an "infinite" amount of data (the "Universe" of data), the mean becomes the true central value, since on our postulate of a large number of small errors with random plus and minus signs, the summation of large numbers of readings would give cancellation of random effects.

2-20. The Standard Deviation of the Mean

As we have discussed, the mean of a large number of readings is more reliable than the mean of a small set, and far more reliable than a single reading. This must signify that the S.D. of the mean becomes smaller when the mean is based on a large number of readings. We shall now investigate the basis for this expectation.

The arithmetic mean of the n readings, x_1, x_2, ..., x_n, may be written,

$$\bar{x} = \frac{1}{n}\,(x_1 + x_2 + \;.\;\;.\;\;.\; + x_n) \tag{27}$$

We may find the variance of this quantity, noting that $\partial \bar{x}/\partial x_1 = 1/n$, and similarly for all other derivatives. Therefore, the variance of $\bar{x}$ is given by the sum of the terms,

$$V(\bar{x}) = \left(\frac{1}{n^2}\,V_1 + \frac{1}{n^2}\,V_2 + \;.\;\;.\;\;.\; + \frac{1}{n^2}\,V_n\right), \tag{28}$$

by application of (20). There are n terms in the parentheses, so we may take the total value as n times the mean variance of the single terms, or,

$$V(\bar{x}) = \frac{n\bar{V}(1)}{n^2} = \frac{1}{n}\,\bar{V}(1) \tag{29}$$

where the symbol $\bar{V}(1)$ is used to denote the mean variance of a single reading. Therefore, the variance of the mean of n terms equals $1/n$ times the mean variance of the single terms that enter into the mean.

In terms of the standard deviation, from (29),

$$\sigma(\bar{x}) = \frac{\sigma(1)}{\sqrt{n}} \tag{30}$$

This equation has important experimental implications. It shows that the precision can be increased (σ reduced) by taking more observations, but the improvement is *slow* because of the $\sqrt{n}$ factor. That is, 90 readings are only *3* times as good as 10 readings, not 9 times. This is a "law of diminishing returns" that makes it impractical to secure a large increase of precision simply by taking more readings. If increased precision is required, beyond a certain point, a new experimental plan or new equipment should be used, rather than depending upon extensive repetition of readings.

Actually, there is another objection to extremely long sets of observations, in addition to that of diminishing returns indicated by (30). As a set of readings is extended over a long period of time, the difficulty of holding conditions steady becomes increasingly great, so additional uncertainties usually are introduced in spite of all precautions. For this reason, the gain in precision by an increased set of readings may be materially less than would be indicated by the mathematical relationship in (30).

2-21. Extension to Finite Data

So far we have assumed generally that we were dealing with large sets of data, or that the Universe mean is known in computing deviations, so that we can apply the random-chance ideas inherent in the Normal law. In any actual investigation we deal with a finite amount of data, which may be regarded as a few readings drawn by chance from the large "Universe," or "Population." Some items of handling must be modified, and some new ideas introduced, when we consider the finite case.

As proved in par. 2-19, the arithmetic mean of a set of observations is the most probable value of the unknown, and this fact holds true regardless of the extent of the data — for a given set of data, the mean is the best estimate we can make. The computation of variance and standard deviation are somewhat different in the case of finite data.

One question that arises when we begin to analyze a set of data concerns the nature of the Universe from which it is drawn. Does it follow the Normal law, or some other distribution? This question is of some importance, since the theoretical relationships are based upon Normality. We cannot answer the question with certainty, particularly if the sample is small. We frequently assume that the distribution is Normal; there is a presumption in favor of this assumption, since an approximation to Normality is met in many physical situations. However, in an unknown type of variation the assumption should not be made without some study. For large samples, Normality may be checked by fitting an equation to the data. A shorter way, for at least a rough check, is to plot the cumulative probability data on probability paper. Probability paper is designed with a special abscissa scale distributed so that Normal data result in a straight line — the S.D. can be determined from the intercepts for the 16 per cent and 84 per cent ordinate values. (84 − 16 = 68 = percentage of cases for ±1σ.)

Some distributions are encountered that are only roughly Normal in form; there is a central maximum, with values falling off fairly smoothly on either side, but not of exact Normal form. Many of the tests and conclusions based upon normality can be used in these cases without being seriously invalidated.

2-22. Standard Deviation for Finite Data

The individual deviations used in computing the S.D. should be measured from the true mean, that is, from the universe mean, giving values such as

$$(x_1 - \mu), (x_2 - \mu), . \circ \; ,$$

resulting in a sum-of-squares,

Sum of squared deviations from universe mean
$$= (x_1 - \mu)^2 + (x_2 - \mu)^2 + . \; . \; . + (x_n - \mu)^2 \quad (31)$$

However, in handling a limited set of data, we do not know the universe mean, and must compute deviations from the mean of the set, as, $(x_1 - \bar{x})$, $(x_2 - \bar{x})$, ..., so that the sum becomes,

Sum of squared deviations from mean of set
$$= (x_1 - \bar{x})^2 + (x_2 - \bar{x})^2 + . \; . \; . + (x_n - \bar{x})^2 \quad (32)$$

We proved in par. 2-19 that the sum of squared deviations is a minimum when measured from the mean of the set; that is, (32) is smaller than the correct value, (31). We remove the "bias" introduced in this way by using a smaller divisor in computing the mean square; the best estimate of the standard deviation of a set of n readings is obtained by using $(n-1)$ as the divisor,

$$\text{Variance} = s^2 = \frac{1}{n-1}[(x_1 - \bar{x})^2 + (x_2 - \bar{x})^2 + . \; . \; . + (x_n - \bar{x})^2] \quad (33)$$

(The S.D. of a limited set is called s to distinguish it from the S.D. of the Universe, denoted by σ. However, this distinction is not always made, and σ is used for finite as well as infinite data. The difference between them, in a numerical sense, becomes small for large sets of data.)

The standard deviation as computed here gives a measure (r.m.s. value) of the dispersion of the data — that is, it gives a measure of the chance deviation of *any single point* of the data. (For $\pm s$, 68 per cent of the readings are included on the average, or approximately a 2-to-1 chance that any reading taken at random will have a deviation no greater than $\pm s$.) The *average* of a *set* of readings can be expected to be closer than this, so, as proved in (30), the mean may be expected to diverge from the true value by the smaller deviation,

$$\text{S.D. of mean of } n \text{ readings} = s_n = \frac{s_1}{\sqrt{n}} \quad (34)$$

The term *degrees of freedom* is used a great deal in statistical computations. Degrees of freedom (d.f.) may be regarded as the number of observations which may vary independently of each other without changing the values of the constants estimated from the data. For n observations, the set of data in itself possesses n degrees of freedom, in that any reading may have any value, as governed by the probability curve. In the use of the data to determine the mean and the S.D., one relationship connecting the readings is imposed in finding the mean, leaving $(n - 1)$ d.f. for the S.D. This is the divisor as we found it in (33).

A general rule may be stated for degrees of freedom: The number of degrees of freedom equals the number of independent observations minus the number of constraints imposed in determining the parameters. As we have seen, the determination of the mean of a set of data imposes one constraint. As another example, if we are fitting the equation of a straight line to a set of n readings, *two* constraints are imposed (two constants for the line), so the d.f. remaining for the computation of the variance is $(n - 2)$.

"Degrees of freedom" seems at times a somewhat elusive concept. It may appear more reasonable if we consider the effect in (33) if n were used as divisor, instead of $(n - 1)$. Suppose that we go to the extreme of having just *one* reading. In this case the average is the same as the reading, and the deviation is zero, so with n as divisor, the result is,

$$\text{S.D.} = \frac{0}{1} = 0 \quad \text{(incorrect)}$$

This cannot be correct, as with only one reading we have no certainty that the reading is correct, and so cannot say that the standard deviation is zero. However, if we use $(n - 1)$, the result is,

$$\text{S.D.} = \frac{0}{1 - 1} = \frac{0}{0} = \text{indeterminate} \quad \text{(correct)}$$

This result is correct, since with only one reading we are in no position to estimate the deviation of the result from the true value.

The use of degrees of freedom in finding S.D. may be regarded as compensating, in a small set, for the uncertainty of the mean, which is used in finding the deviations of the individual observations. The use of d.f. rather than the number of observations, is very important in a small group of observations; in a large set the difference is inconsequential.

Example:

We shall take a set of five readings and determine the mean, the variance, and the S.D. The variance will be computed by two methods. One method is a literal application of (33), by first finding $\bar{x}$, then subtracting it from each x value in turn, and squaring the deviations so obtained. The numbers are tabulated, showing 0.268 for the sum of the squared deviations. The other method depends upon the relationship,

$$\sum (x - \bar{x})^2 = \sum x^2 - \frac{S^2}{n} \quad *$$ (35)

where $S = \sum x$ = sum of the x values. This form is particularly convenient if the calculations are made on a computing machine with cumulative multiplication, in which case the individual entries in the x^2 column need not be made; also, by this method the individual deviations need not be computed, which is a saving of labor, particularly if the mean is carried out to several decimal places. The method would not be good for slide rule work unless, as in the table, each number is reduced by a constant. Otherwise, squares accurate to five significant figures would be needed in this problem.

x	$\begin{array}{c} x - 24 \\ = x' \end{array}$	$\begin{array}{c} (x - 24)^2 \\ = x'^2 \end{array}$	$x - \bar{x}$	$(x - \bar{x})^2$
24.2	0.2	0.04	− 0.02	0.0004
24.3	0.3	0.09	0.08	0.0064
23.9	− 0.1	0.01	− 0.32	0.1024
24.6	0.6	0.36	0.38	0.1444
24.1	0.1	0.01	− 0.12	0.0144
121.1	$S = 1.1$	$\sum x'^2 = 0.51$		0.2680

$$\bar{x} = \frac{121.1}{5} = 24.22$$

$$\left(\text{or, } \bar{x} = 24 + \frac{1.1}{5} = 24.22\right)$$

$$\sum x'^2 \qquad = 0.51$$

$$\frac{S^2}{5} = \frac{1.1^2}{5} = 0.242$$

$$\sum (x - \bar{x})^2 \; = 0.268$$

The sum-of-squares of the deviations is seen to check by the two methods to give 0.268

---

*This relationship may be proved as follows:

$$\sum (x - \bar{x})^2 = \sum (x^2 - 2x\bar{x} + \bar{x}^2)$$
$$= \sum x^2 - 2\bar{x} \sum x + n\bar{x}^2$$
$$= \sum x^2 - n\bar{x}^2, \text{ since } \sum x = n\bar{x}$$
$$= \sum x^2 - \frac{S^2}{n}, \text{ where } S = n\bar{x}$$

Therefore,

$$\text{Variance} = V = s^2 = \frac{0.268}{4} = 0.067$$

$$s = \text{S.D.} (x) = \sqrt{0.067} = 0.259$$

$$\text{S.D. of mean} = \frac{0.259}{\sqrt{5}} = 0.115$$

2-23. Summary

The quantities discussed in the foregoing paragraphs may be summarized for reference, with distinction between Universe (Population) and finite data cases:

Universe	*Finite Sample* *(n readings)*
Mean = true value = μ	Mean = most probable value $= \bar{x}$
Variance = $V_u = \sigma^2$ = mean of deviations measured from μ; $V_u = \dfrac{\Sigma (x - \mu)^2}{N}$	Variance = s^2 $= \dfrac{\Sigma (x - \bar{x})^2}{n - 1}$
S.D. of individual readings $= \sigma = \sqrt{V_u}$	S.D. of individual readings $= s = \sqrt{V}$
S.D. of mean $= 0$	S.D. of mean $= s_n = \dfrac{s}{\sqrt{n}}$

2-24. Combined Errors — Effect of Component Errors

In par. 2-18 we studied the combination of variances for a function composed of several component effects, applying it to the statistical situation. We shall now return to this subject to give it further treatment, and to see other places in which the same forms may enter. The component terms have direct physical significance, useful in assessing many processes in measurement in which the desired result depends on several factors. For example, we may compute R from E/I, where both E and I are subject to error.

As written in par. 2-18, if the combined variation may be expressed as a function of several variables,

$$X = f(x_1, x_2, \ldots x_n) \tag{36}$$

then the change in X effected by the variation of x_1 is, in differential notation,

$$dX_{z1} = \frac{\partial X}{\partial x_1} dx_1 \tag{37}$$

and similarly for x_2 and other components, as expressed in (18). When these variations are summed up in the *random* case, the resulting scatter in X is as written in (22).

However, we shall now investigate (37) to derive a general meaning that may be applied to other situations as well. Suppose that we consider the area of the rectangle shown in Fig. 2-7(a), with increments indicated in both x and y directions. The area is given by,

$$\text{Area} = A = xy \tag{38}$$

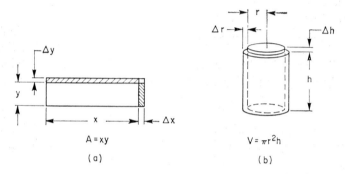

$A = xy$

(a)

$V = \pi r^2 h$

(b)

Fig. 2-7. Graphical interpretation of the effect of component increments upon a combined function.

The effect of an incremental change, Δx, in the length is an increment of area,

$$\Delta A_x = \frac{\partial A}{\partial x} \Delta x = y \Delta x \tag{39}$$

which expresses the small shaded rectangle at the right of the figure. Similarly,

$$\Delta A_y = \frac{\partial A}{\partial y} \Delta y = x \Delta y \tag{40}$$

which gives the area of the small shaded rectangle across the top. The two increments together express the area that would be added by simultaneous addition of the Δx and Δy increments, except for the small corner piece, which is the product of two small quantities, and so is unimportant. Note that for the proportions shown in the figure, an increment Δy causes a larger error in area than does an equal Δx.

As a second illustration, consider the circular cylinder of Fig. 2-7(b). The volume is expressed,

$$\text{Volume} = V = \pi r^2 h \tag{41}$$

The effect on the volume of a change of radius is

$$\Delta V_r = \frac{\partial V}{\partial r} \Delta r = 2\pi r h \, \Delta r \tag{42}$$

Note that $2\pi r$ is the perimeter of the added layer (inside and outside perimeters are nearly the same for a thin shell), so the volume of the shell is represented by (42). Likewise,

$$\Delta V_h = \frac{\partial V}{\partial h} \Delta h = \pi r^2 \, \Delta h \tag{43}$$

gives the volume of a disc added by an increase Δh in height. Again, the component expressions have geometric significance.

The expressions for the components have many uses. We shall apply them mainly to the analysis of errors, but we should not think that they are peculiarly restricted to errors. For example, if we have a metal cylinder and apply a copper plating Δr thick to the cylindrical surface, we may obtain the volume by (42).

It should be emphasized that the expressions are valid if the components act *independently*, otherwise statements such as (37) are not true. Also, they are true only for *small* increments (except linear functions), so that we may neglect increments of second, and higher, order.

The way that we *combine* the increments depends on their use.

2-25. Combined Errors — Normally-distributed Functions

The basic point is that the variances due to different component errors are additive for the deviations arising in dispersed data, with multiplying factors such as $(\partial V/\partial x_1)^2$ to express the influence of each term upon the combined result. We derived the expressions in par. 2-18; what we are doing here will not change those results, but merely add comparisons and explanation. We derived in equation (20),

$$V(X) = \left(\frac{\partial X}{\partial x_1}\right)^2 V(x_1) + \left(\frac{\partial X}{\partial x_2}\right)^2 V(x_2) + \cdots \tag{44}$$

We obtain $\sigma(X)$ as the square root of $V(X)$, and may, if we wish, replace $V(x_1)$ by $\sigma_{x1}{}^2$, giving (22),

$$\sigma(X) = \sqrt{\left(\frac{\partial X}{\partial x_1}\right)^2 \sigma_{x1}{}^2 + \left(\frac{\partial X}{\partial x_2}\right)^2 \sigma_{x2}{}^2 + \cdots} \tag{45}$$

We may write this in notation similar to (37) as,

$$\sigma(X) = \sqrt{(dX_{x_1})^2 + (dX_{x_2})^2 + \cdots}, \tag{46}$$

where

$$dX_{x1} = \frac{\partial X}{\partial x_1} \sigma_{x_1}$$

$$dX_{x2} = \frac{\partial X}{\partial x_2} \sigma_{x_2}$$

. . . . . .

Note, in (44), that variances add directly in the case of *dispersed* data of Normal form. This leads to (45) or (46), which show that component errors add as *square-root-of-sum-of-squares* in *normally-dispersed* data. (This applies also to P.E.'s, since they are directly related to S.D.'s by the 0.6745 factor.)

The reason for this form of addition can be justified in a general way, apart from the mathematical derivation. The sum-of-squares addition gives a *smaller* result than would direct addition of the component terms, and this is reasonable, since in a combination of random effects it is *not probable* that the component errors would all be at maximum value simultaneously. *Scatter* makes it probable that a maximum deviation of one factor will be combined in many cases with a small deviation of another factor, or even a deviation of such sign as to counteract part of the effect.

The sum-of-squares combination carries a lesson for us in planning experiments to achieve precision. If we find that one of the terms in (45) or (46) is much smaller than the other, it will have only minor influence upon the result. Accordingly, it is not efficient to expend a great deal of time and effort in improving the precision of the small part. It is much more to the point to try to improve the large term.

Special Cases:

(a) Sum of Quantities.
   If X is the sum of two functions,

$$X = x_1 + x_2,$$

the partial-derivative terms are unity, and (44) becomes,

$$V(X) = V(x_1) + V(x_2) \tag{47}$$

$$\sigma(X) = \sqrt{\sigma_{x_1}^2 + \sigma_{x_2}^2} \tag{48}$$

(b) Difference of Quantities.
   Note that if $X = x_1 - x_2$, the variance and S.D. of X are the same as written in (47) and (48) for the sum, since variance is a squared quantity, and hence positive. The fact that the S.D. is the same for the sum or the difference of two quantities enters in an important way in some statistical combinations. The S.D. is the same in *numerical* value in the two cases, but is of course larger in proportion for the difference. This is another

example of the decrease in relative precision suffered in taking the difference of quantities.

2-26. Guarantee Errors (Limiting Errors)

Makers of measuring equipment give a statement and a guarantee of the degree of reliability of their product. If a 1000-ohm resistor is guaranteed to have an error no greater than 0.1 per cent, it may be specified in the catalog as 1000 ohms $\pm$ 0.1 per cent, or, 1000 ohms $\pm$ 1 ohm. This notation should not be confused with the plus-or-minus notation met for S.D. (or P.E.). The manufacturer is *not* specifying a S.D. or P.E. of 0.1 per cent — it would certaintly not be satisfactory to the buyer to know that he had only a statistical chance of being within the specified limit. The maker *promises* that the error is *no greater* than the limit set; a reputable and experienced maker stakes his reputation on the truth of the promise, and usually tries to stay well within the limit, frequently issuing shop instructions for adjustment to half the guarantee in order to have a margin against errors in production.

A 0.1 per cent guarantee on a decade resistor box (usually a larger tolerance on the small steps) carries the promise that *no* setting on the dials will be in error by more than the given limit (within specified temperature and time limits). There is no expectation of a Normal-law distribution of values — in fact, Normal law permits some large deviations, which a guarantee does not.

The same idea of a guarantee limiting the *worst possible case* applies also to measurements involving several components, each of which is delimited by a guarantee error. For example, suppose that three decade boxes, each guaranteed to ± 0.1 per cent, are used in a Wheatstone bridge circuit to measure an unknown resistor, X. The operating process is to adjust the decade resistors until the galvanometer shows no deflection, then read resistance values from the dials and compute X from the bridge equation. Balance is a matter of the true values (subscript t) of the bridge arms, which may differ from the nominal or marked values (subscript n). That is, the true value of X is

$$X_t = \frac{R_t A_t}{B_t}$$

The true values of the bridge-arm resistances may be written, for ± 0.1 per cent guarantee, as,

$$R_t = R_n(1 \pm 0.001),$$

with similar expressions for A_t and B_t. Accordingly,

$$X_t = \frac{R_n A_n}{B_n} \frac{(1 \pm 0.001)(1 \pm 0.001)}{(1 \pm 0.0001)}$$

We must recognize the possibility of cumulative effects in the $\pm$ signs. The two terms in the numerator *may* both be plus to the full limit, and the denominator minus, giving a resultant error of practically 0.003 or 0.3 per cent. Thus the limits on X, as set by the equipment used in the experiment, is

$$X_t = X_n(1 \pm 0.003)$$

(The partial-derivative approach, as in (37), may also be used in the analysis of this problem.)

Note that the ± 0.3 per cent is not a probable error in the technical use of the term, nor is it "probable" in the general sense of "likely." (Usually the component errors are below the limit, and the signs may not match the worst combination.) The 0.3 per cent error is *possible* in the sense that it is *not impossible* by the guarantees on the components.

It is seen that the guarantee value is obtained by taking a *direct sum* of the possible errors, adopting the algebraic signs that give the *worst possible case*. The setting of guarantee limits is necessarily a pessimistic process. This is true from the standpoint of the maker of the equipment in his promise to the buyer, and it is also true of the user in setting accuracy limits in results of measurements. If you make technical measurements for a client, and submit a report over your signature, stating the results and the accuracy of the results, you are not in a position to promise more than this type of combination, unless you actually conduct a closer calibration of the equipment used. Or, if you appear as expert witness in a law suit which is to decide whether certain material meets the specifications under which it was sold, you will be decidedly off-base if you assert greater accuracy than the value computed by algebraic combination. (This is so unless you conduct a new calibration — and even then you must recognize limits to *your* calibration.)

We shall cite one further example, and make computations by evaluation of component errors as in (37). Suppose power to a resistor is being determined by,

$$P = I^2R$$

and both I and R have certain *guaranteed* limits to their specified values. Within what limits can we guarantee the computed value of power? For a numerical case, suppose that we know,

$$I = 4.00 \text{ amperes} \pm 0.5\% = 4.00 \pm 0.02 \text{ amp}$$
$$R = 100.0 \text{ ohms} \pm 0.2\% \quad = 100.0 \pm 0.2 \text{ ohm}$$

Then, for the component errors in power,

$$\Delta P_I = \frac{\partial P}{\partial I} \Delta I = 2 IR \cdot \Delta I = 2 \times 4.00 \times 100.0 \times (\pm 0.02)$$

$$= \pm 16 \text{ watts}$$

$$\Delta P_R = \frac{\partial P}{\partial R} \Delta R = I^2 \cdot \Delta R = 4.00^2 \times (\pm 0.2)$$

$$= \pm 3.2 \text{ watts}$$

The indicated power is

$$P = 4.00^2 \times 100.0 = 1600 \text{ watts,}$$

to which we must append the uncertainties on the guarantee basis as,

$$P = 1600 \pm 16 \pm 3.2 \text{ watts}$$
$$= 1600 \pm 19.2 \text{ watts}$$
$$= 1600 \text{ watts} \pm 1.2\%$$

Note that the two uncertainties in watts are added directly for the guarantee case. Also, the percentage error in the result might have been obtained as $(2 \times 0.5 + 0.2)$, but the above type of analysis may be useful in situations with more involved relationships of the components.

In Summary:

(1) Component effects may be evaluated in the same way, by (37), for either random or guarantee cases.

(2) The method of combination of components differs in the two cases:

RANDOM EFFECTS — SQUARE ROOT OF SUM OF SQUARES

GUARANTEE LIMITS — DIRECT ARITHMETIC SUM, INTER-
PRETING SIGNS TO GIVE THE WORST POSSIBLE DISCREPANCY

2-27. Some Applications of Statistical Methods

Statistical methods cannot create facts not contained in the experimental data, but they can help to reveal relationships that otherwise might be obscured by random fluctuations. They can give some measure of the reliability of information, and can help in comparisons of materials and processes:

(1) In measurements of basic and derived quantities, sometimes involving rather long, interrelated series of tests.

Computations are made to determine the desired quantity with the greatest precision inherent in the test data, and also to obtain a measure of the precision of the result, which frequently involves a combination of components, as we have discussed in recent paragraphs. Sometimes a quantity is measured by two methods, and if the two methods agree closely, we can feel happy and fairly secure. However, if each method yields a closely grouped set of data, but the difference between the averages is more than can be accounted for by the small scatter, a discrepancy is discovered; we must then conduct additional tests to determine which method is at fault (both may be). Thus, the measure of precision is important in detecting inaccuracy.

(2) The design of a system, or the investigation of conditions within an operating mechanism, involving a large number of factors.

It may be impossible, or at least impractical, to evaluate analytically the effect of each factor, and to select the combination that gives the best operating conditions. A series of tests is needed to evaluate the effects of the components, and to determine the direction and amount of change in the result attributable to a change in one item. The situation would be relatively simple if tests and data were unlimited, but this is never the case since tests involve both time and money. It is thus desirable to extract the greatest amount of information, with the greatest possible reliability from a minimum of data. Statistical studies make very important contributions in this area. At one time statistical methods were considered applicable only to large masses of data, but recent developments permit increased reliability of results derived from "small data" experiments. Agricultural research into the yields of different varieties of grain and the effects of soil treatments gave great impetus to the handling of data of limited amount and of large dispersion. The difficulties of controlling "external conditions" and the impracticability of providing a large number of plots for each treatment and sub-treatment made it necessary to plan experiments and handle the results in such way that valid conclusions could be drawn from limited data in spite of masking random effects. Several terms in common use, such as "treatment," "plot," and "split-plot" give indication of the origin. However, it has been found that experimental work in other technical fields can profit from the same ideas. It is generally possible to keep external conditions under close enough control, so that the scatter is proportionally smaller, but the problem remains of separating significant trends from random effects.

(3) Sampling, inspection, and quality control.

It is necessary to inspect the output of manufacturing processes to insure that the product meets the requirements, but in quantity production it is not practical or economical to test each unit. Information must be obtained by testing a few samples, in order to give reliable measure of the whole "population." Such problems arise in factory production and in acceptance tests by purchasers. Quality control is a branch of statistics that applies the results of sampling to effect better control of processes, so that rejected material will be at a minimum. A great deal of specialized study has been given to sampling theory, so we shall not attempt to cover it here. Some of the statistical tests that we discuss will be useful in forming a basis for judging several applications.

2-28. An Introduction to Statistical Tests

Every measurement is subject to experimental error of random nature, the scatter being due either to small unknown causes or to a residue from incomplete correction of known disturbing factors. We cannot expect that the errors in a small set of data will balance out to give the mean the correct value it would have if disturbing elements could be ruled out completely, or, what amounts to the same thing, if an extremely large amount of data could be taken. Accordingly, we are disturbed if we see a great deal of variation between readings, as it casts doubt upon the reliability of the mean. For measurement of a single quantity in a homogeneous set of tests, we may use Variance or Standard Deviation as a measure of the regularity of the data, and hence of the degree to which random effects were kept under close control. Therefore, the smaller the S.D. in comparison with the mean, the greater the confidence we have in the measurements (at least with respect to the *precision* of the determinations). The S.D. of the mean is $s/\sqrt{n}$; this is the *average* expectation for a group of n determinations. For a *single set* of n readings the deviation may be greater; this matter will be considered below in the "*t*-distribution."

We frequently make comparisons of some property of two materials or of two treatments in a given process, or of two methods of measuring a quantity. With the scatter that may be expected in a small set, the means and variances of two sets will differ, even if drawn from the *same* test conditions on the *same* material. Accordingly, when we conduct tests on two materials and find a difference in the results, the question arises whether the observed difference shows a real superiority of one material, or whether it is the scatter inherent in small samples. Unless the improvement is conspicuously greater than the variation that might occur by chance in two random samples drawn from the same population, the case is not proved for the merits of the new material or treatment. Differences must be investigated to determine whether they have real meaning, so we have "Tests of Significance." Statistical methods are necessary, since there is no other way to distinguish between chance variations and the real effects of physical relationships. Actually, we cannot disprove the possibility in any case that the entire observed variation occurred by chance, but we may be able to show that the odds against chance occurrence are unacceptably great.

The basis of analysis is the "Null Hypothesis," which means, simply, that we begin by assuming that there is no real difference between the two sets of measurements, and that they may, in fact, come from the same population. We compute the appropriate statistics (means, or variances, as required by the problem in hand), and make comparisons between them. We then consult a table that has been prepared for chance occurrences in accordance with the Normal law. If the table indicates a very low probability of chance occurrence of the observed ratio, we conclude that the Null Hypothesis is false, and that a real difference does exist between the two sets. For example, if we find from the tables for our test conditions that the probability figure is 10 per cent, it means there is a 1-in-10 chance that there is no essential difference in the systems from which the two sets of data came. Said in another way, it means that if similar tests were to be repeated many times, that one time in ten a difference as great as observed could be expected to occur, on the average, with data drawn from the same population. Odds of 1/10 sound fairly impressive, but we may not wish to accept the results as conclusive on this basis, as 1-in-10 shots *do* occur at times. If the odds of chance occurrence are longer, as 1/20 or 1/100, we must conclude either that a very improbable event has occurred, or that the hypothesis of no real difference must be rejected. If the odds are sufficiently long, the latter conclusion is the logical one.

A 1 per cent probability of the observed sets of data occurring from random sampling, with no real difference in the systems from which they came, gives 99 per cent "confidence" that there is a real difference. A *low* chance probability value corresponds to a *high confidence level:* that is, a high confidence that a real difference exists. A 1 per cent probability (of the Null Hypothesis being true) gives 99 per cent support for the existence of a significant difference in the materials or treatments.

A probability level above 10 per cent is usually not considered significant statistically; from 5 per cent to 10 per cent, possibly significant; 1 per cent to 5 per cent significant; below 1 per cent highly significant. The tables usually show values of the statistical function for 10 per cent, 5 per cent, and 1 per cent levels. The probability level that we require before accepting the validity of a new material or treatment depends upon the application. The confidence level that we require depends on the consequences that might attend a mistaken adoption of a particular course of action — that is, on how much a mistake may cost.

The difference in results between two treatments, in order to be regarded as significant, depends not only upon the ratio of the values, but also upon the amount of data upon which the comparison is based. With more data, greater confidence may be placed in means or variances, so the ratio of values in two sets need not be as great to give a prescribed confidence level. The extent of the data is expressed in terms of the degrees of freedom of each set, and tables are available which show probabilities in terms of the degrees of freedom.

2-29. Tests Applied to the Mean

There are many cases in which it is necessary to estimate the reliability (precision) of a mean, or to determine the significance between the results of different materials or processes. Statistical analysis helps define the types of comparison needed, and provides tests for the purpose. The basis for the test is the "*t*-distribution," and is frequently called "Student's test" from the nom de plume of the originator, W. S. Gosset. This test extends to small samples, many of the comparisons previously conducted only on large groups of data, and thus may be considered the beginning of small-sample testing. The test takes several different forms, depending upon the type of use.

(1) *To estimate the precision of the mean of a set of readings.*

Many practical situations require decisions based upon amounts of data that are small in the statistical sense, so it is important to be able to estimate the degree of precision for the results of limited sets. The mean and the standard deviation, *s*, computed from a small sample possess uncertainties due to random effects. We consider that in taking a small set of readings we are drawing a few values by chance from the large universe of readings that might reside in the situation. Random variations do not, in general, cancel completely in a small group, so the computed results may lie on either the low or high sides of the values that would come from a large set.

Degrees of freedom, rather than the number of observations was used in (33) in finding *s*, to allow, *on the average*, for the bias introduced by the use of a small sample. We need now to estimate the limits within which the *mean* of a set may lie. We cannot state a hard and fast limit, but we can determine from the laws of probability the limits beyond which there is only a $\frac{1}{10}$ chance (or $\frac{1}{20}$, $\frac{1}{100}$, etc.) that the true (universe) value will lie. For this purpose, the quantity *t* is defined as,

$$t = \frac{\text{difference of sample mean from universe mean}}{\text{S.D. of mean of } n \text{ determinations}}$$

$$= \frac{\bar{x} - \mu}{s/\sqrt{n}} \tag{49}$$

where *s* is the S.D. of the *n* readings, based on $n - 1$ degrees of freedom, and so *t* is based upon the same d.f.

The expression is of somewhat limited usefulness in this form, since the universe mean, μ, is not generally known, except in some theoretical cases, or where a large amount of data has been accumulated on a process. One use of *t* is suggested if (49) is rewritten,

$$(\bar{x} - \mu) = t \frac{s}{\sqrt{n}} \qquad (50)$$

where t is obtained from a table of the t-distribution, which is available in many books on statistical methods, and in handbooks.*

Example:

Five observations of a quantity are, 4.56, 4.43, 4.61, 4.58, and 4.47. By calculations as shown in par. 2-22,

$$\bar{x} = 4.53, \quad \sum (x - \bar{x})^2 = 0.0234, \quad V = \frac{0.0234}{4} = 0.0058$$

$$s = \sqrt{0.0058} = 0.077, \qquad \text{S.D. of } \bar{x} = \frac{0.077}{\sqrt{5}} = 0.034$$

The S.D. as obtained here signifies, on the basis of *average expectations*, that 68% of the time the mean of a set of five readings will not differ from universe mean by more than 0.034, or, for 95% of the time by more than $1.96 \times 0.034 = 0.067$. However, realizing that we are basing our predictions on a *single* set of five, we make allowance for a somewhat greater deviation. We refer to a table of t-distribution, of which a small sample section is reproduced for illustration:

| | | *Mult. Factor for the S.D. of the Mean* | | | | | |
| | | *Degrees of Freedom* | | | | | |
Probability (Chance Occurrence)	*Significance Level*	2	3	4	5	10	∞
10%	90%	2.92	2.35	2.13	2.02	1.81	1.64
5%	95%	4.30	3.18	2.78	2.57	2.23	1.96
1%	99%	9.93	5.84	4.60	4.03	3.17	2.58

From the table above, for 4 d.f., if we wish a 95 per cent assurance of being correct, the true mean of the quantity being measured lies between the limits

$$4.53 \pm 2.78 \times 0.034 = 4.53 \pm 0.09.$$

That is, we can expect, if we take many groups of five readings, that the means of the groups will lie within these limits (4.62 to 4.44) in 95 per cent of the tests. That is, the odds are 19 to 1 that these limits bracket the true value. Note that for the given five readings, the mean, 4.53, is the best estimate we can give of the mean. The added uncertainty of a small sample (as compared with infinite data) is compensated by the use of 2.78 σ instead of 1.96σ to include 95 per cent of the cases. (The ∞ column in the table is the same as the Normal law.)

---

*See *Handbook of Chemistry and Physics,* 41st Ed. pp. 218-19.

The "confidence level" is high if the probability is small that the given departure from the mean could occur through chance for samples drawn from the same universe. Note, from the table, that if we require a higher degree of certainty, we must allow a greater range of doubt in estimating the possible departure of sample mean from the universe mean.

(2) *Test of differences.*

Suppose that we wish to test the relative merits of two processes for making a product. We divide each batch of material into two parts, use Process A on one and Process B on the other, and measure the pertinent property of the product. We do this n times, and then examine the data to determine if there is a significant difference in the results by the two methods.
Let

x_1 = measurement on product from Batch 1 by Process A

y_1 = measurement on product from Batch 1 by Process B

$d_1 = x_1 - y_1$ = difference, Batch 1 (similarly through d_n)

The mean difference,

$$\overline{d} = \frac{(\sum d)}{n}$$

The S.D. of the differences is,

$$s = \sqrt{\sum (d - \overline{d})^2 / (n - 1)}$$

and the S.D. of the mean difference is $s/\sqrt{n}$.
In this case we define t as,

$$t = \frac{\overline{d}}{s/\sqrt{n}} = \frac{\text{mean of the differences}}{\text{S.D. of the mean of the differences}} \qquad (n - 1 \text{ d.f.}) \quad (51)$$

We can calculate this quantity for the given data, and then find in the table, for the given degrees of freedom, the probability that this difference could occur by chance, if there is really no difference between the results by the two processes. This answers the question, "Is the difference of results real, or does it appear only because of random effects in a small sample?" If the probability of chance occurrence of the observed ratio is only 5 per cent or 1 per cent, we would probably reject the "null hypothesis," and decide that there *is* a real difference between the two processes. If the chance probability is as great as 10 per cent, we may require further tests before we decide definitely on one process or the other.

(3) *Two independent samples of any size, to determine whether their means differ significantly.*

Suppose that we have two independent sets of readings,

$$\text{Set } X: \quad x_1, x_2, \ldots \ldots x_n$$

$$\text{Set } Y: \quad y_1, y_2, \ldots \ldots y_m$$

We take the tentative attitude that there is no difference between the sets (null hypothesis). On this basis, we pool all the data, and compute the combined variance as indicated by (33) for each set,

$$V = s^2 = \frac{\sum (x - \bar{x})^2 + \sum (y - y)^2}{(n - 1) + (m - 1)} \tag{52}$$

On the basis of the common variance,

$$\text{Variance of } \bar{x} = V_{\bar{x}} = \frac{V}{n}$$

$$\text{Variance of } \bar{y} = V_{\bar{y}} = \frac{V}{m}$$

If we write for the difference of the means,

$$d = \bar{x} - \bar{y},$$

Then, the variance of d is,

$$V_d = \left(\frac{\partial d}{\partial x}\right)^2 V_{\bar{x}} + \left(\frac{\partial d}{\partial y}\right)^2 V_{\bar{y}} = \frac{(1)^2 V}{n} + \frac{(-1)^2 V}{m}$$

$$= V\left(\frac{1}{n} + \frac{1}{m}\right) \tag{53}$$

then, S.D. of d = S.D. of $(\bar{x} - \bar{y}) = \sqrt{V}\sqrt{\frac{1}{n} + \frac{1}{m}}$ \hfill (54)

$$= s\sqrt{\frac{1}{n} + \frac{1}{m}} \tag{55}$$

where $s = \sqrt{V}$, with V as given in (52). From this, t is defined as,

$$t = \frac{\bar{x} - \bar{y}}{s\sqrt{\frac{1}{n} + \frac{1}{m}}} = \frac{\text{difference of the means}}{\text{S.D. of the difference of means}}$$

$$(n + m - 2 \text{ d.f.}) \tag{56}$$

Again, we can judge from the results and the tables whether the observed difference between the means is greater than we would reasonably expect by chance sampling.

It is assumed in this case that the two samples have the same S.D.

within the limits of random sampling, and strictly, the test is valid only under these conditions. However, the test is seriously invalidated only if the difference is very marked. An exact test exists, but is more difficult, and is avoided if the above test is reasonably adequate, or can be made so by a change of variable (such as using log x in place of x).

2-30. Analysis of Variance

Analysis of the variance of sets of observations is a powerful method of extracting information from test data, and one of the most important branches of statistical work. It is essential in separating regular variations from random effects in limited sets of data. In experimental programs that investigate the influence of several factors upon a system, the analysis permits estimates of the effect of each factor, and the remaining part due to error. We shall take a brief look at the planning of experiments to indicate how it is possible, by careful arrangement of a test program, to take advantage of possibilities of statistical analysis.

The "Ratio Variance Test" applies to the "F-distribution" (from the name of the originator, R. A. Fisher). The quantity F is defined as,

$$F = \frac{V_1}{V_2}, \quad \text{where } V_1 > V_2 \tag{57}$$

A large V_1/V_2 ratio means a small probability that the two results could be drawn from the same universe by random sampling, and that a real difference is therefore involved. Tables of this distribution have been prepared. The table is a multiple array, in that different numbers of degrees of freedom must be listed for V_1 and for V_2, and at different probability levels.

An example will be given to illustrate the application of the test to a simple case. The following table shows the results of tests of the dissipation factor of samples of electrical cable, as conducted by two different test procedures. The samples were cut from the same production run, and hence are subject only to such variations as occur in processing a continuous length from the same batch of material. The data are to be examined to determine the regularity of results by the two methods of test.

Nine readings are available from one method and five from the other. While more information would be helpful, there are many situations in which conclusions must be drawn from relatively limited data. The dissipation factors were numbers such as 0.00612; the entries in the table were formed by subtracting 0.006 and moving the decimal point four places, to give simpler numbers to handle. The means and variances were computed as in an earlier example, in par. 2-22.

DISSIPATION FACTOR OF CABLE SAMPLES

Sample	Method A $(D - 0.006) \times 10^4$	d^2	Sample	Method B $(D - 0.006) \times 10^4$	d^2
1	1.2	1.44	1	3.9	15.21
2	− 0.4	0.16	2	0.4	0.16
3	2.2	4.84	3	4.3	18.49
4	0.5	0.25	4	1.7	2.89
5	2.7	7.29	5	0.7	0.49
6	1.6	2.56			
7	2.1	4.41			
8	0.9	0.81			
9	2.4	5.76			
	$S = 13.2$	27.52		$S = 11.0$	37.24

$$\bar{d}_A = 13.2/9 = 1.47$$
$$\bar{D}_A = 0.006147$$
$$\Sigma\, d^2 = 27.52$$
$$13.2^2/9 = 19.36$$
$$\Sigma\,(d - \bar{d})^2 = 8.16$$
$$V_A = 8.16/8 = 1.02$$

$$\bar{d}_B = 11.0/5 = 2.20$$
$$\bar{D}_B = 0.006220$$
$$\Sigma\, d^2 = 37.24$$
$$11.0^2/5 = 24.20$$
$$\Sigma\,(d - \bar{d})^2 = 13.04$$
$$V_B = 13.04/4 = 3.26$$

$$F = \frac{V_B}{V_A} = \frac{3.26}{1.02} = 3.20$$

From the tables: For $V_B = 4$ d.f., $V_A = 8$ d.f.,

$$F = 2.81 \quad \text{for } 10\% \text{ probability}$$

$$F = 3.84 \quad \text{for } 5\% \text{ probability}$$

(By interpolation, $F = 3.20$ for approximately 8 per cent probability)

The variance by Method A is 1.02, and by Method B, 3.26. It thus appears that A has a much greater degree of regularity and precision than B, but the question arises, "May this difference occur by chance, due to the relatively small number of readings?" The ratio of variances is 3.26/1.02 = 3.20. If we now go to an F-distribution table for four degrees of freedom for B and eight for A, we find a ratio of variances of 2.81 for a 10 per cent probability level, and 3.84 for 5 per cent. Our value is between these figures, and by interpolation we may determine that a variance ratio of 3.20, for the given degrees of freedom, corresponds to a probability of about 8 per cent (confidence 92 per cent). That is, there is an 8 per cent chance (1 in 12) that the two methods may be essentially the same in precision, and that the observed difference is due to chance variations in the readings or in the samples. Odds of 12 to 1 that there is a difference are fair evidence of a real difference, although, if there is a great deal at stake, we may wish to make further tests before accepting A over B.

This study, it should be noted, has to do with the regularity of results by the two methods. Such tests cannot detect a *systematic* error in either method, and so we cannot say that A is more *accurate* than B. Some sort of basic calibration of the two pieces of equipment would be needed to determine accuracy. There is some difference in the averages by A and B in the above table, but the difference is bracketed by a $\pm 1\sigma$ range, and so is not conclusive of error.

2-31. The Planning of Experiments

Careful planning of an experimental program, particularly one of large scope, is essential if definite answers are to be obtained. Planning contributes to the precision of the results, and helps to get the maximum information from the minimum of experimental work. Some experiments may be relatively simple, but others involve the combination of a large number of factors, each of which must be evaluated in its effect on the functioning of the complete system. The first necessity in planning is a clear understanding of the objectives, otherwise much data may be taken that do not contribute usefully to the desired result. In a situation with many new features it is difficult to appraise the importance of some factors accurately. Some false starts are unavoidable, but they may be reduced by care in planning. Sometimes a short exploratory set of measurements is useful.

It is very important that the planning be done *in advance*, and not improvised from step to step. Moreover, in a program of any size, it is important that this be done with the statistical aspects in mind. This means, in general, that a qualified statistician should aid in the planning. It happens all too often in poorly organized experiments that much data are found at the end to be useless, or nearly so, and much time and labor has been wasted. It is too late to consult a statistician at this stage; his help should have been sought at the start.

The means of increasing the precision and accuracy of the experiment must be considered. One of these is the repetition ("replication," in statistical terms) of a complete set of experimental trials. Replication serves (1) as protection against faulty material and out-of-line readings, (2) to provide an estimate of precision, (3) to increase precision (it can increase precision of results, but does not eliminate systematic errors). Another method of improving experimental results is the technique of planning to balance out undesired variables, or to separate the effects of different variables. A little later we shall consider several types of balanced design that permit a number of factors to be included, with provision for the statistical separation of the effect of each factor.

A number of variables may enter into the design of a device or system. We wish to evaluate the influence of the factors, so that we select the optimum combination for our purpose. This requires a series of experiments in which the factors are changed in an orderly manner, and the results measured in each case. An important consideration in arranging the test schedule is the presence or absence of *interaction* between the factors. If the effects of the factors on the result are additive, each one acting independently of the others, then there is no interaction. That is, if an increment of factor A produces the same result regardless of the level of factor B, these two factors are free of interaction. Freedom from interaction simplifies statistical combinations. The situation is analogous to the property of linearity in electrical circuits, which makes superposition of voltage and current components possible. Where interaction exists, more data are required, and more involved treatment of the data, in order to separate the effects of the factors.

There are many, many possible test arrangements that may be used, depending upon the conditions. The subject fills many books and technical articles, so we will take only a short look at the principles of a few of the basic schemes. In doing so we shall refer to "treatments" (methods, processes, etc.) being applied to "experimental units" (materials, batches, plots, groups of vacuum tubes, and so on).

2-32. Randomized tests

In a randomized test the different treatments that are to be compared are allotted to the experimental units entirely by chance. There is one merit here, in that the chance arrangement is in accord with analysis based upon laws of probability. Complete flexibility is allowed, in that any number of treatments and replicates may be used, with no fixed pattern required. Also, the statistical analysis is easy. However, there is some objection on the grounds of accuracy; since the allocation of treatments is entirely random, there is no assurance that the units which receive one treatment are similar to those which receive another treatment, so that the whole of the variation among the units enters into the experimental error. A considerable number of tests may be required before reduction of sampling error is effected. "Partially randomized" experiments are also used, where experimental units are assigned to treatments in as homogeneous a way as possible. "Block designs" are arranged to balance out uncertainties of the randomized design.

2-33. Factorial tests

In a factorial experiment, tests are made with all combinations of the different factors. If 3 materials are to be processed by 3 different methods at one temperature, 9 tests are needed, (with replications of each). If there are 3 materials, 3 methods, 3 temperatures, then 27 test conditions are required. The total runs up rapidly with an increasing number of factors at several different levels. In a program of measurement of gear noise, consideration was given at first to testing a gear transmission at 2 speeds, 3 loads, 3 rates of lubricant feed, 3 temperatures and 4 directions of introducing lubricant, giving

$$2 \times 3 \times 3 \times 3 \times 4 = 216 \text{ test conditions.}$$

(The number of tests was later reduced by a split-block design.) The factorial design gives thorough range, but a great deal of testing, including replications.

2-34. Latin square designs

The Latin square is an arrangement which permits at least two factors *other* than the one being studied to vary during an experiment, but excludes the principal part of their variation from the error of the experiment. A

Group on First Factor

	I	II	III	IV
1	A	B	C	D
2	B	C	D	A
3	C	D	A	B
4	D	A	B	C

Group on Second Factor

Fig. 2-8. A Latin square arrangement with 4 × 4 blocks.

diagram of a 4 × 4 Latin square is shown in Fig. 2-8, where the letters A, B, C, D represent one factor or treatment. Note that each letter appears once, and only once, in each row and in each column. (There are several arrangements, or permutations, of letters that satisfy this requirement.) The name "Latin square" derives from the appearance of the letters in the squares. Note that all row totals are equally affected by variations of the first factor, and all column totals by variations of the second factor. The double grouping eliminates from the errors all differences among rows, and equally all differences among columns. While different arrangements may be made, the experimental material should generally be arranged so that differences along rows and columns represent the major sources of variation. The systematic arrangement of variables in the Latin square permits reliable information to be obtained from a much smaller number of tests than would be required with a completely randomized plan.

As an example, we shall suppose an experimental investigation of the effect of four different types of cathode construction upon the mutual conductance of a vacuum tube. We recognize that variations may enter due to peculiarities of the machine used to process the tube and to the abilities of the operator. We arrange for four machines and four operators to construct the tubes for the test. If all of tube Type I were made by Operator A on Machine 1, any bias arising from machines and operators would be included in the test results, so we want to avoid this possibility by using a combination of the factors. If we plan to test tubes of all four types made by each of four operators on each of four machines (a complete

Cathode Types

		I	II	III	IV	Row Total
Machines	1	A	B	C	D	
	2	B	C	D	A	
	3	C	D	A	B	
	4	D	A	B	C	
Col. Total						

A, B, C, D = Operators

Fig. 2-9. A Latin square arrangement in a vacuum-tube development program.

factorial arrangement) the number of tests becomes $4 \times 4 \times 4 = 64$. As an alternative to so many tests we outline the program of Fig. 2-9. It is not known in advance whether machine or operator differences exist, but this program guards against possible introduction of errors from these sources, and permits evaluation of the errors. All column totals are affected equally by machine and operator variations (assuming no interactions), so the results may be subjected to an analysis of variance to give a comparison of tube types, to determine whether there is a significant difference for any type.

It may be of interest to list the degrees of freedom entering in this design of test. We may separate the sources of variations, and the d.f. of each:

Source	d.f.	As applied to 4 × 4 square
Rows	$r - 1$	3
Columns	$r - 1$	3
Treatments	$r - 1$	3
Errors	$(r - 1)(r - 2)$	6
Total:	$r^2 - 1$	15

The size of Latin squares may range from small, as 3×3, to higher values, with 10×10 as about the upper limit. Small squares suffer in accuracy, and require replication; large squares become unwieldy.

2-35. Latin Square, Split-block (Split-plot) Design

Sub-treatments may be introduced in each block of a Latin square, to provide for testing under a greater assortment of variables. We may take as an illustration the noise measurements of a gear transmission (Reference 9), mentioned in par. 2-33, in which it was desired to measure the gear noise under various conditions of load, speed, temperature, rate of lubricant feed, and direction of lubricant feed. The main divisions are shown in Fig. 2-10(a), which shows the combinations of the main blocks, but not the order in which the tests are run; the order is randomized to avoid systematic changes. Each block represents a day of testing, and is done at one temperature, since temperature adjustments are slow. Note, that when the data is arranged according to the diagram, each column contains tests ("treatments") at 3 loads, 3 temperatures, and one rate of lubricant feed. Each row contains tests at 3 lubricant feeds, 3 temperatures, and one load. Therefore, comparison by columns balances out temperature and load (assuming no interaction effects), giving a means of assessing lubricant effects. Similarly, comparison by row totals gives a measure of load effect, eliminating temperature and lubricant. The same data may be tabulated by temperatures, in which case each temperature group contains readings at all three loads and lubricant rates. Analysis of the data consists in tabulating the noise measurements by temperature, by load, etc., and computing values of the mean and variance for the noise readings.

Lubricant Feed — Gal/Min

Load	0.5	1	2
25%	120°	90°	160°
100%	90°	160°	120°
125%	160°	120°	90°

(a) General plan

90°–1 GPM		
Tang.,	bottom,	300 rpm
Inmesh,	top,	1200 rpm
Inmesh,	bottom,	300 rpm
Inmesh,	top,	300 rpm
Tangent,	top,	1200 rpm
Tangent,	top,	300 rpm
Tang.,	bottom,	1200 rpm
Inmesh,	bottom,	1200 rpm

(b) Subdivision of one block in (a) (different orders used in other blocks)

Fig. 2-10. A 3 × 3 split-block design used in a gear-test program.

Figure 2-10(b) indicates division into four directions of lubricant feed and two speeds (these additional readings may be taken rather quickly, once temperature conditions are established). Precautions are taken to randomize the order of the tests, so that no systematic variation or "bias," is introduced by changes of machine conditions during the tests. The split-block design illustrates how more than the three main variables may be introduced in the Latin square. Also, it illustrates that a great deal of information is obtained by a relatively small number of tests. The scheme shown here required a total of 72 determinations, as compared with 216 for a complete factorial set.

2-36. Incomplete Latin Squares; Youden Squares

The Latin square requires an equal number of classes in the three factors. This may be inconvenient, or undesirable, or the facilities in one factor may be insufficient. It is possible to get around this difficulty in some instances, and have a non-square array that still possesses balancing features of the complete blocks. "Youden squares" are one form of balanced incomplete block that possesses the double-control of the Latin square without requiring equality of numbers of classes in the three factors. This form of "square" originated with the work of W. J. Youden at the National Bureau of Standards.

The idea of this form of array may be illustrated by some tests that were conducted at the Bureau on special cells, investigated as a way of setting up a definite temperature*. The cell, containing a pure substance, maintains the melting point of that substance for at least 24 hours; the temperature is measured by a resistance thermometer left in the cell over-night to attain thermal equilibrium. A study was made to determine the reproducibility of temperature when a cell is used repeatedly, and to check agreement between cells filled with the same substance. A separate thermometer must be used with each cell being measured on any one day. If the same thermometer is used for all tests, the measurements require several days, possibly introducing day-to-day errors (called "day errors"), due to imperfect control of all conditions. If several thermometers are used, thermometer differences are introduced, in spite of care in calibration ("thermometer errors").

The following schedule of tests is a way of eliminating thermometer errors and day errors. There are 7 cells, denoted by A, B, C, D, E, F, G. Three thermometers, marked a, b, c, are used. The tests cover a period of 7 days, as follows:

	Days						
Thermometer 1	2	3	4	5	6	7
a	*A*	*B*	*C*	*D*	*E*	*F*	*G*
b	*B*	*C*	*D*	*E*	*F*	*G*	*A*
c	*D*	*E*	*F*	*G*	*A*	*B*	*C*

Note that each cell is measured by each thermometer on one day or another. Also, each cell is measured on one of the days with each of the other cells (once, in this schedule). Thermometer and day effects are balanced out, as may be shown by studying one cell, say A. If, for Day 1, we take twice the reading of A and subtract the sum of B plus D, we have, using A_1 to represent the result (subscripts denote thermometers):

---

*Reference 10,

$$\text{Day 1:} \quad 2A_a - (B_b + D_c) = A_1$$

$$\text{Similarly, Day 5:} \quad 2A_c - (E_a + F_b) = A_5$$

$$\text{Day 7:} \quad 2A_b - (C_c + G_a) = A_7$$

$$\text{Identity:} \quad A - A \qquad = 0$$

$$\text{Sum:} \quad 7A - (\text{all cells}) = A_1 + A_5 + A_7$$

or,

$$A = \text{Average of all} + \tfrac{1}{7}(A_1 + A_5 + A_7)$$

Similar equations may be written for each of the other cells.

Note that the differences, A_1, A_5, A_7 are free of day effects (A_1 is based on readings taken all one day, etc.). Also the sum of these quantities is free of thermometer errors since, as shown in the total, each thermometer appears in two plus terms and also in two minus terms. Thus the differences of the cell averages is known in terms such as $(A_1 + A_5 + A_7)$ which are free of both day and thermometer errors. The comparisons of each thermometer are obtained directly from the average for each thermometer, since each thermometer was used with each cell, and was used every day. If one thermometer is known as a standard, the other thermometers and the cells can be evaluated in terms of the standard.

Only certain numbers of cells, days, and thermometers can be combined in a scheme of this sort. Each cell must be found with each other cell on one or another day, an equal number of times. In addition to the arrangement above, we may use 4 thermometers, 7 cells, each cell being associated with each other cell twice. Other numbers are 4 and 13, 5 and 11, ... It will make an interesting exercise for the reader to study the requirements of these sets of numbers, and devise a test schedule.

This example was given to illustrate the possibilities of balancing out errors by careful advance planning of experiments. While the particular situation presented here is quite special, the same type of analysis may be useful in many places. It seems appropriate to close with a quotation from the final paragraph of the reference given above.

"The experimental arrangements given here show that a thought-out scheme for pairing thermometers and cells, combined with a well-planned time schedule for making the measurements, improves the precision of the comparisons that the investigator has in mind. ... The arrangements do not introduce any complications in the experimental work or call for any changes in technique. Indeed, insofar as these arrangements obviate the necessity for reproducing and maintaining the environmental conditions, the burden upon the investigator is lightened as the precision of the measurements improves."

REFERENCES

1. Davies, Owen L., *Statistical Methods in Research and Production*, London: Oliver and Boyd, 1949.

2. Youden, W. J., *Statistical Methods for Chemists*, New York: John Wiley & Sons, Inc., 1951.

3. Waugh, Albert E., *Elements of Statistical Method*, New York: McGraw-Hill Book Co., 1943.

4. Brownlee, K. A., *Industrial Experimentation*, Chemical Publishing Co., 1948.

5. Cochran, William G. and Gertrude M. Cox, *Experimental Designs*, New York: John Wiley & Sons, Inc., 1957.

6. Kempthorne, Oscar, *The Design and Analysis of Experiments*, New York: John Wiley & Sons, Inc., 1952.

7. Freeman, H. A., *Industrial Statistics*, New York: John Wiley & Sons, Inc., 1942.

8. Villars, Donald Statler, *Statistical Design and Analysis of Experiments for Development Research*, Dubuque, Iowa: Wm. C. Brown Co., 1951.

9. Day, B. B. and F. R. Del Priore, "The Statistics in a Gear-Test Program," *Industrial Quality Control*, **IX**, No. 5 (March 1953).

10. "Latin and Youden Squares: Statistical Designs for Greater Precision," *NBS Tech. News Bul.*, **34**, No. 5, (May 1950).

11. Rider, Paul R., *Modern Statistical Methods*, New York: John Wiley & Sons, Inc., 1939.

PROBLEMS

2-1. Ten determinations of the resistance of a resistor gave the values 101.2, 101.7, 101.3, 101.0, 101.5, 101.3, 101.2, 101.4, 101.3, and 101.1. Assume that only random errors are present. Compute: (a) the arithmetic mean, (b) the standard deviation of the readings, (c) the standard deviation of the mean of the readings.

2-2. Six determinations of a quantity, as entered on a data sheet, and presented to you for analysis, are the following: 12.35, 12.71, 12.48, 10.24, 12.63, 12.58. Examine the data, and on the basis of your conclusions compute the arithmetic mean, the standard deviation of the readings, and the standard deviation of the mean of the readings.

2-3. Each letter of the word "errors" is written on a separate card, and the cards are placed in a hat. Compute the average probability that the first four cards, drawn by chance, and placed on a table in the order of drawing, will spell the word "rose."

2-4. Same statement as in problem 2-3, but to form "dart" from "standard."

2-5. Same statement as in problem 2-3, but to form "pose" from "possesses."

2-6. Eight perfect pennies are tossed as a group, with care to avoid bias (each coin has equal chance, head or tails). Determine the average probability in a large number of tosses of having exactly five heads showing in a toss. Determine the average probability of having five or more heads.

2-7. Consider the eight pennies of problem 2-6. Calculate the arithmetic mean, the variance, and the standard deviation of the discrete probability distribution which describes the possible number of heads in a long series of throws. Calculate the combined chance of having either three, four, or five heads.

2-8. Four perfect dice are tossed at random in a long series of throws. Determine the average probability of four aces showing per throw, then of 3, 2, 1, and zero. Calculate the mean and the standard deviation of the distribution.

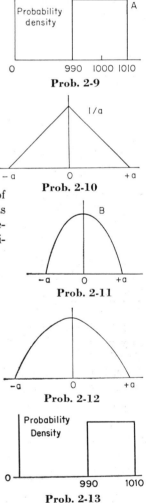

2-9. Suppose that the resistors of a group may have with equal probability any value between 990 and 1010 ohms, as expressed by the rectangular frequency-distribution diagram ("probability-density" curve). Compute the mean square (variance) of the deviations, and the standard deviation.

Prob. 2-9

2-10. If the curve showing the probability of occurrence of deviations of different amounts has a triangular form, as shown, find the mean square (variance) of the deviations, and the S.D.

Prob. 2-10

2-11. If the curve showing the probability of occurrence of deviations of different amounts has a parabolic form between $-a$ and $+a$ (zero elsewhere), as indicated in the sketch, compute the variance and the S.D.

Prob. 2-11

2-12. If the equation,

$$y = \cos(\pi/2a)x$$

expresses the probability of occurrence of deviations between $-a$ and $+a$ (zero elsewhere), compute the variance and the S.D.

Prob. 2-12

2-13. Variables B and C are each distributed according to the rectangular probability-density curve shown in the sketch. For the function defined by their product,

$$F = B \times C$$

find the following quantities,
  (a) The form of the probability-density curve of F
  (b) The probable value of F
  (c) The variance and S.D. of F.

Prob. 2-13

2-14. The three variables A, B, and C are each distributed as shown in problem 2-13. Calculate the arithmetic mean and the standard deviation of the probability distribution which describes the function,

$$M = A \log_{10} (B \times C)$$

2-15. A d-c load draws measured values of 112.3 v and 6.2 amp, with standard deviations of ± 0.2 v and ± 0.05 amp, respectively. Find the indicated number of watts, and its standard deviation.

2-16. A rod of circular cross section has a diameter of 1.020 in., with a S.D. of 0.0015 in. The length is 5.12 in. with a S.D. of 0.01 in. Find the probable volume of the rod, and the S.D. of this value.

2-17. The probable resistances of two resistors, and their standard deviations are specified as follows:

$$R_1 = 18.62 \text{ ohms}, \qquad \text{S.D.} = 0.025 \text{ ohm}$$
$$R_2 = 74.48 \text{ ohms}, \qquad \text{S.D.} = 0.050 \text{ ohm}$$

Find the probable value and the S.D. for the two resistors, (a) in series, (b) in parallel.

2-18. A steel ball is measured by micrometer calipers, and found to differ slightly in different directions. The probable mean radius and its S.D. as determined by extensive measurements are:

$$r = 0.500 \text{ in.}, \qquad \text{S.D.} = 0.002 \text{ in.}$$

The material has a unit weight:

$$w = 0.280 \text{ lb per cu in.},$$
$$\text{S.D.} = 0.005 \text{ lb per cu in.}$$

Find the probable weight and the S.D. in weight, knowing that the volume of a sphere is $4\pi r^3/3$.

2-19. Three resistors, according to an extensive set of measurements, have probable resistances and standard deviations:

$$R_1 = 10.00 \text{ ohms}, \qquad \text{S.D.} = 0.01 \text{ ohm}$$
$$R_2 = 25.00 \text{ ohms}, \qquad \text{S.D.} = 0.02 \text{ ohm}$$
$$R_3 = 50.00 \text{ ohms}, \qquad \text{S.D.} = 0.10 \text{ ohm}$$

Determine the resistance and the standard deviation when these three resistors are connected in parallel.

2-20. The arms of a Wheatstone bridge, marked in order around the bridge, are B, A, X, and R. The three known arms have the constants,

$$A = 1000.3 \text{ ohms}, \qquad \text{S.D.} = 0.1 \text{ ohm}$$
$$B = 100.12 \text{ ohms}, \qquad \text{S.D.} = 0.008 \text{ ohm}$$
$$R = 425.5 \text{ ohms}, \qquad \text{S.D.} = 0.48 \text{ ohm}$$

Find the probable value of X, and the S.D. of X.

2-21. A resistance is measured by the voltmeter-ammeter method. The voltmeter reading is 96.6 v on a 300-v scale. The ammeter reading is 3.22 amp on a 10-amp scale. Both voltmeter and ammeter are guaranteed to be accurate within ±0.5 per cent of the full-scale indication. Find the indicated value of resistance, and the limits within which you can guarantee the result. Consider power losses in the instruments to be negligible.

2-22. The current flowing in a resistor was measured as 4.00 amp on a 10-amp ammeter on which the error is guaranteed by the maker not to exceed ±0.5 per cent of the full-scale indication. The resistor is guaranteed to be 50.0 ohms within ±1.0 per cent. What power was consumed, and to what percentage can you guarantee the result?

2-23. The power factor and phase angle in a circuit carrying sinusoidal current and voltage are determined by measurements of current, voltage, and power. The voltage is read as 125.0 v on a voltmeter of 150-v scale; the current 3.00 amp on a 5-amp ammeter; the power, 225 watts on a 500-watt wattmeter. The ammeter and voltmeter are guaranteed to be accurate within ±0.5 per cent of full-scale value, and the wattmeter within 0.25 per cent of full scale.

(a) To what percentage accuracy can you guarantee the power factor, as obtained from these readings?

(b) For the possible error in (a), show the amount contributed by each instrument.

(c) What is the possible error in phase angle?

2-24. A Wheatstone bridge has ratio arms which are guaranteed to be accurate within ±0.025 per cent, and a variable arm ("rheostat arm") guaranteed to ±0.05 per cent. What is the possible error in a resistance measurement on this bridge, as determined by the guarantees?

2-25. (a) During a test of an electric motor the input was read as 4120 watts and the output as 3730 watts, with each value in doubt by ±10 watts. What is the percentage uncertainty in the losses of the motor as determined by the difference of input and output readings?

(b) The same as (a), except that we are testing a transformer and have an input of 3070 watts and an output of 3000 watts, each with an uncertainty of ±10 watts.

2-26. A resistance $R_1 = 10$ ohms is paralleled by $R_2 = 40$ ohms. Find the change in resistance of the parallel circuit if R_2 is increased by an increment of 1 ohm.

2-27. State the number of significant figures in each of the following numbers:

(a) 302   (b) 302.10   (c) 0.00030
(d) 0.00003   (e) 5.01×10^4 (f) 50100

2-28. Three resistors have values of 62.3, 2.73, and 0.612 ohms, respectively, with an uncertainty of one unit in the last figure given in each case. Find the resistance of the three in series, giving only significant figures in the answer.

2-29. If in a d-c circuit I = 2.75 amp and E = 117.5 v, each with an uncertainty of one unit in the last place given, compute the watts and give only significant figures in the answer.

2-30. Each of the following numbers has an uncertainty of one in the last figure. Determine the number of significant figures in the product.

(a) 46.2 × 3.15, (b) 4246.2 × 3.15, (c) 46.2 × 2.15.

Compute the percentage uncertainty in each product caused by a change of one unit in the last figure of each number.

CHAPTER 3

Aids to Computation

3-1. Accuracy in Computation

In measurements work we generally stress the idea of accurate results, so we must make computations with an accuracy at least as good as the basic data and preferably somewhat better, to avoid adding an appreciable computational error to a possibly unavoidable experimental error. It is possible to keep the computational error to as low a value as we desire by checking carefully and by retaining a large number of figures in the computations. However, this may easily be overdone and be carried to a silly extreme. It is important to develop judgment as to what is important and what unimportant, so that satisfactory results may be obtained with a minimum of effort.

It is not particularly laborious to carry a large number of figures if a computing machine is available. This is often not the case, particularly for student work. There are frequent situations in which tiresome long-hand work can be avoided by the use of simple mathematical principles without the sacrifice of the desired accuracy, and these shorter methods are useful also for making approximations by mental arithmetic. Computations can be made difficult or easy, depending on technique. Mathematical methods are tools; the skillful craftsman in any field can secure good results with minimum effort by the intelligent selection and use of available tools.

3-2. Binomial Expansion

The binomial expansion has a number of uses. We shall write it first in general terms before making specific applications:

$$(a + b)^n = a^n + na^{n-1} b + \frac{n(n - 1)a^{n-2}}{2!} b^2 + .\quad .\quad . \tag{1}$$

If we make one of the terms in the parentheses unity, as we always can by factoring, we may write:

$$(1 + d)^n = 1 + nd + \frac{n(n - 1)}{2!} d^2 + .\quad .\quad . \tag{2}$$

The situations in which this is particularly helpful are those for which d is very much smaller than unity. For these cases we may make the following approximations, neglecting the terms of the second or higher power in d. (The first *two* terms are generally used. The first neglected term is shown in brackets, for possible use in checking error.) If $d \ll 1$,

$$\frac{1}{1 \pm d} \simeq 1 \mp d + [d^2] \tag{3}$$

$$\sqrt{1 \pm d} \simeq 1 \pm \tfrac{1}{2} d + [-\tfrac{1}{8} d^2] \tag{4}$$

$$\frac{1}{\sqrt{1 \pm d}} \simeq 1 \mp \tfrac{1}{2} d + [\tfrac{3}{8} d^2] \tag{5}$$

Whether the approximation is satisfactory or not depends on how small d is and also on the required accuracy. It is possible to improve the accuracy by the computation of the term in brackets, but the short-cut method appears less attractive when this added step is required. There are many cases in which we meet correction terms very close to unity, and then the two terms are satisfactory. (For the square root of a number of any size, see par. 3-5.)

Examples:

$$\frac{1}{0.99} \simeq 1.01 \qquad \text{(True value 1.0101 ...)}$$

$$\frac{1}{0.9835} \simeq 1.0165 \qquad \text{(True value 1.01677 ...)}$$

$$\frac{1}{9.835} \simeq 0.10165$$

$$\frac{1}{1.0052} \simeq 0.9948$$

$$\sqrt{1.0028} \simeq 1.0014 \qquad \text{(True value 1.001399 ...)}$$

$$\frac{1}{\sqrt{1.016}} \simeq 0.992 \qquad \text{(True value 0.992095 ...)}$$

3-3. Right Triangle with One Short Side

Solution can be made for the hypotenuse of a right triangle with one side much smaller than the other by use of the binomial expansion. This has electrical application in a-c circuits in finding impedances, adding right-angle components of voltage, and so on.

Let a = longer side, b = shorter side, c = hypotenuse. Then

$$c = (a^2 + b^2)^{1/2} = a\left(1 + \frac{b^2}{a^2}\right)^{1/2} \tag{6}$$

$$\simeq a + \frac{b^2}{2a} \quad \text{if } b \ll a$$

Example:

$$a = 100, \quad b = 8, \quad c = 100 + \text{64}/\text{200} = 100.32$$
$$\text{(Exact value, } c = 100.3194 \ldots)$$

The "correction" term in such a case is small, and can be computed with enough accuracy by slide rule, or even by mental arithmetic. The error by this method becomes 0.1 per cent when b/a is about 0.3, but is only 0.02 per cent for b/a of 0.2.

A similar expression can be used in the reverse direction, if c and b are known, to find a:

$$a \simeq c - \frac{b^2}{2c} \quad (b \ll c) \tag{7}$$

Another computation that may be needed in the right triangle is the finding of an angle, which frequently causes trouble in the case of angles near zero or 90°. Very small angles are frequently met in measurement work, as in a-c bridges, and cannot be disregarded. Any angle can be found in trigonometric tables, but there is difficulty at times when a slide rule is used. The ordinary tangent scale reads tangents from 0.1 to 10, or angles from about 5.7° to 84.3°. Scales are provided for smaller angles, as on the ST scale of the Trig or Decitrig rules which read from tangent of 0.01 to 0.1. Provision is made for still smaller angles, but the ordinary user does not make sufficiently frequent application to feel assurance in the process. A very simple way to determine the small angles is based on the fact that below about 0.1 the sine, tangent, and the angle itself (in radians) are practically equal. Degrees are obtained by multiplying radians by 57.3. The following examples are stated in terms of a right triangle having sides a and b.

Examples:

$$a = 100, \quad b = 8, \quad \theta = \tfrac{8}{100} \times 57.3 = 4.58°$$

$$a = 6.50, \quad b = 0.012, \quad \theta = \frac{0.012}{6.50} \times 57.3 = 0.106°$$

$$a = 0.012, \quad b = 6.50, \quad \theta = 90 - 0.106 = 89.894°$$

3-4. Improved Accuracy with the Slide Rule

We frequently wish to make computations to 0.1 per cent or better, and would like to use a slide rule for speed and convenience. However, 0.1 per cent is about the limit for a single setting on a 10-in. slide rule. Even simple computations will have several times this error, probably about 0.5 per cent. Sometimes a long-hand computation may be the easiest way out. Other times it is possible by some ingenuity to use the slide rule in such way that greater precision is attained. This may take the form of using the slide rule to compute a difference or correction term, rather than the whole quantity. Some examples are given in the following paragraphs; no doubt the reader can exercise his ingenuity and extend the list. (The use of the binomial expansion, already given, belongs in this group, for it frequently permits the finding of a result to a high degree of precision by the slide-rule computation of a correction term, as in finding the hypotenuse of a right triangle.)

(a) *Quotient near Unity (or Ten, etc.).*

Use the slide rule to compute only the *difference* from unity, as:

$$\frac{10542}{10311} = \frac{10311 + 231}{10311} = 1.0224$$

The last four is written in large type, as it can be read fairly definitely. By direct slide-rule computation $10542/10311 = 1.022$, with the last figure somewhat in doubt. There is thus better than a 10 to 1 increase in precision in this case.

Other examples are

$$\frac{0.9223}{0.9535} = 1 - \frac{0.0312}{0.9535} = 0.9673$$

and

$$\frac{81.76}{7.83} = 10 + \frac{3.46}{7.83} = 10.442$$

(b) *Correction Terms to Simplify a Denominator.*

Cases not covered in (a) may be helped by rounding out the denominator to 10, 5, or other number that may be divided mentally. It depends on the fact that if additions are made to the numerator and denominator of a fraction in the same proportion as the original numerator and denominator, the value of the fraction is not changed. Stated mathematically:

$$\frac{a}{b} = \frac{a + d\dfrac{a}{b}}{b + d}$$

Examples:

$$\frac{621}{982} = \frac{621 + 18\frac{621}{982}}{1000} = \frac{621 + 11.4}{1000} = 0.6324$$

$$\frac{307}{475} = \frac{307 + 25\frac{307}{475}}{500} = \frac{307 + 16.16}{500} = \frac{323.16}{500} = 0.64632$$

(c) *Multiplication.*

The accuracy of multiplication can be increased if the first digit is multiplied accurately, and the slide rule is used only for the less important numbers. This is easy when the first digit is one, but the method may be extended to other cases with some increase in work.

Example:

$$862.3 \times 1.027 = 862.3 + 862.3 \times 0.027$$
$$= 862.3 + 23.3 = 885.6$$

(Direct slide-rule multiplication of the original numbers gives 885.)

(d) *Differences.*

The difference of nearly equal numbers must be computed carefully if accuracy is to be achieved. In finding the difference of two nearly equal fractions, do *not* try to evaluate each fraction individually by ordinary slide-rule division. Reduction to a common denominator and subtraction of terms in the numerator is one method, as indicated here:

$$\left(\frac{501}{2501} - \frac{1000}{5000}\right) = \frac{2505 - 2501}{2501 \times 5} = \frac{4}{2501 \times 5} = 0.000320$$

Note that slide-rule precision is adequate *after* the subtraction is made. Computations of this sort will be encountered in problems of unbalanced bridges.

Another possibility is to make use of methods (a) or (b) above, which may be adequate if the difference is not too small.

$$\left(\frac{1762}{1689} - \frac{1184}{1160}\right) = \left(1 + \frac{73}{1689}\right) - \left(1 + \frac{24}{1160}\right)$$
$$= 0.0432 - 0.0207 = 0.0225$$

Note that by direct slide-rule division the result would be

$$1.043 - 1.021 = 0.022$$

under the best conditions, and might easily be

$$1.042 - 1.022 = 0.020,$$

which is an error of 10 per cent.

3-5. Extraction of Square Root

Sometimes we want to find the value of a square root to greater accuracy than can be obtained by direct reading on a slide rule. There are several ways to do this. One method is to use logarithms; however, we may not have a table of logarithms, or we may not want to do the interpolation necessary when high precision is required. There is also a longhand method of root extraction, sometimes taught in elementary school; this is a bit tedious, and the details of computation may easily have been forgotten.

There is a simple alternative method which is less well-known than it should be. The process is this:

(1) Divide the given number by an approximate square root (obtained by slide rule, or even by mental estimate.)

(2) Take half the sum of divisor and quotient.

The scheme depends on the idea that if the divisor is a little below the true value, the quotient is a little high, so the average tends to cancel the discrepancies. However, we can show the mathematical basis in a more exact way. Suppose the number to be represented by A^2. The true square root is A, but our first approximation presumably differs from this by a small amount, d, which may be either plus or minus. Division of A^2 by $(A - d)$ may be accomplished by the binomial theorem as in par. 3-2, or by direct longhand division as shown here:

$$A^2 \qquad\qquad \underline{/A - d} \tag{8}$$

$$\frac{A^2 - Ad}{+ \, Ad} \qquad A + d + \frac{d^2}{A} + \ldots \tag{9}$$

$$\frac{Ad - d^2}{+ \, d^2}$$

$$\frac{\text{Sum}}{2} = \frac{(8) + (9)}{2} = A + \frac{d^2}{2A} + \ldots \tag{10}$$

$$\simeq A, \quad \text{if } d \ll A \tag{11}$$

Note that the first-order error terms cancel, leaving only second- and higher-order terms. If the approximate divisor is fairly close to the true value, these error terms are very small. For example, if d is 1 per cent of A (for comparison, call $A = 1$, $d = 0.01$), then the error is,

$$\text{Error} = \frac{d^2}{2A} = \frac{0.0001}{2} = 0.00005 = 0.005\%$$

That is, a trial divisor within 1 per cent of the true value gives, by this method, a square root accurate within 0.005 per cent. A trial divisor obtained by slide rule should be closer than 1 per cent, so the result would be even better than the 0.005 per cent indicated above. If the first divisor happens to be selected roughly (mental arithmetic), the division may indicate a rather large spacing between divisor and quotient; the process may then be repeated to obtain a closer approximation.

To illustrate by a numerical example, suppose the square root of 21,649 is desired. That is,

$$A^2 = 21,649$$

Approximate A by slide rule $= 147$

By division, $21,649/147 = 147.272$

$$\overline{\text{Average} = 147.136}$$

(As a check, we find that a further stage of division gives $A = 147.135992$, so the second division is not needed unless extreme accuracy is required.)

The method described above is especially easy if a motorized desk computing machine is available. (Manuals for computing machines frequently describe a method of square-root extraction that is much slower than the division process above.) Even by longhand division the process is relatively easy.

PROBLEMS

3-1. Find the hypotenuse of a right triangle whose sides are 250 and 25.

3-2. A complex impedance expressed in Cartesian form is $Z = 2.0 + j120$. Convert to the polar form.

3-3. Convert $1000 + j2.5$ to polar form.

3-4. Perform the following operations by approximate methods and secure results to an accuracy of at least 0.1 per cent.

(a) $\dfrac{1}{1.0169}$ (b) $\dfrac{1}{97.82}$ (c) $\dfrac{1}{0.001015}$

(d) $\dfrac{895}{964}$ (e) $\dfrac{258}{1096}$ (f) $\dfrac{583}{796}$

3-5. Compute the following by the binomial expansion approximation.

(a) $\sqrt{1.0182}$ (b) $\sqrt{102.44}$

(c) $\dfrac{1}{\sqrt{1.0096}}$ (d) $\dfrac{1}{\sqrt{101.64}}$

(e) $\dfrac{1}{\sqrt{0.9862}}$ (f) $\dfrac{1}{\sqrt{0.00964}}$

(g) $\sqrt{1.058}$ to 0.1 per cent; to 0.01 per cent

3-6. Compute the following differences so that the result is correct to one part in a hundred.

(a) $\dfrac{284}{560} - \dfrac{1}{2}$

(b) $\dfrac{2000}{3050} - \dfrac{300}{459}$

(c) $\dfrac{1}{\sqrt{1.0028}} - \dfrac{1}{\sqrt{1.0102}}$

3-7. Compute the square roots of the following quantities by the method of par. 3-5, with results accurate within 0.01 per cent.

(a) 637.29   (b) 0.0002838

3-8. Compute the square roots of the following quantities by the method of par. 3-5, with results accurate within 0.01 per cent.

(a) 736,242   (b) 13.386

3-9. Two resistors of nominal 100-ohm value have resistances which may be represented by $R_1 = 100 + a$, $R_2 = 100 + b$, where a and b are the *small* discrepancies (positive or negative) from the exact 100 ohms.

(a) Derive an expression for the ratio of the resistance given by these two resistors in series to their resistance in parallel, to show how this ratio differs from the value of four which exists for exactly equal resistances.

(b) To gain an appreciation of magnitudes, evaluate (a) for $a = 0.1, b = -0.2$.

CHAPTER 4

Measurement of Resistance

4-1. Foreword — Bridges, Galvanometers and Indicating Instruments

Most resistance measurements — practically all precision d-c measurements — are made by some type of bridge circuit, and these circuits require the use of a galvanometer as balance indicator. The subjects of bridges and galvanometers are accordingly tied together very closely. It is thought better, however, to consider them in adjacent chapters rather than to mix them. The author has elected to treat the bridges first, as preliminary laboratory measurements can then be begun with an elementary knowledge of galvanometer characteristics. On account of the close connection, however, it is recommended that reading be carried on in parallel in chaps. 4 and 5.

The subject of electrical indicating instruments is also closely allied to some parts of resistance measurements, but is presented here in a later chapter, since in many curricula this material has been treated in earlier courses. Chapter 17 is made available as reference material for those who have covered the subject before; it may be introduced at this point for those for whom the arrangement is appropriate.

4-2. Resistor Materials

Early experimenters used iron wire for making resistors, not because of good characteristics but because it was easily obtained. Later, a platinum-silver wire was developed for the British Association unit. The alloy known as German-silver was used for some resistors but has some undesirable characteristics. These two materials were great improvements over iron with respect to resistance change with temperature, but still the temperature effect was much greater than is desirable in a measuring device.

In 1884 Edward Weston discovered the two alloys now known as constantan and manganin. Constantan has a small resistance-temperature coefficient, but an objectionably large thermal emf against copper. Manganin, consisting of copper, nickel, and manganese, has to a large degree the qualities that are important in a resistor for measurement work, and is still used for very many precision resistors. The foremost feature is a temperature coefficient that can be made almost zero in the range of the usual room temperatures. The situation may be better understood by reference to Fig. 4-1, which shows the general form of the resistance-

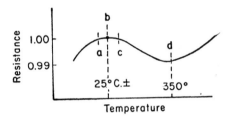

Fig. 4-1. Qualitative resistance-temperature curve for manganin.

temperature curve of manganin. There is a section from a to c in which the resistance change is very small. The position of the flat section can be controlled to some extent by the composition and heat treatment. For standard resistors, decade boxes, etc., the flat section is placed near 25°C. A somewhat different composition is used for ammeter shunts to give the flat region near 50°C.

Another property of manganin that is of very great importance is its low thermoelectric emf against copper. In a resistance unit the ends of the resistance wire are soldered to copper terminals, and copper wires are employed for external connections. Two thermocouples are thus formed, and if the two junctions are not at the same temperature there is a net emf set up around the circuit. A manganin-copper thermocouple has the low value of about 2 $\mu v/°C$, whereas other materials are in general much higher (constantan has about 40 $\mu v/°C$ against copper).

Other important properties of material for precision resistors is a high specific resistance (to avoid bulkiness) and permanence of value. Manganin is satisfactory in both respects. The resistivity is about 290 ohms for a circular-mil foot, which is a little more than 25 times the value for copper. A manganin resistor is quite permanent in value if it is heat-treated and aged to remove the strains incident to winding it to the required shape, and if the winding form and insulation are designed and treated to minimize the effects of temperature and humidity.

Several nickel chromium alloys have been developed and marketed under various trade names ("Nichrome," "Chromel," etc.) and have found extensive use in electrical heating devices due to their resistance to oxidation at high temperatures. They have the merit of an electrical resistivity somewhat more than twice that of manganin, but have not been much used for precision resistors due to rather large values of temperature coefficient and thermal emf against copper. Two typical alloys of this type are included in Table 2. Modifications of the nickel chromium alloys by the addition of small amounts of aluminum and copper have been developed recently with the objective of keeping the high resistivity, yet reducing the temperature coefficient. High resistivity is valuable, in requiring less material and in making possible the design of a resistor of smaller physical size. This is particularly important in resistors for a-c bridges in which we wish to keep residual capacitances to ground to as low a value as possible. (This is discussed later in chap. 13.) Another merit of the new alloys is a high tensile strength, that permits winding resistors from wire as small as 1 mil (0.001 in.) in diameter. They have excellent properties with respect to resistivity, temperature coefficient, thermal emf, stability, and resistance to corrosion.

An alloy of gold and chromium has been studied, and has had several applications. It has a nearly zero temperature coefficient over a wider range than any of the other alloys. The resistance-temperature curve is somewhat similar to Fig. 4-1, but the section *b-d* of the curve is much shorter and can be made to have nearly a zero slope by proper heat treatment, giving constant resistance from 20 to 30°C. Its thermoelectric effect against copper is a little larger than for manganin.

Information about these, and other, alloys is given in references **1, 6, 7, 8, 11,** and **13,** listed at the end of the chapter.

4-3. Resistance Standards

After 1884 the unit of resistance was based formally on the mercury ohm, but this was used only as a periodic check. The working standards were wire resistors, which since 1892 have been manganin. At this time several 1-ohm manganin resistors were constructed at the Reichsanstalt in Germany. Values were assigned to them in 1897 in terms of the mercury ohm as defined by the Congress of 1893. The average of the group has in the main been assumed to remain constant, but periodic checks have been made.

The international ohm was maintained at the National Bureau of Standards after 1910 by a group of 1-ohm resistors of a type developed there by E. B. Rosa.* Beginning in 1931 a new, double-walled type,† developed by J. L. Thomas, was introduced and is now used entirely for maintenance of the ohm.

All of these resistors were designed for immersion in a constant-temperature oil bath. The Reichsanstalt type has the resistance wire in a perforated container that is lowered into the oil, so that the oil circulates directly over the resistance element. This type has been found to give seasonal variations due to humidity, absorbed by the oil, and transferred to the insulation surrounding the wires. There is also a slow change, not consistent among individual units, over a period of years. In the NBS Rosa type the resistance wire is in a sealed container holding oil, which is thus protected from atmospheric effects. The long-time stability, however, is about the same as for the Reichsanstalt type. The Thomas type uses a double-walled construction, with the resistance wire wound about the inner cylinder, and the space between cylinders filled with dry air and sealed. The resistance wire is first wound to shape, then annealed on a mandril of the same size as the final mounting. Due to the care in annealing and mounting, these resistors have proved to be highly stable. Their load rating is somewhat less than for the other types, due to poorer heat transfer to the oil bath, but this is not serious. In other respects this type is markedly superior. Figures indicating stability are given in the reference.‡

Figure 4-2(a) shows the outer appearance of one of the double-walled resistors, which is constructed as a 1-ohm unit. The heavy terminals to right and left are the main current connections, and the binding posts to the rear are the potential connections. This is the 4-terminal construction used on all low-resistance standards for reasons that are discussed in par. 21 of this chapter. Figure 4-2(b) shows the double-walled construction, with a few wires indicated in the space between walls. The wire is in contact with the inner wall, except for a thin layer of insulating material.

The Rosa-type resistor, while not used as the primary standard for the maintenance of the ohm, does find application in many laboratories as a secondary standard which can be sent to Washington every few years for recertification at the Bureau. Its stability and freedom from seasonal humidity effects suit it very well to such service. Resistors of this type are customarily made in the decimal multiples, 1, 10, 100, 1000, and 10,000 ohms.

---

*See reference 3.
†See references 4 and 5.
‡See reference 5.

Fig. 4-2. Double-walled standard resistor, Thomas type. (a) View of 1-ohm standard. (b) Cross-sectional view. (*Courtesy of the National Bureau of Standards.*)

(a)

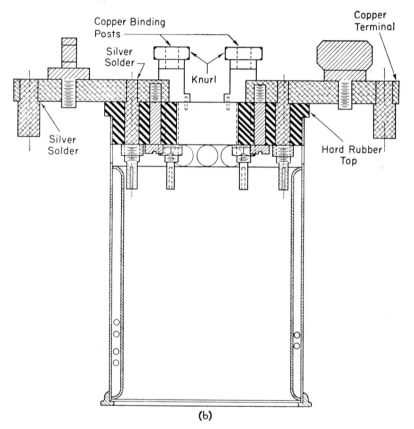

(b)

4-4. Methods of Measurement

Resistors are used in many places in electrical circuits to perform a variety of useful functions. Also, the property of resistance enters as an important element in the behavior of circuit elements such as coils, transformers, and condensers. It is important in many cases to have information regarding the magnitude of the resistance and to be able to rely upon the information to a degree of accuracy adequate to the particular purpose. Measurement of resistance is, accordingly, basic to many working circuits, meters, and machines. Resistors are used, in addition, as standards for the measurement of other resistors and for the determination of inductance and capacitance. Measurements of resistance are made most commonly by the following methods:

1. Ohmmeter
2. Voltmeter-ammeter method
3. Potentiometer method
4. Wheatstone bridge
5. Kelvin bridge

These methods will be discussed in some detail, particularly 4 and 5. A method for the precision comparison of resistance will also be considered.

4-5. The Series-type Ohmmeter

The ohmmeter is a convenient direct-reading device for the indication of resistance to a rather low degree of accuracy. The statement regarding accuracy is not intended in an unfavorable sense, for there is a large field of application for an instrument that gives even an approximate value by a direct meter reading with practically no adjustment required by the operator. An ohmmeter is useful for getting the approximate resistance of circuit components such as heater elements or motor field coils, measuring and sorting radio resistors, checking circuit continuity, and so on. It is also useful in a measurement laboratory as an adjunct to a precision bridge, for if the resistor to be measured is truly an "unknown," time is saved in bridge balancing if an approximate resistance value is first obtained by an ohmmeter.

The ohmmeter consists basically of a sensitive d-c instrument connected in series with a resistance and a battery to a pair of terminals to which the unknown is connected. The current through the instrument depends on the unknown, and so may be used as an indication of the unknown, provided that calibration problems are taken into account adequately.

Figure 4-3 shows the elements of a simple single-range ohmmeter. Before readings are taken, the terminals A-B must be shorted together, and the shunt adjusted so that the meter pointer comes to the top mark on the scale; this is the zero-resistance point. When resistance is inserted at X, the current through the circuit is reduced and the pointer drops lower on the scale, which thus has "0" at the extreme right and "∞" at the extreme left (for the usual instrument having the zero-current position at

TABLE 2

PROPERTIES OF MATERIALS FOR PRECISION RESISTORS

Name	Composition %	Resistivity		Temperature coefficient per °C	Thermal emf against copper $\mu v/°C$	Tensile strength lb/in.²	
		Microhms for cm cube	Ohms for cir. mil ft			Max.	Min.
Manganin	Cu 82–84 Mn 12–13 Ni 4–5	48	290	±0.00001*	1.7	90,000	40,000
Constantan	Cu 60–55 Ni 40–45	49	294	−0.00001	40	135,000	60,000
Gold-Chromium	Au 97.9 Cr 2.1	32	193	Negligible 20–30°	7		
Therlo	Cu 85 Mn 9.5 Al 5.5	47	282	0 at 25°	0.3		
Evanohm	Ni 74.5 Cr 20.0 Al 2.75 Cu 2.75	133	800	0.00002 −50 to +100°	2	180,000 at 20% elon.	
Karma	Ni 73 Cr 20 +Al +Fe	133	800	0.00002 −50 to +150°	2	180,000	130,000
Nichrome Chromel C Tophet C	Ni 60 Cr 16 Fe 24	112	675	0.00026 0 to 100°	20	200,000	95,000
Nichrome V Chromel A Tophet A	Ni 80 Cr 20	108	650	0.00023 0 to 100°		200,000	100,000

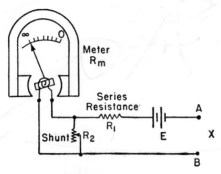

Fig. 4-3. Simple series-ohmmeter circuit.

the left end of the scale). If E, R_1, and R_2 are constant, markings can be placed on the scale for the position of the pointer corresponding to different values of R_x, and the values read from the instrument can be as accurate as the calibration process and the repeating-accuracy of the meter movement. The chief difficulty is the fact that ohmmeters are usually powered by batteries, and the battery voltage changes gradually with use and age. It is necessary to provide an adjustment to counteract the effect of the battery change; this is the purpose of the adjustable R_2 in Fig. 4-3. If R_2 were not present, it would be possible to bring the pointer to full scale by adjustment of R_1, but this would change the calibration all along the scale and cause a large error. Adjustment by means of R_2 is much better, for the parallel resistance of R_2 and the instrument coil is always low in comparison with R_1, and hence the change in R_2 needed for adjustment purposes does not change the calibration very greatly. The circuit of Fig. 4-3 does not compensate completely for ageing in the battery, but it does a reasonably good job within the expected limits of accuracy of the ohmmeter.

As the foregoing discussion indicates, all ohmmeters of this type have "0" at one end of the scale and "∞" at the other, regardless of the constants of the circuit. The values of the constants do, however, affect the distribution of the scale markings between these two extremes. A convenient quantity to use in design is the value of R_x for the half-way mark on the scale. (Call this value R_h). If the full-scale current and resistance of the meter, the battery voltage, and R_h are specified, the circuit can be "designed," that is, values can be found for R_1 and R_2. Design can be approached by recognition of the fact that, if introduction of the unknown reduces the current to half value, the unknown must be equal to the internal resistance consisting of R_1 in series with the parallel combination of R_2 and R_m. The battery current at half scale is thus $E/2R_h$, and at full scale E/R_h. With the battery current and the sensitivity of the instrument movement known, R_2, and then R_1, can be computed.

Example: Assume the following values:

$$\text{Full-scale current of instrument} = 1 \text{ ma}$$
$$R_m = \text{Resistance of instrument} = 24 \text{ ohms}$$
$$R_h = \text{Half-scale resistance} = 2000 \text{ ohms}$$
$$E = 3 \text{ v}$$

From these values and the discussion above, for R_x shorted,

$$\text{Battery current} = 3/2000 = 0.0015 \text{ amp}$$
$$\text{Instrument current} = 0.001 \text{ amp}$$
$$\text{Shunt current} = 0.0015 - 0.001 = 0.0005 \text{ amp}$$
$$R_2 = 24 \times 0.001/0.0005 = 48 \text{ ohms}$$

For R_m and R_2 in parallel,

$$R_p = (24 \times 48)/(24 + 48) = 16 \text{ ohms}$$
$$R_1 = 2000 - 16 = 1984 \text{ ohms}$$

The circuit could be designed for other values of R_h, within limits. However, if R_h were taken as 3000 ohms, with other data as above, the battery current would be the 0.001 amp required by the instrument and would leave no current for the shunt, thus no provision for adjustment, and no leeway for weakening of the battery. For an R_h equal to 3000 ohms or higher it would be necessary to use either a more sensitive instrument or a higher battery voltage.

A lower R_h can be arranged either by redesign or by provision of a shunt as shown in Fig. 4-4, the latter arrangement being particularly useful in the construction of a multirange ohmmeter. If R_3 in Fig. 4-4 is made $\frac{1}{9}$ of the unshunted internal resistance (R_1 plus R_2 and R_m in parallel), the internal resistance is made $\frac{1}{10}$ as great as before, also $\frac{9}{10}$ of the current passing through the unknown flows through the shunt and $\frac{1}{10}$ through the old path. If such a shunt is added to the example above and the unknown is made 200 ohms, the battery current is ten times as great as before, but the instrument current is the same. Therefore, the scale marking that was

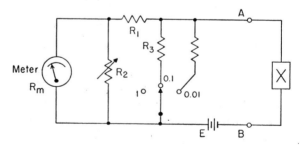

Fig. 4-4. Shunted series-type ohmmeter.

2000 ohms now represents 200 ohms, so we can use the former scale and apply a multiplier of 0.1 to the readings. We could similarly provide a multiplier of 0.01, but if we go much lower by this method we find that the current through the unknown is objectionably large for some purposes and the battery drain is too great for a small battery. (We should consider the magnitude of current passed through the unknown when we measure anything of a delicate nature, such as a galvanometer or microammeter. Unthinking use of an ohmmeter may result in serious damage in such cases.)

Some ohmmeters use rectified alternating current as the power source. In this case the voltage may be regulated to give a constant supply and taps can be provided to give definite fractions of the total voltage. Scale-changing can be accomplished by switching that makes simultaneous changes of E and R_1. Combination of this idea with the use of shunts makes possible an ohmmeter that gives readings over a very wide range.

The scale arrangement on an ohmmeter is so different from the usual ammeter or voltmeter that it may be helpful to make a study of it, both to understand the scale markings and also to help in computation of errors in the readings. We shall use the circuit of Fig. 4-3, with a resistance R_x connected to the terminals A-B. An equation for the current in the meter may be obtained in the following form:

$$I_m = \frac{R_p E}{R_m(R_p + R_1 + R_x)} \tag{1}$$

where the symbol R_p is used for convenience for the parallel combination of R_2 and R_m. With terminals A-B shorted we have the same equation, except with R_x placed equal to zero. Equation (1) may be rearranged:

$$I_m = \frac{R_p E}{R_m(R_p + R_1)} \times \frac{R_p + R_1}{R_p + R_1 + R_x} \tag{2}$$

The first fraction equals I_{FS}, the full-scale current ($R_x = 0$) to which the instrument should always be adjusted before measurements are made. Therefore

$$I_m = I_{\text{FS}} \frac{R_p + R_1}{R_p + R_1 + R_x} \tag{3}$$

If we divide through by I_{FS} and use

$$\frac{I_m}{I_{\text{FS}}} = s$$

which is the current as a fraction of full-scale value, and solve for R_x, we obtain

$$R_x = (R_p + R_1)\frac{1 - s}{s} = R_h \frac{1 - s}{s} \tag{4}$$

As an alternative, solution of this equation for s gives

$$s = \frac{R_h}{R_x + R_h} \tag{5}$$

which explains the distribution of markings along the scale of the ohmmeter.

Equation (4) may be useful for computation of errors in R_x due to scale errors in the instrument. Following the discussion in par. 2-24,

$$dR_x = \frac{\partial R_x}{\partial s} ds \tag{6}$$

Some ohmmeters are constructed with an adjustable soft-iron shunt across the pole pieces of the meter movement, as indicated in Fig. 4-5.

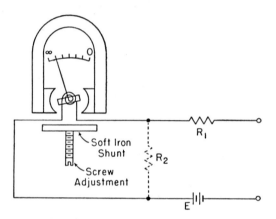

Fig. 4-5. Series-type ohmmeter with magnetic adjustment to compensate for changes in battery voltage.

The iron piece serves to modify the air-gap flux of the magnet, and hence controls the sensitivity of the movement. The pointer can thus be set on the full-scale mark in compensation for battery changes, without changing the electrical circuit. The superiority of this method may be understood from eq. (3), which shows that the *ratio* of I_m to I_{FS} depends on resistance values, which stay constant in the magnetic adjustment. With the arrangement of Fig. 4-3, however, R_p is changed to some extent by change of R_2.

4-6. The Shunt-type Ohmmeter

An ohmmeter may be constructed on the idea shown in Fig. 4-6. If terminals A-B are shorted ($R_x = 0$), the meter current is zero, and if R_x is removed, the pointer rises on the scale and can be adjusted to the full-scale point by proper selection of the resistor R_1. This ohmmeter, accordingly, has "0" at the left and "∞" at the right of the scale, in contrast with

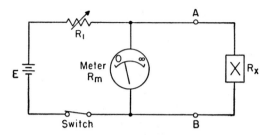

Fig. 4-6. Shunt-type ohmmeter.

the type previously studied. The shunt type is particularly suited to low resistance values. Note that a switch is needed to keep the battery from running down when the instrument is not in use.

The circuit of Fig. 4-6 may be analyzed in the following way:

With X open,

$$I_{FS} = \frac{E}{R_1 + R_m} \tag{7}$$

With X connected,

$$I_m = \frac{E}{R_1 + \dfrac{R_m R_x}{R_m + R_x}} \frac{R_x}{R_m + R_x} \tag{8}$$

The meter current in the latter case, expressed as a fraction of the full-scale value, is

$$s = \frac{I_m}{I_{FS}} = \frac{R_x(R_1 + R_m)}{R_1(R_m + R_x) + R_m R_x}$$

$$= \frac{R_x(R_1 + R_m)}{R_x(R_1 + R_m) + R_1 R_m}$$

$$s = \frac{R_x}{R_x + R_p} \tag{9}$$

where

$$R_p = \frac{R_1 R_m}{R_1 + R_m}$$

Equation (9) checks the fact that the half-scale point occurs for $R_x = R_p$. The distribution of scale markings is nearly linear in the lower part (for $R_x \ll R_p$), but becomes progressively more crowded as R_x is increased.

4-7. Crossed-coil Ohmmeter (Ratio Meter)

The ratio meter is so named because the indication depends upon the *ratio* of the currents in two coils and not upon the individual values. It has many applications in addition to its use as an ohmmeter. The same principles may be utilized in two forms of construction:

(a) A fixed permanent magnet, as in most d'Arsonval instruments (except for different pole construction), with a moving system consisting of two coils mounted on opposite sides of the common axis. The spirals carrying current to and from these coils must be very flexible in order to produce negligible restraining torque, so that the coil position is determined entirely by the magnetic forces of the two coils. The central iron piece is shaped as shown in Fig. 4-7(a) and mounted off-center to give a tapering air gap. Currents are passed through the coils to produce torques in opposite directions and in the directions indicated by the arrows. If, from initial equilibrium, the current in one coil increases relatively to the other,

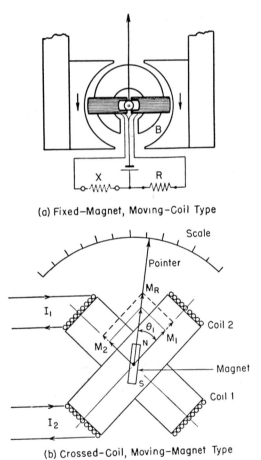

(a) Fixed–Magnet, Moving–Coil Type

(b) Crossed–Coil, Moving–Magnet Type

Fig. 4-7. Construction of d-c ratio meters. [*Part (a) courtesy of Daystrom, Inc., Weston Instruments Division.*]

that coil moves to a position having a longer air gap and hence a lower flux density, while the other coil is moving at the same time to a position of higher flux density. Equilibrium is reached when the resulting torques are equal in magnitude. (NOTE: Fig. 4-8(c) shows a variation of this type of construction.)

If the coils are in the equilibrium position and both currents are increased in the same proportion, both torques increase equally and equilibrium is not disturbed. The dependence of the indication on the relative values of current, and not on the actual magnitudes, is useful in many applications. Figure 4-7(a) indicates the use of this movement as an ohmmeter, with X as the unknown resistance and R as the standard resistance that establishes the scale of the instrument. A change of battery voltage affects both coils in the same proportion and hence does not affect the calibration, as discussed above.

(b) The crossed-coil type, made up of two fixed coils at 90°, with the moving system consisting of a small permanent magnet fastened to the pivoted shaft that carries the pointer. No current-carrying springs or spirals are needed in this type. (Damping may be provided by placing a copper cup around the magnet.)

Type (b) is shown in Fig. 4-7(b), which indicates the two crossed coils and the small pivoted permanent magnet. The arrows M_1 and M_2 indicate the magnetomotive forces of the two coils, and M_R their resultant. The small magnet tends to move into alignment with the resultant field, since it is subject to no spring restraint. The angle θ depends on the ratio of M_1 and M_2, and hence on the ratio of I_1 and I_2. Note that changes of the two currents in the *same proportion* leave the resultant, as indicated by the dotted lines, in exactly the same position and cause no change in indication. The action is different in detail from that in Type (a), but the behavior is similar in the types insofar as proportional changes of current in the two coils do not disturb the equilibrium position. The two types may be used interchangeably, in general, in circuit applications.

Figure 4-8(a) shows the crossed-coil ratio meter as an ohmmeter. Coil 1 is connected in series with a fixed resistor to the power supply, and hence I_1 is proportional to voltage E. I_2 carries the current flowing in the resistance R_x. The reading of the meter depends on the ratio of the two current strengths, and therefore on the resistance R_x. Note that a change of the supply voltage E affects both coils in the same proportion and, hence, does not change the reading. It is thus unnecessary to use a regulated power supply or to provide an adjustment for voltage change.

Figure 4-8(b) shows the modifications needed in the measurement of very low resistances. The voltage for the potential coil of the ohmmeter is obtained from a second set of terminals, marked "P," to avoid errors caused by resistance of the current leads or by contact drop at the current binding

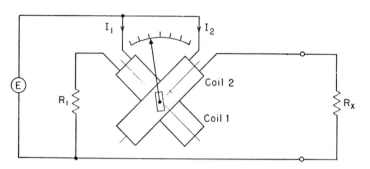

Fig. 4-8(a). Application to measurement of high resistance.

posts. The use of a shunt is indicated in the current line, for low-valued unknowns require such large currents to give enough voltage drop that it is not practicable to pass the entire current through the ohmmeter coil.

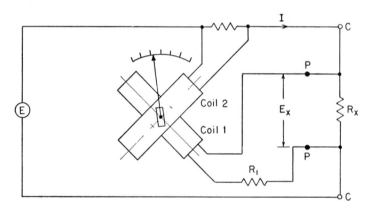

Fig. 4-8(b). Application to measurement of low resistance. Separate potential terminals are used on R_x to avoid errors due to lead resistance and contact resistance at the current terminals.

Figure 4-8(c) shows the circuit of a multirange instrument that is based on the principles of Fig. 4-8(b), although the meter movement is of a different type. The "Ducter" ohmmeter shown here has the same form of coil and magnet construction shown later (Figs. 4-30 and 4-31) for the "Megger" insulation testers. "Ducters" are made with various full-scale ranges from 5 ohms down to 100 microhms, one scale division in the latter case representing 1 microhm. The instrument of Fig. 4-8(c) has a scale 0-1000 microhms (each division = 10 microhms), with multipliers of 1, 10, 100, and 1000, giving a top reading of 1 ohm.

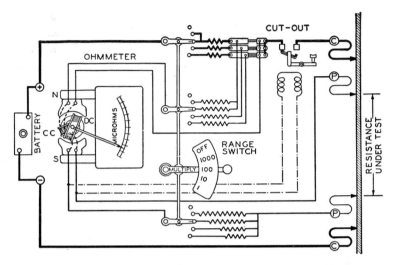

Fig. 4-8(c). Circuit of 'Ducter' low-resistance ohmmeter. (*Courtesy of the James G. Biddle Co.*)

4-8. Voltmeter-Ammeter Method

The voltmeter-ammeter method is a direct application of Ohm's law by measurement of current flowing through the unknown and the voltage drop across it. The circuit may be connected as in Fig. 4-9, (a) or (b).

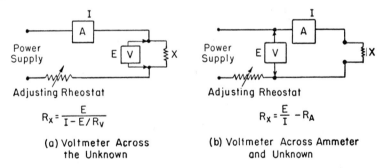

$$R_x = \frac{E}{I - E/R_V}$$

$$R_x = \frac{E}{I} - R_A$$

(a) Voltmeter Across the Unknown   (b) Voltmeter Across Ammeter and Unknown

Fig. 4-9. Measurement of resistance by voltmeter-ammeter method.

If the effect of the instruments may be neglected, the resistance of the unknown is E/I. In (a) the voltmeter is directly across the unknown and receives the correct voltage, but the ammeter reading includes the current drawn by the voltmeter. In (b) the current is correct, but the voltmeter reading is too large by the amount of the ammeter IR drop. Correction may be made in either case if the constants of the instruments are known. In many cases one or other of the two connections may be used without correction. Circuit (a) is better for a low-resistance unknown (E small, I large), while (b) is better for a high resistance (E large, I small).

This method is capable of fair accuracy, depending on care in taking the readings and on the quality of the instruments. If the two instruments are of "$\frac{1}{2}$ per cent accuracy," and are read near full scale, the instrumental error in the result may be anything from 0 to 1 per cent. If read near half-scale, the percentage error may be twice as great and for lower readings may mount correspondingly higher. With less accurate instruments the possible error is of course increased. It is difficult to secure much better than 1 per cent accuracy in a resistance value under usual conditions, and the error may be considerably higher. However, the method is useful in some laboratory work in which high accuracy is not required.

The form of instrument guarantee should be reviewed in connection with the discussion of the previous paragraph. The limiting errors of indicating instruments are customarily specified as a percentage of the *full-scale* reading. That is, a voltmeter that is referred to briefly as a $\frac{1}{2}$ per cent instrument actually carries this specification, "The error at any point of the scale will not exceed 0.5 per cent of the full-scale reading." The *magnitude* of the error may be as great by the guarantee at half scale as at full scale, but is twice as great a percentage of the actual reading. For this reason the percentage accuracy of a reading taken on a low part of the scale may be quite poor.

4-9. Potentiometer Method

The potentiometer method of measuring resistances may be regarded as an outgrowth and refinement of the voltmeter-ammeter method. The scheme, as shown in Fig. 4-10, is to place a known resistor in series with the

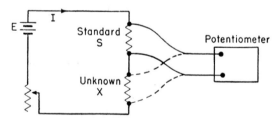

Fig. 4-10. Measurement of resistance by potentiometer.

unknown and then to measure the voltage drops across each by a potentiometer. The voltage across the standard may be considered as giving the current, the combination of resistor and potentiometer thus serving as a more accurate substitute for an ammeter. More directly, for computation,

$$\frac{X}{S} = \frac{E_x}{E_s}$$

provided that the current is kept constant during the two voltage measurements. This method is capable of considerable accuracy if the voltages are large enough to permit accurate determination. However, a current large enough to give adequate voltage drops may cause serious heating of the resistors, particularly with low values of resistance. This is one objection and the difficulty of holding constant current is another, so the method is not generally as desirable as a bridge. It is not much used, except under special conditions.

4-10. Wheatstone Bridge

The circuit arrangement commonly known as the Wheatstone bridge is the most widely used method for the precision measurement of resistance. The Wheatstone bridge consists of four resistance arms, together with a source of current (battery) and a detector (galvanometer). The circuit is shown in Fig. 4-11. The current through the galvanometer depends on the

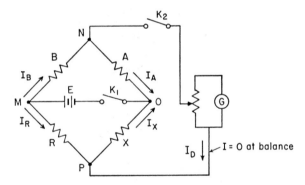

Fig. 4-11. Wheatstone bridge circuit.

difference of potential from N to P, or in other terms, on the difference in the potential drops from M to N and M to P. The current may flow in either direction, depending on which of these potential drops is the greater. A "balanced" condition can be attained in which no current flows through the galvanometer, provided that the drop in potential M to N equals the drop M to P. That is

$$I_B B = I_R R \qquad (10)$$

If the galvanometer current is zero,

$$I_B = I_A = \frac{E}{A + B} \qquad (11)$$

and
$$I_R = I_X = \frac{E}{R + X} \tag{12}$$

Substitution of (11) and (12) in (10) gives the well-known relationship,

$$B \cdot X = A \cdot R \tag{13}$$

or
$$\frac{X}{R} = \frac{A}{B} \tag{14}$$

If three of the resistances are known, the fourth may be determined from (13) or (14)

$$X = \frac{A}{B} R \tag{15}$$

The measurement of the unknown is made in terms of three known resistances, and is independent of the characteristics or calibration of the deflection instrument (galvanometer), provided only that it has sufficient sensitivity to permit the balance point to be found to the required degree of precision. The fact that adjustment is made for zero current in the detector at balance causes the term "null method" to be applied to this arrangement.

4-11. Errors in a Wheatstone Bridge

The Wheatstone bridge is a convenient and accurate method for measuring resistance. The main, or limiting, source of error is inaccuracy in the three known resistors making up the bridge. This, and other possible sources of error may be listed as follows:

(1) Discrepancies between the true and marked values of resistance of the three known arms.

(2) Personal errors in finding the balance point, in taking readings, and in making computations.

(3) Thermal emf's in the bridge or galvanometer circuit.

(4) Changes of resistance due to self-heating (I^2R) of one or more arms of the bridge.

(5) Inaccuracy of balance point, due to insufficient sensitivity of the balance detector.

(6) Errors due to the resistance of leads and contacts. This effect is usually negligible, except for unknowns of low resistance.

With respect to item (1) it may be noted that since X is measured in terms of the other three resistances, it is thus known to approximately the same order of accuracy. If A, B, and R are decade resistance boxes guaranteed to be accurate within 0.1 per cent of the marked values and if the marked values are all higher than the true values to the limit of the guarantee, then the resistance determined for X from the marked values will be

0.1 per cent higher than the true value (assuming that the balance is perfectly made). However, if A and R should each be 0.1 per cent high, and B 0.1 per cent low, then the result for X will be in error by 0.3 per cent. This is the *limiting* value of error for this case as determined by the equipment (not including personal errors, and others). The *probable* error is much smaller.

Thermal emf's in the bridge arms may cause considerable trouble, particularly in low-resistance circuits, but are not so important in measurements of high resistances, particularly if the impressed voltage is made higher. A thermal emf in the galvanometer circuit may be serious in some cases, so care should be taken to minimize the effect in precision measurements. Some sensitive galvanometers have all-copper systems (i.e., copper suspensions as well as copper coils) so that there is no junction of dissimilar metals to produce an emf. A galvanometer may be placed in a box, and surrounded by cotton, to avoid temperature inequalities caused by drafts. The effect of thermal emf's can be balanced out in practice by interposing a reversing switch between the battery and the bridge, making a bridge balance for each polarity, and averaging the results. A test for thermal emf may be made by closing the galvanometer key with the battery switch open. A steady deflection denotes a thermal emf which may be compensated by double balance, as above, or by using the false (thermal) zero for bridge balancing.

4-12. Loading of Precision Resistors

Great care should be taken to keep the I^2R loss in the bridge arms to a safe amount. Moderate heating may simply introduce a small error in a particular measurement, due to a temporary change in resistance. Excessive heating, however, may cause a *permanent* change in resistance value. The voltage applied to a bridge should be kept low to avoid the possibility of such an occurrence, otherwise a valuable piece of precision equipment may no longer be suitable for precise work. The damage, moreover, may not be discovered immediately. A few minutes of careless use may thus cause subsequent erroneous results and, eventually, a large expense for repairs. The current, or wattage, should be computed in advance, particularly when low resistance values are used, to determine whether the proposed setup is safe.

Instrument makers specify in their catalogues the safe current or power values for their precision resistors and decade boxes. Experimenters should know these values before attempting to use the equipment. Some decade boxes have the safe current value marked on the panel for each decade, and this is an excellent idea, particularly for general laboratory use.

4-13. Unbalanced Conditions in Wheatstone Bridge

The computation of unbalanced conditions in a bridge circuit is frequently of interest, although the resistance determination, which is the desired end result, comes from the equation of the balanced condition. The computation of the current flowing through the galvanometer for a small discrepancy in the resistance of a bridge arm, as compared with the true balance value, is a necessary first step in determining whether the galvanometer has sufficient sensitivity to detect that amount of unbalance. Galvanometers differ not only in the amount of current required for a given deflection, but also in their resistance, so it is often not possible to tell by inspection which of two galvanometers will give the greater sensitivity with a particular bridge. Solution of the bridge circuit for a small unbalance yields this information.

Equations have been derived for the galvanometer current of an unbalanced bridge in terms of the resistances of the branches. The equations have the disadvantage of being decidedly unwieldy. A different, and better, approach to the problem is by means of a circuit theorem generally referred to as Thevenin's theorem. This method has several advantages. The computation is shorter and simpler. But more important, it replaces a formula by an idea, and it makes possible the visualization of the effect of circuit parts, the change from one galvanometer to another, and so on.

4-14. Thevenin's Theorem

Thevenin's theorem may be stated as follows: Any linear and bilateral network with two accessible terminals may be replaced by an emf acting in series with an impedance: the emf is that between the terminals when they are unconnected externally, and the impedance is that presented by the network to the terminals, when all sources of emf in the network are replaced by their internal impedances.

This means, in effect, that for a member we wish to connect to two points in a circuit, the whole circuit, no matter how complicated, may be replaced by two simple elements, an emf and an impedance. The emf is the open-circuit voltage across the two points, before the new member is connected. The impedance is that which is seen looking into the circuit from the two points (all emf's set equal to zero, but internal impedance of generators retained). First, we shall illustrate the theorem by the simple d-c circuit, shown in Fig. 4-12. With S open:

$$I = \frac{125}{1 + 4 + 20} = 5 \text{ amp}$$

$$E_0 = 5 \times 20 = 100 \text{ v}$$

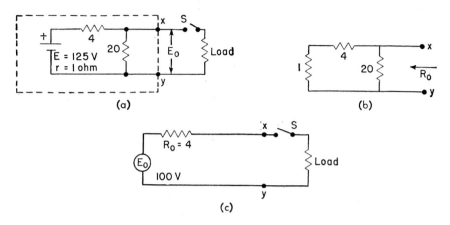

Fig. 4-12. Illustration of Thevenin's theorem. (a) Original circuit. (b) Internal resistance as seen from terminals x-y. (c) Equivalent circuit by Thevenin's theorem.

The resistance x-y in diagram (b) is

$$R_{xy} = R_0 = \frac{5 \times 20}{25} = 4 \text{ ohms}$$

Now, by Thevenin's theorem, we may replace the original circuit by the circuit of Fig. 4-12(c), which is *equivalent for any load we wish to connect from x to y*. This may be tested by trying any convenient load, say 6 ohms, connected first to the original circuit and then to the equivalent circuit.

4-15. Application of Thevenin's Theorem to Wheatstone Bridge

We shall select the galvanometer corners of the bridge as the "two accessible terminals" of Thevenin's theorem, since we are interested in the current flowing in the galvanometer. We shall take the following problem to illustrate the method. In this case the battery will be assumed to have negligible resistance. This is permissible in most bridge problems, and though not essential to the method, it does produce simplification of computation. The effect of resistance in the battery will be considered later.

The bridge would be balanced if the OP arm were 400 ohms. The diagram shows it to be $400 + x$, so x represents the unbalance, which we shall consider to be a *small* amount; that is, $x \ll 400$. It would be equally permissible to take a small fixed unbalance, such as 1 ohm, which would give the same information, since, as we shall see, the galvanometer deflection is proportional to the unbalance for small unbalances. We are interested generally in the case of small unbalances, so that we may determine the possibility of balancing to the limit of precision of the bridge. The Thevenin method is, of course, not limited in this way, but the computations are simplified if we drop terms of insignificant value. The engineer must frequently distinguish between important and unimportant factors in a situation or become hopelessly mired; the present case is one that is aided by the development of the same type of judgment. Minor terms may be included when they are needed.

The first step in the application of Thevenin's method to the circuit of Fig. 4-13(a) is to find the voltage from N to P with K open. This is the difference of the voltages MN and MP (or, PO and NO), and may be expressed as

$$E_0 = 4\left[\frac{1000}{5000} - \frac{100}{500 + x}\right]$$

$$= \frac{4(500 + x - 500)}{5(500 + x)}$$

$$\simeq \frac{4x}{2500} = 1600 \ x \ \mu v, \quad \text{if } x \ll 500$$

Note that the increment, x, has been dropped from the denominator *after* reduction to a common denominator. The discrepancy is only 1 in 500, if $x = 1$ ohm, and may be much less. Galvanometer deflections need not be computed to such close limits.

The internal resistance of the bridge circuit *as seen from the detector terminals* is, from Fig. 4-13(b),

$$R_0 = \frac{1000 \times 4000}{5000} + \frac{100(400 + x)}{500 + x}$$

$$\simeq 800 + 80 = 880 \text{ ohms}, \quad \text{if } x \ll 500$$

This gives the equivalent circuit of Fig. 4-13(c), which really summarizes the essential matters of the solution. What we do beyond this point depends on the given conditions. If we know the amount of unbalance, we can compute E_0 and I_g, and from this and the galvanometer sensitivity (expressed in microamperes required for a 1 mm deflection) we can find the deflection. Or, if the galvanometer sensitivity is given, we can solve for the unbalance needed to give a deflection of 1 mm; this value of unbalance is of interest in approximating the lower limit of precision, with respect to *detector sensitivity* — we can work with fair certainty to somewhat less than 1 mm, but 1 mm is probably a good criterion to use.

It should be emphasized that the Thevenin method is directed particularly to the finding of the detector response, and that this is a major virtue of the method. The detector response is the item of interest most of the time, and Thevenin finds it easily, directly, and accurately, and is much

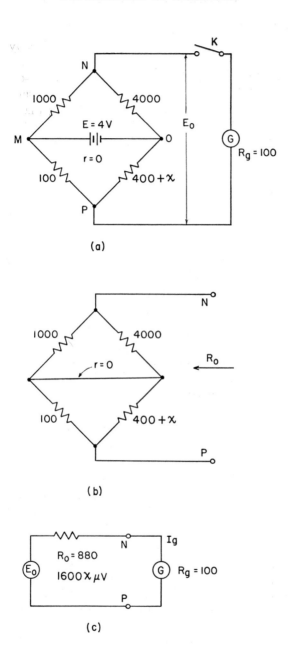

Fig. 4-13. Application of Thevenin's theorem to a Wheatstone bridge circuit. (a) Wheatstone bridge circuit. (b) Internal resistance for the Thevenin circuit. (c) Thevenin's equivalent circuit.

superior to a formula approach, or to other circuit theorems. Moreover, the method may be easily extended, if needed occasionally, to find the currents in the bridge arms. The bridge-arm currents have already been expressed, in (11) and (12), for the balanced case, or key-open case. The change by closing the galvanometer key is so minute, for small unbalances, that (11) and (12) are entirely adequate for finding the heating in the resistors. However, the effect of closing the galvanometer key may be found easily by superposition, if desired, as follows, for symbols and directions as shown in Fig. 4-11:

$$I_R = \frac{E}{R+X} - I_D \frac{X}{R+X} \tag{11a}$$

$$I_X = \frac{E}{R+X} + I_D \frac{R}{R+X} \tag{12a}$$

with similar expressions for I_A and I_B.

For the present problem, suppose that $x = 1$ ohm. Then we have

$$E_0 = 1600 \ \mu v$$

$$I_g = \frac{1600}{880 + 100} = 1.63 \ \mu a$$

$$\text{If Galv. Sens.} = \frac{0.1 \mu a}{mm}, \quad \text{then}$$

$$\text{Deflection} = \frac{1.63}{0.1} = 16 \ \text{mm}$$

We should note, as a merit of the Thevenin method, that a change from one galvanometer to another of different resistance and sensitivity may be tried very simply, as only the last step in the above computation needs to be repeated. The equivalent circuit is useful also if we wish to "match" the galvanometer to the bridge so that it will absorb the maximum power from the bridge; the criterion for this case is to make the galvanometer resistance equal to R_0.

One detail of computation of E_0 is worth some attention. The expression is formed as the difference of two fractions, and for small unbalances these two fractions are nearly equal. Therefore, for reasonable precision in a numerical case, *do not attempt to evaluate each fraction separately by slide rule.** It is a simple process, *first*, to reduce the quantity to a common

---

Example: Suppose that E_0 is found to be

$$E_0 = 4\left[\frac{1000}{1100} - \frac{4000}{4401}\right] = 4(0.909091 - 0.908884) = 828 \ \mu v$$

The two quotients are indistinguishable by slide rule, but the result corresponds to a reasonable 1-ohm unbalance. Reduction to a common denominator saves handling

denominator, then, *after subtraction* of the two terms in the numerator, slide-rule accuracy is ample for the remaining computations. This a case where intelligent handling of a computation saves long-hand divisions that would be needed for evaluation of the separate terms. This matter was discussed in chap. 3.

Partly for general interest and partly for comparison with the next problem, we shall compute the power consumed in the bridge arms for an unbalance of 1 ohm in the 400-ohm arm. It is not necessary, in general, to compute the power for all arms, but only for the one with maximum power. This can be seen by inspection of the above circuit to be the 401-ohm arm. (Of the two parallel paths across the battery, the current, and hence power, is greater in the (100 + 401) ohm branch, and of these two units, the power is of course greater in the 401-ohm resistor). The power dissipated is

$$W_{max} = W_{401} = \left(\frac{4}{501}\right)^2 \times 401 = 0.0256 \text{ watt}$$

This value is true, strictly speaking, when the galvanometer key is open, but the change due to the closing of the galvanometer branch is negligible for so small an unbalance.

4-16. Bridge with Detector and Battery Interchanged

The interchange of battery and detector in the circuit of Fig. 4-13 gives new values in the equivalent circuit

$$E_0 = 4\left(\frac{1000}{1100} - \frac{4000}{4400 + x}\right)$$

$$\simeq 826x \ \mu\text{v}, \quad \text{if } x \ll 4400$$

$$R_0 = \frac{1000 \cdot 100}{1100} + \frac{4000(400 + x)}{4400 + x}$$

$$\simeq 90.9 + 363.6 = 454.5 \text{ ohms}, \quad \text{if } x \ll 400$$

For $x = 1$ ohm, for comparison with the previous case,

$$E_0 = 826 \ \mu\text{v}$$

$$I_g = \frac{826}{454.5 + 100} = 1.50 \ \mu\text{a}$$

---

six-digit numbers, as illustrated by the comparable calculation by reduction to a common denominator,

$$E_0 = 4000 \frac{4401 - 4400}{1100 \times 4401} = 4000 \frac{1}{1100 \times 4401} = 826 \ \mu\text{v}$$

and this more accurate result may be obtained by slide rule.

$$\text{Deflection} = \frac{1.50}{0.1} = 15 \text{ mm}$$

$$W_{max} = W_{1000} = \left(\frac{4}{1100}\right)^2 \times 1000 = 0.0132 \text{ watt}$$

We may note that the galvanometer current in this case is a little less than for the previous arrangement, so we may say that the second bridge is less sensitive to unbalance. Further than this, generalized proof can be given[*] for the statement that the sensitivity is greater when the galvanometer is connected from the junction of the two high-resistance arms to the junction of the two low-resistance arms. This statement regarding sensitivity has received undue attention in the past, for it is of limited value, for two reasons. First, the difference in sensitivity is usually not great—not great enough to have appreciable effect on the detection of the balance point. In the second place the statement may even be misleading. It is based on the *same voltage* in both cases. A more adequate comparison would be on the basis of the *same power* dissipation in the arm that has the greatest heating. The second bridge has a maximum power of 0.0132 watt instead of 0.0256 watt, or about one-half, for 4 v impressed. For the *same wattage* in the second case, the impressed voltage may be raised to

$$E = 4\sqrt{\frac{.0256}{.0132}} = 5.56 \text{ v}$$

and the galvanometer current is increased to,

$$I_g = 1.50 \times \frac{5.56}{4.0} = 2.09 \text{ μa}$$

Note that on the basis of equal wattage for the most heavily loaded arm the *second* arrangement is *more sensitive* than the first. This reverses the sensitivity statement given above. The answer for maximum sensitivity thus depends on the conditions laid down, whether constant voltage or constant maximum wattage. It makes a difference, also, whether the detector resistance is fixed or whether the best value may be selected for a given case. The two circuit arrangements give the same power to the detector on the basis of the same maximum bridge-arm wattage *if* the optimum detector resistance is used for each case.

In summary, the difference in sensitivity (at constant voltage) for the two arrangements is important only when there is a very large ratio of values in the bridge, and for this case the so-called maximum sensitivity connection places the two lower resistances in series across the battery, so the discrepancy in heating is extreme. However, if the comparison is based

---

[*]Reference 10, p. 179.

on equal maximum wattage the common criterion should be reversed if the detector resistance is low. (It is correct for a high-resistance detector. The cross-over between "low" and "high" resistance for this purpose occurs at the geometric mean of the internal bridge resistances for the two circuit arrangements.)

The relationships stated above are developed in Appendix I, to which the reader is referred for a more complete discussion. Even aside from this study, however, it seems desirable to emphasize again the importance of computing in advance the power dissipated in low-resistance bridge arms in order to avoid serious heating.

4-17. Resistance in the Battery Circuit

The general equations become quite involved for a bridge with any amount of unbalance and with resistance in the battery branch. The Thevenin method is much more desirable, and in spite of the added resistor is only slightly longer than in the cases given above. If the resistances of

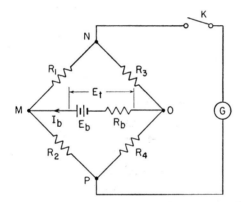

Fig. 4-14. Wheatstone bridge with resistance in the battery circuit.

the four bridge arms are specified, as shown in the circuit of Fig. 4-14, the Thevenin circuit may be formed in the following manner:

(1) Find E_t, K open.

$$R_{\text{bridge}} = \frac{(R_1 + R_3)(R_2 + R_4)}{R_1 + R_2 + R_3 + R_4} \tag{16}$$

$$E_t = E_b \frac{R_{\text{bridge}}}{R_{\text{bridge}} + R_b} \tag{17}$$

(2) Use E_t from eq. (17) as the voltage impressed on the bridge in order to compute the E_0 of the Thevenin circuit, as in previous problems.

(Note: If solution is being made for the amount of resistance unbalance required to produce a given small galvanometer deflection, the same method may still be used, as E_t and R_0 are changed only slightly by a small unbalance. E_0 may be found from the E_t computed for the balanced case.)

(3) Find R_0 for the circuit as shown in Fig. 4-15. This may be done directly by means of a delta to wye transformation of three of the resist-

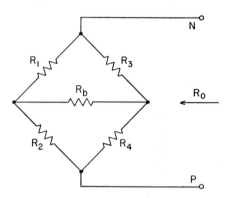

Fig. 4-15. Internal resistance for the circuit of Fig. 4-14.

ances, followed by parallel and series combinations. However, practically the same result can be obtained for *small* unbalances by ignoring R_b and computing the resistance from N to P as $(R_1 + R_2)$ in parallel with $(R_3 + R_4)$. Near balance, the resistance N-P is changed very little whether the R_b branch is opened or closed; that is, R_b may have any value from zero to infinity with very little effect on the resistance NP. For small unbalances, accordingly, we obtain a satisfactory approximation from either of these extremes. The delta-wye transformation should be used only for very large unbalances, as it increases the labor considerably, and in the usual case is more likely to introduce additional computational error than to give greater accuracy.

4-18. Limitations of the Wheatstone Bridge

There are limitations at both the high-resistance and low-resistance ends of the scale to the values that can be measured accurately by the Wheatstone circuit. The high-resistance limit is set partly by the reduced sensitivity to unbalance. (In terms of the Thevenin circuit, this is due to the high value of R_0 in series with E_0 and the galvanometer.) The limit can be extended to a certain extent by increasing the voltage applied to the bridge, provided that care is taken to avoid overheating of any arm of the bridge. Inaccuracy may also be introduced by leakage over insulation of the bridge members when measuring a high resistance. The upper limit of resistance measurement ranges from a few hundred thousand ohms to several megohms or so. The limit can be extended by special design features.

The lower limit for measurements is set by the resistance of the connecting leads and by contact resistance at the binding posts. The error due to the resistance of the leads may be corrected fairly well, but contact resistance presents a source of uncertainty that is difficult to overcome. The lower limit for accurate measurements is in the neighborhood of 1 to 5 ohms. The limit may be lowered somewhat by particular care in the connections,* but in general the Kelvin bridge is preferable for resistances less than 1 ohm.

4-19. Arrangement of Ratio Arms

Built-up Wheatstone bridges have special arrangements of the ratio arms for convenience and, in precision bridges, for aid in checking. The precision bridges usually have two arms each with values of 1, 10, 100, 1000 and 10,000 ohms with plug arrangements so that any resistor may be used in either ratio arm. One of the best methods is the Schöne scheme shown in Fig. 4-16. (One resistor of each value is marked with a prime to

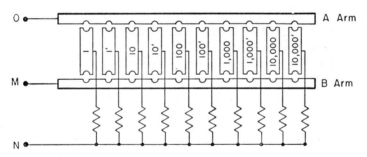

Fig. 4-16. Schöne arrangement of ratio arms. The terminals are lettered to correspond with Fig. 4-11.

distinguish it from the other.) Any resistor can be connected to either the A or the B side of the bridge by placing the plug in the proper position.

This arrangement of the resistors makes it possible, when using equal ratio arms, to interchange the resistors. For example, if we are using the ratio as 1000/1000, we can make one balance using the ratio A/B as 1000/1000′, and then a second balance 1000′/1000. If there is a small difference between the 1000′ and 1000 resistors, the two balance values are

---

*Reference 14, para. 2.23 and 2.25.

not exactly the same. One value is a little larger than it should be, but the balance with the inverted ratio is correspondingly less. The average of the two values eliminates the effect of errors in the ratio arms for any reasonable discrepancy in their resistances. This is important in precision work because it eliminates one of the potential sources of error. The only remaining calibration error in the bridge is in the variable arm (often called the "rheostat arm").

The direct interchange of arms cannot be made if other than equal arms are used. However, the arrangement of Fig. 4-16 is still valuable, as it makes available two arms of each value to use as a check on each other. We might, for example, take balances with the ratios set at 1000/100, 1000/100', 1000'/100, and 1000'/100'. If the four results are close together, we at least gain confidence in the result. At any rate, if one resistor is materially in error, we become aware of the fact and can take steps to have it corrected.

4-20. Commercial Wheatstone Bridges

The Wheatstone method is so useful that a great many forms of bridge are available commercially, with varying design features, accuracy, and cost. The complete bridge units include at least the ratio arms, the rheostat arm, and battery and galvanometer keys; the portable bridges also include a battery and a galvanometer.

Laboratory-type bridges furnish the three bridge arms and connections, and are used with sensitive external galvanometers and external batteries that may be selected in accordance with the measurement being made. They are available with resistors adjusted within ± 0.01 or ± 0.02 per cent for the most accurate work. The highest-accuracy bridges usually have the Schöne arrangement of ratio arms which permits selection of individual arms and interchange of arms when a one-to-one ratio is being used; this is important in the most accurate measurements, as it eliminates ratio-arm errors from the results. Other bridges have a single dial which selects *ratio*, but not the values of the individual ratio arms; a schematic diagram of this sort is shown in Fig. 4-17. Bridges in the highest-accuracy class must use either a plug arrangement for the resistors or extremely well-made dial switches, so that switching uncertainties do not affect accuracy.

Figure 4-18 shows a precision bridge. It has the Schöne arrangement of ratio arms and a six-dial rheostat arm (resistance steps from 10,000 ohms to 0.1 ohm). There are two ratio arms each of 10,000 ohms and 1000 ohms, which permits interchange of arms on 10,000/10,000 and 1000/1000 ratios. One additional arm each of 1, 10, 100, and 100,000 ohms makes possible step-up or step-down ratios. The limit of error is 0.01 per cent for resistances below the megohm range, with larger tolerances for higher

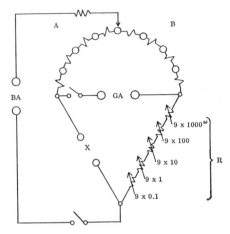

Fig. 4-17. Schematic diagram of a general-purpose Wheatstone bridge.

Fig. 4-18. Precision Wheatstone bridge.   (*Courtesy of Leeds & Northrup Co.*)

resistances. The complete metal enclosure provides electrostatic shielding, which is important, particularly for measurements of high resistance. The circuits have guard electrodes to protect against current leakage, an important consideration for operation under high-humidity conditions.

Small portable bridges include the entire bridge circuit, and require only the connection of the unknown for the measurement to be made. The battery usually consists of one or two cells of flashlight size, with provision for external battery connections when a higher voltage is needed. The galvanometer must be rugged, and is consequently of rather low sensitivity. The performance of such bridges is limited by the galvanometer and by the moderate accuracy of the components, but they are very convenient where the bridge must be carried to the work, and are much more accurate than an ohmmeter.

4-21. Purpose of the Kelvin Bridge

The Kelvin double bridge, or simply Kelvin bridge for short, may be regarded as a modification of the Wheatstone bridge to secure increased accuracy in the measurement of low resistance. The term double bridge is used due to the presence of a double set of ratio arms.

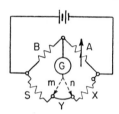

Fig. 4-19. Wheatstone bridge with lead resistance.

An understanding of the Kelvin arrangement may be obtained by a study of the difficulties that arise in a Wheatstone bridge in the measurement of resistances that are low enough for the resistance of leads and contacts to be appreciable in comparison. Consider the bridge shown in Fig. 4-19, where Y represents the resistance of the lead that connects from S to X.

Two possible connections for the galvanometer are indicated by the dotted lines. With connection to m, Y is added to X, so that the computed value of the unknown is higher than X alone, if Y is appreciable in comparison with X. On the other hand, if connection is made to n, X is in fact computed from the known value of S only, is accordingly lower than it should be.

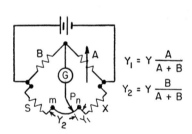

$$Y_1 = Y \frac{A}{A + B}$$

$$Y_2 = Y \frac{B}{A + B}$$

Fig. 4-20. Wheatstone bridge, illustrating theoretical correction of lead resistance.

Suppose that instead of using point m, which gives a *high* result, or n, which makes the result *low*, we can slide the galvanometer connection along to any desired intermediate point, as shown in Fig. 4-20. If at point p the resistance of Y is divided into two parts such that

$$\frac{Y_1}{Y_2} = \frac{A}{B}$$

then the presence of Y *causes no error* in the result. From the usual balance relationship, we may write

$$\left(X + \frac{A}{A + B} Y \right) = \frac{A}{B}\left(S + \frac{B}{A + B} Y \right) \tag{18}$$

From this equation,
$$X = \frac{A}{B} S \tag{19}$$

indicating that Y does not affect the result for X.

The process described here is obviously not a practical way of achieving the desired result, as we would have trouble in determining the correct point for the galvanometer connection. It does, however, suggest the simple modification, that we connect two actual resistance units of the correct ratio between points m and n, and connect the galvanometer to their junction. This is the Kelvin bridge arrangement, which is shown later in Fig. 4-22.

One complication, not mentioned so far, is the fact that in actual practice the resistance Y includes not only the ohmic resistance of the connecting wire, but also the contact resistance between wire and binding post. Contact resistance is a variable and uncertain element, as it depends on such things as the cleanness of the surfaces and the amount of pressure between them. The resistance between clean surfaces should be small, but may nevertheless not be negligible if the measured resistance is only a fraction of an ohm. The uncertainty must be removed before definite values can be given for low resistances. This is accomplished by the construction described in the next section.

4-22. Construction of Low-resistance Standards

Low-resistance standards are constructed with four terminals as shown in Fig. 4-21. One pair of terminals is used to lead the current to and from the resistor. The voltage drop is measured between the other pair of terminals. The voltage E, indicated in Fig. 4-21, is thus I times the resistance from B to B', and does not include any contact drop that may be

A, A' – Current Terminals
B, B' – Potential Terminals

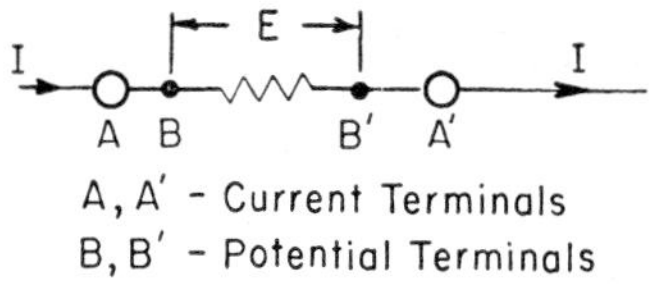

Fig. 4-21. Low-resistance standard.

present at terminals A and A'. This construction is used for ammeter shunts for the same reason.

Resistors of low values are measured in terms of the resistance between the potential terminals, which thus becomes perfectly definite in value and independent of contact drop at the current terminals. (Contact drop at the potential terminals need not be a source of error, as the current crossing these contacts is usually extremely small — or even zero for null methods.)

4-23. The Kelvin Bridge

The Kelvin bridge incorporates the idea of a second set of ratio arms and the use of four-terminal resistors for the low-resistance arms for the reasons discussed in the two preceding sections. The diagram of the bridge is given in Fig. 4-22. Arms a and b are the new ratio arms, used to give a point p

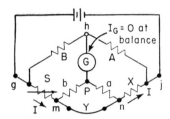

Fig. 4-22. Circuit of the Kelvin bridge.

at the proper potential between m and n to eliminate the effect of Y, as discussed above. The ratio a/b should be the same as A/B, but separate symbols will be used in the derivation below, so that it will be possible to see the error that is introduced in case the second ratio does not quite duplicate the first.

The derivation will be made here by equating the voltage drops g-h and g-m-p, which must be equal in order that the condition of zero galvanometer deflection be realized. The voltage g-h is given by

$$E_{gh} = \frac{B}{A+B} E_{gj} = \frac{B}{A+B} I\left[S + X + \frac{(a+b)Y}{a+b+Y} \right] \quad (20)$$

since no current is flowing in the galvanometer branch. Similarly,

$$E_{gmp} = I\left[S + \frac{b}{a+b} \cdot \frac{(a+b)Y}{a+b+Y} \right] \quad (21)$$

If these two values are equated and solution made for X, the result is

$$X = S\frac{A}{B} + \frac{bY}{a+b+Y}\left[\frac{A}{B} - \frac{a}{b} \right] \quad (22)$$

Now, if $A/B = a/b$, the equation becomes

$$X = \frac{A}{B} S \quad (23)$$

Equation (23) is the usual working equation for the Kelvin bridge. It indicates that the resistance of the yoke has no effect, provided that the two sets of ratio arms have equal ratios. Equation (22) is useful, however, as it shows the error that is introduced in case the ratios are not exactly equal. It indicates that it is *desirable to keep Y as small as possible* in order to minimize the error in case of discrepancy between the ratios.

The Kelvin bridge is used to measure resistances less than an ohm; with smaller resistances it can measure to increments of the order of a millionth of an ohm or less. Determination is made in all cases of the resistance between the *potential terminals* of a *four-terminal* resistor; only in this way can a definite value be specified for a low resistance, as we discussed in par. 4-22.

4-24. Construction of Kelvin Bridge

The construction frequently used in precision Kelvin bridges to satisfy the balance equation, (23), is to make S the continuously variable element, and to provide a limited number of ratio values. Figure 4-22 indicates a variable S, but does not show how the variation is to be accomplished. It is not permissible to have a sliding contact carrying the main current *between the connection points to the ratio arms,* as inclusion of contact drops would produce large error. The arrangement of Fig. 4-23 avoids this

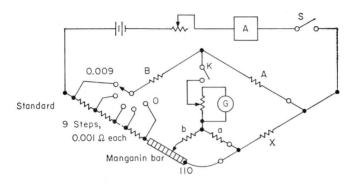

Fig. 4-23. Circuit diagram of a precision Kelvin bridge

difficulty. Here the standard consists of 9 steps of 0.001 ohm each and a calibrated manganin bar of 0.0011 ohm with a sliding contact. The switching by the contacts on B and b changes the drop between ratio arm connection points, although it does not change the resistance around the battery circuit. By this arrangement, the contact resistance is in series with the comparatively high resistance of the ratio arms, and thus has no appreciable effect; also, the sliding contact carries only the *ratio-arm current,* of milliampere magnitude or less. It is important that the standard be oriented as shown in the figure, with the manganin bar at the end near the yoke. In this way the resistance introduced into the yoke is only the unused portion of the bar, and hence small; any unused steps at the other end of the standard are simply in series in the battery circuit, and cause no difficulty. The ratio should be selected, when possible, so that a large part of the standard is used in the measuring circuit. In this way the resistance is read to the largest possible number of significant figures, and accuracy is improved by the better guarantee on the main steps as compared with the bar.

Figure 4-24 represents a bridge embodying the ideas of Fig. 4-23. The low-resistance standard and the ratio box are constructed as separate units. Connections to the 0.001-ohm steps on the standard are controlled by a dial switch. The manganin bar of 0.0011 ohm has a scale divided into 110 divisions, with a vernier that assists in reading to 0.1 of a division. The smallest readable increment is thus 1 microhm in a total resistance of 0.0101 ohm, which is a readability of 1 part in 10,000. The ratio box has two resistors each, closely matched to equality, of 10,000, 1000, 400, 300,

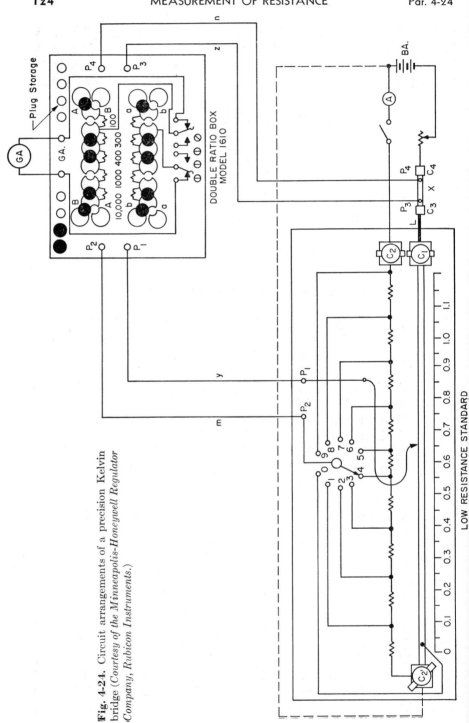

Fig. 4-24. Circuit arrangements of a precision Kelvin bridge (*Courtesy of the Minneapolis-Honeywell Regulator Company, Rubicon Instruments.*)

and 100 ohms, with a plug arrangement which permits the setting up of several ratio values. Interchange of the corner plugs changes the ratio from step-up to step-down without changing any wire connections, which is a novel and convenient feature of this ratio box.

Some Kelvin bridges have been constructed using a fixed standard resistance and variable ratio arms. On bridges of moderate accuracy, the ratio arms may be twin circular slide wires. Variable ratio arms of decade construction are used on some foreign precision Kelvin bridges, but this entails expense for many resistances adjusted to close tolerances.

Kelvin bridges are available also in portable form, including all bridge components and the galvanometer. The sensitivity and accuracy are lower than for the laboratory type, but portability is essential for the measurement of many things that cannot be moved to the laboratory. As examples we may mention the series field coil, or the commutating field coil, of a large motor or generator; or a large metering shunt mounted on a switchboard or in a bus–bar structure.

The Kelvin double bridge principle has been found useful in applications other than the measurement of low resistances. The idea has been used to eliminate the effect of lead and contact resistances in strain-gage bridges. Also, the temperature of the field winding of large generators is measured in terms of their resistance; the Kelvin circuit is used to avoid errors from the voltage drops at the slip rings (small auxiliary brushes are used for the potential connections).

Samples of electrical conductors must be measured to determine their conductivity, and whether the composition and processing has met specifications. A holder is constructed with potential connections an exact distance apart. Measurement on a Kelvin bridge determines the resistivity and, from the measured length and crosssection, the conductivity.

4-25. Adjustment and Checking of Kelvin Bridge

An important matter to check for the Kelvin bridge is the equality of ratios A/B and a/b. This can be determined by making a balance first with the regular Kelvin bridge circuit of Fig. 4-23. Now the yoke connection Y should be opened to see whether the galvanometer still indicates balance; if so, the ratios are equal. It can be seen that, with Y open, the circuit becomes a Wheatstone bridge with the lower arms equal to $(S + b)$ and $(a + X)$. Since X and S are very small in comparison with a and b in the first place and have been adjusted to the A/B ratio in the second place, we are practically making a Wheatstone comparison of A/B and a/b. It is important that the ratios be very close to equality so that their difference, as shown in eq. (22), shall not contribute an appreciable error to X, even when we are measuring to microhms.

It is necessary in measuring low resistances to pass fairly large currents through S and X in order to give enough voltage drop across them. It is important to observe the rated values for the resistors, to avoid damage by overheating. For example, a 0.1-ohm air-mounted standard has a rated current of 15 amp, a 0.01-ohm standard 100 amp, and a 0.01-ohm adjustable standard 30 amp.

Even with the large currents the voltage drops are small, so a sensitive galvanometer must be used with a Kelvin bridge. As another consequence of the low voltages and low resistances, the presence of thermal emf's may cause considerable error. Readings should be taken with both battery polarities to guard against this trouble.

4-26. Unbalanced Kelvin Bridge

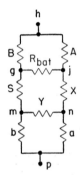

Fig. 4-25. Circuit for finding R_0 of the Kelvin bridge.

The galvanometer current of an unbalanced Kelvin bridge can be found by the same Thevenin method we used with the Wheatstone bridge. The finding of E_0, the open-circuit voltage across the galvanometer terminals, is slightly more involved than for the Wheatstone bridge, but may be found by the ideas indicated in eqs. (20) and (21) above.

The circuit for the internal resistance of the bridge as seen from the galvanometer terminals is shown in Fig. 4-25, lettered as in Fig. 4-22. The solution for the resistance h to p in the general case apparently requires two delta to wye transformations. However, this is rarely necessary, as S, X, Y, and R_{bat} are usually *very* small compared with A, B, a, and b and may be neglected with no appreciable loss in accuracy. We then have A and B in parallel, plus a and b in parallel.

4-27. Measurement of High Resistance

High resistances of the order of hundreds or thousands of megohms are often encountered in electrical equipment, and frequently must be measured. Common examples are the insulation resistance of machines and cables, the leakage resistance of capacitors, and the high resistances used in vacuum-tube circuits.

One difficulty in the measurement of such resistances is the leakage that occurs over the circuit components, binding posts, and so on, in the equipment itself or in the measuring set. Unless such effects are controlled, our measurements give us the combined effect of the unknown with the leakage paths in parallel and not the unknown itself. The leakage paths are not only an undesired part of the result, but they are generally variable from day to day, depending on humidity conditions. The effect of the leakage paths on the measurements can be removed by use of some form of guard circuit.

Figure 4-26(a) shows a battery and microammeter connected in series with a high resistance which is mounted by binding posts on a strip of

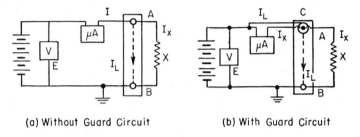

(a) Without Guard Circuit   (b) With Guard Circuit

Fig. 4-26. Application of guarded terminal for measurement of high resistance.

insulating material. The current through the meter includes the leakage current I_1 as well as I_x, and accordingly the resistance computed from E/I is not the true value of R_x. In (b) a guard terminal has been added, surrounding terminal A, and connected to the battery side of the meter. The leakage current flows from this terminal, and since it does not flow through the meter it causes no error in the computed R_x. Terminals A and C are at nearly the same potential, so there is little tendency for current to flow between them.

High resistances may be measured in several ways, of which we may list the following as the best known: (a) direct-deflection method; (b) loss-of-charge method; (c) bridge method; (d) megohmmeter.

(a) *Direct deflection method.*

The direct deflection method is basically that of Fig. 4-26, which is in effect the voltmeter-ammeter method we mentioned in an earlier section. For high resistances a sensitive galvanometer is used instead of a micro-ammeter, as shown in Fig. 4-26. A calibrated Ayrton shunt is included to protect the galvanometer and to provide several scale ranges. A diagram is shown in Fig. 4-27. A "Test-Short" switch is added so that a capacitive type of specimen, such as a capacitor or a length of cable, can be discharged after the test measurement.

Figure 4-27 indicates a measurement being made on a sample of insulating material in sheet form. A full-sized metal disk is used as the electrode on one side of the sheet, and on the other a smaller disk with a "guard ring" around it, separated by a small spacing. With connections as shown, any surface leakage around the edge of the sheet goes to the guard ring and does not enter the measuring circuit. The reading of the

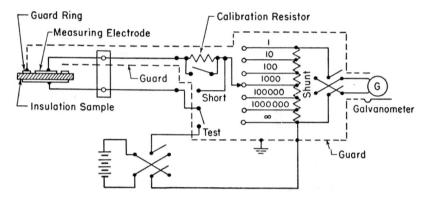

Fig. 4-27. Measurement of high resistance by direct-deflection method.

galvanometer is thus a measure of the *volume* resistivity of the sheet material, independent of surface leakage.

Contact is usually made with the surfaces of insulating materials by means of *mercury* electrodes to insure that contact is made over the entire surface. The lower electrode connection is formed by floating the specimen in a flat dish containing mercury. Rings of insulating material are placed on the top surface to confine the mercury to the desired forms for the main and guard rings. Care must be taken to eliminate air bubbles from the surface of the specimen.

A capacitor with solid dielectric when connected to a steady d-c supply shows an initial rush of current, followed by a small current which decreases gradually for a considerable period of time before it reaches what can be called a d-c steady-state value. This slow charging is due to the "absorbed charge" in the dielectric. For this reason some time must elapse between the closing of the switch and the taking of the reading. The shunt should be in the "∞" position when the battery is first connected to protect the galvanometer from the large initial rush of current.

(b) *Loss-of-charge method.*

In this method a capacitor is allowed to discharge through the resistor to be measured, and the charge remaining at the end of a measured time interval is determined by a ballistic galvanometer. Since the discharge of an R-C circuit follows an $\epsilon^{-t/RC}$ law, R may be computed if the other constants are known. It is desirable to make measurements for several time intervals (with several trials for each), and then to plot log E (where $E = Q/C$) against time to find the value of the resistance. This is tedious and time-consuming, so other methods are used when possible. The method may be applicable to the measurement of some very high resistances, but requires a capacitor of extremely high leakage resistance — at least as high as the resistance being measured. The method may be more attractive if the resistance to be measured is the leakage resistance of a capacitor; in this case the rate of discharge is a measure of the resistance, without the use of auxiliary R or C units. Several modifications are possible; one convenient form uses an electrostatic voltmeter to follow the decrease of capacitor voltage (or a quadrant electrometer for greater sensitivity). Further information can be found in several reference books.*

(c) *Bridge method.*

The bridge method utilizes a Wheatstone bridge circuit, modified by the addition of guard electrodes and the substitution of a vacuum-tube detector for the ordinary galvanometer. Figure 4-28 shows the circuit of a completely self-contained megohm bridge which includes power supply, bridge members, amplifiers, and indicating meter. It has a range from 0.1 megohm to 1,000,000 megohms. The accuracy is within 3 per cent for the lower part of the range to possibly 10 per cent above 10,000 megohms.

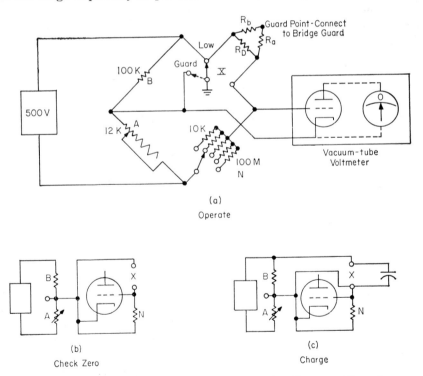

Fig. 4-28. Circuit arrangement of a megohm bridge (*Courtesy of General Radio Company*)

---

*Reference 10, p. 200; reference 18, p. 252.

Sensitivity for balancing against high resistances is obtained by the combination of a high applied voltage (500 v) and the use of a vacuum-tube detector. The dial on A is calibrated 1–10–100–1000 megohms, with the main decade 1–10 occupying the greater part of the dial space. The N arm gives five multipliers, 0.1, 1, 10, 100, and 1000. The A arm (note that A is a divisor, $X = BN/A$) is tapered, so that the dial calibration is approximately logarithmic in the main decade, 1 to 10. The unknown may be measured as a two-terminal resistor connected to the X posts, or as a three-terminal resistor with its guard point connected to the guard point on the bridge. In the latter case, R_b parallels arm B, and affects the calibration unless it is 100 or more times B. R_a parallels the detector, and does not alter the balance point, but does reduce the detector sensitivity to some extent, depending on magnitudes.

A guard circuit is essential for measurements of high resistance by bridge as by other methods in order that the reading shall represent the desired resistance, and not be affected by leakage over a stray path that is not part of the desired measurement. The guard may be constructed as in Fig. 4-27 in case sheet insulating material is being tested. Figure 4-29(a) represents a method of applying a guard if a test is being made of

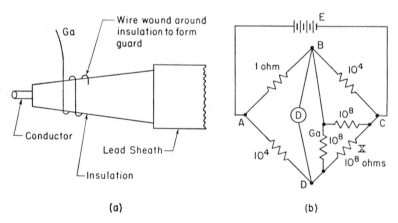

(a)   (b)

Fig. 4-29. Measurement of a three-terminal resistance on a Wheatstone bridge. (a) Method of guarding a cable. (b) Three-terminal resistor in a Wheatstone bridge.

the insulating material in a power cable. It is desired to determine the resistance *through* the cable material, independent of leakage over the surface at the ends. Surface leakage could cause serious error, particularly with a short sample and particularly if humidity conditions are bad at the time the measurements are made. The lead sheath is cut off for a short distance at each end of the cable, and a guard is formed by wrapping bare copper wire in close contact with the insulation. This guard can be used in the direct-deflection method just as in Fig. 4-27.

It is possible in measuring a guarded resistance on a bridge to provide a special guard circuit that holds the guard at the correct potential. More will be seen of this idea later in connection with a-c bridges. However, a simpler expedient is usually adequate for the d-c case. The resistances between the main terminals and between each main terminal and the guard form a *three-terminal resistance*, which is shown connected in a bridge in Fig. 4-29(b). Some numerical values are shown on the figure to indicate the general nature of the problem, it being assumed here that the unknown is 100 megohms, and that the resistance from each terminal to the guard is likewise 100 megohms. If the guard were not connected, the resistance measured from C to D would be 67 megohms, an error of 33 per cent. With the guard connected as shown, the resistance from Ga to D shunts the detector; thus it has no effect on the balance and only a small effect on the sensitivity of the detector. The resistance from Ga to C shunts the BC arm, and for the values shown, gives an error of 0.01 per cent; this is entirely negligible in a measurement of this sort.

A question arises in connecting the cable to this bridge regarding which terminal, C or D, should be connected to the sheath (which should be at ground potential). This may be important, as a rather high voltage may be applied to the bridge to improve the sensitivity. Note that the voltage drop from A to B or D is very small, and that the voltage B to C or D to C is practically equal to the applied voltage, E. If we connect the sheath to C and ground this point, arms AB and AD are much above ground potential, and hence must be protected against leakage. However, if we connect the sheath to D, the AB and AD arms are nearly at ground potential, and conditions are little different for BC and DC than before, except for reversal of grounding. In this latter method of connection neither terminal of the battery is grounded; however, terminal A is so nearly at ground potential that no difficulty need be experienced in insulating it.

(d) *Megohmmeter.*

The megohmmeter consists of a permanent-magnet crossed-coil meter and a power supply that may be either a hand-driven generator or a rectified a-c supply. The two torque-opposing coils are mounted rigidly together on a staff that carries the pointer. The spirals that conduct current to and from the coils are made to have as little torque as possible. One of the moving coils is a current coil in series with the unknown and the generator, and the other coil is connected in series with a fixed resistor across the generator terminals. For any value of the unknown the coils and pointer take up a position such that the torques of the two coils are equal and balanced against one another, and thus the position of the coils is

(a)

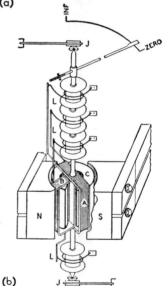

A = POTENTIAL OR DEFLECTING COIL

B = CURRENT OR CONTROL COIL

LLL = LEADING-IN LIGAMENTS HUNG LOOSELY AROUND INSULATING SPOOLS.

JJ = SPRING-SUPPORTED JEWEL BEARINGS.

NS = POLE PIECES OF PERMANENT MAGNET.

C = C-SHAPED IRON CORE MOUNTED IN A FIXED POSITION COAXIALLY WITH THE MOVING SYSTEM

(b)

Fig. 4-30. A 'Megger' insulation tester.   (a) External view.   (b) Diagram showing crossed coils, ligaments, and magnetic system.   (*Courtesy of James G. Biddle Co.*)

independent of the generator voltage because a change of voltage affects both coils in the same proportion. The scale of the meter can be calibrated directly in resistance value, with zero at one end and infinity at the other. The scale distribution depends on the series resistor, and on the shape of the magnet poles. Multi-range instruments can be constructed.

Figure 4-30(a) shows a "Megger" resistance-measuring instrument of the hand-cranked type, and (b) is a diagram of its construction, showing the magnet, crossed coils, and flexible spirals. Figure 4-31 is a wiring

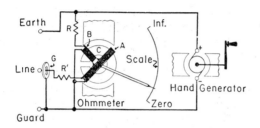

Fig. 4-31. Circuit diagram of a 'Megger' insulation tester. (*Courtesy of James G. Biddle Co.*)

diagram of a single-range instrument, indicating the coil arrangement, generator, and guard. The "Megger" is used very commonly to measure the insulation resistance of electrical machinery, insulators, insulating bushings, and so on. The test is made with a moderately high voltage stress on the insulation, as the generators (or rectified a-c supply) are made, in different models, for 100, 500, 1000, 2500, or 5000 v. Experience has shown that careful periodic measurements are a valuable indication of the condition of insulation and may make it possible to anticipate a serious breakdown.

4-28. Precision Comparison of Resistances

The precision comparison of standard resistors is a specialized process that is carried out mainly by the manufacturers of such equipment and by the national laboratories. The equipment and techniques involved are somewhat apart from the usual type of laboratory measurement, yet are of interest both in themselves and also for the application of some of the methods to other types of work.

Precision standards are usually made in decimal multiples and sub-multiples, as 0.1, 1, 10, 100 ohms, etc. Measurement of a resistor is made on an equal-arm bridge by comparison with a previously calibrated standard resistor of the same nominal value. How a precise set of standards is built up in the first place is an interesting story,* but the process can be indicated only in outline in the space available here. The build-up from one decade to another can be accomplished for resistors, as in other applications, by a set of standards with values of 1, 1, 2, 2, and 5 units (or by ten 1-ohm units). Suppose that we have a 1-ohm unit whose value we know accurately in some manner, as by the absolute method outlined in par. 30. We can calibrate a second 1-ohm unit in terms of the first. Then the two 1-ohm units can be placed in series and used to assign values to the 2-ohm units, taken one at a time. Now we place 1, 2, and 2 in series to establish the 5, and after that take 1, 2, 2, and 5 (or two 5's) to calibrate the 10-ohm resistor. We can repeat the process to build up the next decade and so on. It is interesting that values can be pyramided in this way with no appreciable increase in the percentage error.

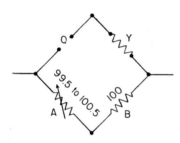

Fig. 4-32. Basic comparison bridge.

Figure 4-32 illustrates the principle of the comparison of nearly equal resistors, as applied to the higher (two-terminal) resistors. Arm B has a fixed value of 100 ohms and A may be varied over a small range, say from 99.5 to 100.5 ohms, which covers a range of ±0.5 per cent and so is more than ample to take care of any normal variations encountered in the bridge. Arm Y is a resistor approximately equal to the unknown, but its precise value does not need to be known. We may first place the standard at Q and balance the bridge, then replace by the unknown and rebalance. The difference in ratios gives the percentage difference between the standard and the unknown. Thus, if the two balances are different by one step on the 0.001-ohm (per step) dial, then one resistor is higher than the other by 0.001 per cent. Note that we not only do not need the value of Y, but we need not know the exact value of either ratio. We are concerned only with the *difference* of the ratios, which measures the difference between standard and the unknown. The unknown can thus be known to the same accuracy as the standard, within very close limits. (If a suitable resistor is not available for Y, we can make two balances, interchanging X and S in positions Q and Y. Then the difference of X and S is one half the difference of the two balance readings.)

Figure 4-33(a) shows the general arrangement of arm A, which has a large fixed resistor in series with three dials, going down to 0.001-ohm steps. The fixed resistor is so chosen that arm A has a resistance of 100 ohms when

---

*References 9 and 11.

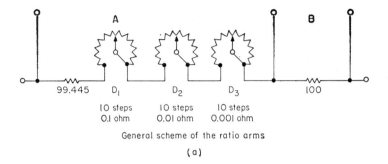

General scheme of the ratio arms

(a)

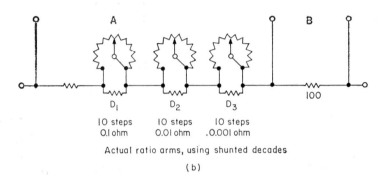

Actual ratio arms, using shunted decades

(b)

Fig. 4-33. Ratio arms for Comparison Set

the dials are set at the center position. A step on $D1$ raises or lowers A by 0.1 ohm, and so on.

The arrangement of Fig. 4-33(a) gives the desired combination of resistances, but in practice it would not be satisfactory as shown because of contact-resistance difficulties. Even a well-made switch gives contact resistance of the order of one or two thousandths of an ohm. Steps of 0.001 ohm on $D3$ would be meaningless if there were so much uncertainty as to contact resistance. For this reason a *shunted decade** arrangement is used. As shown in Fig. 4-34, 30 ohms has been chosen as the shunt resistance, and the dial resistances are computed so that the total effect ranges from 24.5 to 25.5 ohms, with 25.0 ohms at the center position. Note that the small switch contact resistance is now in series with a resistance of from 133 to 170 ohms and so produces negligible effect. The same scheme is used on the $D2$ and $D3$ dials of Fig. 4-33 with different resistance values. The dial shown in Fig. 4-34 gives *steps* of 0.1 ohm, as desired, but does not reduce to zero on the first contact. The minimum resistance value is of no harm in Fig. 4-33, however, as the resistance shown in (a) as 99.445 can be reduced

---

*Reference 11.

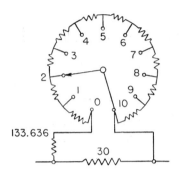

On 0, 133.636 and 30 in parallel = 24.5 ohms
On 1, 136.666 and 30 in parallel = 24 6 ohms
On 2, 139.811 and 30 in parallel = 24.7 ohms

etc.

On 10, 170.000 and 30 in parallel = 25.5 ohms

Fig. 4-34. Shunted decade, 0.1-ohm steps.

a corresponding amount. The only requirement is that the total resistance of arm A equal 100 ohms with all switches set at the center positions.

The shunted-dial construction is used also in bridges designed for measurements with resistance thermometers where it is desired to have dial resistances with very small steps.

4-29. Substitution Methods on a Standard Bridge

It sometimes happens that a measurement of a resistor is desired to closer limits than the accuracy tolerance of the available measuring devices. Close measurement may be made on a quite ordinary bridge if a standard resistor of the same nominal value is available, and if the technique of measurement is considered carefully.

Standard resistors of even decimal values are available with guarantees from 0.1 per cent to 0.01 per cent, according to the choice and the budget of the buyer. Resistors of suitable quality may be sent to the National Bureau of Standards for certification to ±0.01 per cent, when the needs of the situation call for this degree of accuracy. With such a standard resistor available, alternate measurements of standard (S) and unknown (X) on the bridge can give a close measurement of X, provided that balancing is done in such way that minimum changes are made in important bridge constants from one balance to the other. The loss of accuracy in transferring from S to X originates in the tolerances of the bridge as applied to the *difference* in settings in the two cases, so these differences must be kept to a small amount for best results.

For example, suppose the rheostat arm, R, has the value 1001.7 for one balance and 1003.2 for the other; then the uncertainty in ΔR is the tolerance applied to two steps on the units dial and five on the tenths dial, which forms a very small fraction of the total resistance. The difference of X and S, with its uncertainty, is given by ΔR (including its uncertainty) multiplied by the ratio (with its uncertainty). The added uncertainty, over and above that of S itself, can thus be kept very small, so that X may be known to nearly as close limits as S.

By way of contrast, suppose that the first balance gives $R = 1001.7$ and the second 999.2. The difference of *nominal* values is the same as before, but *the uncertainty is much greater*, since a 1000-ohm unit has been removed, and 990 added; the plus-or-minus nature of the guarantees gives a possible error originating in the upper dials equal to their tolerance percentage applied to 1990 ohms — this is in addition to the uncertainty in eight steps on the units dial and five on the tenths dial. Such operation would be highly undesirable, and would vitiate most or all of the advantage in the substitution method. If such change of dials appears in the first measurement attempt, a revised procedure should be followed, such as the addition of a small fixed resistor in the unknown arm for both balances, so that both readings on R are above the 1000-ohm setting.

The gain of accuracy in the substitution method is not in the elimination of errors of the equipment, but in balancing them out between two readings by keeping them as nearly the same as possible. The equipment used as the base of the measurements thus needs only short-time stability plus moderate accuracy. The same principle will be seen later to be very useful in measurements of inductance and capacitance.

4-30. Special-purpose Resistance Measurements

The equipment considered so far (except in par. 27) may be regarded as being suited to measurements of a rather general type. There are several bridges which, though somewhat special in nature, have come into sufficiently wide use to deserve some mention. In some cases the bridge is designed for a particular type of measurement and in others for convenient and rapid routine tests.

(a) *Comparison set.*

Makers of decade resistance boxes frequently use for adjustment of resistors a type of comparison bridge similar in basic idea to Figs. 4-32 and 4-33, but minus some of the refinements needed for extremely precise comparison. The bridge consists of a fixed and a variable ratio arm, plus terminals for connection of battery, galvanometer, standard, and unknown. For readings of ±0.5 per cent by 0.01 per cent steps only two dials are needed in the variable arm. For medium resistance values, the fixed arm might be 1000 ohms, and the variable arm 995 ohms in series with a 1-ohm decade and a 0.1-ohm decade (per step). The merit of this type of bridge is that the dials may be marked directly in the percentage difference between standard and unknown, and the same calibration holds good when used with standards of different amount.

(b) *Limit bridge.*

A convenient bridge for the rapid checking of resistors in quantity production is shown in Fig. 4-35. The limit sliders L_1 and L_2 may be set

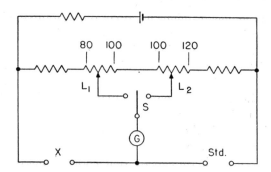

Fig. 4-35. Diagram of a limit bridge.

for the desired lower and upper tolerance limits for the resistors. A sample of the correct value is connected as the standard. A resistor to be tested is connected at X, and the switch S is thrown to left and to right. If the galvanometer deflects in opposite directions for the two positions of S, then the unknown is within the limits; if the galvanometer deflects in the same direction both times, the unknown is outside the limits. This type of test does not tell the exact value of the unknown, but it is fast, as the balancing process is eliminated. Another method of production testing that can be arranged for some sorting into values may be accomplished by use of a fixed bridge and calibration marks on the galvanometer (microammeter) scale, so that the galvanometer deflection indicates the difference of value between standard and unknown.

(c) *Mueller bridge.*

The Mueller bridge* is a special form of Wheatstone bridge used mainly for the measurement of temperature by a platinum-resistance thermometer. The range is frequently from zero to 70 or 110 ohms, reading by 0.0001-ohm steps. To attain such precision and to permit checking, an equal-arm bridge

---

*References 16, 18.

is used, and the decades are of the shunted design to eliminate errors due to contact resistance. The thermometers are of the four-lead type, so that the effect of lead resistance can be balanced out. A more detailed explanation of the Mueller bridge may be found in reference 11.

(d) *Ground-resistance measurements.*

An important type of measurement made in power systems and industrial plants is that of the resistance of ground connections and ground rods that play an essential part in the protection of equipment and personnel against overvoltages due to lightning discharges or accidental faults on the lines. The resistance of the ground connection should be low to give good protection and must be checked, as conditions in the soil cause great variability in resistance in different locations.

A convenient method of making ground-resistance measurements is by means of a special Megger, similar to the ones studied in a previous section, but modified to meet the requirements of this application. A Megger made for this purpose and the method of using it are indicated in Fig. 4-36. It is

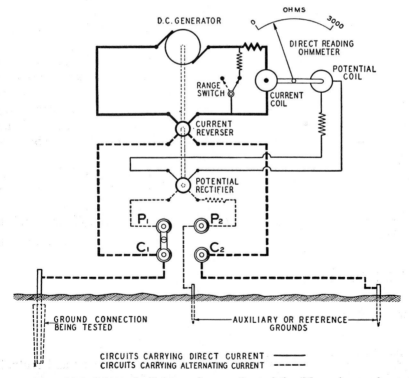

Fig. 4-36. Schematic diagram of connections of the 'Megger' ground tester, showing the principle of operation. (*Courtesy of James G. Biddle Co.*)

necessary to use separate potential and current reference grounds, so that the indication of the instrument represents only the resistance associated with the ground rod under test. The commutating devices are needed to give alternating current in the earth in order to avoid the effects of electrolytic polarization at the electrodes that would be produced by the moisture in the earth if undirectional current were used.

4-31. Absolute Measurement of Resistance

Mention was made in chap. 1 of the measurement of resistance in terms of the basic units in the absolute system. Several methods that have been proposed and used for this purpose are discussed in some detail in references 2 and 15. We do not have space to consider them all but shall look briefly at one scheme, devised by Dr. Frank Wenner and used in recent years at the Bureau of Standards. It involves the commutation of currents in a mutual inductance. The basic idea is shown in simplified form in Fig. 4-37(a), which omits a number of the refinements that are essential for accurate results, as will be discussed later.

Voltage is induced in the secondary of the mutual inductance as a result of the reversals of I by the primary commutator C_p. The secondary commutator C_s is needed so that the secondary voltage is always in the same direction and can be balanced against the voltage across the primary resistor. The galvanometer must have a long period to average the pulsating secondary effect against the steady primary voltage. Means are provided to hold the primary current constant; theoretically this could be a very large inductance, but practically, a better job can be done by electronic regulators.

The instantaneous voltage induced in the secondary coil is $M\ di/dt$. The *average* value, which is of importance in balancing the circuit, may be obtained by noting that we have a change in the primary winding from $+I$ to $-I$ and back from $-I$ to $+I$ in time $1/f$ sec, where f is the number of complete cycles of change per second. Therefore

$$E_{2\text{average}} = 4IfM$$

This is balanced against

$$E_1 = IR$$

If we equate these values we obtain

$$R = 4fM$$

That is, resistance is determined in terms of frequency and mutual inductance. The mutual inductance is a matter of the geometry and can be computed from the dimensions of the coils. The determination of resistance therefore is made in terms of the basic units of time and length.

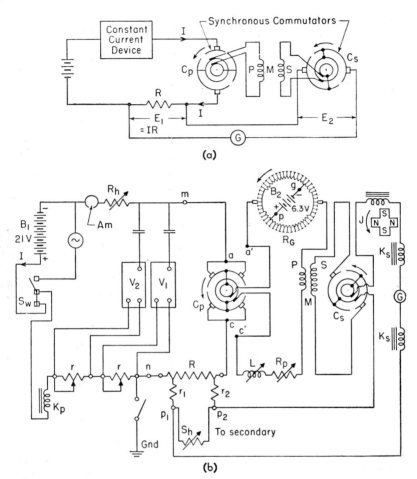

Fig. 4-37. Diagram of the Wenner method for the absolute determination of resistance. (a) Basic circuit. (b) Schematic diagram showing circuit arrangements.

Frequency is based on quartz crystal oscillators which may be relied on to better than one part in ten million. The primary coil is wound on a large and substantial form of porcelain, which is ground accurately to dimensions and has helical grooves ground and lapped in the surface for the single-layer wire winding. All dimensions are measured as accurately as possible with micrometer microscopes. The coil geometry is such that the mutual inductance can be computed to a high degree of certainty.

Some details and refinements of the method are shown in Fig. 4-37(b). (See p. 26 of reference 15, for a short account, or reference 17 for a complete treatment.) V_1 and V_2 are electronic regulators that hold the current steady throughout the cycle. C_p and C_s are the primary and secondary commutators as before, and R_g is another commutator running synchronously with them. The battery is divided in two parts: B_1 supplies the drop in the primary circuit up to points a-c, and B_2 and the rotating resistor system provide the voltage needed in the inductor primary to make the primary current change in an approximately trapezoidal pattern. In this way the difference of potential from a to c is practically zero at all times, and the short-circuiting and reversal can take place at C_p without any disturbance in the circuit. The trapezoidal primary current induces, in the secondary, flat-topped pulses which are made unidirectional by C_s. The magneto-generator, J, introduces an alternating emf (*average* value $= 0$) to balance the pulsations of the secondary voltage and so to reduce the averaging required of the galvanometer.

It is believed that determinations by this method are accurate within a fairly small number of parts per million (ppm), probably less than 10 ppm, and can be repeated within 1 or 2 ppm. Measurements at the Bureau by two methods and at the British National Physical Laboratory by two other methods give an over-all difference from smallest to largest of some 30 ppm, with the above method 6 ppm below the average of the four. The uncertainty in the value of the ohm has thus been reduced to a small amount, and will probably be further reduced in the future.

The following quotation is taken from the paper by J. L. Thomas,* in which he discusses the stability of the new double-walled resistors, and the means of checking them by absolute measurements:

On the basis of relative values, standards of this group are about one order better than well-aged standards of the Rosa type. It does not necessarily follow that their average value is more stable. Unfortunately, there has been no satisfactory method of judging the quality of resistors used in maintaining the unit except their ability to maintain relative values. A group that keeps the same relative values could all be changing, but at the same rate.

Before the war, some preliminary results had been obtained from a commutating method of measuring resistance in terms of inductance, which in turn could be calculated from the measured dimensions of a mutual inductor. By this method it was found possible to duplicate measurements of a 1-ohm resistor to 1 part in a million in terms of the calculated mutual inductance. It is expected that such measurement will in the future give data on the stability of groups of resistors over long periods of time. The stability of the inductor can be checked from time to time from its measured dimensions, and it is believed that this can be done to 1 part in a million. Measurements of the resistors in terms of the inductance would then disclose changes as large as 1 or 2 parts in a million in the resistance of the group being used to maintain the unit. It is not expected that resistance can be measured in absolute units to 1 part in a million by this method. It does appear, however,

---

*Reference 5.

that systematic errors will be sufficiently constant to permit checks of the stability of the unit of resistance to 1 or 2 parts in a million.

4-32. Resistance Measurements Applied to Instrumentation

Resistance measurements, particularly by bridge methods, are used in many applications in which the technique of measurement enters, but in which the objective is the determination of a physical quantity, rather than an interest in the resistance as such. The use of electrical means for the indicating and recording of physical quantities is frequently called instrumentation. A few samples of the things to be measured are: strain in structural or machine members, pressure, torque, speed, acceleration (linear or rotational), temperature, fluid flow, position, and weight. The reasons for the use of electrical methods are many, and differ in different applications. In some cases the superior speed of electrical methods, as contrasted with mechanical means is a compelling reason. Sometimes it is a matter of convenience, and frequently the possibility of remote indication and recording is an important consideration. Some forms of electrical pickup are valuable for the small restraint they place on the measured variable. The possibilities of combining control with indication and recording is another useful property of electrical methods.

Many systems and electrical principles are utilized. Most known electrical phenomena of matter have been used in one way or another. Circuit variations of resistance, inductance, or capacitance are most frequently used, but others, such as thermoelectric effects, the Hall effect (magnetic), and piezoelectric effects, are often employed. The applications using some form of resistance variation are appropriate to the present chapter.

A simple form of resistance variation may be applied in the indication of a mechanical motion or displacement as, for example, the motion of a valve, circuit breaker, liquid level, or the wing flaps or landing gear of an airplane. The pickup, or "transducer," consists of a moving contact on a slide wire resistor of linear or circular form. This type of transducer is especially suited to relatively large motions, and accordingly, rather low speeds. The movement of the contact on the surface of the resistor is attended by some friction, and so places some restraint on the measured quantity, making it unsuitable for delicate elements. The resistance change may be quite large, so that various circuits and instruments are possible. Millions of gages of this type, in rather crude form, are used to indicate gasoline supply in automobiles.

Temperature effects on resistance constitute another class of measuring element. The resistance of most materials is affected to some extent by temperature. Pure metals show, in general, a fairly regular increase of resistance with temperature. This principle is used in resistance thermometry; the most precise form using a platinum thermometer (for linearity and stability) and a Mueller bridge (par. 4-29). Tungsten may also be used. Some elements and compounds show a *negative* resistance–temperature coefficient; i.e., the resistance decreases with an increase of temperature. Carbon has this property, and many of the earth oxides. The magnitude of the change is generally much greater than for metals, which is helpful for instrumentation purposes. The special class of thermally-sensitive resistors is frequently given the name "thermistor." Any of the heat-sensitive devices may be used to measure the temperature of their surroundings; a bridge circuit is generally used, to give zero meter current at the chosen initial temperature and a good scale spread above or below that point. The bridge circuit may consist of one variable arm and three fixed arms, as shown in Fig. 4-38(a); or two variable and two fixed arms; or four variable

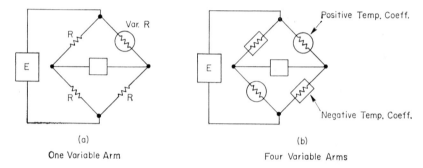

(a)
One Variable Arm

(b)
Four Variable Arms

Fig. 4-38. Bridge circuits with temperature-sensitive members.

arms, as in (b) if both positive- and negative-coefficient elements are available. Any of these bridge configurations may, of course, be analyzed by the methods which we developed earlier.

If the temperature change is produced by *self-heating*, the bridge may be used for voltage measurement or voltage control. If a bridge of either type shown in Fig. 4-38 is connected to the output of a power supply, the bridge may be balanced at one value of voltage. If the voltage increases, the heating of the resistors increases, and the bridge becomes unbalanced. The detector voltage or current may be used for control measures to return the voltage to its desired level.

Another resistance-variation device, the resistance-wire strain gage, will be discussed in the next paragraph.

4-33. Resistance Strain Gages

The resistance of many conductors is affected to some extent by strain in the material — this is true even in the winding of precision resistors.

It has been found possible, under carefully controlled conditions, to use the resistance change as a measure of the amount of strain. Strain is of direct interest in structural members; also, a strain gage on a flexure member can serve for force measuring, weighing, and other purposes. A strain gage is an important basic element in many instrumentation problems.

Part of the change of resistance of a conductor with strain may be explained by the geometric factors of length and cross-sectional area. Figure 4-39 indicates a wire (1) in an unstressed condition, and (2) under

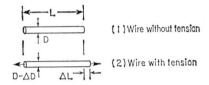

Fig. 4-39. Illustration of the effect of tension on the dimensions of a wire.

tension. The tension causes an increase ΔL in length, and at the same time a decrease ΔD in diameter. By the expression for the resistance of a conductor of uniform section,

$$R_1 = \rho \frac{\text{Length}}{\text{Area}} = \rho_1 \frac{L}{\frac{\pi}{4} D^2} \tag{24}$$

Under tension,

$$R_2 = \rho_2 \frac{L + \Delta L}{\frac{\pi}{4}(D - \Delta D)^2} \simeq \rho_2 \frac{L\left(1 + \frac{\Delta L}{L}\right)}{\frac{\pi}{4} D^2\left(1 - 2\frac{\Delta D}{D}\right)} \tag{25}$$

If we write $(\Delta D/D)/(\Delta L/L) =$ Poisson's ratio $= \mu$, and simplify the fraction, as in chap. 3,

$$R_2 = \frac{\rho_2}{\rho_1} R_1\left[1 + (1 + 2\mu)\frac{\Delta L}{L}\right] \tag{26}$$

Now, if we work on a *geometric basis only*, and neglect for the moment possible differences of ρ_1 and ρ_2,

$$R_2 = R_1 + \Delta R = R_1\left[1 + (1 + 2\mu)\frac{\Delta L}{L}\right] \tag{27}$$

From this, we may express the increment of R as compared with the increment of L,

$$\frac{\Delta R/R}{\Delta L/L} = \text{Strain sensitivity factor} = 1 + 2\mu \tag{28}$$

If μ has a value of about 0.3, as met with some materials, then the factor in (28) equals 1.6.

Several materials have strain sensitivity factors of about 1.6 to 2, which lends some support to our derivation. Other materials range from near zero to about 5, which indicates that molecular strain also affects the resistivity. For strain gage purposes, a high sensitivity is of course desirable, but other considerations enter in addition, particularly the matter of temperature. Many materials have been examined with the view of selecting the best combination of properties. The copper–nickel alloy variously known as Advance or Constantan has a factor slightly above 2, and good temperature characteristics. Another material, originally developed for spring scales, has a strain factor of 3.6, but is less desirable than Advance on a temperature basis; it is used for short-time tests in which possible zero shifts caused by temperature changes do not enter.

In use, the strain wire is usually cemented to the surface to be tested; this is generally accomplished as a preformed gage, consisting of fine resistance wire arranged back and forth in many hairpin loops, and bonded between two layers of paper. The completed gage, about the size of a postage stamp, may then be cemented to the surface of the structural member, and the strain wire must then elongate or contract with the surface to which it is fastened. Gages are made with "gage lengths" from about $\frac{1}{8}$ in. to several inches. The "Gage Factor" of a gage is defined in the same manner as (28), that is, as $(\Delta R/R)/(\Delta L/L)$, but the factor is somewhat less than for a straight piece of wire, owing to inactive portions at the hairpin loops. Some gages use the resistance wire in open construction, without being bonded to a paper base.

Basing the gage factor on the $(\Delta R/R)/(\Delta L/L)$ ratio is useful since it permits strain to be found directly from observed resistance changes. For relation to stress, we know that

$$\frac{\Delta L}{L} = \frac{s}{E} \tag{29}$$

where E = modulus of elasticity = Young's modulus

     s = stress, pounds per sq in.

Equation (29) permits us to obtain the order of magnitude of resistance changes with resistance-wire strain gages. Suppose that we have a gage on a steel member stressed to 30,000 psi. Since E for steel is approximately 30×10^6 lbs/in.^{2}, the unit elongation is

$$\frac{\Delta L}{L} = \frac{30,000}{30 \times 10^6} = \frac{1}{1000} = 0.1\%$$

If the gage factor is 2, then

$$\frac{\Delta R}{R} = \frac{2}{1000} = 0.2\%$$

Since the resistance change is only 0.2 per cent for the fairly high stress of 30,000 psi, gage measurements will be made a great deal of the time at even lower values. A bridge circuit is essential to display so small a change, and to obtain stable indication. Moreover, the bridge circuit must be made very carefully, with solid low-resistance connections. It is important to avoid, or to balance out, the effect of temperature on the resistance of gage and leads, as the temperature increment of resistance may well be as great as the part due to stress. Figure 4-40(a) indicates a bridge with one

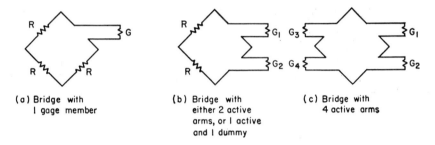

(a) Bridge with 1 gage member

(b) Bridge with either 2 active arms, or 1 active and 1 dummy

(c) Bridge with 4 active arms

Fig. 4-40. Strain gage bridge arrangements.

gage arm, and the other arms as fixed resistors. This would be a bad arrangement, particularly if the gage must be located at some distance from the bridge. To avoid temperature effect on gage and leads, an inactive ("dummy") gage may be added as in (b), with the inactive gage close to the active, and mounted on a piece of material of the same type. Sometimes gages may be mounted on the opposite faces of a beam, so that one gage is in tension and one in compression; the circuit of (b) may be used, with twice the unbalancing effect. Finally, a four-arm bridge may be used as in (c), with two tension gages and two compression; this might be on a beam, or on a shaft in torsion.

The electrical output of any of these bridge arrangements may be analyzed most conveniently by the Thevenin method. Static strain measurements may be conducted by the null method. For varying strains, and sometimes for static tests, the bridge is balanced for zero-stress condition, and an indicating or recording device is connected to the detector terminals. The indication of this device gives a measure of the strain, once calibration has been accomplished. If the detector is a high impedance device (vacuum-tube amplifier) only the Thevenin E_0 need be computed. If the detector is a current-operated device, such as a microammeter, the internal resistance, R_0, must also be found.

In the following expressions it is assumed that all four gages have initial values R, and that the strain adds an increment $+\Delta R$ in tension gages, and an increment $-\Delta R$ in compression gages. The voltage applied to the bridge is E. The derivation is left to the reader.

$$1 \text{ active gage:} \quad E_0 = E\,\frac{\Delta R}{4R}$$

$$2 \text{ active gages:} \quad E_0 = E\,\frac{\Delta R}{2R}$$

$$4 \text{ active gages:} \quad E_0 = E\,\frac{\Delta R}{R}$$

Gage-circuit bridges are frequently operated on a "carrier" alternating current of a frequency considerably higher than any of the strain variations to be recorded. This simplifies amplification, as compared with d-c operation, and avoids trouble from stray thermal voltages. Special circuits may be used for recording fast transients; the gages themselves are capable of very fast response.

REFERENCES

1. Goodwin, W. N., Jr., "Electrical Resistance Alloys," *Weston Engineering Notes*, **2**, 5 (Oct. 1947).

2. Curtis, Harvey L., *Electrical Measurements*, New York: McGraw-Hill Book Co., Inc., 1937.

3. Rosa, E. B., "A New Form of Standard Resistance," *NBS Sci. Technol. Papers*, **5**, 413 (1908–9). S 107.

4. Thomas, James L., "A New Design of Precision Resistance Standard," *NBS J. Research*, **5**, 295 (1930). RP 201.

5. Thomas, James L., "Stability of Double-Walled Manganin Resistors," *NBS Research Papers* **36**, 107 (Jan. 1946). RP 1692.

6. Thomas, James L., "Gold-Chromium Resistance Alloys," *NBS Research Paper* No. 737, **13**, 681 (1934).

7. Godfrey, Theodore B., "Further Data on Gold-Chromium Resistance Wire," *NBS Research Paper* No. 1206, **22**, 565 (1939).

8. Stack, S. S., "Resistance-Thermometer Bridge," *Gen. Elec. Rev.*, **51**, 7, 17 (July, 1948).

9. Wenner, F., "Methods, Apparatus and Procedures for Comparison of Precision Standard Resistors," *NBS Research Paper* (Aug. 1940). RP 1323.

10. Laws, Frank A., *Electrical Measurements*, New York: McGraw-Hill Book Co. Inc., 1938.

11. Thomas, James L., "Precision Resistors and Their Measurement," *NBS Circular* 470 (Oct. 1948).

12. Dike, P. H., "The Effect of Atmospheric Humidity On Unsealed Resistors, Causes and Remedy," *Rev. Sci. Instruments*, **7**, 278 (1936).

13. Thomas, James L., "Electrical-Resistance Alloys of Copper, Manganese and Aluminum," *NBS J. Research* **16**, 149 (1936). RP 863.

14. "Master Test Code for Resistance Measurement," *A.I.E.E.* No. 550 (May, 1949).

15. Silsbee, F. B., "Establishment and Maintenance of the Electrical Units," *NBS Circular* 475 (June 1949).

16. Mueller, E. F., "Wheatstone Bridges and Some Accessory Apparatus for Resistance Thermometry," *NBS Bul.* **13**, 547 (1916). S288.

17. Thomas, James L., Peterson, Chester, Cooter, Irvin L., and Kotter, F. Ralph, "An Absolute Measurement of Resistance by the Wenner Method," *J. Research NBS*, **43**, 4, 291–353 (Oct. 1949). RP 2029.

18. Harris, Forest K., *Electrical Measurements*, New York: John Wiley & Sons, Inc., 1952.

PROBLEMS

4-1. It is desired to construct a series ohmmeter as in Fig. 4-3 with a midscale indication of 5000 ohms. The instrument to be used requires 0.50 ma for full-scale deflection and has a resistance of 50 ohms. The battery voltage is 3.00 v.

(a) Find the values of the resistors needed in the ohmmeter circuit.

(b) Find the range of values of R_2 needed to zero the ohmmeter if the battery voltage may range from 2.70 to 3.10 v. (Use R_1 as in (a)).

(c) If the instrument accuracy is 0.5 per cent of full scale, what is the possible inaccuracy in the resistance value indicated at 30 per cent scale deflection due to the instrument alone?

4-2. A series ohmmeter designed to operate on 4.50 v has a circuit diagram as in Fig. 4-3. The indicating instrument used in the ohmmeter has a resistance of 120 ohms and requires 0.1 ma for full-scale deflection. R_1 is 37,400 ohms. Instrument accuracy = 1% of full scale.

(a) What value of R_2 is needed to zero the ohmmeter?

(b) What value of R_x will cause the instrument to deflect 30 divisions from the left end of the scale if the scale has equal-current 100 divisions? Similarly, find R_x for deflections of 10, 20, and 80 divisions.

(c) Find the possible inaccuracy of the indicated resistance due to the instrument only at the scale points mentioned in part (b). Compute each value both in ohms and as a per cent of the indicated resistance.

(d) Prove that the minimum percentage inaccuracy due to the instrument occurs at the half-scale point.

4-3. A shunt-type ohmmeter, with circuit as in Fig. 4-6, uses a milli-ammeter with full-scale current of 2 ma, a resistance of 25 ohms, and a guaranteed accuracy of 1 per cent of full scale. The battery has an emf of 1.50 v.

(a) Find the value of R_1.

(b) At what point (in percentage of full scale) will 100 ohms be marked on the scale?

(c) What is the possible inaccuracy at the 100-ohm reading, due to the instrument alone?

(d) What addition can be made to the circuit if it is desired to have 100 ohms as the half-scale value?

4-4. The arms of a Wheatstone bridge have resistance of 600, 60, 101, and 1000 ohms, taken in sequence around the bridge. A galvanometer (resistance = 100 ohms) is connected from the junction of the 600- and 1000-ohm resistances to the junction of the 60 and 101. A battery of 4-v emf and negligible internal resistance is connected to the other corners.

(a) Using Thevenin's theorem, find the emf and resistance to be used in an equivalent circuit for computation of galvanometer current.

(b) Find the current through the galvanometer.

(c) Find the power dissipated in the bridge resistor that has the maximum power consumption.

4-5. Repeat problem 4 with battery and galvanometer interchanged in position. Compare the sensitivity (as indicated by galvanometer current) with problem 4, also the maximum power dissipation.

4-6. A resistor being measured on a Wheatstone bridge has a resistance of approximately 1000 ohms. The ratio arms are 1000 ohms each (assumed exact). The galvanometer has a resistance of 50 ohms and requires a current of 0.05 μa to give a deflection of 1 mm on the scale. The battery has an emf of 4 v and negligible internal resistance. How much difference in resistance between the unknown and the rheostat arm is needed to give a galvanometer deflection of 1 mm? (Make the rheostat arm 1000 ohms and call the unknown $(1000 + x)$. Solve for x. Neglect second-order effects in x.)

4-7. Same as problem 6, except that the unknown is approximately 100 ohms.

4-8. An "unknown" resistor being measured on a Wheatstone bridge has actually a resistance of 351.6 ohms. The bridge ratio arms are nominally 1000 ohms each, but one is 1001.6 ohms and the other 999.1 ohms.

(a) If the 999.1-ohm arm is adjacent to the unknown, find the apparent value of the unknown, as indicated by the rheostat arm setting for which balance is secured. This arm can be read to the nearest 0.1 ohm.

(b) Find the new balance value, if the ratio arms are interchanged.

(c) Average the results of (a) and (b) and compare with the resistance of the unknown.

4-9. Does the interchange of ratio arms as in problem 8 make complete correction for errors in the ratio arms of *any* amount? Preferably work analytically, using the ratio resistors as $M = 1000(1 + m)$ and $N = 1000(1 + n)$ — or else use numerical values with extremely large discrepancies.

4-10. Two resistors have values of 100.6 and 99.8 ohms. Find the ratio of their resistance in series to their resistance in parallel.

4-11. Resistor units A, B, and C have nominal values of 100 ohms, and exact values which may be represented as,

$$A = 100 (1 + a)$$
$$B = 100 (1 + b)$$
$$C = 100 (1 + c)$$

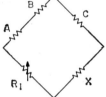

Prob. 4-11

where a, b, and c are the *small* fractional differences (+ or −) from the nominal value. The first balance is taken as shown, with A and B used against C; call the balance value of the rheostat arm R_1. A second value R_2 is found using $A + C$ against B, and a third value R_3 using $B + C$ against A.

(a) Derive an expression for the mean of the three results for X,

$$\frac{R_1 + R_2 + R_3}{3} \times \frac{100}{200}$$

(b) Determine whether the mean reduces the error in X, and if so, to what extent.

4-12. Three nominal 100-ohm resistors have values, $R_1 = 100.6$, $R_2 = 100.4$, and $R_3 = 99.8$ ohms. Two of these resistors are placed in series as one ratio arm of a Wheatstone bridge, and the third is used as the other ratio arm. Three combinations are used; 1 and 2 against 3; 1 and 3 against 2; and 2 and 3 against 1. Find the mean value of the ratios thus secured.

4-13. A resistance of approximately 3000 ohms is needed to balance a bridge. It is obtained on a 5-dial resistance box (steps of 1000, 100, 10, 1, and 0.1 ohms) guaranteed to 0.1 per cent. For this accuracy, how many of these dials would it be worth adjusting?

4-14. A coil of copper wire is measured on a Wheatstone bridge and specified by the operator to have a resistance of 255.2 ohms. The measurement was made at a temperature of about 20°C. To what degree of exactness must the temperature be known if all figures in the resistance value are to be regarded as significant? (Copper has a coefficient of resistance change with temperature of 0.00393/°C, beginning at 20°C.)

4-15. A resistance $R_1 = 10$ ohms is shunted by $R_2 = 90$ ohms. If a sliding contact in the second branch introduces an increment $\Delta R_2 = 0.01$ ohm, by how much is the resistance of the parallel combination changed? [Suggestion: Review the differentiation idea of eq. (37) of chap. 2.]

4-16. Galvanometer A has a sensitivity of 0.05 μa/mm and a resistance of 50 ohms; galvanometer B has values of 0.005 μa/mm and 1000 ohms. One of these galvanometers is to be used in a Wheatstone bridge with equal 100-ohm ratio arms to measure an unknown of about 200 ohms. The 3-v battery connects from the junction of the ratio arms to the junction of the variable arm and the unknown. Determine the deflection which each galvanometer gives for a difference of resistance of 0.05 per cent between the unknown and the variable arm.

4-17. A Wheatstone bridge has ratio arms of 1000 and 100 ohms and is being used to measure an unknown of 25 ohms. Two galvanometers are available. Galvanometer "A" has a resistance of 50 ohms and sensitivity of 0.005 μa/mm, "B" has values 600 ohms and 0.002 μa/mm. Which galvanometer is more sensitive for a small unbalance on the above bridge, and what is the ratio of sensitivities? The galvanometer is connected from the junction of the ratio arms to the opposite corner.

4-18. The arms of a Wheatstone bridge are, in sequence, the ratio arms of 100 ohms and 1000 ohms, the unknown, and the adjustable arm set at 150 ohms. The battery of 3-v emf and negligible internal resistance connects from the junction of the ratio arms to the opposite corner. The galvanometer has a resistance of 90 ohms and a sensitivity of 0.05 μa/mm.

(a) If the galvanometer gives a deflection of 60 mm for the above conditions, by how much does the adjustable arm differ from its true balance value?

(b) What maximum battery voltage may be used with this bridge without exceeding 0.25 watt in any arm?

4-19. A Wheatstone bridge marked A-B-C-D consecutively around the corners has ratio arms AB 1000 ohms and BC 1000 ohms. CD is an unknown of approximately 500 ohms, and DA is the rheostat arm. A 1.5-v battery of negligible internal resistance is connected from A to C. The galvanometer has a sensitivity of 0.005 μa/mm, a resistance of 50 ohms, and an external critical damping resistance of 350 ohms. It is used with an Ayrton shunt of 350 ohms, with relative sensitivity settings of 1, 0.1, 0.01, and 0.001.

(a) What unbalance of the unknown is needed to give a galvanometer deflection of 1 mm, with the shunt on the most sensitive setting?

(b) The rheostat arm is reduced 20 per cent below the balance value. Find the galvanometer deflection with the shunt on the 0.001 position.

4-20. A Wheatstone bridge marked A-B-C-D consecutively around the corners has ratio arms, $AB = 100$ ohms and $BC = 10$ ohms. CD is the unknown of approximately 90 ohms, and DA is the rheostat arm. A battery of negligible internal resistance is connected from A to C. The galvanometer has a sensitivity of 0.005 μa/mm, and an external critical damping resistance of 350 ohms. It is used with a 350-ohm Ayrton shunt with relative sensitivity settings of 1, 0.1, 0.01 and 0.001. $R_G = 50$ ohms.

(a) What is the maximum battery voltage that may be used without exceeding 0.25 watt in any arm?

(b) The rheostat arm is reduced 20 per cent below the balance value. Find the deflection of the galvanometer with the shunt set on the 0.001 position, using the voltage of part (a).

4-21. The arms of a Wheatstone bridge, taken in order around the bridge, are first, the ratio arms of 1000 ohms and 100 ohms, and then the unknown of approximately 500 ohms, and the rheostat arm. The galvanometer connections are made from the junction of the ratio arms to the opposite corner. The galvanometer has a sensitivity of 0.005 μa/mm, a resistance of 100 ohms, and is used with an Ayrton shunt of 500 ohms resistance. The battery has an emf of 3 v and negligible internal resistance.

How much increase of resistance of the rheostat arm above the balance value is needed to give a galvanometer deflection of one millimeter when the shunt is set at 0.01 of the maximum sensitivity?

4-22. The arms of a Wheatstone bridge are set at 1000 ohms to 10 ohms when measuring an unknown of approximately 3.50 ohms. The rheostat arm is variable from zero to 10,000 ohms by 0.1-ohm increments. The galvanometer has a resistance of 40 ohms, and is used with an Ayrton shunt of 360 ohms, set at maximum sensitivity. The maximum allowable power dissipation in any bridge arm, including the unknown, is 0.5 watt. The battery may be connected to either diagonal.

(a) Find the applied bridge voltage for the two battery positions, for the 0.5-watt limit.

(b) Find the relative sensitivities of the bridge and galvanometer to small unbalances in the two cases, using the applied voltages of (a).

4-23. A Wheatstone bridge with ratio arms set at 1000 to 10,000 ohms is used to measure an unknown of 90,000 ohms. The battery has an emf of 3 v and negligible internal resistance and is connected from the junction of the ratio arms to the opposite corner. The galvanometer has a sensitivity of 0.05 μa/mm, and a resistance of 100 ohms. How much change of setting of the rheostat arm is needed to give a galvanometer deflection of 1 mm?

4-24. A Wheatstone bridge with ratio arms of 10,000 ohms and 10 ohms and a rheostat arm variable from zero to 10,000 ohms is used to measure an unknown of approximately 4 megohms. The galvanometer has a resistance of 250 ohms, and is used with an Ayrton shunt of 2250 ohms, set on maximum sensitivity. The maximum allowable power dissipation in any bridge arm is 0.25 watt. The battery may be connected to either diagonal.

(a) Find the applied bridge voltage for the two battery positions, for the 0.25-watt limit.

(b) Find the relative sensitivities of bridge and galvanometer to small unbalances in the two cases, using the applied voltages of part (a).

4-25. A regular Wheatstone bridge is used to measure high resistances (in the megohm range). The bridge has two ratio arms each of the following values, 10,000, 1000, 100, 10, and 1 ohms. The adjustable arm has a maximum value of 10,000 ohms. A battery of 10-v emf and negligible resistance is connected from the junction of the ratio arms to the opposite corner.

(a) What is the maximum resistance that can be measured by this arrangement? Draw a sketch showing the resistance of each arm.

(b) If the galvanometer has a sensitivity of 0.005 μa/mm and a resistance of 25 ohms, how much unbalance (Δr) of the unknown is needed to give a galvanometer deflection of 1 mm for the maximum resistance of part (a)?

(c) If the galvanometer is replaced by one having a sensitivity of 0.0005 μa/mm and resistance of 550 ohms, how much unbalance is needed for 1-mm deflection?

4-26. Three arms of a Wheatstone bridge have resistances of 100.0 ohms, and the fourth arm 100.1 ohms. A galvanometer of 50-ohms resistance and 0.05 μa/mm sensitivity is connected across one diagonal of the bridge. A battery of 3-v emf is connected in series with a resistance of 200 ohms to the other diagonal. Find the deflection of the galvanometer.

4-27. A Wheatstone bridge marked A-B-C-D consecutively around the corners has resistances, AB, 1000 ohms; BC, 100 ohms; CD, 200 ohms; DA, the rheostat arm. The galvanometer, connected from B to D, has a resistance of 50 ohms and a sensitivity of 0.5 μa/mm. A battery, in series with a resistor of 300 ohms, is connected from A to C. Find the battery voltage required so that an unbalance of 1 ohm in the rheostat arm causes a galvanometer deflection of 1 mm.

4-28. A Wheatstone bridge has resistances: AB, 100 ohms; BC, 200 ohms; CD, 2001 ohms; DA, 1000 ohms. A galvanometer with a sensitivity of 0.005 μa/mm and resistance of 50 ohms is used with an Ayrton shunt of 450 ohms, connected from B to D. A battery of 4.50 v is connected in series with a resistor of 200 ohms from A to C.

(a) Find the galvanometer deflection for the above conditions if the shunt is set at 0.1 of maximum sensitivity.

(b) Find the power in the arm that has the maximum wattage.

4-29. A Wheatstone bridge marked A-B-C-D consecutively around the corners has ratio arms AB, 1000 ohms and BC, 10 ohms. CD is the unknown of approximately 40.5 ohms, and DA is the rheostat arm. A 1.5-v battery in series with 250 ohms is connected from B to D. The galvanometer has a sensitivity of 0.005 μa/mm and a resistance of 50 ohms. It is used with an Ayrton shunt of 450 ohms. Find the difference of the rheostat arm from the balance value if the galvanometer gives a deflection of 1 mm with the shunt set at 0.1 of maximum sensitivity.

4-30. A Wheatstone bridge has resistances: AB, 100 ohms; BC, 250 ohms; CD, 2501 ohms; DA, 1000 ohms. A galvanometer of sensitivity of 0.004 μa/mm and 50 ohms resistance is connected with an Ayrton shunt of 200 ohms resistance to bridge terminals B and D. A battery in series with a resistor of 300 ohms is applied from A to C.

(a) Find the battery voltage required to give a galvanometer deflection of 10 mm, with the shunt set on 0.1 of maximum sensitivity.

(b) Find the power dissipated in the bridge arm that receives the greatest power.

4-31. The ratio arms of a Wheatstone bridge are guaranteed to be accurate to ± 0.05 per cent, and the rheostat arm to ± 0.1 per cent. The ratio arms are set at 1000 to 1000 ohms, and the bridge is balanced with the rheostat arm reading 2324 ohms. What are the upper and lower limits of the unknown, on the basis of the guarantees?

4-32. The ratio arms of a Wheatstone bridge are 100 to 10 ohms, and are guaranteed to ± 0.05 per cent. The bridge is balanced with the rheostat arm set at 428.6 ohms. It is guaranteed to ± 0.1 per cent, except the units dial to ± 0.25 per cent and the tenths dial to ± 1 per cent. The contact resistances of the dial switches may be 0.008 ohm (total). Find the limit of error of the unknown by the guarantee and express as a percentage of the apparent value (42.86 ohms).

4-33. A balanced Wheatstone bridge, with circuit as in Fig. 4-11, has values, $A = 10$, $B = 1000$, $R = 3750$ ohms. A and B are guaranteed to ± 0.05 per cent and R to ± 0.1 per cent. Assuming that a 1-ohm change in R can barely be detected, what is the nominal value and the uncertainty of X?

4-34. A Wheatstone bridge has ratio arms, $AB = 10,000$ ohms, $BC = 100$ ohms. The galvanometer indicates balance with the rheostat arm set at 9518.8 ohms. The rheostat arm, D to A, is guaranteed to ± 0.1 per cent, except the units dial to ± 0.25 per cent, and the tenths dial to ± 1 per cent. The ratio arms are guaranteed to ± 0.05 per cent.

Find the upper and lower limits of the unknown on the basis of the guarantees.

4-35. A Wheatstone bridge is used for a resistance measurement by the substitution method, by taking a first balance on a standard resistor, S, which has a value of 10.048 ohms, guaranteed to ± 0.02 per cent. Under these conditions, the other arms of the bridge are, $A = 10$, $B = 1000$, $R = 1006.0$, all guaranteed to ± 0.1 per cent. The unknown, X, is substituted for S; arms A, B, and R are kept at their original values and balance is secured by adjustment of a decade box, $R_2(\pm 0.1$ per cent), shunted across X. Balance is given for $R_2 = 510$ ohms. Specify the value of X, and the limits to which it can be guaranteed.

4-36. A Wheatstone bridge has ratio arms of 1000 and 500 ohms nominal value, respectively, guaranteed to ± 0.05 per cent. The rheostat arm is guaranteed to ± 0.1 per cent, except the 1-ohm steps to ± 0.25 per cent, and the 0.1-ohm steps to ± 1 per cent. A standard resistor guaranteed to be 1000.1 ohms ± 0.01 per cent is first connected to the "unknown" terminals of the bridge, and balance is secured with the rheostat arm set at 2001.6 ohms. The standard is removed and replaced by an unknown, and balance is now secured with the rheostat arm set at 2003.2 ohms. Specify the resistance of the unknown and the percentage of uncertainty in its value. The galvanometer has sufficient sensitivity to permit balancing to the limit of the rheostat dials. Assume, for simplicity in this problem, that the readings do not require interpolation between the 0.1-ohm steps of the rheostat dial.

4-37. The ratio arms of a Wheatstone bridge, A-B and B-C, are 1000 ohms and 100 ohms, respectively, and are guaranteed to ± 0.1 per cent. When a standard of 100.15 ohms, guaranteed to ± 0.02 per cent, is connected in arm C-D, the rheostat arm must be set to a marked value of 1003.0 ohms (± 0.1 per cent) to give a null on the galvanometer. When the standard is replaced by an unknown, a resistance of 2.5 ohms (± 0.1 per cent) must be added in series in arm A-B to restore balance with the settings of the other arms not disturbed. Specify the value of X as closely as possible, and find the limits within which you can guarantee this result.

4-38. A Wheatstone bridge is used for resistance measurements by the substitution method. The ratio arms are marked $A = 100$ ohms, $B = 1000$ ohms, each guaranteed to ± 0.1 per cent. When used to measure a standard resistor, S, certified to be 250.8 ohms ± 0.04 per cent, the galvanometer indicates balance when the rheostat arm is set at a marked value of 2511 ohms. The galvanometer has a resistance of 50 ohms and a sensitivity of 0.005 μa/mm (± 5 per cent). The battery has an emf of 2.00 v, and negligible internal resistance, and is connected from the junction of A and B to the opposite corner.

(a) When an unknown resistor, X, of slightly higher resistance is connected to the bridge in place of S, arms A, B, and R being the same as before, the galvanometer deflection is 78 mm. Find the difference in resistance between X and S.

(b) Within what limits can the resistance of X be guaranteed, for the conditions above?

4-39. The ratio arms of a Kelvin double bridge are 1000 ohms each. The standard resistance is set at 0.1000 ohm; the unknown is 0.1002 ohm. The galvanometer has a resistance of 500 ohms and a sensitivity of 0.005 μa/mm. A current of 10 amps is passed through S and X from a 2.2-v battery in series with a rheostat. The resistance of the yoke may be considered negligible.

(a) What galvanometer deflection is produced?

(b) From the results of (a), what is the resistance unbalance to give a galvanometer deflection of 1 mm?

4-40. The ratio arms of a Kelvin double bridge are 1000 ohms each. The galvanometer has a resistance of 100 ohms, and a sensitivity of 0.003 μa/mm. A current of 10 amps is passed through the bridge from a 2.2-v battery, in series with a rheostat. The resistance of the yoke of the bridge may be considered negligible. With the standard resistance set at 0.1000 ohm the galvanometer gives a deflection of 25 mm. By how much does the unknown differ in resistance from the standard?

4-41. The arms of a Kelvin double bridge are 1000 to 10,000 ohms, and the adjustable standard is set at 0.0100 ohm in measuring an unknown of 0.0997 ohm. A battery of 2.25-v emf and internal resistance 0.0053 ohm is connected across the bridge terminals. The yoke has a resistance of 0.01 ohm. The galvanometer has a resistance of 100 ohms and a sensitivity of 0.005 μa/mm. Find the deflection of the galvanometer.

4-42. A Kelvin bridge has two arms each of 1000 ohms (S side) and 100 ohms (X side). The standard resistance is set at 0.1000 ohm, the unknown is 0.0101 ohm, and the yoke 0.1 ohm. A current of 10 amps is passed through the standard from a 6.2-v battery in series with a rheostat. The galvanometer has a sensitivity of 0.05 μa/mm and a resistance of 50 ohms.

(a) Compute the galvanometer deflection for the above conditions.

(b) If an Ayrton shunt of 300 ohms is used across the galvanometer, with the shunt section set at 3 ohms, compute the galvanometer deflection.

4-43. A Kelvin double bridge has a variable standard resistance consisting of 9 steps of 0.001 ohm each and a variable resistance bar of 0.001 ohm. The ratio arms are 100 ohms (on the standard side) to 1000 ohms (on X side). The galvanometer has a resistance of 50 ohms and a sensitivity of 0.005 μa/mm, and is shunted by its CDRX of 450 ohms. The resistance of the yoke may be considered negligible. The bridge is in balance with $S = 0.008602$ ohm.

(a) Draw a sketch of the circuit and compute X.

(b) The standard is changed to 0.008612 ohm. Find the deflection of the galvanometer. A current of 20 amp is passed through S and X from a battery of 2.20-v emf, with a rheostat in the circuit.

4-44. The adjustable standard of a Kelvin double bridge consists of 9 steps of 0.001 ohm each, guaranteed to ±0.04 per cent, and a 0.001 ohm bar with 100 scale divisions, on which readings are guaranteed to be accurate to ±1 microhm. There are two ratio arms each of 100, 1000, and 10,000 ohms, guaranteed to ±0.04 per cent. A 75-amp, 50 millivolt ammeter shunt is to be measured by this bridge.

(a) Make a sketch of the bridge circuit, showing the ratio arms you would select. Include in the diagram *all* the equipment you would need for an actual laboratory setup.

(b) For the arrangement you select, compute the maximum possible error of the measurement, based on the guarantees. (Consider that the ratios A/B and a/b are matched to negligible difference.)

(c) If 20 amps are passed through S and X in the bridge circuit, and a galvanometer of 500 ohms resistance and 0.0004 μa/mm sensitivity is used, find the resistance increment in X, from the balance value, needed to give the galvanometer a deflection of 1 mm. The yoke resistance is 0.01 ohm.

4-45. A Kelvin bridge has two arms each of 1000 ohms and 100 ohms. The resistance being measured has a value of 0.00100 ohm, and the standard is set at the slightly unbalanced value of 0.00980 ohm. A current of 30 amps is passed through S and X from a 2-v battery, in series with a rheostat. The yoke has a resistance of 0.0024 ohm. The galvanometer has a resistance of 80 ohms and a sensitivity of 0.0035 μa/mm. It is used with an Ayrton shunt of 400 ohms, with selector switch set to give 4 ohms in the shunting portion.

Find the galvanometer deflection under these conditions.

4-46. A Kelvin double bridge has two ratio arms available of each of the following values, 100, 500, 1000, and 10,000 ohms, guaranteed to ±0.05 per cent. The adjustable standard has 9 steps of 0.001 ohm each (±0.05 per cent), and a calibrated bar of 0.0011 ohm with a 110-division scale (any setting guaranteed to ±2 microhms). A d-c ammeter shunt, rated 50 millivolts, 200 amps full scale, is to be measured on this bridge.

(a) Select the ratio arms you would use.

(b) Make a complete circuit diagram, showing *all* equipment needed for the test. A 2.2-v storage battery is available.

(c) For the settings you have selected, find the possible uncertainty in X caused by the uncertainty in the standard, and express as a percentage of X. What additional error may be caused by errors in the ratio arms? (Consider that the two arms of each value are matched to a negligible difference.)

4-47. A Kelvin bridge is balanced with the following constants (notation as in Fig. 4-22): $S = 0.003770$, $A = 100.00$, $B = 1000.0$, $a = 99.92$, $b = 1000.6$, $Y = 0.1$.

(a) Compute X, neglecting the second term in the equation.

(b) Compute X, using the complete equation.

(c) Repeat (b), except that $Y = 0.01$ ohm.

4-48. It is desired to use a Kelvin bridge to adjust a shunt for a 50-mv 50-ohm millivoltmeter (± 0.5 per cent full scale) to form a 5000-amp ammeter (± 1.0 per cent full scale).

(a) Find the resistance and accuracy tolerance of the shunt.

(b) The lowest bridge ratio available is 0.01 accurate to ± 0.02 per cent. The yoke resistance, exclusive of the unused portion of the bar, is 25×10^{-6} ohm. The standard resistance is a manganin bar totaling 0.0011 ohm and divided into 110 divisions, connected as in Fig. 4-23. It is accurate to $\pm 1 \times 10^{-6}$ ohm. Where must the slide be placed on the bar? What is the total yoke resistance? What is the uncertainty of the measurements if the galvanometer permits an uncertainty of setting of no more than $\pm 1.0 \times 10^{-6}$ ohm?

4-49. A Kelvin bridge has a variable standard consisting of 9 steps of 0.01 ohm each, guaranteed to ± 0.05 per cent and a 100-division slide wire of 0.01 ohm, guaranteed to ± 0.00002 ohm. The ratios may have values of 0.01, 0.1, 1, 10, and 100, with accuracy of ± 0.05 per cent. Y, exclusive of the unused portion of the slide wire, has a resistance of 0.005 ohm.

(a) $A/B = a/b = 0.1 \pm 0.05$ per cent. (This means that $(A/B - a/b) = 0 \pm 10^{-4}$.) S is set on 0.005 $\pm$ 0.00002 ohm on the slide wire and zero-fixed steps. Use the complete equation for the Kelvin bridge and find the value and the uncertainty of X.

(b) A/B and a/b are changed to 0.01, with the same percentage accuracy as before. S is set at 0.04 ohm $\pm$ 0.05 per cent on the steps, plus 0.01 ohm $\pm$ 0.00002 ohm on the slide wire. Find the uncertainty of X. Show the sources of uncertainty in X that have been decreased.

4-50. The following is quoted from "Master Test Code for Resistance Measurement" (Ref. 14, sec. 2.25):

As an expedient, when a Kelvin bridge is not available, it is possible to compare the resistance of an unknown four-terminal resistor, X, with a standard four-terminal resistor, S, by using the rheostat arm and one ratio arm of a Wheatstone bridge. In this method S and X are connected in series by a link which connects to one current terminal of each and are supplied with rated current from a battery. The rheostat arm and ratio arm are connected in series with each other across the circuit between the two potential terminals which are most remote from the link. The galvanometer is bridged successively from the junction of the rheostat and ratio arms to first one and then the other of the two potential terminals nearest the link, and the resulting Wheatstone bridge is balanced for each connection. In one connection the link resistance is effectively added to X and the setting of the rheostat arm adjacent to X is R_1. In the other connection the link resistance is added to S and the setting of the rheostat arm is R_2. If B denotes the fixed resistance of the arm adjacent to S, the solution of the bridge equation gives

$$X = S \frac{R_2(R_1 + B)}{B(R_2 + B)}$$

Draw the circuit and prove this equation.

CHAPTER 5

Galvanometers; Shunts

5-1. Construction

Galvanometers are used to indicate or measure small currents in bridge circuits, potentiometers, and other measuring equipment. In many of the applications the circuit is adjusted to give a zero deflection or "null" on the galvanometer, so the function of the galvanometer is to show the presence or absence of current, with no requirement to measure the actual magnitude. It is merely necessary for the instrument to show a readable deflection for the smallest current that is significant in the particular application. There are a few places in measurement work in which deflections of a galvanometer are read, but such use is avoided as much as possible, for the accuracy in reading a deflection is seriously limited.

Galvanometers of many kinds have been devised, including moving-coil and moving-magnet types. Special instruments are needed for some purposes, but the great majority of the galvanometers in use now are the permanent-magnet moving-coil type (frequently called the d'Arsonval type). We shall devote our attention to this one kind for d-c work. The principle is the same as for d-c ammeters and voltmeters with which the student is familiar. In fact, if little sensitivity is needed, a milliammeter may be adequate, or for smaller currents, a microammeter. The latter instrument is called a microammeter if it is calibrated in actual current units, or a "pivoted galvanometer" if the scale markings are arbitrary. A sensitive galvanometer is different only in the modifications that are made to secure greater sensitivity. Pivots are replaced by filamentary suspensions, and the pointer by a mirror and light-beam arrangement.

Figure 5-1 shows diagrammatically the construction used in a high-sensitivity galvanometer. The permanent magnet has soft-iron pole pieces attached to it and a cylinder of soft iron between poles so that a radial magnetic field is set up in the coil space. The coil fits over the cylinder, and is supported by a flat ribbon suspension which also carries current to the coil. The other current connection in a sensitive galvanometer is a coiled wire, called the lower suspension, which has negligible torque effect. This type of galvanometer must be leveled carefully so that the coil hangs straight and centrally in the air gap without rubbing on the poles or cylinder. Some galvanometers, intended to be portable and not to require exact leveling, have "taut suspensions" consisting of straight flat strips, top and bottom, kept under tension.

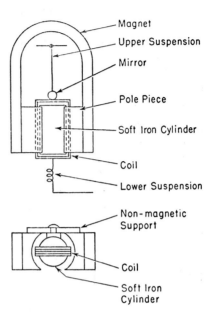

Fig. 5-1. Construction of a d-c moving-coil galvanometer.

A small mirror is mounted on the coil for indication of coil position. It may be used in two ways. A telescope and scale may be mounted in front of the galvanometer, so that the observer reads the scale reflected by the mirror. The more usual modern arrangement, less tiring to the user, consists of a lamp and scale device in which a spot of light is projected on the mirror and reflected to the scale. The scale is usually mounted 1 m from the mirror, although $\frac{1}{2}$ m is sometimes used for greater compactness.

The combination of flexible suspension and long light-beam "pointer" makes possible the construction of a very sensitive current indicator. Commercial galvanometers are available that give a deflection of 1 mm on the scale for a current of 0.0001 microampere, or even 0.00001 in particularly sensitive types. More will be said of the sensitivity and other characteristics of a galvanometer in a later section.

The upper suspension consists of a gold or copper wire of one or two thousandths of an inch in diameter rolled into the form of a ribbon. This is not very strong mechanically, so the galvanometer must be treated carefully to avoid jolts. Sensitive galvanometers are provided with coil clamps to take the strain from the suspension while the galvanometer is being moved.

The following information regarding suspensions is quoted from an L&N booklet,* "Notes on Moving Coil Galvanometers."

The choice of suspension material is based upon the following considerations: Copper is used where the important requirements are low resistance, freedom from thermal electromotive forces, and maximum sensitivity. It has been found that pure copper meets these requirements better than any other material.

Where greater torsion is required, together with fairly low resistance and stable zero, twenty-four-carat gold is used. Fourteen-carat gold has a greater torsion than the twenty-four-carat gold and has a considerably higher resistance. Fourteen-carat gold, which is used in place of phosphor bronze formerly used, has about the same torsion as phosphor bronze. The twenty-four-carat gold has about the same torsion as silver, formerly used, and has the important advantage of not being affected by atmospheric conditions which affect the silver.

Flat rolled suspensions give a low suspension resistance for the desired restoring constant, coupled with a higher tensile strength than would result from a round wire of the same torque.

5-2. Torque and Deflection of a Galvanometer

Figure 5-2 shows the dimensions that enter in the development of the torque relationship of a galvanometer.

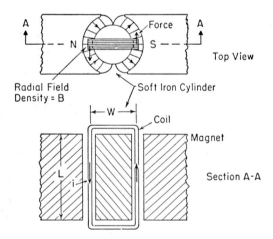

Fig. 5-2. Diagram of air-gap arrangements of a d-c moving-coil galvanometer.

*Reference 1.

Symbols:

L = length (active, of coil sides)
W = width of coil
n = number of turns
B = flux density at the coil position
i = current
S = suspension torque per radian deflection

The force* acting on a coil side is

$$F = nBiL \tag{1}$$

The torque* is

$$T = FW = BnWLi \tag{2}$$
$$= K_1 i \tag{3}$$

The deflection of the coil in radian measure is

$$\theta = \frac{T}{S} = \frac{K_1}{S} i = K_2 i \tag{4}$$

The deflection observed on a millimeter scale has the value, if θ is small and if the scale is 1 m from the mirror, as indicated in Fig. 5-3,

$$d = 1000 \times 2\theta = 2000 K_2 i \quad \text{mm} \tag{5}$$

For a large angle (2θ above 5 to 10°) eq. (5) is not accurate, and we should write

$$d = 1000 \tan 2\theta \quad \text{mm} \tag{6}$$

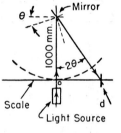

Fig. 5-3. Scale reading of galvanometer.

Equation (5) is correct for any angle if a curved (circular) scale is used and *if* the flux density is *uniform* along the magnet gap. A curved scale was frequently used with the telescope-and-scale, but is not practical on the modern lamp-and-scale devices in which the scale consists of an etched strip of ground glass. The departure from linearity on a straight scale is of

---

*Either MKS or CGS units may be used, as preferred. The MKS system is, of course, the present standard, but CGS will be found in older references and in much current engineering practice.

	MKS	CGS
L, W	m	cm
B	webers/m²	gausses
i	amp	abamp
F	newtons	dynes
T	newton-m	dyne-cm
K_1	newton-m/amp	dyne-cm/abamp
K_2	radians/amp	radians/abamp

no importance for null methods and is appreciable only with large deflections when readings are taken on the scale.

5-3. Definitions of Galvanometer Sensitivity

The sensitivity of a galvanometer may be expressed in several different ways, the first of which expresses the basic current relationship.

(a) *Current sensitivity.*

The current sensitivity is the current (in microamperes) required to give a deflection of one scale division. This is usually a 1-mm division on a scale at a distance of 1 m, but for galvanometers with attached scales the definition is in terms of one scale division, regardless of size of the division.

$$\text{Current Sensitivity} = \text{Microamperes per division} \qquad (7)$$

(b) *Voltage sensitivity.*

The voltage sensitivity is the voltage (in microvolts) that must be impressed on the galvanometer *in series with the critical damping resistance* to give a deflection of one scale division.

$$\text{Voltage Sensitivity} = \left(\begin{array}{l} \text{Microvolts on galvanometer} \\ \text{and CDRX* per division} \end{array} \right) \qquad (8)$$

Foreign manufacturers usually give the voltage at the galvanometer terminals and do not include the critical damping resistor. In this country the usual practice is to include the CDRX, in accordance with the recommendation of the Bureau of Standards. The applications of the voltage sensitivity are usually such that the inclusion of damping resistance is more significant.

(c) *Megohm sensitivity.*

The megohm sensitivity is the resistance of the circuit (in megohms) so that the deflection will be one division with 1 v impressed. As the coil resistance is negligible in comparison, the megohm sensitivity represents the resistance that must be put in series with the galvanometer so that 1 v gives a deflection of 1 division. This is the reciprocal of the current sensitivity.

$$\text{Megohm Sensitivity} = \text{Resistance in Megohms to give 1-mm deflection}$$
$$\text{with 1 v impressed} \qquad (9)$$

---

*CDRX or ECDR = Critical Damping Resistance External.

(d) *Ballistic sensitivity.*

The ballistic sensitivity is the quantity of charge (in microcoulombs) to give a deflection of one division when the galvanometer is used ballistically. This will be discussed in chap. 16.

5-4. Measurement of Galvanometer Sensitivity

The sensitivity of a galvanometer may be measured as indicated in Fig. 5-4. The galvanometer current must be very small, so R_1 is small,

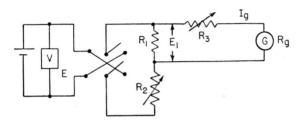

Fig. 5-4. Determination of galvanometer sensitivity.

probably 1 or 2 ohms, and R_2 large, possibly several thousand ohms. R_3 and R_2 can be varied to give convenient deflections. The galvanometer current is computed from the voltage and the resistances. Several readings should be taken, and the results of each expressed as microamperes per millimeter of deflection and the average taken for the galvanometer current sensitivity.

(NOTE: If, as is usually the case, $R_1 \ll R_2$, the following approximations are satisfactory.

$$E_1 \simeq E \frac{R_1}{R_1 + R_2} \simeq E \frac{R_1}{R_2} \quad \text{and} \quad I_g = \frac{E_1}{R_3 + R_g} \tag{10}$$

If R_1 should be too large for this approximation to be satisfactory, the battery plus R_1 and R_2 can be reduced to a simpler circuit by use of Thevenin's theorem.)

The galvanometer current may be computed easily if the resistance of the galvanometer is known. However, if information about R_g is not given, the question arises as how to measure it. The usual methods, including a Wheatstone bridge, pass more current through the unknown than can be tolerated for a galvanometer. The suspension of a sensitive galvanometer is a wire of about 1-mil (0.001-in.) diameter, rolled to the form of a ribbon, so it can easily be damaged by excessive current. It is possible to measure R_g by a Wheatstone bridge *if* computations are made in advance and the circuit so arranged that the current through the unknown does not exceed a small fraction of a milliampere. (The coil should be locked to prevent a violent swing, even with this apparently small value of current.)

A safer method of measurement is by means of deflections. If two readings are taken in the circuit of Fig. 5-4 and deflection is assumed directly proportional to current, it is possible to solve for both R_g and the galvanometer sensitivity. This does not give a very accurate value for R_g, but an approximate value is adequate for most purposes. Also, since the galvanometer circuit is mainly or entirely copper, its resistance changes about 1 per cent for each 2.5°C temperature change in the usual room temperature range, so resistance values closer than 1 or 2 % have no particular meaning unless the temperature is controlled.

While any two deflections can be used to determine the galvanometer resistance, a particularly convenient arrangement is to make the deflections in the proportion of 2:1. This is called the "half-deflection method." Suppose that we arrange the circuit as in Fig. 5-4, and adjust the resistances to give a convenient and fairly large deflection. Read R_3. Now, keeping R_1 and R_2 constant, we readjust R_3 to give a deflection just half as large as before. Call this value R_3'. Assuming a linear deflection-current relationship, it can be shown for this circuit that

$$R_g = R_3' - 2R_3 - R_1 \tag{11}$$

The last term in (11) should be the parallel value of R_1 and R_2, but this is practically R_1 if $R_1 \ll R_2$.

5-5. Dynamic Behavior of Galvanometers

We have considered so far the relationship of current, turns, and so on, for the steady deflection of a galvanometer. There are, in addition, features connected with the dynamic behavior that are important in the use of a galvanometer. The study we shall make has, however, a much wider application than to laboratory-type galvanometers. The same elements enter into most electrical indicating and recording instruments, to electro-mechanical oscillographs, and to many instruments of strictly mechanical actuation. In all these cases we are interested in such matters as the speed of response, overshoot, and damping. In indicating instruments the chief concern is in obtaining a prompt and reliable reading. Recording devices leave a written record of the motion of the moving element, and this inevitably includes the response characteristics of the element as well as the intended record; knowledge of these characteristics is thus highly important in the design and use of instruments, and in interpretation of the records.

If current is passed through the coil to give it a deflection and then the circuit is opened, the coil swings back toward the zero position. As the result of inertia, it overswings and travels nearly as far to the opposite side of the zero. The oscillations decrease slowly and last for a considerable time unless something is done to provide a damping effect.

There are three quantities that determine the characteristics of the motion of the moving coil. They are:

$$(1) \quad \text{Moment of inertia} = J$$

This is the moment of inertia about the axis of rotation of the coil. In units, it is of the nature of mass times square of radius.

$$(2) \quad \text{Stiffness of suspension} = S$$

S is the torque developed by the suspension per radian of coil deflection.

$$(3) \quad \text{Damping constant} = D$$

Damping is provided by motion of the coil in the air and by induced electrical effects if a closed circuit is provided. Damping torque is assumed to be proportional to the coil velocity. This is true for electrical damping, and is at least a fair approximation for air damping (which is usually a minor part of the total). We shall study damping action in greater detail in a later section, but for the present will lump the total effect under the one symbol. $D = $ damping torque for a coil velocity of 1 radian/sec.

If the coil is deflected to an initial angle θ_0 and then allowed to swing freely, we may write for the torques acting on it

Torque of acceleration + Damping torque
$$+ \text{ Suspension torque} = 0 \quad (12)$$

or in terms of the angle θ measured from the rest position

$$J \frac{d^2\theta}{dt^2} + D \frac{d\theta}{dt} + S\theta = 0 \tag{13}$$

This equation is satisfied by solutions of the form

$$\theta = K\epsilon^{mt} \tag{14}$$

where ϵ is the base of the natural-log system and m and K constants. This may be verified by substitution of (14) in (13) which gives the equation

$$Jm^2 + Dm + S = 0 \tag{15}$$

Solution of (15) by the quadratic formula gives the two roots

$$m_1 = -\frac{D}{2J} + \sqrt{\frac{D^2}{4J^2} - \frac{S}{J}}$$

$$m_2 = -\frac{D}{2J} - \sqrt{\frac{D^2}{4J^2} - \frac{S}{J}} \tag{16}$$

The solution for θ is of the form

$$\theta = K_1\epsilon^{m_1 t} + K_2\epsilon^{m_2 t} + \theta_{\text{steady}} \tag{17}$$

(m_1 and m_2 different) where K_1 and K_2 must be evaluated to suit the initial conditions. Note that $\theta_{\text{steady}} = 0$ for the condition assumed in (12), when the coil is given an initial deflection, and is then allowed to return toward zero. This condition is of interest in many bridge measurements; closing the key in the unbalanced condition results in a deflection; we then release the key and wait for the galvanometer spot to return to zero before making the next trial.

The equations for the opposite condition, starting from zero and *closing* the key, may be obtained, if desired, from our derivation by the appropriate changes in the equations. The deflecting torque would appear on the right side in (12) and (13), and the corresponding θ_{steady} in (17). The initial conditions are different in the two cases, and K_1 and K_2 have different values.

We shall give some attention to the complete equations for θ a little later. For the present our interest centers on the roots in (16), for we can tell the *type* of behavior from the form of the roots. There are three possible cases, depending on the quantity under the radical in (16).

Case I. $\qquad\qquad\qquad\qquad \dfrac{D^2}{4J^2} > \dfrac{S}{J}$ (roots real and unequal)

In this case θ is the sum of two terms with real and unequal negative exponents, as in (17). The form of solution is shown by curve I of Fig. 5-5, which indicates that the coil returns very slowly to the rest position without oscillations. This is the *overdamped* case. The curve is plotted from the equation:

$$\theta = \frac{\theta_0}{m_2 - m_1}[m_2\epsilon^{m_1 t} - m_1\epsilon^{m_2 t}] \tag{18}$$

which satisfies the initial conditions, $\theta = \theta_0$ at $t = 0$, and $d\theta/dt = 0$ at $t = 0$. (Also, $\theta_{\text{steady}} = 0$ for our case, as cited above.)

The overdamped case is of minor interest in the use of a galvanometer as a bridge detector, or for current or voltage measurements, as we prefer to operate under Case II or Case III for such operations. However, overdamped operation is useful in measurements of magnetic flux, as discussed in chap. 16.

Case II. $\qquad\qquad\qquad\qquad \dfrac{D^2}{4J^2} < \dfrac{S}{J}$ (roots conjugate-complex)

The quantity under the radical in (16) is negative in this case, but the roots may be changed to the form:

$$m_1 = -\frac{D}{2J} + j\sqrt{\frac{S}{J} - \frac{D^2}{4J^2}} \tag{19}$$

$$= -\frac{D}{2J} + j\beta \tag{19'}$$

It can be shown that roots of this sort give damped sinusoidal oscillations, as shown in curve II of Fig. 5-5, and computed from the equation

$$\theta = \frac{\theta_0}{\beta}\sqrt{\frac{S}{J}}\,\epsilon^{-(D/2J)t}\,\sin\left(\beta t + \sin^{-1}\beta\sqrt{\frac{J}{S}}\right) \tag{20}$$

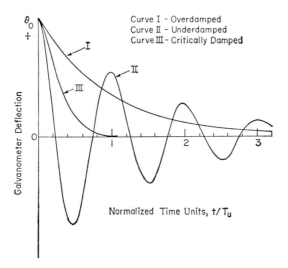

Fig. 5-5. Dynamic behavior of a galvanometer.

The radical in (19) determines the frequency of oscillation. For the undamped case (D approx. zero), the frequency is

$$f_{\text{undamped}} = \frac{1}{2\pi}\sqrt{\frac{S}{J}} \tag{21}$$

or the time of one complete oscillation is

$$T_u = \text{Undamped period} = 2\pi\sqrt{\frac{J}{S}} \quad \text{sec} \tag{22}$$

The rate at which the oscillations die away is determined by the factor

$$\epsilon^{-(D/2J)t}$$

Curve II is for a small value of D and is called the *underdamped* case.

Case III.　　　　　　　　　　$\dfrac{D^2}{4J^2} = \dfrac{S}{J}$　　(roots real and equal)

A new form of solution must be found for this case as (17) is not adequate, for it reduces to only one constant. The solution may be found in mathematics texts and gives a graph of the form shown by curve III in Fig. 5-5. This is called the *critically damped* case. The curve is computed from the equation

$$\theta = \theta_0 \epsilon^{-(D/2J)t} \left(1 + \frac{D}{2J} t \right)$$ (23)

The coil is seen to return rapidly to the rest position, but without overshooting. The critical case is thus important in a physical sense, and not merely as the crossover between cases I and II. It is the arrangement that brings the coil back to the zero point in the minimum time *that is possible without some measure of overshooting*, and it is important for this reason. A galvanometer that behaves as in curve II causes a great waste of time in waiting for the coil to settle to zero. Curve I represents very sluggish behavior and a loss of time. The best operating conditions are attained with a galvanometer that is either critically damped, or slightly underdamped — which may give an even faster return, as we shall see shortly.

If the damping is slightly less than the critical value, the galvanometer swings a small amount beyond the final position on the first swing, and soon comes to rest. It can be shown that if the damping constant D is two-thirds of the critical value, each swing is about 6 per cent of the preceding one. That is, the first overswing is 0.06 of the final displacement, the second overswing 0.0036, and so on. The galvanometer is thus very close to its final position at the end of one complete period. The speed of operation is excellent for any damping from two thirds of critical up to critical.*

5-6. Galvanometer Equations in Operational Constants

The galvanometer equations have been derived in terms of the constants, J, D, and S. These are basic physical constants, and are of interest to the designer, but they are not known by the user, and are rather difficult to evaluate. The quantities that the user is interested in operationally, and that he can determine readily, are the sensitivity, the critical damping resistance, and the period. As the first of the new set of constants, the damping of galvanometer motion is expressed most conveniently with reference to the critical case; that is,

$$\text{Relative damping constant} = \gamma = \frac{D}{D_c}$$ (24)

The value of D for the critical condition was defined in Case III as $D_c{}^2/4J^2$ $= S/J$, from which

$$D_c{}^2 = 4JS$$ (25)

---

*Reference 2, p. 56.

and
$$D_c = 2\sqrt{JS} \tag{26}$$

Another change is to express the frequency of oscillation in terms of the undamped period, from (22),

$$\beta = \sqrt{\frac{S}{J}}\sqrt{1 - \frac{D^2}{4JS}} = \frac{2\pi\sqrt{1 - \gamma^2}}{T_u} \tag{27}$$

With these substitutions, (20) becomes

$$\theta = \frac{\theta_0}{\sqrt{1 - \gamma^2}}\,\epsilon^{-2\pi\gamma t/T_u}\sin\left(2\pi\sqrt{1 - \gamma^2}\,\frac{t}{T_u} + \sin^{-1}\sqrt{1 - \gamma^2}\right) \tag{28}$$

The equation in this form is convenient to use, and helpful for visualization of galvanometer operation. The time scale is normalized in t/T_u units, and so can be applied to a galvanometer of any period. The abscissa scale of Fig. 5-5, for example, is marked in these units, so 1 on the scale would be one cycle with zero damping. Curve II was computed for $\gamma = 0.1$, and for this small amount of damping the period differs so slightly from T_u (0.5 per cent longer), that the difference is not noticeable on the graph.

Figure 5-6 shows to expanded scale the lower portions of curves computed from (28) for $\gamma = 0.6$ and 0.8, with the critical curve, $\gamma = 1$, for

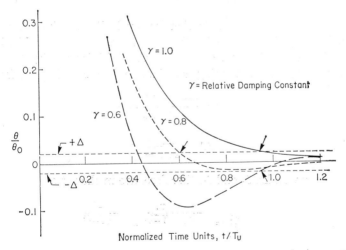

Fig. 5-6. Illustration of the influence of damping upon the time required for a galvanometer to approach the final deflection.

comparison. The 0.8 curve has only a small negative loop, and the 0.6 curve an overshoot of about 10 per cent. These curves are of interest in showing how the galvanometer coil approaches the rest position after deflection. We sometimes speak of the time for a deflection, but this is loose usage, since in theory the deflection only *approaches* the final value asymptotically, and never reaches it. We can specify, however, the time to approach within any desired increment of the final value, *and stay within that increment*, plus or minus. A band of ±1 per cent is sometimes indicated; however, ±2 per cent is used in Fig. 5-6 for easier visualization. The 0.8 curve is seen to enter the band in about 0.6 of a time unit (and *stay* within), whereas the critical curve requires about 0.94. The 0.6 curve cuts across both limits quite early, but enters finally at about 0.94. The arrows on the graph indicate the final entry points. If ±1 per cent limits had been used, the three curves would have nearly the same times, but with a small lead for the 0.6 curve.

As shown in Fig. 5-6, the time of approach to the final deflection is satisfactorily short for any γ from 0.6 to 1, so the operator has considerable leeway in selecting damping arrangements. Some underdamping, with γ from 0.6 to 0.8, actually has some advantage over the critical condition, partly in a small speed margin, but even more in the fact that the small overswing indicates to the operator that the coil is swinging freely, without rubbing on the poles. The same damping principles apply to indicating instruments, voltmeters and the like, and the small overswing associated with underdamping is considered advantageous in giving assurance of free movement of the metering element, and good condition of the pivots and jewels.

The times of maximum deflection (zero velocity) can be determined by differentiating (28), and equating the derivative to zero. Our chief interest, however, is in the ratio of successive peaks, and this is given by the exponential factor for a Δt corresponding to a time-angle of π radians, or from (28),

$$2\pi\sqrt{1 - \gamma^2}\,\frac{\Delta t}{T_u} = \pi$$

Therefore,

$$\Delta t = \frac{T_u}{2\sqrt{1 - \gamma^2}} \tag{29}$$

Two successive peaks, θ_0 and θ_1, therefore, are in the ratio

$$\frac{\theta_1}{\theta_0} = \epsilon^{-\frac{2\pi\gamma}{T_u}\frac{T_u}{2\sqrt{1-\gamma^2}}} = \epsilon^{-\frac{\pi\gamma}{\sqrt{1-\gamma^2}}} \tag{30}$$

In logarithmic form,

$$\frac{\pi\gamma}{\sqrt{1 - \gamma^2}} = \ln\frac{\theta_0}{\theta_1} = L \tag{31}$$

for short. From (31) we may make solution for γ,

$$\gamma = \frac{L}{\sqrt{\pi^2 + L^2}} \tag{32}$$

Equation (32) is valuable in permitting easy evaluation of the relative damping constant of a galvanometer. If the galvanometer is given a deflection, as by the circuit of Fig. 5-4, with $R_1 + R_3$ set for the desired damping, then opening of the battery switch initiates swinging of the light spot. Readings are made of the peak points of successive swings, preferably making several determinations, as accurate reading of the rather fast-moving spot is difficult. The natural logarithm of the ratio is substituted in (32) for evaluation of γ.

5-7. The Mechanism of Damping

Attenuation of motion of a galvanometer coil is basically a matter of dissipating the energy that is stored alternately in the twist of the suspension and in the kinetic energy of the moving coil. The dissipation of energy stems partly from mechanical effects, and usually to a greater extent, from electromagnetic effects from the coil circuit.

(a) *Mechanical damping.*

The major element in the mechanical damping is the motion of the coil through the air surrounding it. A small dissipation of energy is associated with the flexing of the suspension. These mechanical losses are present if the coil is swinging freely with the coil circuit open. The mechanical retardation is, accordingly, often called the *open-circuit damping*, since it is independent of electrical actuation.

The open-circuit damping may be measured as discussed in the preceding section. The coil is given a deflection, as by the circuit of Fig. 5-4, and then the *coil circuit* is opened, and the ratio of successive swings is observed. The relative damping constant is determined from (32) for this condition. This will be denoted by the symbol:

$$\gamma_0 = \text{relative open-circuit damping constant}$$

(b) *Electromagnetic damping.*

Electromagnetic damping is produced by the induced effects when the coil moves in the magnetic field if a closed path is provided including the coil. Figure 5-7 shows the damping circuit produced by a resistor connected across a galvanometer coil. When the coil rotates in the magnetic field a voltage is generated in it, and this circulates a current as shown. The combination of current and magnetic field produces a torque in opposition to the motion, just as in dynamic braking with a d-c motor. For any galvanometer a value of R_d can be found that gives the critical damping condition of curve III. This is called the *critical damping resistance*, or as frequently abbreviated, CDRX (critical damping resistance external). This is one of the important constants of a galvanometer.

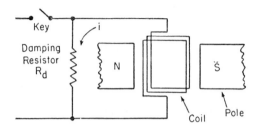

Fig. 5-7. Damping circuit for a galvanometer coil.

The dependence of the damping effect on the other constants of a galvanometer may be shown by the following study. The voltage induced in the coil by its motion in the magnetic field may be written,

$$e_i = \text{rate of change of flux linkages}$$

$$= 2nBL \frac{W}{2} \frac{d\theta}{dt} = K_1 \frac{d\theta}{dt} \tag{33}$$

where K_1 is the same constant used in eq. (3). The current flowing in the circuit of Fig. 5-7 is

$$i = \frac{e_i}{R_{\text{total}}} = \frac{K_1}{R} \frac{d\theta}{dt} \tag{34}$$

where

$R = R_{\text{total}} = R_g + R_d$

$R_g = \text{galvanometer coil resistance}$

$R_d = \text{damping resistance (external)}$

The torque produced by the electromagnetic damping action is

$$T_m = nBiLW = \frac{K_1^2}{R} \frac{d\theta}{dt} = D_m \frac{d\theta}{dt} \tag{35}$$

The dynamic damping torque is thus seen to depend upon the inverse of the total circuit resistance and upon the square of the K_1 constant, which includes the flux density and the number of turns and dimensions of the coil. The torque depends upon the angular velocity of the coil, which is the type of behavior that was assumed in eq. (13).

(c) *Total damping.*

The total damping effect is made up of the two components, open-circuit (air damping) and electromagnetic damping. This may be indicated by the symbols,

$$D = D_0 + D_m \tag{36}$$

or,
$$\gamma = \gamma_0 + \gamma_m \tag{37}$$

The open-circuit component is a fixed amount for any given galvanometer; D_m may be controlled, as derived above, by the constants of the electrical circuit. The subscript c may be added to D_m or D to denote that adjustment has been made for the critical case.

The relative importance of the two components varies to some extent, depending upon the design and type of galvanometer. For an instrument of fairly robust design, the open-circuit damping is generally a small part of the total for usual operating conditions. Very sensitive galvanometers operate with low suspension torques, and smaller forces generally, so the air-damping torque is a larger element proportionately.

5-8. Determination of Damping Resistance

The equations of the last two sections may be combined and extended to illustrate the relationship between the damping resistance and the galvanometer behavior, taking the open-circuit damping into account. Also, the damping resistance may be found for another operating condition if the values are known for one condition.

From (36) and (35),

$$\gamma = \frac{D}{D_c} = \frac{D_0}{D_c} + \frac{K_1^2}{D_c R} \tag{38}$$

or,
$$\gamma = \gamma_0 + \frac{K_1^2}{D_c R} \tag{39}$$

We may compare the damping constants, γ_A and γ_B, given by the damping resistances R_A and R_B, respectively, by writing two equations of the form of (39) with the γ_0 term transposed,

$$\gamma_A - \gamma_0 = \frac{K_1^2}{D_c R_A} \tag{40}$$

$$\gamma_B - \gamma_0 = \frac{K_1^2}{D_c R_B} \tag{41}$$

Division of (40) by (41) gives the result,

$$\frac{R_B}{R_A} = \frac{\gamma_A - \gamma_0}{\gamma_B - \gamma_0} \tag{42}$$

or the alternative form,

$$R_A(\gamma_A - \gamma_0) = R_B(\gamma_B - \gamma_0) \tag{43}$$

(Note that R stands for the *total* galvanometer circuit resistance in both cases.)

The method of evaluating γ was derived in par. 5-6, leading to (32). The open-circuit constant, γ_0, is found from the ratio of successive swings with the coil circuit open. If γ is evaluated for one circuit resistance, then the resistance may be computed from (42) for any other γ, as desired. For a complete test, determinations should be made with several values of resistance; the product $R(\gamma - \gamma_0)$ for all trials should be the same, as indicated by (43). There will, of course, be scatter in the results, so the average of several values should be better than a single value, and out-of-line results can be detected and discarded.

After a satisfactory value of the $R(\gamma - \gamma_0)$ product is established, the total circuit resistance for critical damping may be computed from

$$R_{\text{crit}} = \frac{\text{Mean}[R(\gamma - \gamma_0)]}{1 - \gamma_0}, \tag{44}$$

since γ for the critical case equals one.

The *critical damping resistance* is an item of universal interest that is always specified for a galvanometer, and accordingly, must be determined. One method is visual observation of the action of the spot when a deflection is applied or removed from the coil; beginning with the oscillating condition, decreasing resistance values are tried until a value is found for which the overshoot or negative loop just disappears. Such a determination is not very precise, but is adequate for most purposes. Actually, the curve for $\gamma = 0.9$ has such a minute negative loop that it might easily be mistaken for the critical case, as viewed on the scale of the galvanometer. (Note: The reader may estimate the 0.9 curve by interpolating between the 0.8 and the 1.0 curves on Fig. 5-6. The negative ordinate does not exceed 0.002 of the deflection, and hence would be practically undetectable.)

The question may arise of using the relationships derived above for the determination of the CDRX, using the equations with the substitution of $\gamma = 1$ for the critical case. The cautions enter from the fact that the equations were derived for the oscillating case, based upon a ratio of swings. The critical case gives a second swing of zero, so no ratio of swings exists. However, support for application to $\gamma = 1$ as a limiting case is given by the fact that eq. (23) for the critical case may be obtained from (20) for the oscillating case by allowing β [eq. (19)] to approach zero. So, regarded as a limit, the critical damping resistance (total) may be obtained from the mean $R(\gamma - \gamma_0)$ by division by $(1 - \gamma_0)$.

5-9. Design Features in a Galvanometer

The designer must know the conditions of use of a galvanometer: whether for portable or laboratory use; the degree of sensitivity needed; the damping conditions imposed on the galvanometer by the circuit; the desired period; and so forth. Beyond this he must decide on the dimensions and turns of the coil, the strength and air gap of the magnet, the length and stiffness of the suspension, and various other factors. Equation (2) shows that the torque depends on the product of width by length of the coil — to keep the period low the designer uses a long narrow coil. Equation (2) shows more torque if flux density is increased; this requires either better magnetic material or a shorter air gap. Magnetic materials have been improved in recent years, and this has helped in galvanometer design. A shorter air gap improves the sensitivity but, if too short, makes the galvanometer hard to level to the required accuracy. A more flexible suspension increases the current sensitivity, but also makes the response slower (longer period).

It may be interesting to study the effect of changing the number of turns in the coil winding. Suppose that we have two galvanometers alike in all respects except the coil windings, and suppose that No. 2 has a coil with twice as many turns of wire that is one half as large in cross-sectional area (assuming that the insulation space permits this). We shall make a table with relative values in parallel columns:

	Relative Values	
	No. 1	*No. 2*
Coil turns...............................	1	2
Wire sectional area.........................	1	$\frac{1}{2}$
Current, for same deflection................	1	$\frac{1}{2}$
Coil resistance............................	1	4
Coil loss, I^2R.............................	1	1

It is seen that the two galvanometers are equally good from a power standpoint. In use it is a matter of matching a galvanometer reasonably well to the bridge with which it is used; thus No. 2 galvanometer matches a bridge having four times the best internal impedance for No. 1. The object of this study is to show that equally good galvanometers can be made with different coil windings; which is best in a given case depends on the use to which it is to be put.

It is not the intention to imply that a galvanometer must always match the resistance of the bridge with which it is used. (The term "bridge resistance" as used here means the resistance of the bridge as seen from the galvanometer corners of the bridge. It is the R_0 used in chap. 4.) This is impractical, as a general-purpose bridge is used at various times with different ratio arms and different unknowns. It is necessary in practice to use a general-average instrument. Fortunately, even a considerable mismatch does not cause a great loss in sensitivity.

The question may be raised whether the bridge and galvanometer should have equal resistances for maximum power transfer or whether the galvanometer should be selected so that its CDRX is equal to the bridge resistance, thus giving critical damping. As the CDRX of a galvanometer is generally several times the coil resistance (see Table 3), these two criteria may result in considerably different bridge setups. The answer to the question may depend on the conditions of use. For ordinary bridge measurements the bridge is usually balanced to the nearest unit on the smallest dial. The operator in such adjustments presses the galvanometer key to see whether a deflection results, and then immediately releases it. The damping for the "key-closed" condition is thus of little importance. The important point is the return of the galvanometer to zero when the key is released; this can be accomplished by the use of the appropriate Ayrton shunt, as indicated in Figs. 4-11 and 4-23. Conditions are different in the precision comparison of resistors, in which case the resistance may be specified one place beyond the last dial by reading the galvanometer deflections with the last dial set at the unit just above and just below balance, and interpolating. For this use it is desirable to have the combination of bridge and galvanometer give critical damping (or slight underdamping). It is not practical to change galvanometers frequently as different unknowns are measured, but damping conditions may be controlled by connecting a resistor in series or in parallel with the galvanometer as required.

Galvanometers may be constructed for a wide range of sensitivities. High sensitivity is needed for some purposes, so interest attaches to the question of the highest practical value, and to the factors that limit it. Sensitivity is increased by use of a low-torque suspension, but this introduces some undesirable features, such as a long period and unstable zero — at least, it accentuates the elements that cause instability of the zero. Such instability is disturbing to the user, and interferes with making readings or determining null points to the smallest readable increment. Instability of zero may be caused by several things, including magnetic impurities in the coil structure, nonelastic yield in the suspension, vibration of the support, and stray thermal voltages in the galvanometer or its associated circuit. A limit of sensitivity is set by the small continuous random motions (Brownian motion) caused by bombardment of the coil by air molecules.

5-10. Choice of a Galvanometer

The general conditions of use determine the type of galvanometer to be used, e.g., whether portable or laboratory type. However, after this is settled there are usually several instruments from which to choose, and judgment is required to select the best one for the particular application. A discussion of the main features may be of some aid in making the decision.

The important constants of a galvanometer are the sensitivity, the resistance, the critical damping resistance and the period. These are the chief items that are considered in judging the suitability of a galvanometer for a particular application. A designer may secure a particularly favorable value of one of these constants by certain proportions in his design, but the effect will show in the other constants, possibly in an undesirable way, as the constants are interrelated and dependent on the primary constants and dimensions of the instrument. The sensitivity may be increased by increasing the dimensions of the coil or by increasing the number of turns, but either of these changes increases the moment of inertia and, hence, lengthens the period. A long period is undesirable, as it makes measurements with the galvanometer slower. The user should therefore consider these matters in selecting an instrument. Comparison of galvanometers of different kinds shows a general tendency toward a longer period for greater sensitivity. (This is true if sensitivity is secured by change of suspension or by increase of coil dimensions, and not ordinarily for change of winding for resistance-matching. The latter effect can be ruled out and the merit of different designs better judged if comparison is made on the basis of the *power* sensitivity. The power sensitivity of a galvanometer can be computed as the square of the current sensitivity times the resistance.)

There is sometimes a tendency to consider that the most sensitive galvanometer is the best one. It is, therefore, appropriate to point out that there are drawbacks to excessive sensitivity. An oversensitive galvanometer leads to increased time spent in balancing the circuit to an unnecessary degree of exactness. If this drawback is reduced by shunting, the galvanometer still suffers in comparison with an unshunted, less sensitive, instrument in having a longer period, which means that the operator must wait longer between trials for the light spot to return to zero. The more sensitive galvanometer also costs more, is less stable in zero point, and is more time-consuming to set up and adjust.

Table 3 presents a short tabulation of d-c moving-coil galvanometers, to give a general idea of the types and range of sensitivities available. The pointer types, of which a representative is shown in Fig. 5-8(a), are small portable instruments of relatively low sensitivity and rugged design, useful for a great deal of test work, particularly for portable bridges and potentiometers. Slightly larger instruments are made with a combination of pointer and light spot, so that the pointer may be used for rough balancing. The light spot, with its longer optical arm, gives higher sensitivity for fine balancing; a value as high as 0.04 $\mu a/mm$ may be obtained, as shown in the table. The reflecting type of Fig. 5-8(b) is used with a lamp and scale, frequently mounted at a distance of 1 meter from the galvanometer. This general type is a laboratory standard which is available from several makers, and with a wide range of characteristics, of which a few representa-

(a)

Fig. 5-8. Two typical d-c moving-coil galva-nometers. (a) Pointer-type. (b) Reflecting-type R. (*Courtesy of Leeds & Northrup Co.*)

(b)

tive samples are shown in the table. The galvanometer marked "ballistic" is used for measurements of magnetic flux (discussed further in chap. 16) and is intentionally given a long period, for in its intended use the coil is given an impulse of short duration, and the scale reading must be taken "on the fly" at the limit of the first swing. The entries in the table show the general tendencies discussed earlier; that the more sensitive instruments show, in general, longer periods, higher coil resistances, and higher values of critical damping resistance.

TABLE 3

TYPICAL CHARACTERISTICS OF D-C MOVING-COIL GALVANOMETERS

Type	Sensitivity*	Period (sec.)	Resistance	
			Ext. crit. damping	Coil
Pointer	2 μa	2,5	20	12
	1 μa	3	110	25
	0.25 μa	3	1800	250
Light spot and pointer	0.2, 2.5 μa	2	100	10
	0.04,0.5 μa	2	3500	1000
Reflecting, lamp and scale	0.004 μa	5	900	80
	0.0005 μa	6	10,000	650
	0.0001 μa	14	22,000	500
	0.00001 μa	40	100,000	800
	0.2 μv	5	25	10
(Ballistic)	0.0004 μc	25	30,000	800
Box-type, self-contained	0.005 μa	2.5	400	25
	0.0005 μa	3	25,000	550
	0.5 μv	3	50	17

*Sensitivity per mm at 1 meter or per scale division. Microvolt sensitivities are for critical damping resistance in series.

Another matter of importance is the steadiness of the support available for the galvanometer. Vibration conditions in industrial plants are frequently so severe that sensitive galvanometers can be used only if resort is made to elaborate vibration-absorbing mountings. The less sensitive instruments, and the taut-suspension type in general, are less subject to this difficulty. There are some new detectors for d-c circuits, based on the use of electronic amplifiers with rugged output meters (milliammeters), that are not troubled by vibration conditions. There have been difficulties in the past in making a d-c amplifier sufficiently stable for such service, but recent improvements have done a great deal to remove this limitation.

We may summarize the discussion in the following points:

(1) Select a galvanometer that has enough sensitivity for the job, but not a great deal more.

(2) Choose the galvanometer resistance to make a reasonable match with the circuit in which it is used.

(3) Select the most rugged type of galvanometer that meets the requirements.

5-11. Commercial types

Direct-current galvanometers are made in three general types, not counting some that may be classed as special research tools.

(1) *Portable pointer types* — small instruments, similar in style to portable meters — may be pivoted, but usually have taut suspension.

(2) *Laboratory reflecting types,* made in a variety of styles and characteristics, covering high sensitivity needs.

(3) *Box type, lamp-and-scale,* includes the galvanometer, light source, and scale.

The box-type self-contained galvanometer deserves special mention, for notable improvements have been made in this field in recent years. These instruments include a number of features that make them very convenient in use. They have taut suspensions, so they do not require exact levelling. The box includes the complete optical system, so it is not necessary to make separate mountings for galvanometer, lamp, and scale. High sensitivities are attained in spite of the compact arrangement by means of multiple reflections, to give a long optical arm. One instrument of this type is shown in Fig. 5-9, where part (b) gives a diagram of the optical system. The sensitivities cover most of the range formerly covered only by the reflecting type, as indicated by the few entries in the table.

5-12. Electronic Galvanometers

A vacuum tube or transistor amplifier, followed by a milliammeter may be used as a detector. This combination has advantages of ruggedness and portability that permit its use under conditions that would be extremely difficult for a sensitive suspended galvanometer. Some of these devices have arbitrary scale values, sometimes with variable gain, and are designed simply as balance indicators. The sensitivity may be of the order of a microvolt per scale division or better, and the input impedance a number of megohms, so in most cases the current drawn from the measured source is negligible. Another possibility is a vacuum-tube voltmeter, if sufficiently low scales are provided. Instruments are available with ranges such as 10-0-10 microvolts, giving a fraction of a microvolt per scale division; the input impedance is in the range of megohms. On any of these devices with high input impedance, the indication is practically the Thevenin voltage pro-

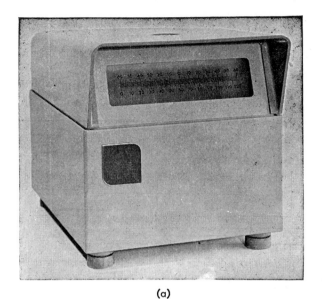

(a)

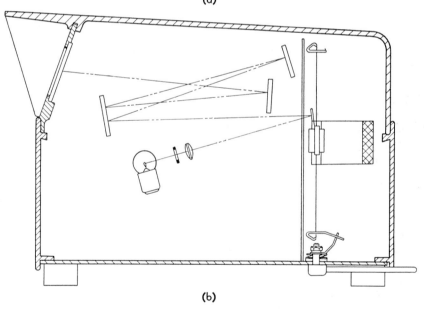

(b)

Fig. 5-9. (a) & (b) Rubicon spotlight galvanometer. (a) External view. (b) Diagram of optical arrangement, showing multiple reflections (*Courtesy of Minneapolis-Honeywell Regulator Co., Rubicon Instruments.*)

duced by the unbalance of the bridge or potentiometer, unless the circuit is itself of very high impedance. Electronic detectors are often useful; however, there may also be difficulties connected with their use. It is desirable to ground the circuit, if possible, at one of the detector terminals, otherwise stray effects and body capacity may cause false indications. Also, a-c pickup must be avoided unless the detector internal circuit is immune to it.

Figure 5-10 shows another idea in electronic detectors; it consists of a moving-coil galvanometer, photoelectric tubes, an amplifier, and an output

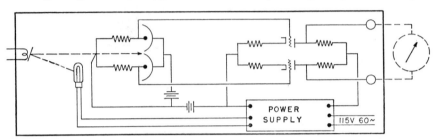

Fig. 5-10. Schematic diagram of photoelectric galvanometer as a null detector. (*Courtesy of Minneapolis-Honeywell Regulator Co., Rubicon Instruments.*)

meter. The galvanometer is of taut-suspension type, of comparatively low sensitivity, but particularly designed for a low period. The unit as a whole gives high sensitivity and fast response.

5-13. Ayrton Shunt

A protective device of some sort must be used with a sensitive galvanometer in the process of balancing a bridge. If the galvanometer has sufficient sensitivity for the final stages of balancing, the large unbalance that probably exists at the start is likely to damage the coil or suspension, or, if not this serious, it sends the light spot so far off scale that the operator cannot tell what to do to improve conditions.

Sensitivity to unbalance can be reduced by several combinations of series or parallel resistors. Figure 5-11(a) shows the arrangement that is most advantageous. If the slider indicated in (a) is near the lower end, the

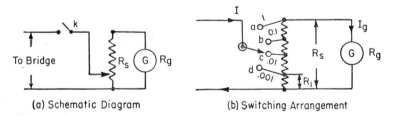

Fig. 5-11. Ayrton shunt.

lower segment of resistance is a shunt, and the upper part is in series with the galvanometer. The current from the bridge divides in two parts, the major part going through the shunt and only a small part through the galvanometer, thus protecting it and giving low sensitivity. As the sliding contact is moved toward the top, the sensitivity is increased, and with the contact at the top the minimum shunting effect is obtained. There is some loss of sensitivity at the top point, as compared with the galvanometer alone, but the loss is not serious if R_s is several times as great as R_g.

When K is open, in Fig. 5-11(a), there is a complete circuit through the galvanometer coil and R_s, and this provides a damping circuit. R_s should, accordingly, be selected to have a value approximately the same as the critical damping resistance of the galvanometer (somewhat higher is satisfactory, as galvanometer response is excellent; if considerably too high, a fixed shunt can be added across R_s). As the CDRX is usually several times the coil resistance (see Table 3), the loss of sensitivity due to using a shunt of this value is not serious.

Figure 5-11(b) represents the usual Ayrton shunt construction, using a dial-type switch. The numbers on the contacts represent the *relative* values of galvanometer current. The resistance of each shunt section can be proportioned to give this result, as shown by the following derivation. For the symbols on the diagram and for the same input current I in all cases, we may write

Contact on a or "1" $$I_{ga} = I \cdot \frac{R_s}{R_s + R_g}$$

Contact on d or "0.001" $$I_{gd} = I \cdot \frac{R_1}{R_s + R_g}$$

therefore $$\frac{I_{gd}}{I_{ga}} = \frac{1}{1000} \quad \text{if} \quad \frac{R_1}{R_s} = \frac{1}{1000}$$

(The subscript a or d denotes the switch position.) Note that the *relative* division of current for the two positions depends only on R_1/R_s and is *independent* of R_g. The shunt thus gives the same *relative* current values for the various steps with different galvanometers, although not the same fraction of the total current.

The Ayrton shunt may be used with exact resistance values for calibrated division of current. For convenience and protection in bridge balancing approximate ratios are satisfactory. The Ayrton shunt is very convenient in bridge measurements as the galvanometer sensitivity can be increased as bridge balance is approached.

REFERENCES

1. Leeds & Northrup Co., "Notes on Moving Coil Galvanometers," *Note Book* ED(1), Philadelphia, 1943.

2. Harris, Forest K., *Electrical Measurements*, New York: John Wiley & Sons, Inc., 1952.

3. Laws, Frank A., *Electrical Measurements*, New York: McGraw-Hill Book Co. Inc., 1938.

4. Whipple, R. S., "The Evolution of the Galvanometer," *J. Sci. Instruments*, **11**, 37 (1934).

5. Wenner, Frank., "General Design of Critically Damped Galvanometers," *NBS Sci. Paper* 273 (1916).

6. Hill, A. W., "The Theory of Moving-coil Galvanometers," *J. Sci. Instruments*, **11**, 309 (1934).

7. Downing, A. C., "Construction of Sensitive Moving Coil Galvanometers," *J. Sci. Instruments*, **12**, 277 (1935).

8. Williams, A. J., Jr., R. E. Tarpley, and W. R. Clark, "D-C Amplifier Stabilized for Zero and Gain," *Trans. A.I.E.E.*, **67**, 47–57 (1948).

PROBLEMS

5-1. A galvanometer gives a deflection of 8.0 cm when a current of 0.48 μa is passed through it.
(a) Find the current sensitivity of this galvanometer.
(b) Find the megohm sensitivity.

5-2. A galvanometer has a sensitivity of 200 megohms. Find the current required to give a deflection of 10 cm.

5-3. A galvanometer has a sensitivity of 2500 megohms, a resistance of 650 ohms, and a critical damping resistance of 9000 ohms.
(a) Find the current sensitivity.
(b) Find the voltage sensitivity.

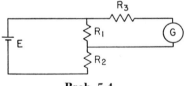

Prob. 5-4.

5-4. In the circuit shown here $E = 1.5$ v, $R_1 = 1$ ohm, $R_2 = 7500$ ohms. When $R_3 = 350$ ohms the galvanometer deflection is 14.0 cm, and when $R_3 = 1080$ ohms, the deflection is 5.0 cm.
(a) Find the resistance of the galvanometer.
(b) Find the current sensitivity of the galvanometer.
(c) For what value of R_3 would the deflection equal 7.0 cm?

5-5. A galvanometer is tested on a circuit as in problem 5-4. $E = 1.5$ v, $R_1 = 1.0$ ohm, $R_2 = 2500$ ohms. When R_3 is 450 ohms the deflection is 150 mm, and when R_3 is 950 ohms the deflection is 75 mm.
(a) Find the resistance of the galvanometer.
(b) Find the current sensitivity of the galvanometer.

5-6. (a) A galvanometer is being tested on a circuit like that shown in problem 5-4. The galvanometer has a sensitivity of 2000 megohms and a resistance of 600 ohms. $R_1 = 1$ ohm and $R_3 = 9000$ ohms. If $E = 1.5$ v, find R_2 to limit the deflection to 150 mm.

(b) After adjustment is made as in part (a), R_1 becomes open-circuited. What is the current through the galvanometer?

5-7. A galvanometer having a resistance of 100 ohms is tested by connecting it in series with 900 ohms. This combination is shunted across a 1-ohm resistor, which is connected in series with 10,000 ohms to a 1.5-v battery. Under this condition the galvanometer deflection is 25 cm.

(a) Compute the current sensitivity of the galvanometer in microamperes per mm.

(b) This galvanometer has a shunt of 1000 ohms when used on a bridge. What is the over-all sensitivity of galvanometer with shunt, with the shunt on maximum-sensitivity setting?

(c) Compute the input point to the shunt of part (b) to give a sensitivity one-tenth as great as in (b).

5-8. The card attached to a galvanometer gives the following data: sensitivity $= 0.004$ $\mu a/mm$, coil resistance $= 50$ ohms, CDRX $= 300$ ohms. It is being used in bridge measurements with an Ayrton shunt of 500 ohms resistance.

(a) Is the galvanometer over-damped or under-damped under this condition of use? Explain briefly.

(b) What is the sensitivity of galvanometer-with-shunt, with the shunt at maximum setting?

(c) Where should the shunt be tapped to give a sensitivity 0.01 as great as in (b)?

5-9. A potential divider consists of a 2-ohm resistor in series with a 5000-ohm resistor, connected across a 1.50-v battery. A galvanometer in series with 1450 ohms is connected across the 2 ohms, and gives a deflection of 10 cm. The galvanometer data card specifies the coil resistance as 50 ohms, and the external resistance needed for critical damping as 300 ohms.

(a) Compute the current sensitivity of the galvanometer.

(b) Compute the megohm sensitivity.

(c) Under the conditions of the test above, is the galvanometer over-damped or under-damped? Explain briefly.

5-10. A potential divider consists of a 2-ohm resistor in series with a 1000-ohm resistor, connected across a 1.50-v battery. A galvanometer in series with 1000 ohms is connected across the 2 ohms, and gives a deflection of 50 mm. When the galvanometer series resistor is changed to 2000 ohms, the deflection becomes 30 mm.

(a) Find the resistance of the galvanometer.

(b) Find the current sensitivity of the galvanometer.

(c) Find the megohm sensitivity.

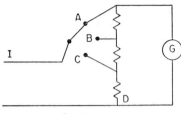

Prob. 5-11.

5-11. Design an Ayrton shunt (that is, specify the resistance of each section) so that the sensitivity on B will be one-tenth, and on C one-hundredth as great as on A. The galvanometer has a sensitivity of 0.05 $\mu a/mm$, a resistance of 50 ohms, and an external critical damping resistance of 350 ohms.

5-12. A galvanometer has a coil resistance of 50 ohms; the resistance needed for critical damping is 500 ohms. It gives a deflection of 60 mm for a current of 0.3 μa when used without a shunt. A 500-ohm Ayrton shunt is available, with taps at 5 ohms and 50 ohms from the lower end (as in Fig. 5-11). Find the current sensitivity of the galvanometer (a) unshunted; (b) shunted, maximum sensitivity; (c) shunted, on each of the tap points.

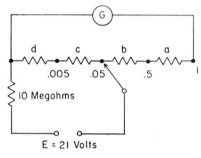

E = 21 Volts

Prob. 5-13.

5-13. A galvanometer has a resistance of 100 ohms, and a sensitivity when used alone of 0.01 $\mu a/mm$. The external resistance needed for critical damping is 750 ohms. It is used with an Ayrton shunt of 2000-ohm total resistance and tap points marked 1, 0.5, 0.05, and 0.005, as shown in the figure.

(a) Determine the values of each of the resistors, a, b, c, d.

(b) What is the galvanometer deflection with the shunt set on the 0.05 tap?

(c) If the galvanometer is not critically damped with the shunt as shown, determine the value of a resistor to be added to give critical damping, and show how it is connected in the circuit.

(d) With the addition made as in (c), find the over-all sensitivity of galvanometer and shunt with connection to the 1 and to the 0.05 taps, and find the ratio of the two values.

CHAPTER 6

Voltage References

6-1. Types of Voltage References

Voltage references have consisted, from the early days of measurements, of voltaic cells of several sorts. The cells were developed over a long period to a high degree of excellence. Techniques have been perfected so that a group of cells of the precision type can be produced with individual differences of a small number of parts per million. In addition, the cells have long life, so in many respects they serve in a most satisfactory way as a standard.

Recent discoveries and developments have provided semiconductor diodes that have definite current-voltage characteristics that can serve as a voltage standard. It cannot be predicted at present to what extent the diodes can replace standard cells. It does not appear that they will be used for the highest precision standardizing work. However, they have ample accuracy for many industrial applications, including recorders and controllers, and have advantages in ruggedness and absence of liquid. Attention will be given in this chapter both to standard cells and to diode standards.

6-2. History of Standard Cells

A working standard of electromotive force was recognized early as a pressing need in electrical measurements, so experimenters naturally turned to the idea of a voltaic cell to provide a reference. It was known that the emf of a cell depends on the electrolyte and electrode materials. Hence there was the natural inference that a given combination of materials can be relied upon to give a definite value. The Daniell cell, consisting of copper and zinc electrodes and copper sulphate and zinc sulphate solutions, was used for a time, but was far from ideal: the emf was not constant and the life short.

There are several properties a cell should have to be a good standard of emf: long life, small temperature effect, and reproducibility. In 1872 Latimer Clark invented a cell that was much better than the Daniell cell. The cell, after some development, consisted of mercury and zinc amalgam electrodes and a saturated solution of zinc sulphate, sealed in a glass enclosure to prevent evaporation of the electrolyte. It had good reproducibility, but a rather large temperature coefficient, and gave trouble by cracking of the glass where the platinum wire was sealed through the glass, owing to alloying of the platinum with the zinc amalgam.

Edward Weston produced the cadmium cell in 1892. It was adopted in 1908 by the London International Electrical Congress with the value of 1.0184 international volts (changed in 1910 to 1.0183). The international volt as a working standard was defined in 1910 as 1/1.0183 of the emf of the Weston Normal (saturated) cell at 20°C. The Weston cell has been the working (or reference) standard of emf since that time, and as such has played a very important part in electrical measurements. It has fulfilled the requirements of a reference standard in a highly satisfactory way.

The cadmium cell is made in an H-shaped glass container as shown in Fig. 6-1. The positive electrode is mercury, with mercurous sulphate as

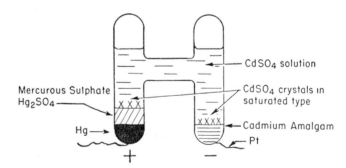

Fig. 6-1. Diagram of the cadmium cell.

depolarizer. The negative terminal is cadmium amalgam, and the solution cadmium sulphate. There are two forms of the cell, the first in which the solution is saturated and a surplus of $CdSO_4$ crystals is placed in the cell to maintain saturation, and the second form in which the solution is unsaturated at ordinary temperatures. These two forms have different properties and uses.

6-3. The Saturated (Normal) Cell

The saturated cell is named the "normal" cell to distinguish it from the unsaturated type and to indicate that it is the basic standard. These cells, when made of highly purified materials, show a high degree of uniformity, individual cells having differences, in general, of a few microvolts from the average of a group, although an occasional cell may be much farther out of line. The emf changes somewhat during the first few months — probably because of the small amount of glass that goes into solution in the electrolyte — and then settles to a comparatively steady value for a long period of time. These cells thus have the properties of reproducibility and permanence to a quite satisfactory degree.

There is an acid form of the cell in which the electrolyte contains added sulphuric acid. The acid cell appears to be better with respect to uniformity and permanence. The National Physical Laboratory in England uses it as their standard, and the National Bureau of Standards in the United States has a group of cells of this type in addition to the so-called neutral type.

The normal cell has a rather large temperature coefficient, about -40 $\mu v/°C$. In fact, it has an effect of about $+0.00031$ $v/°C$ at the positive terminal and -0.00035 at the negative terminal, so it is important to keep the entire cell at a uniform temperature to secure as much cancellation as possible of the plus and minus effects. These cells are always used with close temperature regulation, usually in an oil bath held constant to better than $0.01°C$. Close regulation is needed, first because of the large temperature coefficient, and second because this type of cell is used only where high precision is required.

The equation relating emf to temperature, as adopted in 1911, is

$$E_t = E_{20} - 0.0000406(t - 20) - 0.00000095(t - 20)^2 + 0.00000001(t - 20)^3$$

The emf at 20° was defined as 1.01830 international volts, which equals 1.01864 abs volts.

Normal cells accommodate themselves slowly to changes in temperature and show "hysteresis" if the cell is heated and cooled. The term hysteresis as used here represents the lag in emf value and the generally erratic behavior of the cell while it is coming to equilibrium following a change of temperature. One case has been reported in which cells were subjected to high temperature accidentally, and required two or three months to recover.*

The saturated cell is a standard for the maintenance of the volt, and as such is used only in places such as the Bureau of Standards and the largest commercial laboratories. The value of the cells for this purpose is shown by a quotation regarding their reliability given in par. 5 of chap. 1. Saturated cells are not used for ordinary work, because of the required temperature regulation and the fact that they are not portable. In fact, these cells are not shipped by parcel post or express, but are sent only by special messenger. The unsaturated cell is used for all ordinary measurements.

6-4. The Unsaturated Cadmium Cell

The unsaturated cell is similar in construction to the normal cell, except that it contains no $CdSO_4$ crystals, and the solution is so made that it would be saturated at 4°C, and hence is unsaturated at ordinary room temperatures. In addition, a "septum" or retaining member is used over each electrode to hold the material in place against ordinary motion, so that the cell is portable. (After shipment the cell should be allowed a few days to settle, for best results.)

Figure 6-2 shows external and internal views of an unsaturated Eppley standard cell of the type used for precision laboratory measurements. The cork washers serve to retain the solid material of the electrodes during

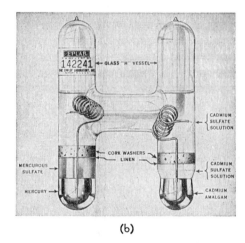

(a)   (b)

Fig. 6-2. Standard cell made by the Eppley Laboratories. (a) Cell mounting. (b) View showing construction of the cell. (*Courtesy of the Eppley Laboratories.*)

---

*Reference 6.

shipment. The bakelite case is lined with copper $\frac{1}{16}$ in. thick to help keep all parts of the cell at a uniform temperature. Cells similar to Fig. 6-2(b) are used in unmounted form in many potentiometric recorders and controllers.

The unsaturated cell differs from the normal cell in some of its characteristics. For one thing, individual cells have slightly different values of emf, so each cell must be measured in terms of a primary standard. The maker supplies a certificate with each cell of the precision type, specifying the value of its emf at a particular temperature. The emf is somewhat higher than for the saturated type. New unsaturated cells ordinarily range between 1.0190 and 1.0194 abs volts.

The unsaturated cell is less constant with age than the normal cell, yet is still remarkably good. Records kept in one laboratory over periods ranging from a few years to about ten of a number of cells carefully used by qualified personnel showed an average decline of about 0.003 per cent (30μv) per year, with a moderate scatter above and below this figure. Accordingly, cells should be checked at intervals of a year, or at most two, so that the uncertainty shall not exceed the usual certification of 0.01 per cent for unsaturated cells. Statistics based on the certification and re-certification of many cells sent to the National Bureau of Standards over a period of years give a much higher value than the one above, but the conditions of use are of course unknown, and undoubtedly cover everything from the most careful use to downright abuse. If a cell shows a change of voltage greater than 30 to 50 μv/year there is a strong presumption that the conditions of use have been far from ideal. *Excessive current drain* (extending to accidental short circuiting) is the most common cause of deterioration of cells. The unsaturated cell has a shorter life than the saturated type, yet experience has shown that some cells last as long as twenty years or longer. The voltage finally becomes unsteady, indicating that it has reached the end of its useful life. The life depends, in addition to age, on the kind of treatment the cell has received.

One important characteristic of the unsaturated cell is its low temperature coefficient, about -10 μv/°C.* It can thus be used for most purposes without temperature correction if the temperature of use does not differ by more than a few degrees from the calibration condition. This cell, also, has a positive temperature effect at one electrode and negative at the other, so it is important to keep all parts of the cell at a uniform temperature. The cells are usually mounted in a bakelite case, lined with copper, to help maintain uniform temperature. The cells should be mounted so that they are not exposed to extremes of temperature, to drafts, steam pipes, or direct sunlight on the case.

---

*Reference 2.

Fig. 6-3. Cut-away view of Weston Standard Cell. (*Courtesy of Daystrom, Inc., Weston Instrument Division.*)

The unsaturated cell is used for a great variety of work in commercial and college laboratories, research projects, and power-system equipment. It is used as the standardizing means in a number of recording meters. The unsaturated form is used for all measurements except those requiring precision of the highest order.

Figure 6-3 is a cut-away view of a Weston standard cell of the laboratory type. The copper shield is shown surrounding the H-shaped glass cell.

6-5. Precautions in the Use of Standard Cells

The principal limitation on the use of standard cells is that nothing more than a minute current should be drawn from the cell at any time and even the small current should flow for only a brief period. It is definitely a potential device and is injured if appreciable current is drawn. It is difficult to set a definite current limit, for damage to the cell is a function of both the magnitude of current and the duration of flow. The makers specify 0.0001 amp as the limiting current, but this should be regarded as an *extreme* figure: the amount should be kept *less* than this, and then only for *momentary* use. The limitation on current means that measurement should be by potentiometer (chap. 7), or other essentially open-circuit device. For a potentiometer, no current is drawn at balance; the momentary flow when tapping the galvanometer key on preliminary balance can be limited to a fraction of a microampere for only a few seconds at most, so there is no need for a large or long-continued drain.

A voltmeter should *never* be used to measure the voltage of a standard cell. The current drain is excessive, and injures the cell. In addition, the reading is meaningless, because of the high internal resistance of the cell. The internal resistance varies from 100 to 500 ohms or more, depending on the dimensions and condition of the cell.

Great care must be exercised *never* to *short-circuit* a cell. The excessive current causes a voltage change that is likely to be permanent. The cell recovers to some degree in a matter of weeks or months, but the recovery probably will not be complete. A cell, once shorted, must be regarded with suspicion and hence is of little value as a standard.

To summarize the precautions, as stated by a maker,*

(1) The cell should not be exposed to temperatures below 4°C nor above 40°C.

(2) Abrupt changes in temperature should be avoided.

(3) All parts of the cell should be at the same temperature.

(4) Currents in excess of 0.0001 ampere should never pass through the cell.

(5) The electromotive force of the cell should be redetermined at periods of a year or two.

These precautions can be accepted fully, except that a lower current limit is safer. A value 0.1 as great — i.e., 10 *microamperes* — is more conservative and more desirable, but even so, being used only for *short periods*, of the order of seconds. A current limit of 10 μa is *ample* for balancing a potentiometer if a galvanometer of any reasonable sensitivity is used, *and if* the galvanometer protection is by a *series* resistor.

A final remark may be quoted from the authority on standard cells, Dr. G. W. Vinal of the National Bureau of Standards,

> "The best answer to the oft-repeated question: 'How much current can I draw from my standard cell?' is 'None'." †

6-6. Silicon Diode Reference Standard (Zener Diode)

The silicon diode is included in the chapter with standard cells since it is an alternative method of providing a reference standard of a degree of accuracy that is acceptable in many measurement processes. It does not supplant standard cells completely, but can be an alternative in some types of use, and opens new possibilities in some metering applications.

Silicon diodes, originally devised as rectifying devices, were found when properly processed to have voltage-current curves similar to Fig. 6-4, characterized by a very small reverse current until a point is reached at which an extremely sharp increase begins. This is the "Zener voltage"; the process is sometimes referred to as a "breakdown" of the diode, but the term is misleading in that the action is not disruptive unless safe current and heating limits are exceeded. With restriction to safe

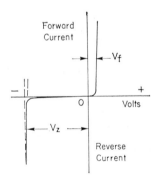

Fig. 6-4. D.C. voltage-current characteristic of a silicon diode, showing the Zener voltage.

---

*From Eppley Laboratories certificate.
†Reference 9, p. 212. Reprinted with permission of the publisher.

values, the process is reversible and repeatable to a high degree of regularity, so the voltage may be used as a reliable standard. The critical voltage value may be controlled over a wide range by the processing of the diode.

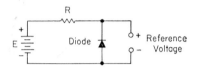

Fig. 6-5. Circuit using a silicon diode to provide a reference voltage.

The slope of the reverse current curve is very steep at the Zener point, corresponding to a low dynamic resistance for operation on this portion of the characteristic. If a circuit is formed as in Fig. 6-5, with a supply voltage much higher than the desired reference voltage, and with a high resistance in series, then if E varies to some extent, most of the change of E is taken up in voltage drop across R (with a small change in diode current), and the reference voltage stays nearly the same. The regulation is good to the extent that the dynamic resistance is very small in comparison with R.

For purposes of analysis, the diode in the Zener region may be represented as a voltage in series with a resistance, as in Fig. 6-6(a), which,

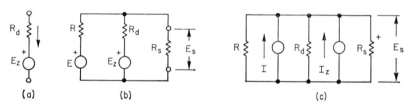

Fig. 6-6. Circuit for the analysis of the voltage-regulating action of a Zener diode.

(a) Circuit to represent a diode in the Zener region
(b) Power supply and diode represented as voltage sources
(c) Power supply and diode represented as equivalent current sources.

combined with the source, gives (b). Conversion from voltage sources (Thevenin) to current sources (Norton), E/R and E_z/R_d, respectively, results in the equivalent circuit of (c). (The reference direction for I_z is taken as shown to put both "sources" on the same basis.) From this circuit,

$$\frac{E}{R} + \frac{E_z}{R_d} = E_s\left(\frac{1}{R} + \frac{1}{R_d} + \frac{1}{R_s}\right) \qquad (1)$$

Solution for E_s results in,

$$E_s = \frac{ER_dR_s + E_zRR_s}{R_dR_s + RR_s + RR_d} \qquad (2)$$

If E is given an increment, so that $E' = E + \Delta E$, then E_s acquires a changed value, $E_s' = E_s + \Delta E_s$, and by subtraction,

$$\Delta E_s = \Delta E \, \frac{R_d R_s}{R_d R_s + R R_s + R R_d}$$

$$= \Delta E \, \frac{1}{1 + \dfrac{R}{R_d} + \dfrac{R}{R_s}} \tag{3}$$

This equation shows that the change in the standard voltage E_s may be very much less than the variation in the supply voltage, so that E_s may be constant within close limits in spite of variations in the supply. The main element in the regulation is the R/R_d ratio, with R large and R_d small. If in addition the supply voltage is moderately well regulated, the remaining variation in the output voltage will be very small. The voltage regulation may be improved even further by the connection of two additional resistors to form a bridge-type circuit (along with the diode and the series resistor) to balance out the effect of the dynamic resistance of the diode.*

One question that always arises in precision equipment is the error that may be caused by changes of temperature. Figure 6-7 shows the effect of temperature, expressed as percentage change of the Zener voltage per degree change of temperature. The coefficient is strongly dependent on the Zener voltage for which the diode is produced, being low (or even negative) for diodes of 4–5 v value, and increasing for higher-voltage diodes. It is possible to select diodes with temperature coefficients very nearly zero,

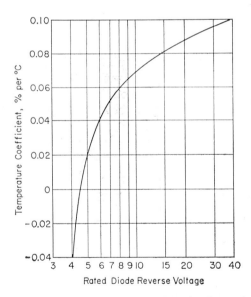

Fig. 6-7. Temperature coefficient of reverse (Zener) voltage of silicon diodes.

---

*Reference 11.

so that the effect of temperature on the standard voltage is quite small. If a standard voltage considerably greater than 5 v must be provided, there is an advantage in using several low-voltage diodes in series, rather than one high-voltage diode. Alternatively, temperature effects can be balanced out by combination of a positive-coefficient diode (about 8 v) with one or two forward-connected diodes (negative coefficient) to cancel the temperature effects. The 8-v diodes have the advantage of greater regularity and predictability than the zero-coefficient diodes.*

The resistance in the Zener region changes also on the basis of the voltage for which the diode is produced. A 5-v diode has a resistance of about 5 ohms, but the value increases to about 1000 ohms for a 40-v diode. This is another advantage of low-voltage diodes, because as shown in (3), low resistance is important to good voltage regulation. Here, again, several low-voltage diodes in series give better results than one high-voltage diode.

Experience with silicon diodes indicates permanence of value to a satisfactory degree. It is not expected that Zener diodes will replace standard cells in work of the highest precision, in standardizing laboratories, but they may serve to maintain a calibrated voltage applied to a potentiometer circuit for thermocouple work, recording potentiometers, and other uses of moderate requirements. Freedom from a standard cell is an attractive feature for a recording potentiometer that may be subjected at times to freezing temperatures. In other fields, the diodes have been applied to indicating instruments such as voltmeters to expand the most used part of the scale, and suppress unneeded parts. Frequency meters of narrow range (58 to 62 cps, for example) have been designed using diodes.

Voltage reference units are now available commercially with operation from a regulated a-c or d-c supply, and with voltage output in the neighborhood of 1.0190 v, so that they may be substituted directly for a standard cell. They are obtainable with accuracy rating good to ±0.01 per cent (which is the usual guarantee on the laboratory-grade standard cell). It is expected that they will be stable and permanent to a high degree, but the limits will be known more surely after a longer experience with them. The early units were considerably more expensive than a standard cell, and required a long warm-up period when first turned on. On the other hand, they are immune to the effects of overload and short-circuit, so the choice between standard cell and diode may be affected by several considerations.

REFERENCES

1. Lamb, F. X., "Weston's Cadmium Cell," *Weston Eng. Notes*, **1**, No. 3 (June, 1946).

*Reference 11.

2. Lamb, F. X., "Weston Standard Cells and Their Applications," *Weston Eng. Notes*, **3**, No. 6 (Dec. 1948).

3. Eppley, Marion, "International Standard of Electromotive Force and Its Low-Temperature-Coefficient Form," *Trans. A.I.E.E.*, **50**, 4, 1293 (Nov. 1931). (Includes bibliography.)

4. Eppley, Marion, "Standard Electrical Cells," *J. Franklin Inst.*, **201**, 17 (1926).

5. Shaw, A. Norman, H. E. Reilley, and R. J. Clark, "The Ageing of Standard Cells," *Phil. Trans. (Royal Soc. London)*, A 229, 125–62 (1930).

6. O'Bryan, L., "Standard Cells," *Gen. Elec. Rev.*, **35**, No. 11, 591–2 (Nov. 1932).

7. Vinal, G. W., "Maintenance of the Volt," *Trans. A.C.S.*, **54**, 247–56 (1928).

8. Smith, D. H., "The Suitability of the Silicon Alloy Junction Diode as a Reference Standard in Regulated Metallic Rectifier Circuits", *Trans. A.I.E.E.*, **73**, Part I, 645, (1954).

9. Vinal, George W., *Primary Batteries*, pp. 160–213, New York: John Wiley & Sons, Inc., 1950.

10. Vincent, George D., "The Construction and Characteristics of Standard Cells," *I.R.E. Transactions on Instrumentation*, *I*-7, 221–234 ,Dec. 1958.

11. Robinson, Paul B., "A Precision, Continuous Voltage Reference for Industrial Recorders," *Proc. 3rd Natl. Conf. on An. and Dig. Instr.*, *A.I.E.E.*, pp 43–48, (April 1959). (Also, *Trans. A.I.E.E.*, No. 60–114, 1960.)

PROBLEMS

6-1. An unsaturated standard cell has an emf of 1.01910 v and an internal resistance of 250 ohms.

(a) If a voltmeter with a 3-v scale and resistance of 300 ohms is connected to the cell, what reading is given? What current is drawn from the cell?

(b) What current is drawn if the cell is short-circuited?

6-2. A standard cell measured by potentiometer (open-circuit) reads 1.01892 v. When a 1-megohm resistor is connected across the cell terminals, the reading becomes 1.01874 v. What is the internal resistance of the cell?

CHAPTER 7

Potentiometers

7-1. Use of the Potentiometer

The potentiometer is a device for the comparison of voltages, based on the use of a standard cell, which gives an accurate foundation for the measurements. The measuring circuit is a resistance network that can also be known to a high degree of accuracy. The potentiometer method is thus capable of very accurate results.

The potentiometer may be used directly for the measurement of voltages, and by measuring the voltage drop across a shunt of known resistance can serve also for the determination of current. It may make these measurements of current and voltage in a working circuit of some sort, or it may be used for the calibration of ammeters and voltmeters. Through the latter use it becomes the standard for our indicating meters for current, voltage, and power, and thus occupies a very important place in the scheme of electrical measurements.

7-2. Principle of the Potentiometer

Figure 7-1 represents the circuit of a simple potentiometer of the slide-wire type. The method of measurement depends upon finding a position for the slider such that the galvanometer shows no deflection when the key is closed. Zero current through the galvanometer means that the emf E is balanced by a voltage drop of equal magnitude along the slide wire, or $E = E'$ in Fig. 7-1.

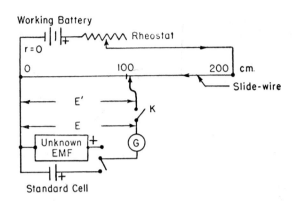

Fig. 7-1. Basic slide-wire potentiometer circuit.

Determination of the unknown emf E now is a matter of evaluation of E'. This may be found from the proportional length along the slide wire, as, for example, if we have established 2.00 v for a 200-cm length of slide wire we can evaluate the voltage for any intermediate point (assuming uniformity of resistance along the wire). The 2.00 v could be adjusted by the use of a voltmeter. However, this is not generally desirable, as the accuracy of a voltmeter is not high enough for many purposes. The usual method of establishing the voltage scale is by means of a standard cell and is accomplished by the following steps. Suppose the emf of the standard cell to be 1.018 v. We place the slider at 101.8 cm on the scale and adjust the rheostat shown in Fig. 7-1 until the galvanometer shows no deflection when the key is closed. The potentiometer standardized in this way may be said to be direct reading, as the voltage at any balance point can be obtained without any computation other than the placing of the decimal point in the proper place (e.g., 101.8 cm = 1.018 v).

One feature of the potentiometer method that should be emphasized is the fact that *no current* is drawn from the source being measured when the balance is attained. The determination of the emf of a source is thus independent of its resistance. The only effect of resistance, if large, is to reduce the sensitivity of the galvanometer to unbalance.

The fact should be noted that a potentiometer is a *ratio* instrument. An unknown voltage is measured in terms of a reference voltage by means of the *ratio of the resistances* required to give the particular potential drops needed to balance the two voltages. It is important that the various resistors in the circuit have correct *relative* values, but the general level of resistance is of secondary interest. In a potentiometer having a dial with a number of 0.1-v steps, it is essential that all the resistance steps be alike within a very small tolerance, although the nominal value may be decidedly different in different circuit designs.

The battery that supplies current to the slide wire is called the working battery, and the current in this circuit is the working current. The working current is adjusted, or "standardized," by means of the rheostat by reference to the standard cell, with the slider set at the point appropriate to the emf of the standard cell. Reference to the standard cell should be made periodically so that the working current may be maintained at the correct value.

7-3. Practical Construction of Potentiometers

The slide-wire type of potentiometer used as illustration in Fig. 7-1 is not a practical form of construction. The long slide wire is awkward, and even for the length shown cannot be read to a very great degree of precision. This difficulty is overcome in practical potentiometers by making part (sometimes all) of the variation by dial resistors and by the use of circular slide wires of one or more turns, thus achieving a considerable length of slide wire in a small space.

Potentiometers have been developed in various forms by the instrument makers. Some of the variety has been due to the special conditions for which particular instruments are built. Other refinements have been added to increase the accuracy and the convenience of measurement. We shall study several typical potentiometers to see the features incorporated in their circuits.

7-4. Simple Potentiometer — Student Type

The circuit of the student-type potentiometer is shown in Fig. 7-2. There is one dial switch with 15 steps of 0.1 v, and a one-turn circular slide wire of 0.1 v. A double-throw switch is provided to connect either the

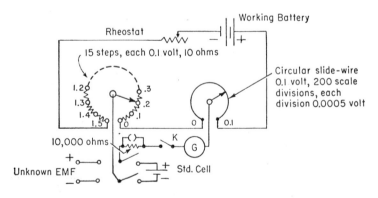

Fig. 7-2. Circuit diagram of the student potentiometer (single range).

standard cell or the unknown to the circuit. The galvanometer circuit includes a key and a protective resistance of 10,000 ohms. The slide wire has 200 divisions, so each division corresponds to 0.0005 v and readings can be estimated to about 0.0001 v. The working current of this potentiometer is 0.1 v/10 ohms, or 0.010 amp, in order to give 0.1 v steps on the main dial. (The slide wire is sometimes designed for 0.11 v, to give some overlap on the 0.1-v dial steps.)

The procedure in making measurements with the potentiometer requires the following steps:

(1) Set the dials to the emf of the standard cell, throw switch to standard cell position, *tap* the key (10,000 ohms in circuit), adjust the rheostat for zero galvanometer deflection. Cut out the 10,000 ohms as balance is approached.

(2) Throw the switch to the unknown and balance by the main dial and slide wire. Begin with the 10,000 ohms in circuit and cut it out as balance is approached. Read the unknown emf directly from the dials.

(3) Check the working current by referring back to the standard cell. On this potentiometer we must reset the dials whenever we check the working current.

It may be interesting to compare this compact arrangement of a dial switch (15 steps) and a single-turn slide wire with the length a plain slide wire would require for equal precision of reading. If we assume the slide wire in Fig. 7-2 to have a circumference of 21 inches (diameter about 7 inches), an equivalent "slide-wire-only" design would be 16 $\times$ 21 inches, or 28 feet! Moreover, the dial arrangement has the advantages of greater accuracy and convenience, in addition to compactness, in comparison with sole dependence upon a slide wire. The dial resistors may easily be adjusted to high accuracy, whereas there are linearity limitations on a slide wire.

7-5. Multiple-range Potentiometers

The student potentiometer of Fig. 7-2 is shown as a single-range device, and as such, is frequently constructed to cover voltages up to about 1.6 v — though, of course, it can be designed for any desired value within reason. It is possible to modify the circuit in a simple way to add a second range, which is usually related by a decimal factor, such as 0.1 or 0.01, in order that the direct-reading feature may still be utilized.

Figure 7-3 (a) indicates how a second range may be incorporated into the potentiometer of Fig. 7-2 by the addition of two resistors and a switch. The current in the 10-ohm dial resistors must be 10 milliamperes on Range 1.0, in order to give 0.1-v steps, and 1 ma on Range 0.1, to give voltage steps just one-tenth as great. It is essential in designing the circuit to be able to change ranges without readjustment of the rheostat, and without

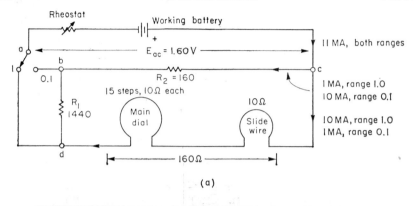

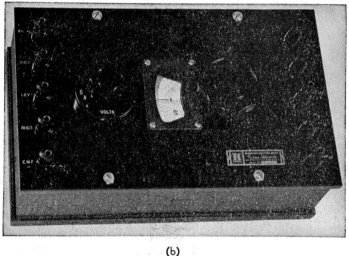

(b)

Fig. 7-3. (a) General scheme of a 2 range circuit (1.6 and 0.16 volt ranges shown). (b) A "student" potentiometer (1.6 and 0.016 volt ranges). (*Courtesy of Minneapolis-Honeywell Regulator Co., Rubicon Instruments Division.*)

requiring a special value of working-battery voltage. This requires that the voltage E_{ac} stay the same automatically for both positions of the switch, which result is obtained if the *total* battery current is kept the same. Therefore, the current through R_2 should be 1 ma on Range 1.0 and 10 ma on Range 0.1, to give a total battery current of 11 ma on either range. With the same voltage E_{ac}, resistor R_1 must equal $9(R_{\text{dial}} + R_{\text{sl.w}})$ to give 1 ma through the measuring circuit on Range 0.1; also, R_2 must equal $(R_{\text{dial}} + R_{\text{sl.w}})$ so that it will now carry 10 ma.

For the circuit assumed in Fig. 7-3 (a), voltages up to 0.16 v may be measured on Range 0.1. Low voltages are balanced in the usual way on

Range 0.1, and then the dial readings are multiplied by 0.1. There are two advantages in having the low range for small voltages: first, the precision of reading is increased by one decimal place, and second, a greater part of the reading is made on the dial resistors which have inherently greater accuracy possibilities than the slide wire. The illustration of Fig. 7-3 (a) was stated in terms of a 10/1 ratio; similar arrangements can be used for any other ratio, and 100/1 is sometimes encountered (used for millivolt ranges met in some temperature measurements by thermocouple).

A "student" potentiometer which embodies a circuit of this type is illustrated in Fig. 7-3 (b). In this instrument the ranges are 0-1.6 v and 0-0.016 v, the latter range being useful in the measurement of voltages of thermocouples. This potentiometer is a versatile and accurate device within the scope of its design, and has a usefulness greater than might be inferred from the "student" appelation. The resistor adjustment to 0.03 per cent, and the readability of 0.0001 v (representing 0.01 per cent at 1.0 v), give it an accuracy adequate for many types of measurement. While it does not have some convenience features incorporated in the more elaborate (and more expensive) potentiometers, and is slightly lower in accuracy, it is capable of giving high-grade results.

The necessity in the potentiometer of Fig. 7-3 of returning to Range 1.0, and resetting the dials, when standardizing, is avoided in some other potentiometers by special circuit arrangements. A standard-cell dial is provided in the precision potentiometers to permit setting to the certified value of a specified standard cell. These features will be illustrated in the potentiometers to be studied in later sections.

The precision potentiometers frequently have three ranges, usually 1.0, 0.1 and 0.01. The method of incorporating the third range will be considered later in connection with specific potentiometers.

7-6. Features of Precision Potentiometers

The student-type potentiometer of pars. 4 and 5 is a simple device, capable of a precision of 10^{-4} v for readings up to about 1.6 v. (The accuracy is less than this, due to tolerances on resistor values, and in part to possible non-linearity of resistance along the slide wire.) The potentiometer is useful in many classes of work. However, for some applications higher precision and accuracy are needed, requiring changes in design features. Other changes and additions are made in the interest of greater speed and convenience in operation.

The readability can be improved by making the slide wire longer — but there are limits to practical dimensions. Another solution is a multi-turn slide wire, with ten (or eleven) turns, giving an additional decade of readability, but requiring considerable "cranking" when turning from one end to the other.

Another arrangement consists of a single-turn slide wire with an intermediate dial. The main dial has 0.1 v steps, the intermediate dial 0.01 v steps, and the slide wire divides 0.01 v further by 1000 (200 divisions, interpolated to fifths). This gives a readability of 10^{-5} v in 1.6 v, and the attainable accuracy is also increased, in comparison with the student type of potentiometer, since the slide wire determines a smaller portion of the reading, and the relatively more accurate dial resistors are used for a larger part. The means of incorporating the intermediate dial between the two "traveling" contacts on main dial and slide wire may be worked out in several ways, some of which we shall meet in the following sections.

Still other potentiometers provide all balancing on decade dials, without recourse to a slide wire. These are usually five-dial instruments, giving readings to 0.00001-v increments in a total of 1.6 v — occasionally six dials are available on very high quality instruments. (Two or more ranges may be provided, to extend the measurements to lower voltages.) These potentiometers have very pleasant operating characteristics, but are less common than the slide-wire type, owing, in considerable part, to the expense of close adjustment of many dial resistors. Several circuit arrangements have been devised to permit the all-dial balancing.

Another feature, incorporated in most high-grade potentiometers, is a standard-cell dial. This dial can be set to the voltage of the cell being used as reference, and left at that setting throughout the measurements, since the dial is used for this purpose alone. The main dials can be left at the setting of the voltage being measured while standardization is being checked, thus increasing the convenience and speed of measurements. The galvanometer is connected in either measuring or standardizing circuit by a double-throw switch. The circuit details of the standard-cell dial and its relationship to range-changing circuits are worked out differently in various potentiometers, and supply some interesting studies of circuit relationships.

A precision potentiometer which reads to increments of 0.00001 v (10 μv) on Range 1.0 has *readability* of 0.1 μv on Range 0.01. This does not mean, unfortunately, that small voltages can be read with assurance to 0.1 μv. Measurements are subject to stray thermal and contact voltages in the potentiometer, galvanometer, and the measured circuit, of the order of one to several microvolts. Such voltages are difficult to locate and control, and can be minimized, in fact, only by special construction, involving the selection of metals for resistors, binding posts and connecting wires, and by the use of thermal shields. Advantage cannot be taken in the usual general-purpose potentiometer of readability to 0.1 μv, but the 0.01 range is nevertheless useful in moving the measurement further up on the main dials. There are special "thermo-free" potentiometers for measurement of very low voltages; we shall study the principles of these circuits later.

7-7. Potentiometer Sensitivity — Study of Unbalanced Conditions

The current flowing in the galvanometer when the circuit is unbalanced can be computed by the use of Thevenin's theorem. Figure 7-4 shows a simple slide-wire type of potentiometer for purposes of illustration. If the "two accessible terminals" are taken as x and y in Fig. 7-4(a), then the voltage, E_0 in the equivalent circuit is the voltage x-y with K open. This is the scale-reading, in case the working current is correctly adjusted. The internal resistance, R_0, as indicated in Fig. 7-4(b), is the parallel combination of R_1 and $(R_2 + R_3)$. The complete Thevenin circuit is shown in Fig. 7-4(c). The resistance marked "r" in this diagram should include the resistance of the lower battery and the galvanometer protective resistance.

Figure 7-4 indicates that current flows in the galvanometer circuit as determined by the difference of the emf E and the potentiometer setting E_0, acting in a circuit having the

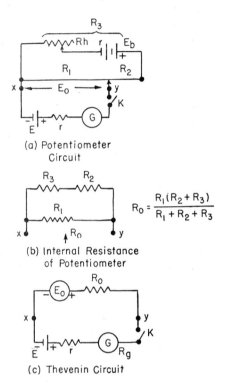

(a) Potentiometer Circuit

$$R_0 = \frac{R_1(R_2 + R_3)}{R_1 + R_2 + R_3}$$

(b) Internal Resistance of Potentiometer

(c) Thevenin Circuit

Fig. 7-4. Application of Thevenin's theorem to potentiometer circuit.

resistance shown in the diagram. This makes possible the computation of the galvanometer current produced by any specified discrepancy in the setting of the potentiometer dials. We can compute, as an extreme case, the current that would flow in case either the working battery or the cell E is connected to the circuit with the wrong polarity. The required galvanometer-protecting resistance may also be determined for any specified case.

7-8. Galvanometer Protection in a Potentiometer Circuit

The problem of protecting the galvanometer in a potentiometer circuit differs from that in a bridge. The difference lies in the fact that the standard cell must also be guarded against excessive current. It is true that the galvanometer can be protected by an Ayrton shunt, but if a shunt is used the resistance of the circuit through the galvanometer and standard cell is made low, as can be seen in Fig. 7-4(c). In case of large unbalance or reversed battery connection the current through the standard cell is objectionably large, even though the galvanometer is protected. The possibility of damage may be avoided by the simple expedient of using a *series* resistor instead of a shunt. The series resistance may be switched in steps, as 100,000 ohms, 10,000, and zero.

Precision potentiometers usually have a series of three galvanometer buttons and self-contained resistors to be used progressively in balancing the circuit. A damping resistance is connected across the galvanometer to an upper contact on the maximum-sensitivity key.

RULE: *Use a series resistor and not a shunt to protect the galvanometer in a potentiometer circuit.*

7-9. Measurement of High Voltage by Potentiometer

Potentiometers of the general-purpose type usually cover a range of about 1.5 to 2 v. If higher voltages are to be measured, it is necessary to use a scale-multiplying device; this is an accurately built voltage-dividing resistor, frequently constructed with multiple ranges. The circuit diagram of a usual form of divider is shown in Fig. 7-5; range-selection is sometimes

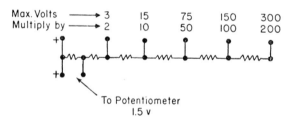

Fig. 7-5. Circuit diagram of the volt box.

accomplished by choice of binding posts, as indicated here, and sometimes by means of a selector switch. The name "volt box" is applied to this device.

It is no longer true, when the volt box is used, that no current is drawn from the source that is measured, although the potentiometer itself draws no current at balance from the volt box. The current drain can be made quite small, however, by the use of a high-resistance divider system. The choice of resistance values involves a compromise. High values are desirable in reducing the current drain on the measured source and the power to be dissipated in the volt box. Low values would be preferred for other reasons: (1) low resistors are, in general, more stable than high resistors, (2) low values give better galvanometer sensitivity, (3) low values minimize the effects of leakage. Volt boxes are commonly constructed with resistances of 100 ohms/v of nominal voltage range, giving a maximum current of 10 ma from the measured source, or 200 ohms/v, giving a maximum of 5 ma. Some foreign instrument makers, notably Otto Wolff in Berlin, use a value of 10,000 ohms/v, associated with high-resistance types of potentiometers.

7-10. The Leeds & Northrup Type K-2 Potentiometer

The Leeds & Northrup Type K-1 and K-2 potentiometers may be regarded as the outgrowth and development of the earlier design which is now carried on as the "student" potentiometer. Several features have been added for increased precision and convenience, including the following for the K-2:

(1) An eleven-turn slide wire, for increased precision

(2) A standard cell dial

(3) A three-range scheme, controlled by a selector switch, giving multipliers of 1.0, 0.1 and 0.01 on the basic range of 0–1.61 v

(4) Built-in galvanometer resistances and working-circuit rheostats. Galvanometer damping resistances of 200 and 2000 ohms are provided, and three keys for graded galvanometer sensitivity.

The slide wire consists of a manganin wire wound in helical grooves on a form of insulating material. It is covered by an inverted cup or drum which carries the sliding contact. The number of turns is read on a vertical glass scale, and the fraction of a turn on a scale engraved around the periphery of the drum. The 11-turn slide wire covers 0.11 v, so with 200 divisions per turn, each division is 0.00005 v, and readings may be estimated to 0.00001 v. The 0.11-v slide wire gives some overlap on the 0.1-v steps on the main dial, which is convenient in balancing a voltage close to one of the 0.1-v points.

The circuit diagram of the K-2 is shown in Fig. 7-6. The range-changing network and the standard cell connections are of particular interest. The 1.0 and 0.1 multipliers are obtained in a manner similar to the two-range circuit of Fig. 7-3. On the 0.01 multiplier, connection is made into this resistance loop at a point that gives the correct dial current ($\frac{1}{10}$ of the total); at this point the parallel resistance is lower than on 1.0 or 0.1, so a series resistance is connected between the loop and the 0.01 contact on the range switch (shown in two parts in Fig. 7-6, to provide a tap point for the standard cell connection.) On the 1.0 multiplier the standard cell is connected to the 0.9-v stud of the main dial, so that 0.9 v plus 0.11 v gives

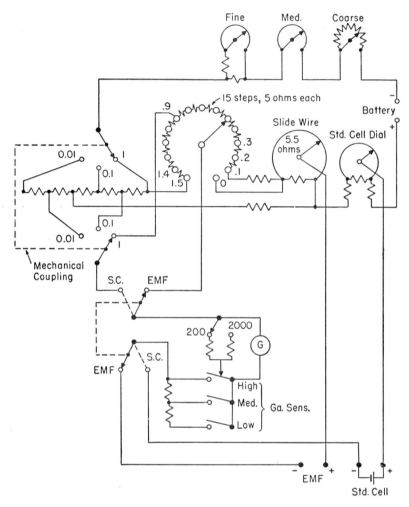

Fig. 7-6. Circuit diagram of type K-2 potentiometer.

1.01 v — the small remaining voltage needed to balance the standard cell is obtained from the standard cell dial. In this way, the same resistors are used, in the main, for standardizing and for measuring, which helps minimize the effect of slight discrepancies in the resistance values. The standard cell connections are tapped at the points on the 0.1 and 0.01 ranges to give 1.01 v from the end of the slide wire.

The L&N potentiometer is a comparatively low-resistance instrument. Each 0.1-v step has a resistance of 5 ohms, so the working current is 20 ma. Low resistance contributes to sensitivity to unbalance, so an extremely sensitive galvanometer is not needed. Another reason for low resistance in this design is the necessity of keeping the slide wire sufficiently large and robust that the wear due to the sliding contact shall not make an appreciable change of resistance.

Another feature worthy of note is the fact that there is no sliding contact which carries the *working current* within the limits of the measuring part of the circuit. This is important in avoiding contact drop and is practically imperative in a potentiometer with so large a working current. (The K-1 potentiometer is similar to the K-2 in its main features, but differs in details of the range-changing circuits and the standard-cell dial.)

7-11. Rubicon Type B and No. 2781 Potentiometers

The Rubicon Type B potentiometer uses a main dial with 0.1-v steps, an intermediate dial with 0.01-v steps, and a single-turn slide wire. The slide wire divides 0.01 v further by 1000 (200 divisions, interpolated to fifths). This gives a readability of 10^{-5} v in 1.6 v, and the attainable accuracy is also good, since the slide wire determines only a small part of the reading, and the relatively more accurate dial resistors are used for a large part. The circuit diagram, Fig. 7-7, shows the main dial followed by the intermediate dial with a two-arm arrangement to which the slide wire is connected. The slide wire may be considered a means of interpolating between the steps of the intermediate dial, or as an alternative viewpoint, the intermediate dial may be regarded as a means of moving the entire slide wire "up" or "down" in potential by 0.01-v steps. This action of the intermediate dial is pictured in Fig. 7-8. The combination of intermediate dial and single-turn slide wire gives the same readability as a multi-turn slide wire.

The travelling two-arm arrangement in which the arms span two steps of the dial is called a *Kelvin-Varley slide*. The circuit bridged across the two arms is in some cases a single decade for further voltage reduction, and in other cases a continuation of Kelvin-Varley construction in several decades. The Type B and No. 2781 potentiometers use this circuit as a means of voltage interpolation by a slide wire.

The current in the 10-ohm, 0.1-v, main dial steps is 0.01 amp, and this current flowing in the 1-ohm resistors of the intermediate dial *between zero and the lower arm* gives 0.01-v steps. The 11-ohm slide wire, in series with 7 ohms, is shunted by 2 ohms on the dial, and hence carries one tenth of the current, or 0.001 amp, and has a voltage of 0.011 v. The moving contacts are in a circuit of 18 ohms, and carry only 0.001 amp, so no difficulty is experienced from contact resistance. The total circuit resistance is the same for all switch settings.

The standard cell is connected from the "0" stud on the main dial to a small slide wire connected between the 1.0 and 1.1 studs. The resistors

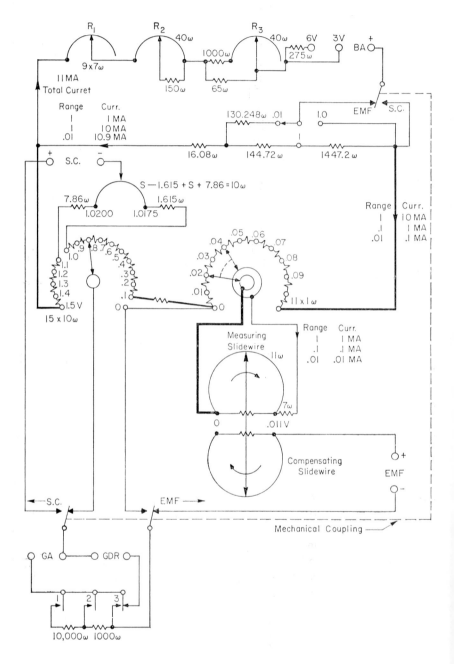

Fig. 7-7. Schematic Diagram of Rubicon Type B Potentiometer *(Courtesy of Minneapolis-Honeywell Regulator Co., Rubicon Instruments).*

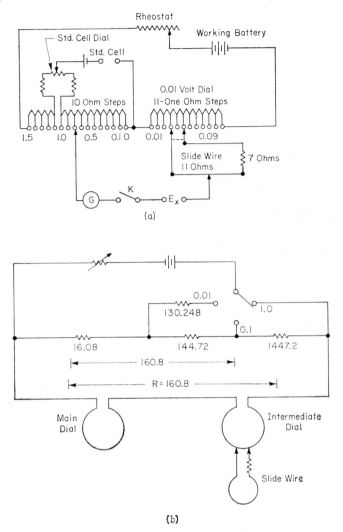

Fig. 7-8. Circuit details of the Rubicon Type B potentiometer.

at the ends of this slide wire serve to take up part of the 0.1 v step so that the calibrated part of the slide wire can be restricted to the limited range, 1.0175 to 1.020 v, since standard cells in good condition have voltages close to 1.0190 v. The use of the main-dial resistors for standardization is a desirable feature, contributing to accuracy, since the common usage for standardizing and measuring tends to balance out the effect of slight discrepancies in the values of the resistors.

The main features of the range-changing circuit of the Type B potentiometer are illustrated in Fig. 7-8(b). The 1.0 and 0.1 multipliers follow the pattern of Fig. 7-3. The connection for the 0.01 multiplier is tapped into the resistor loop to give 0.1 ma in the dial path and 10.9 ma in the shunt; this parallel combination has lower resistance than on the other multipliers, so the series resistor (130.248 ohms) is needed to maintain the same total working current. (The EMF-SC switch is arranged to return the connections to Range 1.0 for standardization.)

The resistors of the Type B are adjusted within 0.01 per cent, and the general accuracy is of this order. The limit of error on Range 1.0 is specified as 30 microvolts, or 0.01 per cent of the reading, whichever is greater. On the lower ranges the percentage limit of error is somewhat greater. Great care has been taken to minimize thermal emf's in the circuit. The last galvanometer key is constructed with spring copper leaves and gold contact points, a combination essentially free of thermal emf. An interesting detail is the arrangement of the compensating slide wire in Fig. 7-7, used to complete the galvanometer circuit and to balance out thermal voltages at the slide-wire contact. If a sliding contact is moved rapidly, some heat is generated, and the resulting thermal emf, although of short duration, causes difficulty in balancing. The compensating slide wire is constructed in the same way and of the same materials as the measuring slide wire, so the thermal effects at the two sliders should be nearly equal, and the directions are such that they cancel out around the galvanometer circuit.

7-12. Feussner-Brooks Potentiometer

The circuit of this potentiometer is a radical departure from the potentiometers studied so far. The circuit was originated by Feussner and later modified by Dr. H. B. Brooks of the National Bureau of Standards. Potentiometers of this design are produced by the Otto Wolff concern in Germany. The Wolff line was originated in Berlin before 1900 by Otto Wolff, and has been produced since that time, except for an interruption caused by World War II. Wolff instruments have held a leading place in the field of fine resistors, bridges, and potentiometers.

In this potentiometer, all balancing is done on decade dials, and no slide wire is used. This is made possible by the Feussner decade, which consists of a double set of resistors and contacts, working together mechanically, but insulated electrically. The Wolff potentiometer of Fig. 7-9 utilizes three of these decades and two plain decades (actually, 14 or 19 steps on the main dial, in various models). Note that the sliding contacts of the 1000-ohm and 100-ohm dials are the two ends of the measuring circuit, much as in the previous potentiometers. The 10-, 1-, and 0.1-ohm decades have two sets of resistors, made as nearly identical as possible. One set in each decade is in the measuring circuit, the other in the working circuit *outside* the measuring part, and connection is made so that resistance

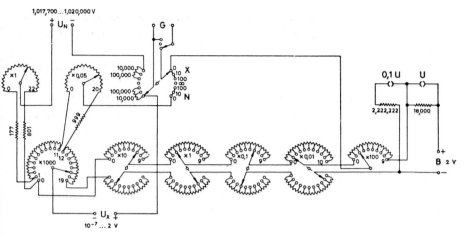

Fig. 7-9(a). Circuit of the Wolff potentiometer, Feussner-Brooks type. (*Courtesy of Physics Research Laboratories, Inc.*)

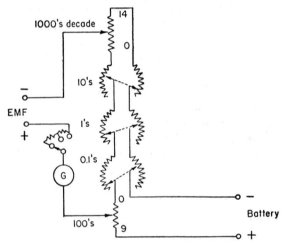

Fig. 7-9(b). Schematic diagram of Fig. 7-9(a).

is reduced in the outside part as it is increased in the measuring section. In this way the working current is kept constant, although the drop within the measuring part is changed.

The double construction is used in the Wolff potentiometer in the three lowest-resistance dials. This obviously is desirable, as any slight inequality between the upper and lower sets of resistances is thus less likely to be serious than it would be in the 100-ohm or 1000-ohm dials.

The instrument shown in Fig. 7-9 has two ranges, with multipliers of 1.0 and 0.1, controlled by the position of a plug. Wolff produces single-range models, also instruments that are combined Wheatstone and Kelvin bridges and potentiometers. The setting for the standard cell is made on two dials, for 0.0001-v and 0.00001-v steps, respectively; no slide wire is used in the circuit.

This is a high-resistance potentiometer and a high-sensitivity galvanometer must be used with it. The 0.1-v steps have a resistance of 1000 ohms, so the working current is 0.1 ma. There are several sliding contacts in the measuring circuit, and this is ordinarily considered undesirable. However, the Wolff potentiometer performs excellently in spite of the contacts, and is convenient and accurate. This is due to the small working current and the high-grade construction of the contact system. Dry cells may be used satisfactorily for the working current, because of the small current drain. Other potentiometers have been built, based on the same principle of construction.*

7-13. Bonn Five-dial Potentiometer

The Bonn five-dial potentiometer of the Rubicon line is another instrument that carries out all balancing on decade dials, without recourse to a slide wire, but the method and the circuit properties are much different than the Feussner type. Use is made in this design of 10-ohm resistors in the main dial and 1-ohm resistors in a current-attenuating circuit in the succeeding dials. These low resistances have the merit of being highly stable, and of giving a low-resistance circuit. The four lower dials use double-dial construction, with nine 1-ohm units on each dial, and the two arms moving together to transfer the connections to the next dial.

The current in the resistors of the main dial is 0.01 amp, which is divided at the next dial between a direct path of 9 ohms and a shunt path of 81 ohms, giving 0.001 amp to the next dial, which flows in 1-ohm resistors to give 0.001-v steps. Further divisions by ten occur at the next two dials. The simplified diagram of Fig. 7-10 indicates how each stage modifies the voltage drop in the measuring part of the circuit, but keeps the working current constant. The effect of switch contact resistance is minimized, since the contacts are in series with a resistance of 81 ohms. Dial II contacts, carrying the largest current (0.001 amp), are made of silver to insure low contact resistance; the following dials are less critical.

Two ranges are provided, 0 to 1.6 v in steps of 10 μv, and 0 to 0.16 v in steps of 1 μv with range-changing resistors as in some of the other potentiometer circuits. The circuit is returned automatically to the high range for standardizing. There are two standard-cell dials, so that settings may be made to 0.00001 v. Another interesting feature is the use of compensation on dial I (as in the Brooks deflection potentiometer) so that the total

---

*Reference 2.

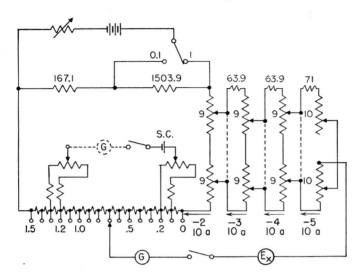

Fig. 7-10. Schematic diagram of the Bonn five-dial potentiometer.

galvanometer circuit resistance is held more nearly constant. (There is a slight variation with the change of setting the remaining dials.)

Care has been taken to minimize thermal voltages in the potentiometer circuit. A Wenner thermo-free reversing key of copper and gold construction is used to permit elimination of the effect of thermal emf's in the galvanometer circuit. In addition, this reversing key doubles the deflection of the galvanometer when approaching final balance, which aids in the most precise setting of the last dial. Three galvanometer keys are provided, the high-sensitivity key being of copper and gold construction to minimize thermal effects.

7-14. Leeds & Northrup Type K-3 Potentiometer

The K-3 potentiometer is a recent development which differs markedly from the potentiometers which we have considered so far, and from previous instruments of this manufacturer. The most radical difference lies in its being a *parallel* circuit device. That is, the main dial of 0.1-v steps forms one path, and a combined intermediate dial and slide wire (single-turn) form the other path. The basic range is 1.611 v, with multipliers for lower ranges. The general plan of the circuit is indicated in Fig. 7-11.

The unknown voltage is balanced by the difference of the voltages of the two parallel paths, as marked in the figure. If n_A represents the number of digits on the A dial (from the bottom end), and using the current values as marked (and which we can verify as correct relative values in terms of the given resistances),

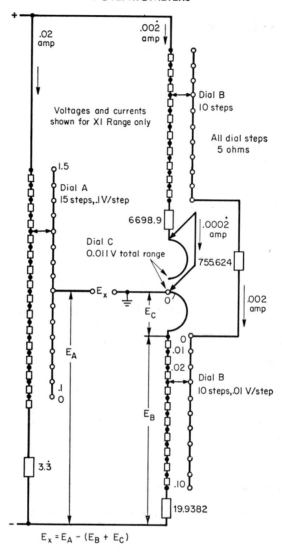

Fig. 7-11. Basic circuit of the K-3 potentiometer, with galvanometer and range-changing circuits omitted. (*Courtesy of Leeds & Northrup Co.*)

$$E_A = 0.020(n_A \times 5 + 5.0 + 3.\dot{3})$$

Note that a dot over the last digit is used in the diagram to indicate that this is a repeating digit, to save writing to awkward lengths. By simplification,

$$E_A = 0.1 \, n_A + 0.16\dot{6}$$

For study of the *B-C* branch,

let n_B = number of digits on dial B, from top

s = slide-wire resistance (lower sector, from upper end to the slider)

Then

$$E_B + E_C = 0.000\dot{2}[(10 - n_B) \times 5 + 19.9382] + 0.000\dot{2}\, n_B \times 5$$
$$+ (51.735 - s) \times 0.000\dot{2}$$
$$= -0.01\, n_B - 0.000\dot{2}s + 0.166914$$

Accordingly,

$$E_x = E_A - (E_B + E_C)$$
$$= 0.1n_A + 0.01n_B + 0.000\dot{2}s - 0.000247$$
$$= 0.1n_A + 0.01n_B + 0.000\dot{2}(s - 1.11)$$

The equation shows 0.1-v steps on dial A, 0.01-v steps on B, and an additional amount along the slide wire, with the zero displaced 1.11 ohms from the end. The calibrated part of the slide wire covers 0.011 v, which requires $0.011 \div 0.000\dot{2} = 49.5$ ohms. The remainder of the 51.735 ohms is used, part at top, and part at bottom (about equally divided, as shown by the 1.11 ohms), for the adjustment of zero. Since E_x is the difference of two branches, a slight drift of the resistance of either path with aging would displace the zero point. The zero can and should be checked, and corrected if necessary; the procedure is to short the input terminals, with dials A and B on zero, and adjust for balance by the slide wire. If the slide wire does not read zero with the galvanometer at zero, the slide-wire dial is loosened and moved to read zero. The small excess resistance at the ends of the slide wire provide for this adjustment in either direction.

The resistance of the B-C branch is needed for analysis of range-changing circuits. The parallel part of B-C consists of $(50 + 6698.9 + 51.735) = 6800.635$ ohms in parallel with 755.624, giving 680.063 ohms. This is in series with $(50 + 19.9382)$ to give a total of $750.001 = 750$ ohms. Branch A has a resistance of

$$(16 \times 5 + 3.\dot{3}) = 83.\dot{3} \text{ ohms}$$

The combined effect of A in parallel with B-C is thus

$$(83.\dot{3} \times 750)/(83.\dot{3} + 750) = 75.0 \text{ ohms}$$

The currents marked on the diagram may be verified by multiplying each value by the corresponding path resistance, as found above, to show equality of voltage drops for the two paths.

This potentiometer has range multipliers of 1.0, 0.1, and 0.01, secured in a manner similar to some of the other potentiometers. The range-changing circuits are well illustrated in Fig. 7-12. There are taps brought from the resistor network to permit standardizing on any range. The

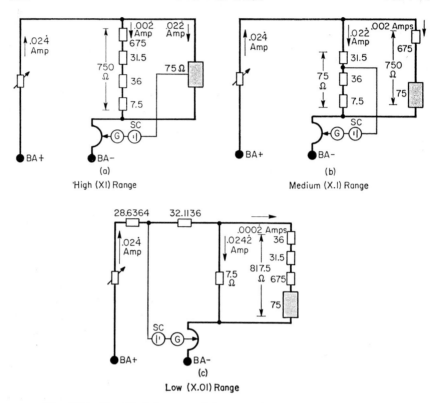

Fig. 7-12. Simplified diagrams of the range-changing circuits of the K-3 potentiometer. (*Courtesy of Leeds & Northrup Co.*)

contacts of dial B carry a current of 2 ma, but this is in a circuit of 755 ohms, so the effect of contact resistance is negligible. The slide-wire contacts carry 0.2 ma in a circuit of 6800 ohms, again making contact resistance negligible.

The potentiometer is mounted in a metal case, providing electrostatic shielding. Binding posts and internal circuits have guards against leakage for operation under high-humidity conditions. Care has been taken to keep thermal voltages to a low level. There is a reversing key in the galvanometer circuit, which permits checking for thermal voltages in this path, and also doubles galvanometer sensitivity in close balancing. Another feature is the read-out dial, with all numbers adjacent in one window, with the decimal point moved automatically as the range switch is changed.

7-15. Special-purpose Potentiometers

The potentiometers described in the foregoing sections may be considered general purpose instruments, in that they may be used for a wide variety of measurements, although their characteristics may not be the best possible for certain kinds of work. There are in addition several potentiometers that are designed either for special ranges, special accuracy, or for convenience in a particular type of measurement.

One special instrument is the deflection potentiometer, which is a combination of potentiometer and deflecting meter. This will be discussed in the next section. There are several low-range potentiometers that are designed for temperature measurements with thermocouples. Some of these are small devices, constructed for portability and convenience rather than extreme accuracy.

There are other low-range potentiometers that are specially designed for accurate measurement of very small voltages. The chief problem in this class of measurement is to keep stray thermal effects to small enough values to permit the accurate determination of the small unknown voltage. "Thermo-free" potentiometers have been devised by several workers, notably Diesselhorst, White and Wenner.* Special precautions are taken to avoid junctions that may introduce thermal emf's in the part of the circuit that is in series with the unknown.

Another special device is the standard cell comparator in use at the Bureau of Standards. It is designed to read the difference between the reference cell and the unknown cell. By measuring only the difference between the cells, the measuring equipment need cover only a small range, and by confining the measurement to the difference a very accurate determination can be made.

7-16. Brooks Deflection Potentiometer

The deflection potentiometer is a combination of potentiometer and deflection instrument, in contrast to the strictly null devices we have studied so far. The major part of the voltage to be measured is balanced against a voltage derived from a potentiometer dial, and the remaining small amount is read by the deflection of a millivoltmeter. One potentiometer dial is used (30 steps, each 5 mv, in the model shown in Fig. 7-13), and intermediate values are read from the meter. The accuracy is not so high as for a straight potentiometer, but operation is much faster, as it is not necessary to hunt for a null point by the adjustment of several dials.

One application of this instrument is to the calibration and checking of d-c ammeters, voltmeters, and wattmeters, for which purpose the accuracy is adequate and the increased speed is very helpful. To check any one point on the meter scale, the potentiometer dial is set on the required step, the power supply is adjusted to bring the pointer of the tested meter to the

---

*See references 4, 5.

Fig. 7-13. Brooks Model-7 deflection potentiometer. (*Courtesy of the Rubicon Co.*)

desired mark, and the deflection is read on the potentiometer deflection instrument. The desired value is thus read directly, with no adjustment of the potentiometer other than the initial setting of the main dial. Another application is to the testing of incandescent lamps. The light output is very sensitive to changes of voltage, so for precision testing the voltage should be adjusted and read more closely than can be done on a voltmeter. The power supply, however, is rarely completely steady, so the balancing on a straight potentiometer is difficult and tedious.

If a potentiometer circuit is not completely balanced, the current through the galvanometer depends on the difference of the unknown voltage and the potentiometer setting for any one position of the potentiometer sliding contact. The unbalance-current can thus be used, at any one setting, as a measure of the voltage unbalance that must be added to or subtracted from the dial indication to obtain the true value of the unknown voltage. However, the proportionality between current unbalance and voltage unbalance depends also on the resistance of the circuit as viewed from the galvanometer. The resistance must be constant for all settings of the potentiometer dial if the scale of the meter is to be read directly in terms of voltage. The resistance is not constant for all positions in the usual potentiometer, as can be seen by inspection of Fig. 7-4(a). If the slider is near the left, the resistance R_1 is small and hence R_0 is small. R_0 increases as the slider is moved toward the right.

The circuit devised by Dr. H. B. Brooks* of the Bureau of Standards overcomes these difficulties, and keeps the galvanometer circuit resistance constant by adding resistance of the correct amount at each switch tap point. The sketch of Fig. 7-14 shows the potentiometer dial. The units

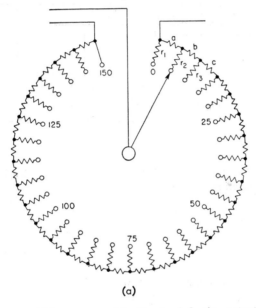

(a)

Fig. 7-14(a). Dial arrangement for Brooks deflection potentiometer.

a, b, c, etc., are the main potentiometer resistors. The tap resistors r_1, r_2, r_3, etc., are adjusted to special values so that the total galvanometer circuit resistance is the same for all settings of the dial. The idea of the "internal resistance" of the potentiometer, and the need to keep it constant, follow from the Thevenin method discussed in par. 7 and illustrated in Fig. 7-4. The parallel circuit of this figure indicates the method of computation for the 30 resistors needed in the taps; the volume of computation may appear

---

*References 6 and 7.

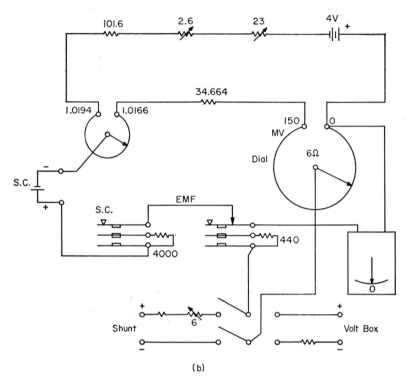

Fig. 7-14(b). Diagram of the Brooks deflection potentiometer; circuit diagram (dial details omitted).

rather large, but the calculations are simple, and can be reduced to routine, once the resistances of the other parts of the circuit are established.

The dial has 30 steps of 5 mv and 0.2 ohm each for a total of 150 mv and 6 ohms; the working current is, accordingly, 0.025 amp. The circuit is designed for a nominal 4 v battery, with rheostat range to cover voltages from 3.56 to 4.2 v. The Thevenin resistance R_0, with the dial set at full scale, is 6 ohms for the dial path, and the remaining circuit resistance for the other path. For the nominal 4 v supply,

$$R_{\text{total}} = 4/0.025 = 160 \text{ ohms}$$

so

$$R_0 = (6 \times 154)/160 = 5.775 \text{ ohms}$$

R_0 computed for lower dial steps would come out a smaller value, so the compensating resistors, r_1, r_2, etc., are added to keep R_0 the same for all positions. The change of rheostat setting to compensate for the battery condition introduces a small change of R_0; however, for the extremes of battery voltage (as above) R_0 changes by about 0.04 ohm, which is only

$\frac{1}{500}$ of the circuit resistance (R_0, 5.775; meter, 7.4; external, 6 ohms,—total = 19.2 ohms, approximately). The meter can be read, at best, to 0.1 of a scale division in 30 divisions, or $\frac{1}{300}$, so the possible extreme change is less than the limit of readability.

The deflection potentiometer is used with a volt box to measure higher voltages, or with a shunt to measure current; however, consideration must be given to maintaining the same resistance in the galvanometer circuit for all ranges. Figure 7-15(a) gives the circuit of a volt box for the deflection

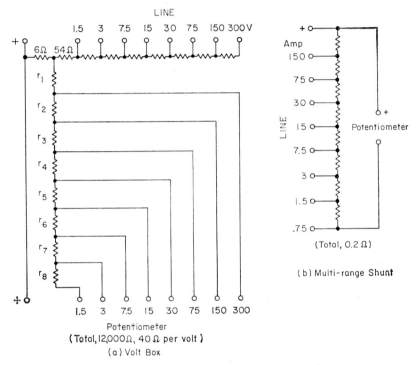

Fig. 7-15. Volt box and shunt for deflection potentiometer.

potentiometer, designed on the basis of 40 ohms per volt. If we suppose for the moment that no resistors r_1, r_2, ... were used, and measure a resistance-less source connected to the 1.5-v terminals (60 ohms), we have for the source resistance from the potentiometer side

$$(6 \times 54)/60 = 5.4 \text{ ohms}$$

Similarly, for a source connected to the 300-v terminals,

$$(6 \times 11994)/12000 = 5.997 \text{ ohms}$$

Corresponding values may be found for the other ratios. We may decide to round out the value to 6 ohms for all positions — we can evaluate r_1, then

r_2, and so on. The total of the string down to the 1.5-v post is 0.6 ohm, which added to the 5.4 computed above, checks the desired total of 6 ohms. (The first two or three steps are small — if ignored, the numerical results are not changed appreciably, but all are indicated on the diagram for completeness of analytical idea.)

For current measurement, any standard shunt of appropriate resistance and current capacity may be used. Figure 7-15(b) shows a multirange shunt. Note that the resistance as seen from the potentiometer posts is 0.2 ohm for all ranges. To read correctly, since the potentiometer is calibrated for a 6-ohm volt box, the shunt should have 6 ohms also; this could be added in the shunt lead, but for convenience a 6-ohm rheostat is included in the potentiometer [see Fig. 7-14(b)]. The rheostat can be set to the resistance needed to bring the resistance to the correct total. The deflection potentiometer has separate binding posts for shunt and volt box, and a double-throw switch, so the throw-over from one to the other is fast. Voltage and current can be read in quick succession in the calibration of a wattmeter.

The deflection instrument has a zero-center scale, 30-0-30 divisions for the 3-0-3 mv range. The smallest divisions are thus 0.1 mv, making estimation possible to perhaps 0.01 mv. The maker specifies a limit of error of 0.02 per cent plus 0.03 mv. The deflection potentiometer is rather expensive because of the large number of special-value resistors that must be adjusted to close tolerances.

7-17. Lindeck Element

The potentiometers considered so far have been of the *constant-current* type — that is, the current through the circuit is maintained at a constant value, and the unknown is balanced against the voltage drop across an adjustable portion of the circuit resistance. An alternative scheme is to use a *constant resistance*, and to vary the current through it to obtain the variable voltage used to balance and measure the unknown.

The circuit scheme is illustrated in Fig. 7-16. The current is adjusted until the galvanometer indicates a null, for which condition E_x must equal the voltage drop across the standard resistor. The current required for balancing is read on a milliammeter. This arrangement is known as a Lindeck element.

The accuracy of measurement is not so high as for the constant-current type, being limited by the indicating instrument used to measure the current. However, for some purposes, such as low-voltage thermocouple measurements, the accuracy is adequate, and there are some advantages in its use. The adjustment is simple, and there are no sliding contacts present in the E_x circuit to introduce spurious thermal emf's. Multiple ranges may

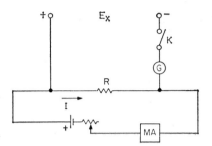

Fig. 7-16. Lindeck potentiometer element.

be provided simply by switching different standard resistors (four-terminal type) into the circuit. Some potentiometers have combined the constant-current construction for the upper decades with a Lindeck element for final balancing; this combination is convenient for some purposes, and gives good accuracy. If the milliammeter is accurate to within one per cent, the Lindeck element may in effect replace two decades of dial balancing.

The Lindeck element has particular advantages in the measurement of very small voltages since it affords excellent protection against parasitic effects. There are no sliding contacts in the measuring circuit, and the few elements operating at low level may be enclosed quite easily in a thermal shield. The Lindeck element has qualities which merit more attention in its field of usefulness than it has received.

7-18. Microvolt ("Thermo-free") Potentiometers

Potentiometers are needed in a number of situations in which the voltages are in a low range, of the order of a small fraction of a volt. While the standard general-purpose potentiometers can be read to low values (usually to 0.1 μv on the lowest range), they are not ideal for the very low part of the scale. The difficulty is in the presence of thermal voltages ranging from a fraction of a microvolt to several microvolts. It is essential for the accurate and dependable determination of small voltages that thermal and contact voltages be kept to very low levels, and this requires special planning and circuit arrangements. The potentiometers designed for the low ranges are sometimes called microvolt, and sometimes thermo-free, since the two ideas are closely associated for low readings.

One principle in design is to avoid sliding contacts in the measuring circuits that operate at very low levels, (1) since contact resistance multiplied by current gives a voltage of uncertain amount, and (2) from the danger of thermal voltages at these junctions. There must of course be variation somewhere in the system, but it should be in a branch of the circuit operating with relatively high voltage and high resistance.

Another thing to be avoided is a combination of metals that have appreciable thermoelectric properties between them. It would be ideal if the entire circuit could all be of one metal, but this is impractical; the wires to the galvanometer and its moving coil are necessarily of copper, and the resistors in the instrument cannot be copper. The galvanometer branch is helped by using an all-copper system — suspension as well as coil and leads (instead of a phosphor bronze or gold suspension). Manganin is a good material to use for the resistors, since it has a relatively low thermal power against copper. Even so, the junctions must be carefully managed, as the thermal voltage of this combination is nearly 2 μv per degree. Inside the potentiometer the entire measuring loop may be of manganin to avoid dissimilar materials in this critical portion of the circuit. The two binding posts for the galvanometer leads should be close together, and enclosed in a thermal shield (a box made of heavy copper or aluminum). All critical portions of the circuit should be in a shield to maintain the entire circuit at the same temperature, and resistors subject to appreciable self-heating should be positioned carefully. The galvanometer may be placed in a padded box to avoid uneven cooling effects of air currents.

"Thermo-free" potentiometers have been devised by many workers, notably Diesselhorst, White, and Wenner.* Various ingenious schemes have been worked out to obtain a variable voltage to balance against the unknown voltage, without having moving contacts in the measuring path. We shall study a few of these principles without going into a complete study of circuit details; one or two decades may serve to illustrate the idea in some cases.

(a) *Diesselhorst potentiometer.*

The Diesselhorst potentiometer uses a split circuit, the current being fed in by the moving arms at the middle (main) dial, and flowing in parallel paths, right and left. The diagram of Fig. 7-17 is simplified to some extent by the omission of less important details and resistor values.

We may study the operation of the circuit in terms of the following items, with reference to the diagram:

(1) The ratio of resistances, left path to right path, is 10/1, so that the current in the right path is 10/11 I, and in the left path 1/11 I.

(2) The double-dial construction maintains the same circuit resistance for any setting of any dial.

(3) The total current is standardized by reference to a standard cell. Multiple ranges are generally provided; the diagram shows only one, for simplicity.

---

*References 3, 4, 5, 8, and 9.

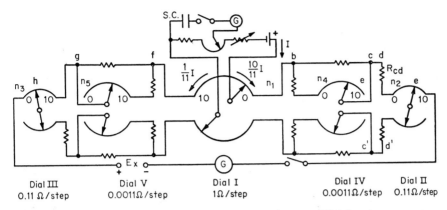

Fig. 7-17. General plan of Diesselhorst potentiometer (Dial decades are used — shown here as smooth arcs for ease in drawing.)

(4) Dials IV and V use shunted construction (chap. 4, par. 28) to minimize contact effects. They provide steps of 0.0011 ohms, beginning from an initial value R_0, so that

$$R_4 = R_0 + n_4 \times 0.0011$$

$$R_5 = R_0 + (10 - n_5) \times 0.0011$$

$$= R_0 + 0.011 - n_5 \times 0.0011$$

where n_4 and n_5 represent the number of digits of dial setting (note direction of calibration on V).

(5) The unknown voltage, E_x, is balanced by the *difference* of the voltages E_{ae} and E_{ah}, because of current directions.

(6) Contact effects are very small. Dial I contacts are in series with the comparatively high resistance of the battery circuit, so contact resistance has negligible influence on current value; any thermal voltage adds to or subtracts from battery voltage, and is negligible. Dial II contact resistance is in a path of the order of 1000 ohms, and Dial III of 1000, and so cause no difficulty. Dials IV and V have a fixed resistance of about 1 ohm shunted by a variable of about 100, so contact resistance is discounted by a factor of $1/100^2$, and thermal voltage by a factor of $1/100$, reducing the effects to negligible levels.

(7) The E_x loop is solid, with no sliding contacts (except in the shunt paths of IV and V). The loop can be made all-manganin except for the galvanometer, and its binding posts can be close together in a thermal shield.

We may write equations for the voltage drops of the right and left branches in terms of the dial settings, n_1, n_2, ... for the galvanometer null:

$$E_{ae} = \tfrac{10}{11} I[n_1 \times 1 + n_2 \times 0.11 + R_0 + n_4 \times 0.0011 + R_{cd}]$$

$$E_{ah} = \tfrac{1}{11} I[(10 - n_1) \times 1 + (10 - n_3) \times 0.11 + R_0 + 0.011$$
$$- n_5 \times 0.0011]$$

The unknown voltage is the difference of these two quantities for the null balance, so performing the subtraction, and combining terms, we have,

$$E_x = E_{ae} - E_{ah}$$
$$= I[n_1 + 0.1n_2 + 0.01n_3 + 0.001n_4 + 0.0001n_5]$$
$$+ \tfrac{1}{11} I[9R_0 + 10R_{cd} - 11.111]$$

The second bracket may be made zero by choice of resistance values. (Let $R_0 = 1$ ohm, then $R_{cd} = 0.2111$ ohms.)

We now have E_x expressed in decade terms, by the first bracket of the equation, based upon the value of the total current I. If $I = 0.1$ amp, then the largest dial steps are 0.1 v, and the smallest 0.00001 v. If a multi-range scheme is arranged, with additional values $I = 0.01$ amp, 0.001 amp, and 0.0001 amp, (as in some Wolff potentiometers), the steps are proportionally smaller, giving readings to 0.01 μv.

The Diesselhorst circuit requires a rather large number of odd-value resistors adjusted to close tolerances. Some other circuits have been devised with the objective of simplifying the resistor arrangement.

(b) *Wenner difference potentiometer.*

Figure 7-18 represents a decade element of a "difference" potentiometer originated by Wenner[*]; it consists of a double decade with a re-

10 x 5 Ω

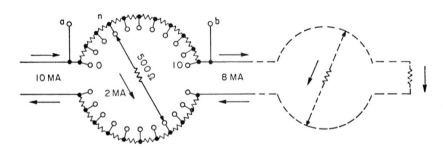

Fig. 7-18. Schematic diagram of a Wenner decade.

sistance shunted across the moving arms. The name derives from the fact that the voltage change in the measuring side of the circuit is due to the difference in current in the dial resistors produced by the shunting path. As the moving resistor always bridges the same number of steps,

---

[*]References 3, 9.

the circuit resistance is always the same, and the total voltage drop is constant for the decade, but voltage drop is shifted from one side of the circuit to the other — one side being the balancing voltage for the unknown. A variable balancing voltage is thus obtained with no sliding contacts in the measuring path, the only contact resistance being in series with the high-resistance shunting element.

The illustrative decade shown in the figure has 5-ohm dial resistors and a shunt of 500 ohms which carries 2 ma out of 10 ma that comes to the decade. The voltage drop a to b, with the dial set at n units, is

$$E_{ab} = n \times 5 \times 0.010 + (10 - n) \times 5 \times 0.008$$
$$= 0.01n + 0.4$$

The voltage between a and b thus changes by 0.01-v increments as the dial is moved. (The 0.4 v can be balanced out.) Note that the 0.01 v equals the product of the step resistance multiplied by the 2 ma diverted by the shunt, $5 \times 0.002 = 0.01$ v.

Decades of this sort can be used in various ways in conjuntion with other elements to form a complete potentiometer circuit. One combination consists of dials at the two ends of a divided circuit, with three difference decades between them. Another possibility is a feed-in dial as in Fig. 7-17, with two Wenner decades on each side to give five-dial readings. Decades of this type have the merit of using equal even-value resistors along the dial.

7-19. Rubicon Thermo-free Potentiometer

The Rubicon thermo-free potentiometer is another approach to the problem of eliminating the effects of thermal voltages from the measuring circuits. A balancing voltage is produced in an all manganin circuit which has no sliding contacts, in this way eliminating the important sources of spurious emf. Adjustment of some sort must be provided, of course, but the circuit can be arranged so that sliding contacts are not in the *measuring* part of the circuit. If they are in the current-supply system, in series with the battery voltage and the fairly high series resistances, their effect on measurements can be kept to a very low level.

Figure 7-19 is a schematic diagram showing the main features of two decades only, for simplicity. The current supplied to the circuit is adjusted to 10 milliamperes, as determined by the standard cell, standard cell resistors, and the galvanometer. The voltage introduced into the E_x loop may be studied first by reference to the left half of the circuit, a to b. If the dial reading is represented by n_1 digits, the division of current between the two parallel paths gives for the measuring branch, a-b, the following value,

$$i_{ab} = 10 \times \frac{n_1 \times 11}{121} \text{ ma}$$

and the voltage drop a-b is,

$$e_{ab} = 10 \times \frac{n_1 \times 11}{121} \times 11 \text{ mv}$$

$$= 10 \times n_1 \text{ mv}$$

A similar analysis holds for the current division of the other dial, but the drop b-c is across 1.1 ohms, so each step is one millivolt. The compensating resistors shown in Fig. 7-19 in the vertical risers to the switch contacts are computed so that the resistance in the working-battery circuit remains constant when the switch arm is moved from one position to another. Range-changing resistors, F and G, are provided to give a 0.1 multiplier when desired.

A four-dial or six-dial potentiometer can be designed utilizing additional elements of the type we have analyzed here, with additional resistors in series in the E_x loop. It is necessary, of course, for each pair of dials to have its own working battery, and the current must be standardized in each circuit.

The basic design of the circuit, as outlined above, gives a variable voltage for balancing against E_x, but removes sliding contacts from the E_x circuit. In addition, other refinements must be made to keep stray thermal emf's below the microvolt level. Copper connections between

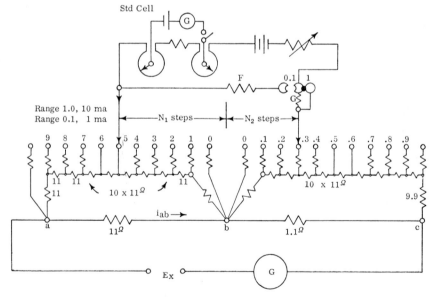

Fig. 7-19. Schematic diagram of two decades of a "thermofree" potentiometer circuit.

resistors and decades are avoided, since they would introduce numerous "thermocouple" junctions, each with a thermal voltage. In this instrument, the connecting leads in the measuring circuit are of manganin, becoming in some places extensions of calibrated resistors. In addition, the entire potentiometer circuit is enclosed in a heavy thermal shield of copper or aluminum, to guard against temperature differences for different parts of the circuit. The essential measuring loop is protected still further by an inner thermal shield.

The connections for the unknown enter the potentiometer as a pair of copper wires. The connection posts are mounted within a thermal shield, and are placed in close proximity to each other to eliminate the possibility of temperature difference between them. The connections for the copper wires to the galvanometer are similarly mounted, and an all-copper galvanometer is used (suspension as well as coil) to avoid junctions of different metals. The galvanometer is often surrounded with cotton or other heat insulation and mounted in a box with only a window for the light beam, to guard against the effects of air currents in the room.

A six-dial potentiometer of this type is made by Rubicon. The first four dials consist of two pairs of dials utilizing the principle analyzed in Fig. 7-19. The last two dials are similar in principle, but may be of somewhat simpler construction owing to the low level of operation and to the reduced percentage accuracy required of these small voltage steps. These dials are supplied by an internal cadmium cell, which is adequate for long operation, since the drain is small (only 12 microamperes) and intermittent. The series resistance in the battery circuit is so large in comparison with the dial resistance that compensating resistors are not needed, as in Dials I to IV. Standardization for the last two dials is made by comparison with a step on Dial IV. This potentiometer can give readings on Range 1.0 from zero to 111,111 microvolts in steps of 0.1 microvolt, and on Range 0.1 to 11,111.1 microvolts in steps of 0.01 microvolt. The presence of six measuring decades does not imply that an accuracy of one part in a million can be attained; however, the sixth dial is useful in measuring small voltages (Dial I on zero), and in following small changes in a voltage.

"Double" thermo-free potentiometers are made, consisting of two sets of dials coupled to the resistor system. This makes possible the measuring of two different voltages in quick succession without resetting of dials, as would be required in a single potentiometer. The double construction is a great convenience in some thermocouple and chemical work.

REFERENCES

1. Stein, I. Melville, "Design of Potentiometers," *Trans. A.I.E.E.*, **50**, 4, 1302 (Nov. 1931).

2. Eppley, Marion, and W. R. Gray, "An Improved Feussner-Type Potentiometer," *Rev. Sci. Instruments*, **2**, 242 (1931).

3. Wenner, Frank, "Note on Potentiometer Design," *Phys. Rev.*, **31**, 94 (1910).

4. Harris, Forest K., *Electrical Measurements*, New York: John Wiley & Sons, Inc., 1952.

5. Laws, Frank A., *Electrical Measurements*, New York: McGraw-Hill Book Co. Inc., 1938.

6. Brooks, H. B., "A New Potentiometer for the Measurement of Electromotive Force and Current," *NBS Bull.*, **2**, 225 (1906).

7. Brooks, H. B., "Outline of Design of Deflection Potentiometers," *NBS Bull.*, **8**, 419 (1912).

8. White, Walter P., "Potentiometers for Thermoelectric Measurements, Especially in Calorimetry," *J.A.C.S.*, **36**, 1868 (1914).

9. Behr, L., "Wenner Potentiometer," *Rev. Sci. Instruments*, **3**, 3, 109–20 (Mar. 1932).

10. Bobier, F. C., and L. O'Bryan, "Precision Potentiometer of Improved Design," *Gen. Elec. Rev.*, **35**, 3, 185–7 (Mar. 1932).

PROBLEMS

7-1. A slide-wire potentiometer is arranged as in Fig. 7-1. The slide wire has a resistance of 100 ohms; the working battery has an emf of 4 v and negligible internal resistance. The galvanometer resistance is 50 ohms. For the standard cell, $E = 1.018$ v, $r = 200$ ohms. The rheostat is adjusted so that the standard cell is in balance with the slide-wire contact set at 101.8 cm.

(a) Find the working current and the resistance of the rheostat.

(b) With the setting as in part (a), the connections to the standard cell are reversed by accident. Find the current through the standard cell when K is closed (no protective resistance in the circuit). What harm would this do?

(c) Find the protective resistance to limit the current to 10 μa.

(d) Suppose an Ayrton shunt of 500-ohms total resistance is used with the galvanometer (instead of the series resistance), with the multiplier switch set at 0.001. Find the galvanometer current and the standard-cell current for the condition of (b).

7-2. The diagram indicates a potentiometer with a range of 1.7 v and a working current of 50 ma. The 16 fixed resistance coils provide discrete steps of 0.1 v each. The emf slide wire has the same resistance as one of the fixed coils and a 100-division scale that may be read to 0.1 division. The standard-cell slide wire permits the potentiometer to be standardized, with a standard cell having a voltage between 1.016 and 1.020.

(a) What is the smallest voltage step that may be read with this potentiometer?

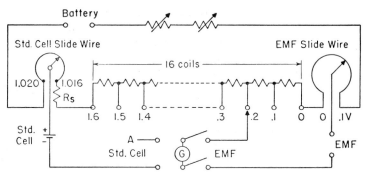

Prob. 7-2.

(b) What is the resistance in ohms of (1) each of the resistance coils, (2) the resistance R_s, and (3) the standard-cell slide wire?

(c) Show (1) where to connect terminal A of the double-pole double-throw galvanometer switch, (2) the polarity of the battery terminals, and (3) the polarity of the emf terminals.

7-3. A potentiometer of the student type (text, Fig. 7-2) is designed to have a dial of 15 steps, 0.1 v and 10 ohms each, and a 10-ohm slide wire. Because of error, the resistance of the step between the zero and the 0.1-v dial contacts is 10.1 ohms. This potentiometer is standardized against a standard cell of 1.0190 v and then is used to measure an unknown voltage that gives balance for a reading of 0.2150 v on the dials. What is the true value of the unknown? Express the error in the observed reading as a percentage.

7-4. A potentiometer of the student-type has 15 steps of 10 ohms and 0.1 v each, and a slide wire of 11 ohms. The working battery has a voltage of 3.0 v.

(a) With the dials set at 1.2460 v, what is the internal (Thevenin) resistance of the potentiometer circuit as seen from the galvanometer position?

(b) If the potentiometer is standardized against a standard cell marked 1.01920 but actually 1.01884 v, what dial reading is given when balancing a source of 0.6362 v?

(c) The potentiometer is as described above, except that the 0.1-to-0.2 v step is 10.2 ohms, and the 0.9-to-1.0 v step is 9.6 ohms. Standardization is carried out in the regular manner against a correct 1.01900-v standard cell. What reading is given when balancing against a 0.2466-v source?

7-5. A student-type potentiometer has a main dial with 15 steps of 10 ohms and 0.1 v each, and a slide wire of 10 ohms. (This potentiometer does not have a standard-cell dial.) It is used with a rheostat and a 3-v battery. The voltage being measured has a value of 0.8 v.

(a) If the standard cell is marked 1.01920 v, but actually is 1.01895, what reading is given by the potentiometer for the 0.8-v source?

(b) If the standard cell is correct as marked, but the resistor between the 0.9 and the 1.0 contacts on the dial has a resistance of 10.1 ohms, find the reading given for the 0.8-v source.

7-6. The main dial of a student-type potentiometer has 15 steps, each 0.1 v and 10 ohms, and a slide wire of 20 ohms, shunted to give it a range of 0.1 v. The potentiometer is used with a 3-v battery and a galvanometer of 30 ohms resistance and a sensitivity of 0.25 microampere per mm.

(a) Find the resistance of the slide-wire shunt.

(b) The working current is standardized correctly and the dials are set at 0.45 v when measuring an emf of 0.449 v with negligible internal resistance. Find the galvanometer deflection. Show your method of analysis clearly.

(c) The switch is thrown to standardizing position without resetting the dials. The standard cell has an emf of 1.019 v and a resistance of 250 ohms. Find the protective resistance to limit the galvanometer deflection to 200 mm, and show how it should be connected.

7-7. A potentiometer has 15 steps, each of 0.1 v and 5 ohms, and a slide wire of 5.5 ohms, in series with the working battery (emf = 2.20 v, r = 0.1 ohm) and a rheostat. The galvanometer has a sensitivity of 0.03 $\mu a/mm$ and a resistance of 50 ohms.

(a) Find the resistance needed in the rheostat.

(b) If the slide wire has 11 turns, 200 divisions per turn, and the reading can be estimated to one-fifth of a division, to what fraction of a volt can the reading be taken?

(c) The potentiometer is being used to measure an emf of 1.20 v with negligible internal resistance. What error of setting (in volts) from the true balance is necessary to deflect the galvanometer spot one millimeter?

7-8. A potentiometer of the student type has a main dial with 15 steps, each 50 ohms, and a slide wire with a resistance of 75 ohms. The working current is adjusted so that each main-dial step represents 0.1 v. The working battery has an emf of 3 v. The galvanometer has a resistance of 50 ohms and a protective series resistance of 10,000 ohms. The standard cell has an emf of 1.0186 v and an internal resistance of 250 ohms.

(a) Find the working current and the voltage range covered by the slide wire.

(b) With the working current properly adjusted, the standard cell is connected to the circuit, but with the dials set by accident at 1.1186 v instead of 1.0186. Find the galvanometer current with the protective resistance included.

(c) Same, without the protective resistance.

7-9. A potentiometer has a dial with 15 steps, each 0.1 v and 5 ohms, and an 11-turn slide wire of 5.5 ohms and 200 divisions per turn. The working current is standardized correctly, using a working battery of 2.2-v emf and a rheostat. The galvanometer has a sensitivity of 0.005 $\mu a/mm$ and a resistance of 50 ohms.

The potentiometer is balanced correctly against a source of 1.50 v and 10-ohms internal resistance. The slide-wire setting is then moved 2 divisions. What galvanometer deflection results? (No series resistance is added to the galvanometer.)

7-10. A potentiometer of the student type has a main dial with 15 steps, each 10 ohms and 0.1 v and a slide wire covering 0.11 v. The working battery has an emf of 3 v and resistance of 0.1 ohm. The standard cell has an emf of 1.0186 v and a resistance of 250 ohms. The galvanometer has a resistance of 50 ohms and a sensitivity of 0.05 μa/mm. With the working current adjusted correctly and the dials set for 0.600 v, the switch is thrown to standard-cell position without resetting the dials.

(a) Find the galvanometer current, with no protective resistance.

(b) Find the protective series resistance to limit the deflection to 50 mm.

7-11. A potentiometer with circuit of the student type has a main dial with 15 steps, each 0.1 v and 10 ohms, and a slide wire of 10 ohms. The working battery has an emf of 3.0 v and negligible internal resistance. The galvanometer has a resistance of 200 ohms and a sensitivity of 0.25 microampere per mm. It is being used to measure an emf of 0.6108 v with internal resistance of 152 ohms. The working current is standardized correctly against a standard cell of 1.0190-v emf.

(a) Draw a sketch of the circuit.

(b) Compute the change of voltage setting on the potentiometer dial to give a galvanometer deflection of 1 mm in this measurement.

(c) Compute the resistance to be placed in series with the galvanometer to limit the deflection to 20 mm for a 0.1 v unbalance of the potentiometer setting in the above case.

(d) If the conditions are otherwise as stated, but the main-dial step between the 0.9 and 1.0-v contacts has a resistance of 9.8 ohms, find the reading given on the dials for the above unknown, after the potentiometer has been standardized in the usual manner.

7-12. A student-type potentiometer has a main dial of 15 steps of 10 ohms and 0.1 v each, and a slide wire of 20 ohms, shunted to give it a range of 0.1 v. (This potentiometer does not have a standard cell dial.) It is used with a rheostat and a 3-v battery. The galvanometer has a resistance of 50 ohms and a sensitivity of 0.5 microampere per mm. It is being used to measure a source of 0.150 v with negligible internal resistance.

(a) Find the resistance of the slide-wire shunt.

(b) Find how much the setting of the potentiometer must be changed from the balance position to give a galvanometer deflection of 1 mm for the 0.15 v source. Show your analysis clearly.

(c) If the resistance step between zero and the 0.1 contact on the main dial has a resistance of 10.1 ohms instead of 10 ohms, what reading is shown on the potentiometer for the 0.15 v source after the potentiometer is properly standardized against a 1.0190-v standard cell?

7-13. The circuit shown here is to be used for measurement of voltages, with 0.1-v steps on a dial switch, and intermediate amounts to be read on the microammeter. The microammeter is to cover a range of 0.150 v (its scale will be calibrated in volts), and is to have the same calibration for all dial switch positions. Determine the resistance value for all resistors.

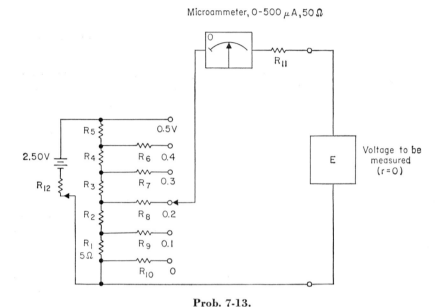

Prob. 7-13.

7-14. It is desired to calibrate a 0–150-v d-c voltmeter with a potentiometer. The manufacturer states that the potentiometer is accurate to ± 0.0001 v, and the 100/1 ratio of the accompanying volt box to ± 0.05 per cent. It is found that, when the voltmeter reads 80 volts, the potentiometer reading is 0.8025 v. Calculate the error of the voltmeter reading in volts and in per cent of the full-scale range of the voltmeter, and specify the uncertainty of the correction, on the assumption that the galvanometer sensitivity is sufficient to permit definite balancing to the last figure given.

7-15. An unknown low resistance is measured by connecting it in series with a standard 0.1-ohm resistor and a battery, and measuring the voltage drops across X and S by a potentiometer. The voltage across X reads 0.02371 v, across S, 0.80135. The battery is then disconnected and the voltage across X is found to be 30 μv, with polarity in the same direction as before; across S, $E = 0$.

(a) What is the source of the 30 μv across X?

(b) Compute the resistance of X to at least 0.1 per cent, on the assumption that the above readings are exact.

(c) What is the possible uncertainty in X, if either of the above readings may be in error by ± 0.01 per cent or ± 0.1 mv, whichever is larger?

7-16. A potentiometer as shown in Fig. 7-2 has a working battery of $E = 3$ v, a standard cell of $E = 1.0180$ volts and $r = 250$ ohms, and a galvanometer of 50 ohms resistance.

(a) At which end of the scale (i.e., voltages near zero or near 1.6 v) is the galvanometer more sensitive to small unbalances of setting?

(b) Find the galvanometer sensitivity needed to give a deflection of one scale division for an error of setting of one-fifth scale division on the slide wire when standardizing the working current.

7-17. The sketch indicates the working-current circuit of a potentiometer. With a plug inserted at A the potentiometer is direct-reading. With the plug moved to B the voltage drop x to y is to be one tenth as great, but the working current drawn from the battery is to remain constant. Compute the correct values for R_1 and R_2.

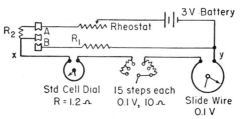

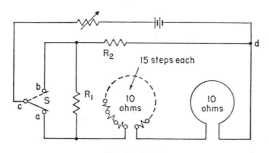

Prob. 7-17.

7-18. A potentiometer has a main dial with 15 steps of 10 ohms each, and a slide wire of 10 ohms. With switch S on the high range (contact a) the current through the dial resistors is to be 10 ma, and on the low range, 1 ma. Find the values of R_1 and R_2 to satisfy these requirements and to give the same total current and the same voltage drop c to d for both positions of S.

Prob. 7-18.

7-19. A student-type potentiometer has a main dial with 15 steps of 10 ohms and 0.1 v each, and a slide wire of 16 ohms, with a shunt, to cover 0.11 v. There are two range multipliers, 1.0 and 0.01 (no standard cell dial). It is used with a rheostat and a 3-v battery. The galvanometer has a sensitivity of 0.5 microampere per millimeter, and a resistance of 100 ohms. The voltage being measured is 1.20 v, with negligible internal resistance.

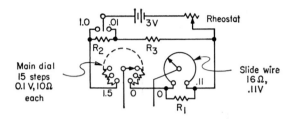

Prob. 7-19.

(a) Compute the values of the resistors R_1, R_2, R_3, and the rheostat resistance.

(b) The 1.20-v source is connected to the potentiometer, after correct standardization. How much change in dial setting from the correct balance point is needed to give a galvanometer deflection of 50 mm, with no galvanometer protective resistance?

7-20. A potentiometer consists of a main dial of 15 steps of 10 ohms, 0.1 v (on Range 1.0) each, and a slide wire of 11 ohms, together with range-changing resistors and switch, to give range multipliers of 1.0, 0.1, and 0.01.

(a) Determine the correct resistances of R_1, R_2, R_3, and R_4.

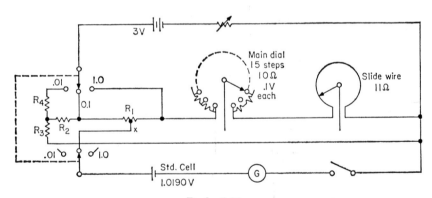

Prob. 7-20.

(b) Determine the tap point "x" on R_1, so that the potentiometer may be standardized on Range 0.1 against a standard cell of 1.01900-v emf.

7-21. Design the resistor system involved in the range-changing circuit of Fig. 7-6. The problem may be simplified by first computing the resistances needed to give the desired currents (20, 2, and 0.2 ma, respectively) through the dial resistances for the 1, 0.1, and 0.01 positions of the range switch, ignoring the tap points for the standard cell. The connection points for the standard cell may be computed later. The total working current on all ranges is 22 ma.

CHAPTER 8

Applications of Potentiometers

8-1. Comparison of Standard Cells

Standard cells are used in numbers in various circuits and instruments, so it is important to check the cells periodically, measuring an old or uncertain cell against a new one. These comparison measurements can be made very well on the general purpose potentiometers described in the previous chapter. These instruments can be read to 0.00001 v by interpolation in the smallest divisions for the slide-wire types, and directly on the last dial of the dial types. Closer determination is not significant unless the temperature is controlled, for the voltage changes by about 0.00001 v for a 1°C change in temperature.

While a standard cell may be measured on a potentiometer just as any other voltage source, a preferred method is to buck it against the cell that is being used as the standard, and measure the *difference* voltage. A general-purpose potentiometer is set up in the usual way, and standardized against any cell of reasonable accuracy. The unknown cell and the cell used as standard are connected in series-opposition to the "Unknown EMF" terminals of the potentiometer, and the difference voltage is measured. The difference subtracted from (or added to, as the case may be) the voltage of the standard, gives the voltage of the unknown. Since only the small difference is being measured, the standardization of the potentiometer is not critical, and the measurement may be made on a low range (Range 0.1, or even 0.01), thus giving definite readability to the last digit of interest in the standard cell voltage. In this way, the voltage of an unknown cell may be determined to practically the same accuracy to which the reference cell is known. There are "standard cell comparators" made, which are convenient if the volume of measurement is large. However, the method outlined above may be carried out on a general-purpose potentiometer to an entirely adequate accuracy, and serves well for the operations of most laboratories.

Precautions must be taken in handling cells to avoid the possibility of short-circuiting. Polarity must be checked carefully in connecting the standard cell and the working battery to the potentiometer. As an added precaution, a high galvanometer series resistance should be used for the first trial. The sequence in handling the connecting leads can also be important. In making connections connect to the cell *last;* in removing connections disconnect from the cell first.

8-2. Measurement of Voltage

The idea of voltage measurements was discussed to a considerable extent in the chapter on the potentiometer, since the instrument is primarily a voltage measuring device. Voltages within the range of the potentiometer are measured by direct comparison with voltages of the resistance-network system, without drawing current from the unknown. Higher voltages require a potential divider network, so that the potentiometer measures a definite fraction of the unknown. The potential divider (the "volt box" of par. 9, chap. 7) draws current from the measured source, so the no-current feature of the usual potentiometer measurement is lost. A straight high-voltage potentiometer could be made, but there is no demand for it, and the design would have several undesirable features. The current drain of the volt box is ordinarily of no importance.

Voltage measurements by potentiometer are made in chemical work, such as thermocouple emf and electrode potentials. In this case both the accuracy and the absence of current drain of the potentiometer are valuable features.

8-3. Measurement of Current

Current is measured by a potentiometer in terms of the voltage drop across a known resistor (a *four-terminal* resistor, for low values) placed in the circuit of the unknown current. From the standpoint of working standards in the laboratory, current is a quantity that is derived from the voltage of a standard cell and known resistances. This is in contrast to the absolute method of measurements, in which resistance and current are the quantities measured primarily; voltage then is derived from them.

8-4. Calibration of a Voltmeter*

A potentiometer provides the means of calibrating a d-c (or electro-dynamometer type a-c) voltmeter in terms of the emf of a standard cell and the known resistances in a potential-divider system. A circuit for the purpose is shown in Fig. 8-1.

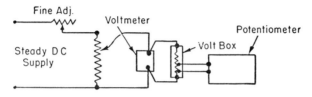

Fig. 8-1. Calibration of a voltmeter by a potentiometer.

If the voltmeter is receiving its initial calibration, the voltage impressed on the instrument is set at one of the major voltage divisions by means of the rheostats. Adjustment is made until the potentiometer indicates that the desired voltage is obtained, and the division mark is placed on the scale. Calibration is carried out at a selected number of cardinal points, and the intermediate marks are interpolated.

If the voltmeter has been calibrated at a previous time, and a check calibration is desired, the pointer of the meter is adjusted by the rheostat to a major point on the scale, and the potentiometer is balanced to give a reading of the correct voltage. The correction for this point is the true value minus the scale reading. When the readings have been taken at the required points with increasing voltage and then with decreasing voltage, a correction curve is plotted, giving correction as ordinate against scale reading as abscissa. The observed points are joined by straight lines, as nothing is known of the variation between points, and there is no reason to expect a smooth variation for the curve as a whole. Note that the correction as defined above is the quantity that must be *added* to the observed value to give the true value.

The process of calibration, particularly the balancing of the potentiometer, is made quite slow and tedious if the d-c supply voltage is unsteady. As the potentiometer can be read to one or two more figures than the voltmeter, it is a help to keep the galvanometer at somewhat reduced sensitivity and not to attempt to read the last figure on the potentiometer.

The potentiometer method is the basis of all voltmeter, ammeter, and wattmeter calibration. However, it is more accurate than needed for ordinary working instruments and slow (unless a deflection potentiometer

---

*See chap. 17, par. 34-36.

is used). Accordingly, the potentiometer is used to calibrate laboratory standard instruments, which are large long-scale well-made meters that are usually guaranteed to be accurate within 0.1 per cent of their full-scale range. Ordinary laboratory and switchboard instruments are calibrated or checked by comparison with a standard instrument. A laboratory standard voltmeter is shown in Fig. 8-2. It has a scale approximately 12 in.

Fig. 8-2. Weston Model-5 Laboratory Standard d-c voltmeter. (*Courtesy of the Weston Electrical Instrument Corp.*)

long with 300 scale divisions. There is a set of six circular arcs and a diagonal line between arcs within each scale division, as an aid in interpolation. If, for example, the instrument has a 150-v range, each division is 0.5 v, with the interpolation marks making readings possible to 0.1 v. The scale is backed by a mirror to help avoid parallax in taking readings.

If a deflection potentiometer is available, it may be used directly for the calibration of each working voltmeter or ammeter. The speed of operation is nearly as great as for the comparison method.

8-5. Calibration of an Ammeter*

The circuit for the calibration of a laboratory standard d-c ammeter is indicated in Fig. 8-3. A standard shunt of suitable resistance value and current-carrying capacity is connected in series with the ammeter, and the voltage across the shunt is measured by the potentiometer. The current can be determined very accurately in this way, as both the resistance of the

---

*See chap. 17, par. 34-36.

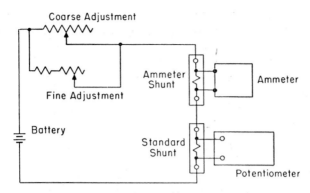

Fig. 8-3. Calibration of a d-c ammeter by a potentiometer.

shunt and the voltage drop across it are measurable to a high degree of accuracy.

The general method of conducting the calibration is similar to that for the voltmeter, and the discussion will not be repeated.

8-6. Measurement of Temperature by Thermocouple

A difference of potential exists between two metals that are placed in contact. If a complete circuit is made of the two metals and if all parts of the circuit are at the same temperature, no current flows, as the emf's at the two junctions are equal and in opposite directions around the circuit. If one junction is heated its emf is increased, and the two junctions no longer balance completely. The current flow may be measured, or the circuit may be opened at one place and attached to a potentiometer to measure the net emf acting around the circuit. This effect may be used to determine the temperature of one junction if the temperature of the other junction is known.

Thermocouples are a very convenient and useful means of measuring temperature in many different applications. They are small, light, relatively cheap, (except for platinum couples), can be mounted at a considerable distance from the indicating instrument, and can be used in places inaccessible to ordinary thermometers. They may be imbedded in the field coils of electrical machines, and may be mounted on bearings or other places, in which the temperature is a matter of importance. They have many industrial applications, such as ovens, furnaces, and so on.

Milliammeters are sometimes used as indicators with thermocouples. However, the accuracy is poor, as the current depends on the resistance as well as the emf. The resistance of the thermocouple circuit may change due to the effect of temperature on the leads, or due to poor contact at binding posts or corrosion of the thermocouple wire. A potentiometer is a much better instrument to use, since it draws no current and hence is independent of the resistance of the circuit. Many special potentiometers are made for this purpose.

Metals differ in their "thermoelectric power." Tables giving data on this subject are available in many books. Information is frequently given in the form of thermal emf's of different metals against one metal, such as platinum or lead, as reference. Metals in the list can then be compared by subtracting the values against the common reference.

Different combinations of metals are used for different purposes and in different temperature ranges. For low-temperature work a couple consisting of copper to constantan or iron to constantan is frequently used. Constantan was mentioned in chap. 4 as having a small resistance change with temperature, but of being disqualified as a precision resistor material due to its high thermal emf against copper. This property, however, makes it useful in thermocouples. It may be used up to about 300°C. The combinations of iron to constantan, chromel to alumel and platinum to platinum rhodium are usable to higher temperatures.

The following table gives approximate temperature ranges. Exact limits cannot be stated, as the upper limit is a matter of giving a reasonably long life, which depends also on other factors, such as the size of wire, the protection given to the wires, and the nature of the surrounding atmosphere. The table also shows the emf of these couples for the cold junction at 0°C and the hot junction at 100°C in order to give some idea of the relative values of emf. Thermocouple tables should be consulted for other temperature ranges since the relationship of emf to temperature is not linear.

Thermocouple	Operating Temperature Range	Millivolts 0 to 100°C
Copper to constantan	−250 to +300°C	4.28
Iron to constantan	−250 to +700°	5.28
Chromel to alumel	−250 to +1000°	4.10
Platinum to platinum-rhodium	0 to +1600°	0.643

The relationship of emf to temperature for a thermocouple is only approximately linear, so tables must be consulted for accurate values. The tables give the emf of a hot junction against a reference junction maintained at a definite temperature, usually 0°C. In some precision measurements the cold junction is kept at 0°C by immersion in ice water, but for most work a computed correction is made by adding the emf corresponding to the room-temperature junction to the emf observed on the potentiometer. The temperature corresponding to this computed emf of the hot junction will be found in the tables. (This discussion is stated in terms of a measured temperature higher than room temperature, but the same method applies equally well below room temperature.) Some thermocouple potentiometers have a manually set correction for reference-junction emf. Others incorporate a temperature-sensitive element in the circuit to compensate automatically for the reference junction.

The general-purpose potentiometers discussed in chap. 7 have low ranges that are used frequently for thermocouple work. These instruments, however, are intended primarily for measurements under laboratory conditions. Figure 8-4 shows a portable potentiometer that may be moved about easily, as it has a self-contained galvanometer, standard cell, and battery. Instruments of this type are used a great deal in industrial plants and for research measurements of many kinds.

The chief use of portable potentiometers is in thermocouple measurements. Some instruments are, in fact, calibrated directly in temperature

Fig. 8-4(a). A portable precision potentiometer.

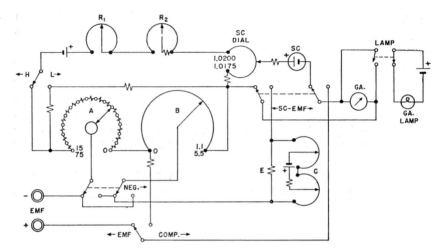

Fig. 8-4(b). Schematic diagram of portable precision potentiometer shown on page 247. (*Fig. 8-4 Courtesy of the Minneapolis-Honeywell Regulator Co., Rubicon Instruments.*)

for a particular type of thermocouple; this is a convenient arrangement when temperature measurements of the same range are being made continually, but it does limit the use of the instrument to one kind of measurement with one kind of couple. The general purpose potentiometer has a scale calibrated in millivolts.

Many small portable potentiometers do all the balancing on a single slide wire. The potentiometer of Fig. 8-4, in contrast, has a main dial of 15 steps and a slide wire, equivalent in effect to a 200-in. slide wire, thus giving high precision of reading. The two-range circuit switching gives scales of 0-16.1 millivolts on the low range, and 0-80.5 millivolts on the high range. Another feature is the "compensating circuit" which can be set to the voltage corresponding to the reference junction temperature; thereafter the measuring dials read the voltage corresponding to the measuring junction temperature.

8-7. Self-balancing Potentiometers

The potentiometer principle has such merit in accuracy and dependability and such wide range of application that instrument makers saw the desirability of producing a potentiometer that could be used in commercial work without requiring the constant attention of an operator. Several makers now produce "self-balancing potentiometers," which operate on the true potentiometer principle with automatic balancing, draw a curve of the quantity being measured, and can be mounted on a switchboard. In some cases the balancing process is initiated by a galvanometer pointer controlling a mechanical drive. In other instruments an electronic amplifier responds to an unbalanced condition and drives a motor to move the potentiometer slider to balance.

Figure 8-5 shows a potentiometer of the electronic type. It employs a

Fig. 8-5. Speedomax G self-balancing potentiometer. (a) External view. (b) Schematic circuit diagram, showing use for temperature measurement. (c) Cover open, showing the slide-wire mounting. (d) Mechanism open, showing motors, standard cell, and amplifier. (*Courtesy of Leeds & Northrup Co.*)

Fig. 8-5(a).

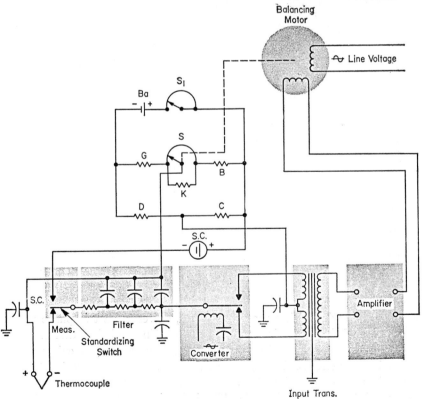

Fig. 8-5(b).

Fig. 8-5(c).

Fig. 8-5(d).

circular slide wire with a moving contact, driven by a two-phase motor. One phase of the motor is supplied from the line and the other from the amplifier, in proportion to any unbalanced voltage that exists in the potentiometer circuit. The motor drives the slider in a direction to eliminate the unbalance and, at the same time, moves a pen on the chart paper and a pointer on a visible scale. The balancing process is quite rapid, so the chart record corresponds to the quantity being measured except for a momentary lag for very rapid variations. The response is fast, the pen being able to move from one side to the other of a ten inch chart in about two seconds.

Instruments of this sort are exceedingly versatile, as they can be used to measure many different kinds of variables. We may mention a few applications, without attempting to present a complete list.

(a) Direct-current, by voltage drop across a shunt.

(b) Direct-voltage, by voltage-dividing resistors.

(c) Revolutions per minute, by tachometer voltage.

(d) Temperature, by thermocouple voltage or by resistance-thermometer bridge.

The instrument can be used not only to indicate a quantity, but also to initiate a control process, such as the control of heat to maintain a constant furnace temperature. Combined instruments can be arranged for totalizing separate instruments, or for reading d-c power, by the addition of a "transmitting slide wire" mounted to move with the potentiometer slider.

These instruments have a standard cell as the basic reference in the measuring process. The working current is furnished by a dry cell, and standardizing of the working current is carried out automatically, usually at approximately 1-hour intervals. The potentiometer may be designed for full-scale deflection for different amounts of voltage; 50 mv is common, but other values may be obtained.

Figure 8-5(a) shows the outside view of a self-balancing potentiometer, which in this case is calibrated for temperature measurements. A visual indication is given by the pointer, and a permanent record from one or more stations is made on the 10-in.-wide chart. View (c), with the cover and chart paper swung open, shows at the center the circular slide wire with dust cover, and to its left the small rheostat that regulates the working current. View (d) shows the entire mechanism open. The large motor drives the chart, and the smaller motor drives the pen, the pointer, and the balancing arm on the slide wire. The standard cell can be seen at the lower right corner, and to its left the dry cell that furnishes the working current. The amplifier and power unit is shown at the back of the case.

Sketch (b) is a schematic diagram showing one of the many uses to which this instrument can be put, in this case temperature measurement by means of a thermocouple. A bridge is formed of two resistors and the slide wire. (The circuit for regulating the working current is not shown here.) The thermocouple is in the detector branch. The system is in equilibrium when the slide-wire contact is in position to give a voltage unbalance just equal to the thermocouple voltage. If the temperature and thermocouple voltage now change, the unbalanced voltage, working through the amplifier, causes the motor to drive the slide-wire contact to the new balance point. The converter is a vibrator device, driven synchronously from the 60-cycle line, which changes the d-c unbalance into an a-c that is amplified and then impressed on the control winding of the motor. The converter makes possible the use of an a-c, rather than a d-c, amplifier.

PROBLEMS

8-1. The calibration of a microammeter (full scale = 20 microamperes) is to be checked in the laboratory, and no standard meter of suitable range is available for comparison. Show in detail in a circuit diagram the set-up you would use for this calibration, drawing upon equipment that may be expected to be present in a well-stocked laboratory. Tell the magnitude and accuracy requirement of each component that enters the calibration in a numerical way. The meter being checked is rated to be in the 0.5 per cent accuracy class, so it is required that the calibration process shall not introduce an uncertainty of more than 0.1 per cent.

8-2. A resistor of approximately 100 megohms is to be measured in the laboratory to an accuracy within 0.25 per cent, with 300 v impressed upon it. A "megohm bridge" is available, but its accuracy rating is 2 per cent. Application of 300 v is not permissible on the standard Wheatstone bridges.

Devise a method of measurement to meet the conditions. Give a diagram of the circuit you would use, and specify accuracy requirements, making use of equipment that may reasonably be expected in a good laboratory, with realistic ratings of accuracy.

8-3. A wattmeter designed for special-purpose measurements has a current-coil rating of 0.10 amp and a potential-circuit rating of 10 v. The full-scale marking is one watt on a 100-division scale. The calibration of this instrument is to be checked in the laboratory, but no standard wattmeter of such low scale is available for direct comparison. The laboratory is well supplied with standard measuring equipment of good quality.

Devise a method of calibration that will not introduce an uncertainty greater than one-tenth of a division on the 100-division scale. Show the complete circuit arrangement that you would use, and indicate numerical values and the accuracy requirements of the components. (Don't specify unusual equipment or accuracies that are beyond reasonable commercial expectations.)

CHAPTER 9

Alternating-Current Bridges:

Principles; Basic Circuits

9-1. Application of A-C to Bridge Measurements

Measurements of inductance, capacitance, and of some other quantities may be made conveniently and accurately by a-c bridge circuits. Some of these measurements may be made by d-c methods, but the a-c bridge has advantages that favor its use. Historically, some of the a-c bridges are the outgrowth of the use of d-c on the same bridge, with an initial balance on steady d-c, and a second balance with "make-and-break" of the current supply to the bridge (with adjustment to eliminate the momentary kick of the galvanometer). Conversion to a-c not only made these bridges more convenient to use, but also opened the possibility of other circuits that could not be balanced by the d-c method. The credit for the definite change from d-c to a-c belongs to Max Wien, who in 1891 laid down the basic principles and published a collection of bridge networks.

The simple form of a-c bridge bears a strong resemblance to the d-c Wheatstone bridge: it consists of four arms, a power supply, and a balance detector. The power source furnishes alternating current of the desired frequency and suitable magnitude to the bridge. Power may be taken from the power line for low-frequency measurements. Higher frequencies are usually supplied by vacuum-tube oscillators or, sometimes, by buzzers or tuning-fork devices. A common form of detector for the audio-frequency range is a telephone headset. Galvanometers, built for low frequencies, are either of the separate field excitation type or the vibration galvanometer. Rugged and convenient detectors can be constructed for low or high frequencies by using a vacuum-tube amplifier followed by a meter or a cathode-ray oscilloscope as indicator.

9-2. Derivation of Bridge Equations

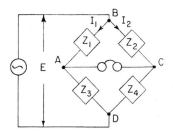

Fig. 9-1. Four-arm a-c bridge.

A bridge is shown in general form in Fig. 9-1, the arms being indicated as impedances of unspecified sorts. The detector is represented as a telephone headset. We shall derive the generalized conditions of balance for this bridge.

Balance is secured by adjustment of one or more of the bridge arms. Balance is indicated by zero response of the detector, which means that points A and C are at the same potential at all instants. We shall use the complex notation in deriving the equations of the bridge, as this form of expression is compact and convenient and lends itself to generalized forms showing the basic conditions of balance. Bold-face type will be used to represent quantities in the complex notation; these may be either impedances and admittances or currents and voltages, for which the complex notation is a means of expressing the relationships of time vectors ("phasors").

The condition of zero difference of potential from A to C requires that the voltage from B to A equal the voltage from B to C both in magnitude and in time relationship, or phase. This may be expressed in the complex notation

$$I_1 Z_1 = I_2 Z_2 \tag{1}$$

where currents and impedances are complex quantities. With no current flowing in the detector at balance

$$I_1 = \frac{E}{Z_1 + Z_3} \tag{2}$$

and

$$I_2 = \frac{E}{Z_2 + Z_4} \tag{3}$$

By substitution of (2) and (3) in (1) we obtain

$$Z_1 Z_4 = Z_2 Z_3 \tag{4}$$

or

$$\frac{Z_1}{Z_2} = \frac{Z_3}{Z_4} \tag{4a}$$

or $$Y_1Y_4 = Y_2Y_3 \tag{4b}$$

This is the general condition for balance of the four-arm bridge. It may easily be shown that the balance equation, just as in the d-c bridge, is not affected by interchange of power supply and detector although the sensitivity may be greater one way than the other. The balance equation may be expressed in any of the above forms, or in still other modifications. Equation (4a) may look more familiar; however, (4) is more convenient in many cases. A combination of admittance and impedance may be employed in one equation, as admittance is more convenient for arms consisting of parallel components, and impedance for series groupings.

Equation (4) states that the product of impedances of one pair of *opposite* arms must equal the product of impedances of the other pair of *opposite* arms, with the impedances expressed as complex numbers. It is instructive to change this equation by conversion of impedances to the polar form. In the polar form,

$$Z_1 = Z_1 \epsilon^{j\theta_1} = Z_1/\underline{\theta_1}$$

for short, where Z_1 represents the magnitude of the complex impedance, and θ_1 its angle. If similar forms are written for all impedances and substituted in (4), the following result is obtained:

$$Z_1Z_4/\underline{\theta_1 + \theta_4} = Z_2Z_3/\underline{\theta_2 + \theta_3}, \tag{5}$$

since the magnitudes are multiplied and angles added in multiplication of complex numbers.

Equation (5) shows two requirements that must be met in the balancing of a bridge. The first and obvious requirement is that the magnitudes must satisfy the relationship $Z_1Z_4 = Z_2Z_3$. The second relationship, which is not always as fully appreciated, is

$$\theta_1 + \theta_4 = \theta_2 + \theta_3 \tag{6}$$

The sum of the phase angles of one pair of opposite arms must equal the sum of the angles of the other opposite pair, for balance. It is necessary, of course, to distinguish between inductive and capacitive impedances by giving the angle a plus sign in one case, minus in the other.

Equation (6) is useful in several ways. It may enter in the derivation of bridge equations or in the calculation of bridge problems. It is very useful in the inspection of a tentative bridge circuit, in telling whether balance is *possible* with the arms as proposed.

We shall apply the above equations to a study of some of the most common and most useful of the bridge circuits. The balance formulas are gathered together in tabular form in Appendix III.

9-3. Capacitance Comparison Bridge

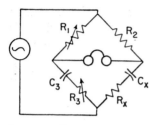

Fig. 9-2. Capacitance comparison bridge.

Figure 9-2 shows a common form of bridge circuit for measurement of an unknown capacitance by comparison with a known capacitance. The resistances shown in the capacitance arms are not, in general, separate resistance units, but represent resistances inherent in the capacitors. As shown below, it may be necessary to add resistance to one capacitor arm to secure balance.

For substitution in eq. (4) we may write the impedances for this case:

$$Z_1 = R_1 \qquad Z_2 = R_2$$

$$Z_3 = R_3 - j\,\frac{1}{\omega C_3} \qquad Z_x = R_x - j\,\frac{1}{\omega C_x},$$

and substitute in (4)

$$R_1\left(R_x - j\,\frac{1}{\omega C_x}\right) = R_2\left(R_3 - \frac{j}{\omega}\frac{1}{C_3}\right)$$

$$R_1 R_x - jR_1\,\frac{1}{\omega C_x} = R_2 R_3 - jR_2\,\frac{1}{\omega C_3} \qquad (7)$$

Equality of two complex numbers requires that the real parts be equal on the two sides of the equation, likewise the j terms. Therefore

$$R_1 R_x = R_2 R_3 \qquad \text{or} \qquad R_x = \frac{R_2 R_3}{R_1} \qquad (8)$$

$$R_1\,\frac{1}{\omega C_x} = R_2\,\frac{1}{\omega C_3} \qquad \text{or} \qquad C_x = C_3\,\frac{R_1}{R_2} \qquad (9)$$

We see that *two* quantities, R_x and C_x, are determined and that there are *two* balance conditions. This requires that two quantities be variable in this bridge as, for example, R_1 (or R_2) and R_3. Suppose that we choose to secure balance by adjustment of R_1 and R_3. Since we assume that arms 3 and X are predominantly capacitive, with minor resistive effect, we give first attention to the capacitive balance, which from (9) requires adjustment of R_1. (Leave R_3 at zero for this stage.) Adjust R_1 to give minimum sound in the phones; this will be a minimum and not a true null because R_3 is presumably not at its correct setting. We now turn attention to R_3 to decrease still further the hum in the phones. If C_3 is a high-grade standard capacitor, its resistance is probably lower than C_x, so we first try added resistance in arm 3; if this makes the balance worse we know added resistance is needed in arm X instead.

We now go back to R_1, for with the decrease in residual tone we should be able to find the balance point for R_1 more precisely than before. Alternate adjustments of R_1 and R_3 should soon yield practically zero response of the detector. (Later we shall discuss grounding and shielding of the bridge elements. This is necessary, in general, for complete and precise balance.)

The process described above is representative of the procedure in balancing most a-c bridges. There are two adjustments, one resistive and the other reactive, that must be made to secure balance. For the usual magnitude-responsive detector these adjustments must be made alternately until they "converge" on the balance point. The rapidity of convergence depends on the circuit arrangements and also on the characteristics of the unknown. We shall discuss this matter more fully later, but for the present it will be sufficient to mention that convergence is best when both adjustments are in the same arm. This could be done in Fig. 9-2 by making C_3 and R_3 variable (R_1 fixed), but in general variable capacitors are not available with as high accuracy as fixed standards, so the fixed standards are preferred.

Beginners frequently have difficulty in securing a good null response in the detector. The reason for the poor balance for the main capacitance or inductance determination is the failure to appreciate the need for a *good resistance balance*. It is only when *both* adjustments are correctly made that a sharply defined null point is secured. Also, it is important to recognize which component of balance is the major one and to make adjustment for it first.

Note in eqs. (8) and (9) above that the frequency does not appear in the final result and, hence, does not need to be known exactly. The balance is independent of frequency, provided that the constants of the elements of the bridge do not change appreciably with frequency.

9-4. Inductance Comparison Bridge

Figure 9-3 shows a bridge for measurement of an unknown inductance by comparison with a standard inductance. The method is basically similar to the capacitance comparison of the preceding section, so derivation of the equation is left to the reader. The balance equation for inductance is

$$L_x = L_s \frac{R_2}{R_1} \qquad (10)$$

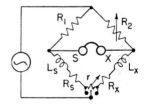

Fig. 9-3. Inductance comparison bridge.

The figure shows a variable resistance "r" connected by a switch so it may be included either with the standard or the unknown, as required for balance. The resistance components are larger in proportion for coils than for capacitors, so the resistance balance is more important here than in the previous bridge. The resistance equations for the two positions are

r with S $$R_x = \frac{R_2}{R_1}(R_s + r) \tag{11a}$$

r with X $$R_x + r = \frac{R_2}{R_1} R_s \tag{11b}$$

If a calibrated variable inductor (an "inductometer") of suitable range is available it may be varied, instead of R_2, as indicated in Fig. 9-3. Inductometers consist of fixed and movable coils, so mounted that the total inductance changes with the position of the movable coils. One well-known type is the Brooks inductometer, which is discussed in chap. 13.

9-5. Measurement of Inductance by Capacitance — Maxwell Bridge

The Maxwell bridge has the valuable feature that it permits measurement of inductance in terms of capacitance. A capacitor has some points of advantage as a standard, in comparison with an inductor, as it gives practically no external field, is more compact, and is easier to shield.

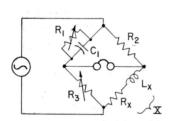

Fig. 9-4. Maxwell bridge for measurement of inductance in terms of capacitance.

The circuit diagram is shown in Fig. 9-4. The impedance of arm 1 may be worked out by the usual formula for parallel impedances and substituted in (4), or admittance may be used in a modified form of (4). Using the latter alternative for illustration,

$$Y_1 = \frac{1}{R_1} + j\omega C_1$$

$$Z_2 = R_2; \ Z_3 = R_3$$

$$Z_x = R_x + j\omega L_x$$

Substitution in the equation $Z_x = Z_2 Z_3 Y_1$ gives

$$R_x + j\omega L_x = R_2 R_3\left(\frac{1}{R_1} + j\omega C_1\right)$$

By separation of real and j terms

$$L_x = R_2 R_3 C_1 \tag{12}$$

$$R_x = \frac{R_2 R_3}{R_1} \tag{13}$$

The units used here are ohms for resistance, henrys for inductance, farads for capacitance.

This is a convenient and useful bridge for the determination of inductance within certain ranges. Difficulty with it arises when the unknown coil has a high value of Q ($Q = \omega L/R$), for then the parallel resistance R_1 becomes very high — frequently higher than available decade boxes — and also the resistance balance is difficult to determine precisely. The need for a high R_1 may be seen from eq. (13), or from the study we made above in terms of phase angles. Since arms 2 and 3 are pure resistances, the angles of arms 1 and X must add to zero; that is, the capacitive angle of arm 1 must equal the inductive angle of X. This requires that R_1 be large in comparison with the reactance of C_1, in case X is a high-Q coil. The Hay bridge, which is a modification of the Maxwell bridge, is more desirable in this case.

The Maxwell bridge works well for low values of Q, except for unusually small values ($Q < 1$), which is a range that may include anything from inductive resistors to coils measured below their usual operating frequency; for example, radio-frequency coils show low Q values when measured on audio frequencies. For this type of measurement the Maxwell bridge has very poor convergence of balance and gives the effect known as "sliding balance" when R_1 and R_2 (or R_3) are used for the balance adjustments. (If balance is made by C_1 and R_1 there is no interaction, but a variable capacitor is sometimes not suitable.) The term "sliding balance" describes a condition of interaction between the controls, so that when we balance with R_1, then go to R_3 and back to R_1, we find a new apparent balance point; that is, the balance point appears to move, or "slide," and settles only gradually to its final point after many adjustments. The general problem of convergence to balance will be considered in a later section (see par. 9-17), but we may give consideration now to the special problem involved in the present case.

For balancing accomplished as in Fig. 9-4, R_3 enters into both balance equations. A preliminary inductive balance is made with R_3, and then R_1 is varied to give a resistive balance which is dependent on the R_3 setting as shown by (13). Accordingly, when R_3 is changed for a second inductive balance, the resistive balance is disturbed, and moves to a new value, giving slow convergence to final balance, particularly for a low-Q coil, for which the resistance effect is prominent. An interesting development* in this connection is made in the General Radio Type 1650-A bridge, in which R_1 is coupled to move with R_3 during inductive balancing to keep their ratio constant. For resistive balancing, R_1 is moved independently of R_3. The "sliding" effect is thus eliminated, and balancing can be accomplished in a small number of operations, even with low-Q coils.

---

*"Orthonull," a patented feature.

9-6. Hay Bridge for Measurement of Inductance

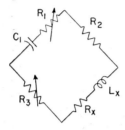

Fig. 9-5. Hay bridge measurement of inductance in terms of capacitance (oscillator and detector omitted from diagram).

The Hay bridge, shown in Fig. 9-5, differs from the Maxwell only in having a resistance in series with the standard capacitor, instead of in parallel with it. For large phase angles this change requires a *low* series resistance instead of a very high parallel resistance. The Hay circuit gives more convenient values of resistance and better balancing for high-Q coils.

The balance equations are altered by the change in the circuit. We may use the same method as before and equate the products of the impedances of opposite pairs of arms.

$$\left(R_1 - j\frac{1}{\omega C_1}\right)(R_x + j\omega L_x) = R_2 R_3 \tag{14}$$

Equation (14) may be separated into the two equations

$$R_1 R_x + \frac{L_x}{C_1} = R_2 R_3 \tag{15}$$

$$\omega L_x R_1 - \frac{R_x}{\omega C_1} = 0 \tag{16}$$

Both (15) and (16) contain L_x and R_x, so we must solve them simultaneously for the constants of the unknown. This gives

$$L_x = \frac{R_2 R_3 C_1}{1 + \omega^2 C_1^2 R_1^2} \tag{17}$$

$$R_x = \frac{\omega^2 C_1^2 R_1 R_2 R_3}{1 + \omega^2 C_1^2 R_1^2} \tag{18}$$

The equations for this bridge differ from previous cases in that they contain ω. It thus appears necessary to know the frequency accurately in order to compute L_x and R_x. However, it may be shown that the term $\omega^2 C_1^2 R_1^2$ is small in comparison with unity in case the coil has a high value of Q. The relationship to the coil Q arises from the fact that the phase angles of arms 1 and X must be equal in magnitude, since arms 2 and 3 are pure resistances. The tangents of the phase angles are

$$\tan \theta_1 = \frac{X_{C1}}{R_1} = \frac{1}{\omega C_1 R_1}$$

$$\tan \theta_x = \frac{X_{Lx}}{R_x} = \frac{\omega L_x}{R_x} = Q_x$$

Therefore $$\omega C_1 R_1 = \frac{1}{Q_x} \qquad (19)$$

The term $\omega^2 C_1^2 R_1^2$ in the denominator of (17) has the value 0.01 in case $Q = 10$ and is smaller for higher values of Q. Thus in some cases it is possible to drop this term without causing excessive error. In other cases we may not wish to ignore this term entirely; however, it is of such minor importance that it may be computed with sufficient accuracy from an approximate value of ω. Note that if the $\omega^2 C_1^2 R_1^2$ is dropped from (17), the equation for L_x is the same as for the Maxwell bridge.

For low Q, on the other hand, the $\omega^2 C_1^2 R_1^2$ term in the denominator becomes very important; not only does it complicate the computation, but it requires that the bridge frequency be known to close limits (to about half the tolerance of the desired L_x measurements). Moreover, the numerator is no longer a close measure of L_x, but may be several times as great. For example, if $Q = 0.5$, the denominator is $(1 + 4)$, so the numerator is five times as great as would be needed for the Maxwell bridge relationship. The Hay bridge is thus poorly suited to finding the constants of a low-Q inductor in series components, with respect to difficulties of computation, and it gives poor convergence if balancing is accomplished as in Fig. 9-5. (The equations are simple in parallel components.)

Figure 9-6 illustrates the possibility of using the same bridge elements and combining them in different ways to achieve bridge circuits for a variety of measurements. Part (d) is the Maxwell circuit, and (e) the Hay arrangement, except that the inductive unknown is shown in parallel components. Parts (a) and (b) are used for capacitance measurements in series or parallel components, respectively, (a) being the same as shown in Fig. 9-2. These circuits are the basis of the General Radio 1650-A bridge.

9-7. Comparison of Maxwell and Hay Bridges

The Maxwell bridge has the obvious advantage of simpler balance equations, and this factor indicates its use for coils of low Q, for which the Hay equations have awkward characteristics. It is not so desirable for high-Q coils, as it requires large values, frequently inconveniently large, for R_1 and does not give a sharp resistance balance.

The Hay bridge is more desirable in the higher ranges of Q as it avoids the difficulties of the Maxwell bridge. In this range the equations may be simplified, or at least do not require laborious computation. No definite dividing line can be drawn between the two arrangements, for there is an area of overlap in which either may be used. If the coil Q is above 10,

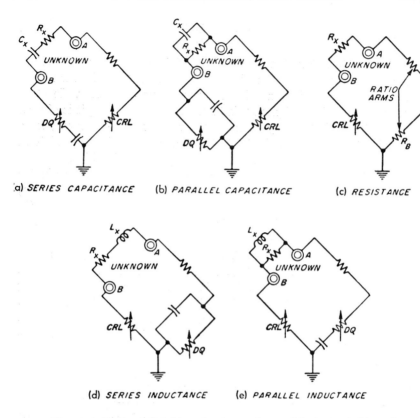

Fig. 9-6. Variety of bridge circuits made possible by switching of components. (*Courtesy of General Radio Co.*)

the advantage is probably with the Hay circuit. It is usually easy to arrange a switch or plug scheme so that the resistance R_1 can be placed either in parallel or in series with the standard condenser, in order to use either form of bridge in a given measurement.

Both bridges give poor convergence to balance for a low-Q inductor if R_1 and R_3 are the variable elements in Figs. 9-4 and 9-5. If a low-Q inductor is to be measured in series components, a good arrangement is the Maxwell bridge with C_1 and R_1 as the adjustable elements; sliding balance is avoided and simplicity of calculation is retained.

The comparison of Maxwell and Hay bridges with respect to the form and simplicity of balance equations is reversed if the inductor is analyzed in terms of parallel components. The reader will find it interesting to try this analysis. We usually specify coil characteristics in terms of series L and R, but the parallel form is useful in some applications.

9-8. Owen Bridge for Measurement of Inductance

The Owen bridge is another circuit for the measurement of inductance in terms of a standard capacitor. One arm consists of the standard capacitor only, and an adjacent arm contains a resistance and capacitance in series, as shown in Fig. 9-7. The possibility of balancing the bridge is indicated by a study of the phase angles of the arms. Arm 1 has a capacitive angle of 90°, and X an inductive angle less than 90° due to the presence of resistance R_x; hence the sum of this pair is slightly on the capacitive side. Arm 2 has an angle of zero, so balance may be secured if arm 3 has a capacitive angle equal to the difference of the angles of arms 1 and X.

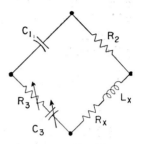

Fig. 9-7. Owen bridge for determination of inductance.

The equations for the Owen bridge may be derived as in other cases. The product of impedances of opposite arms is

$$-j\frac{1}{\omega C_1}(R_x + j\omega L_x) = R_2\left(R_3 - j\frac{1}{\omega C_3}\right)$$

Separation of real and j terms gives

$$L_x = C_1 R_2 R_3 \tag{20}$$

$$R_x = C_1 R_2 \frac{1}{C_3} \tag{21}$$

These equations are interesting in several respects. Equation (20) is identical with the Maxwell relationship. The factors C_1 and R_2 appear in both equations, so in addition to this common part L_x may be said to be measured in terms of R_3, and R_x to be measured in terms of C_3.

The Owen bridge is a useful and convenient method and has had a number of applications. It has one drawback in that it requires a decade capacitor, which is not always available. One point of merit is the fact that the two adjustable elements, R_3 and C_3, are in the *same* arm. This makes the reactive adjustment independent of the resistance adjustment and avoids the interlocking effect or "sliding balance," which sometimes causes difficulty when the two adjustments are in different arms.

The advantages and disadvantages of the Owen bridge may be summarized as follows:

Advantages

(a) R_3, used for determination of L_x, is a high-accuracy decade box (0.1 per cent or better).

(b) C_3, used for determination of R_x, is a decade capacitor. It is not so accurate (about 1 per cent accuracy rating) as R_3, but adequate for R_x, which is usually not so important as L_x.

(c) R_3 and C_3 are in the same arm, thus avoiding interdependence of adjustment.

Disadvantages

(a) A decade capacitor is required.

(b) The decade capacitor, C_3, tends to become rather large in the measurement of high-Q coils.

(NOTE: As an alternative, C_3 may be changed to a *small* value in *parallel* with R_3 if it is inconveniently large in the series arrangement. These two forms of Owen bridge are related much as are Maxwell and Hay with respect to resistance in arm AB. Problem 9-9 deals with the parallel Owen bridge.)

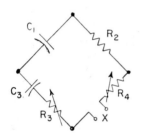

Fig. 9-8. Modified Owen bridge.

A modified form of Owen bridge may be used to avoid the disadvantages mentioned above. As shown in Fig. 9-8, a fixed capacitor (preferably adjustable by steps) is used for C_3, and an adjustable resistance, R_4 is added. It is used in this way:

(a) Short X, set $R_3 = 0$ and balance by adjustment of R_4. For this capacitance bridge

$$R_4 = \frac{C_1 R_2}{C_3} \tag{22}$$

(b) Insert X and rebalance by R_3 and R_4. Call the new value R_4'. For this case, following (21),

$$R_x + R_4' = \frac{C_1 R_2}{C_3}$$

From (22) $$R_x = R_4 - R_4' \tag{23}$$

as before $$L_x = C_1 R_2 R_3$$

Note that by this procedure we have eliminated the need of a decade capacitor and now use resistors for both adjustable elements. The fact that R_x is found by the difference of two readings of R_4 may seriously affect the accuracy, if R_4 is large in comparison with R_x. It is for this reason that optional values of C_3 should be available, so that a desirable combination may be selected.

9-9. Relationship of Maxwell, Hay, and Owen Bridges

The Hay and Owen bridges are modifications of the Maxwell bridge. That this is true may be seen most simply by consideration of the form to which each circuit reduces in the measurement of a pure inductance. They all simplify to a pure capacitance in the arm opposite the pure inductance, and the other two arms become pure resistances. The differences in the three bridges are in the methods used to balance the *resistance component* of the coil. The Maxwell bridge uses a parallel resistance in arm 1, Hay a series resistance in arm 1, Owen a series capacitance in arm 3, using the numbering system of the above diagrams. These differences are of some moment in particular applications, as they give different operating characteristics, different shielding conditions and different features with respect to calibration. The Owen bridge has been used successfully, as by the Bell Telephone Laboratories* in a bridge that has the advantages of having independent R and L adjustments and of being direct reading for inductance.

There is a method of classifying, or cataloguing, bridge circuits into "ratio-arm" and "product-arm" types. We shall talk more of this later, for it has value in some respects. In other respects the classification is artificial, as, for example, when applied to the three bridges discussed above. These bridges are placed in different categories, in spite of the strong family relationship discussed above — Maxwell and Hay are called product-arm bridges, and Owen a ratio-arm bridge.

9-10. Residuals

The arms of the foregoing bridges have been considered, in the main, to be composed of combinations of pure elements, although the presence of resistance in inductors was recognized, and also in capacitors in some cases. Actually, all elements possess something of all three properties, R, L, and C. A resistor, even with careful construction, has some inductive effect and some distributed capacitance which gives a shunting action. Reference to pure circuit elements is an idealization which is convenient and sufficiently accurate in many cases, but which is nevertheless only approximately true. The departure from truth becomes very important in some bridges.

The small amounts of the unwanted quantities that unavoidably remain in an impedance unit are referred to as the "residuals." Thus a resistor has residual L and C, a capacitance has residual L and R, and so forth. A study of residuals and their effect on bridge conditions will be made in chap. 13.

---

*References 6 and 8.

For the present we shall pretty much neglect these effects so that we may concentrate on the broad general principles of the bridges.

9-11. Wagner Earth Connection*

It has been assumed in the discussion so far that the four bridge arms are constant lumped impedances, with no interaction or stray effects between the elements, or between the various elements and ground. Actually, there are capacitances between the elements and from each element to ground. These capacitances shunt the various bridge arms and cause error in the results unless they are determined and included in the calculations. The error depends on the make-up of the bridge and may be negligibly small in some cases and of great importance in others. The stray capacitances are small, in general, and cause little error if their impedances are large in comparison with the impedances of the bridge arms. The stray capacitances are important in the measurement of small capacitances or of large inductances and become increasingly serious at higher frequencies.

The stray capacitances must, for accurate results, either be measured and included in computations or their effect on the bridge balance must be eliminated. It is frequently difficult to measure the stray values, as the paths are indefinite, or change with adjustments of the bridge. A valuable method of control is to surround the arms by shields connected to ground or to points of suitable potential. This does not eliminate capacitance — in fact, it generally causes an increase — but it does make capacitances definite in amount and position and susceptible to control.

Shielding will be studied in greater detail later. Another method, the use of special ground circuits, may reduce the effect of the stray capacitances on the balancing of the bridge. The subject is introduced in a preliminary way at this point to help in the early stages of laboratory work. It will be discussed more fully in a later chapter.

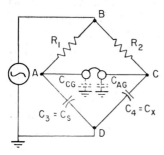

Fig. 9-9. Capacitances of detector to ground.

One troublesome capacitance is between detector and ground. If one side of the oscillator is grounded, as shown in Fig. 9-9, the capacitance from detector terminal C to ground is in parallel with branch CD. A capacitance of equal or unequal amount may exist from A to ground. Unless these two stray capacitances happen to be in the same

---

*Reference 1, p. 542; reference 7, p. 432.

ratio as C_s and C_x, they affect the balance setting and hence give an erroneous result.

The effect of detector capacitance to ground can be eliminated by removing the oscillator ground and using a connection that, in effect, brings the points A and C to ground potential. This is called the Wagner earth connection. Its application to a simple capacitance bridge is indicated in Fig. 9-10. It consists of two auxiliary arms, *similar to two of the bridge arms*, with the midpoint grounded. The bridge is given a preliminary balance, with the detector connected to A, and R_1 is adjusted to give minimum sound

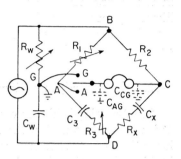

Fig. 9-10. Wagner earth connection on a capacitance bridge.

in the phones. The detector is then switched to G, and R_w is adjusted for minimum detector response. When the detector is switched back to A, some unbalance probably will be shown by the detector. R_1 and R_3 are adjusted for minimum, then the switch is thrown to G, and R_w readjusted. (A variable resistor may be needed in series with C_w for sharp ground-circuit null.) It is possible, by a few adjustments, to obtain practically zero response for both positions A and G. When this result is attained, points A and C are not only at equal potential, but *both* are at *ground* potential. Therefore the capacitances indicated by C_{AG} and C_{CG} have no effect, for there is no difference of potential across them. The ground circuit has a further effect. There are also capacitances from terminals B and D to ground (not shown on Fig. 9-10). When the ground circuit is applied, the capacitance C_{BG} is in effect in parallel with R_w, and C_{AG} in parallel with C_w. These capacitances play some part in the adjustment of the ground circuit, *but do not introduce currents into the bridge network* and hence do not affect balance conditions.

The same idea can be applied to other bridges, such as Maxwell or Hay. The grounding arms must duplicate the impedance and phase effects of one pair of bridge arms — which pair depends on convenience. Note in Fig. 9-10 that if the oscillator were connected from A to C, the ground circuit would then be either two resistors in series or two capacitors (one variable). In the Maxwell bridge the ground circuit may duplicate either the inductive side of the bridge or the capacitive side, depending on which is more convenient (generally the latter).

With many bridges it is not necessary to make as accurate a job of balancing the ground circuit as of balancing the main bridge. Points A and C should be brought essentially to ground potential, but a small differential to ground produces little error. In the same way, the ground circuit may

need to be balanced for the *principal* component only. For example, arm
AD in Fig. 9-10 must have adjustment for resistance to balance R_x, but
in the ground circuit a plain capacitance C_w is frequently satisfactory.
However, sensitive bridges for the accurate measurement of very small
capacitances must have close adjustment of the ground circuit, including
careful control for both components of balance.

9-12. Schering Bridge for Measurement of Capacitance

The Schering bridge is one of the most important of the a-c bridges.
It is used a great deal in the measurement of capacitance in general, and
in particular in the measurement of the properties of insulators, condenser
bushings, insulating oil, and other insulating materials. It is met in several
arrangements, or modifications, to adapt it to special applications. We
shall study only the basic circuit arrangement at this time, but much
detailed study can be given to matters of arrangements, shielding, earth
circuits, and so on.

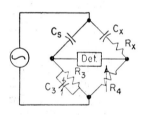

Fig. 9-11. Schering bridge.

The Schering circuit is basically the same as
the capacitance comparison bridge of par. 3. The
difference is in the method of balancing the
resistive component of the unknown. The circuit
is shown in Fig. 9-11, in which R_x represents the
resistance inherent in the unknown capacitor,
portrayed as a series resistance. An adjustable
capacitance C_3 is connected across R_3 for the
measurement of R_x, instead of the resistance in
series with C_s in the comparison bridge of par. 3.

For general measurements, C_s is usually a high-grade mica capacitor
having very low resistance effects. For insulation testing, C_s is generally
an air capacitor, carefully designed to keep insulating material out of strong
electric fields as much as possible. The combination of air dielectric, good
design, and guard electrodes results in practically zero loss in the standard.
In the circuit of Fig. 9-11, accordingly, the sum of the phase angles of arms
1 and 4 is $0° + 90° = 90°$. The angle of arm X is less than $90°$ due to the
(usually small) effect of R_x. Balance can be attained by giving arm 3 a
small capacitive angle by connecting C_3 in parallel with R_3.

The equations for the balanced condition are found by the method used
in previous bridges. If we place $\mathbf{Z}_x = \mathbf{Z}_1 \mathbf{Z}_4 \mathbf{Y}_3$ we obtain the equation

$$R_x - j\frac{1}{\omega C_x} = \frac{R_4 C_3}{C_s} - j\frac{R_4}{\omega C_s R_3}$$

From the j terms

$$C_x = C_s \frac{R_3}{R_4} \tag{24}$$

From the real terms

$$R_x = \frac{C_3}{C_s} R_4 \tag{25}$$

The power factor of the unknown

$$\mathrm{PF}_x = \frac{R_x}{Z_x}$$

can be approximated, for phase angles near 90°, by

$$\mathrm{PF}_x \simeq D_x = \frac{R_x}{X_x} = \omega C_x R_x \tag{26}$$

where D = dissipation factor $(D = R/X = 1/Q)$

By substitution of (24) and (25) in (26)

$$D_x = \omega C_3 R_3 \tag{27}$$

Equation (27) not only is convenient for computation of dissipation factor, but also suggests an arrangement that is used in several bridges. If R_3 is fixed, then for a particular frequency PF_x depends directly on C_3. Accordingly, the dial of C_3 may be calibrated directly in dissipation factor. The calibration holds for the given frequency, but may be used at another frequency if correction is made by multiplying by the ratio of the frequencies.

The "dissipation factor," D, of a series R-C circuit is defined as the cotangent of the phase angle θ, or tangent $(90° - \theta)$. The angle $(90° - \theta)$ is called the "phase-defect angle," as it is the angle by which an actual capacitor falls short of perfection. It is better to mark the loss dial of a Schering bridge as "dissipation factor" (rather than power factor) since this is true regardless of magnitude, whereas it equals power factor only for small defect angles. The difference between the two terms is insignificant for small-defect angles, but becomes appreciable for dissipation factors greater than 0.1.

9-13. Substitution Methods

This section deals, not with a new circuit, but with a method of handling circuits so that the effects of calibration errors and stray couplings in the bridge members are cancelled out to a large degree. It is necessary to have a standard of nearly the same value as the unknown that is to be measured, and accordingly the method is not convenient in general routine measurements involving odd values of the unknown. When a suitable standard is available, however, comparison can be made in this way to a smaller degree of uncertainty than in the ordinary bridge.

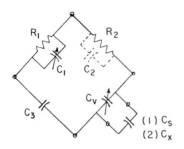

Fig. 9-12. Measurement of capacitance by substitution method.

The method may take several forms, depending on the equipment available. Figure 9-12 illustrates a substitution method in the measurement of capacitance. R_1 and R_2 are decade resistors, C_3 is a fixed capacitor, preferably (but not necessarily) of approximately the magnitude of the unknown. C_v is a variable capacitor, or "vernier," of sufficient range to make up the difference between standard and unknown. C_1 is a small variable capacitor to compensate for losses in C_x as in the Schering circuit. (This correction may be needed across R_2 instead, if C_3 has greater losses than C_s or C_x. Otherwise, we can use a small fixed C_2 plus variable C_1.) The procedure is to connect C_s and balance the bridge by R_1 and C_1, with C_v set at a convenient point on its scale. With R_1 and R_2 held constant we connect C_x in place of C_s and change C_v to secure balance. (Adjust C_1 if necessary for sharp balance.) Since arms 1, 2, and 3 have been kept constant, the *total* capacitance in arm 4 must be the same as before.

$$C_s + C_{v1} = C_x + C_{v2}$$

or
$$C_x = C_s + (C_{v1} - C_{v2})$$

Therefore, the *change* in C_v gives the *difference* between C_s and C_x. Note that we do not depend on the marked values of R_1, R_2, and C_3, but only on their constancy for the period of time needed for the measurements. Also, stray capacitances in the bridge arms cause no error since they are present for both measurements and have the same distribution of voltages both times. C_v is used to measure the difference, and if this is small, accuracy requirements on it are not severe. All told, we can be sure of C_x to nearly the same accuracy as C_s is known. (Note: A very small second-order error is introduced in case S and X differ materially in their conductance components.)

A modified method, somewhat less accurate, may be used if a suitable C_v is not available. In this case we must change the setting of R_1 (or R_2) when we change from C_s to C_x. The main principle is to arrange the first reading so that the changes occur only in the *lower dials* of R_1. Thus if we decide that 1000 ohms is a good value for the arms, we should *not* set R_1 at one unit on the thousands dial, for we may need to drop to a lower value, such as 992, and one unit on the thousands dial may not be duplicated in resistance by ten one-hundred steps. It is much better to start with a value such as 955 ohms, for then we can change a few ohms in either direction without changing the upper dial. If we take two readings on R_1, first with C_s in arm 4, then C_x, we may find the value of C_x by

$$C_x = C_s \frac{R_{1x}}{R_{1s}} \tag{28}$$

where the subscripts of R_1 denote the balance to which they belong. In this method we depend on C_s and the *change* of R_1, but are independent of calibration errors in C_3 and R_2.

Substitution methods may be applied to other measurements, including inductance and resistance and may be used with other types of bridge. The study we have made here, however, will give an idea of the principles involved.

9-14. Series and Parallel Components of Impedance

The arms of some of the bridges we have considered consist of series combinations of impedances, and others of parallel groups, depending on convenience of adjustment and simplicity of formula. It is frequently necessary to change from one form to equivalent values in the other, so it may be convenient to review the connecting relationships.

The series constants consist of R, X, and Z, as pictured in Fig. 9-13, and with relationships as indicated by the impedance triangle in Fig. 9-14. The

Fig. 9-13. Series and parallel representations of a circuit element.

corresponding parallel quantities are G, B, and Y — although the R_p and X_p of Fig. 9-13 find some uses. If one set of constants is given, the other set may be found from it, either by the well-known series and parallel operators, or by the complex notation. The relationships are tabulated in Fig. 9-14.

Either D or Q may be used to express the properties of inductive or capacitive impedances. In common usage Q is more generally used for coils and D for capacitors. However, dissipation factors are generally more convenient for either coils or capacitors combining the loss effects of several parts of a circuit.

9-15. A-C Bridges under Unbalanced Conditions

The indication of the detector when an a-c bridge is unbalanced may be found by the same method used with d-c bridges; that is, by application of Thevenin's theorem. Solution of the unbalanced bridge is important

Series Values Given | Parallel Values Given

$$G_p = \frac{R_s}{R_s^2 + X_s^2}$$

$$B_p = \frac{X_s}{R_s^2 + X_s^2}$$

$$R_s = \frac{G_p}{G_p^2 + B_p^2}$$

$$X_s = \frac{B_p}{G_p^2 + B_p^2}$$

$$R_p = \frac{1}{G_p} = R_s \left(1 + \frac{X_s^2}{R_s^2}\right)$$

$$= R_s (1 + Q^2)$$

$$R_s = R_p \frac{1}{1 + Q^2}$$

$$= R_s \left(1 + \frac{1}{D^2}\right)$$

$$X_p = \frac{1}{B_p} = X_s (1 + D^2)$$

$$X_s = X_p \frac{1}{1 + D^2}$$

$$= X_s \left(1 + \frac{1}{Q^2}\right)$$

$$\text{Dissipation Factor} = D = \frac{R_s}{X_s} = \frac{X_p}{R_p} = \frac{1}{Q}$$

$$\text{Storage Factor} = Q = \frac{X_s}{R_s} = \frac{R_p}{X_p}$$

Fig. 9-14. Series-parallel relationships.

frequently as a measure of the sensitivity needed in the detector in order to determine the balance point to the required degree of accuracy. The only difference in computation of unbalances in the a-c case is the use of the complex notation in representing currents, voltages, and impedances.

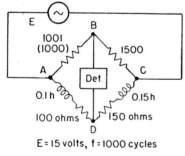

E = 15 volts, f = 1000 cycles

Fig. 9-15. Unbalanced inductance bridge.

The bridge of Fig. 9-15 will be taken as an example. If arm AB were set at 1000 ohms, the bridge would be in complete balance and the detector voltage and current would be zero. We shall change arm AB to 1001 ohms to find whether the detector response is great enough to make possible the finding of the balance point on AB within one ohm. The *open-circuit* voltage from B to D (detector disconnected) may be found with respect to the applied

voltage as reference in the following way:

$$E_0 = 15\left(\frac{1001}{2501} - \frac{100 + j628}{250 + j1570}\right) \tag{29}$$

The second fraction in the brackets is equal to $\frac{1000}{2500}$, which equals the voltage from A to D or A to B under balanced conditions. However, if this fact is not recognized, we can proceed with the above equation by factoring the second fraction as follows

$$E_0 = 15\left[\frac{1001}{2501} - \frac{100(1 + j6.28)}{250(1 + j6.28)}\right]$$

$$= 15\frac{250250 - 250100}{2501 \times 250} = \frac{15 \times 150}{2501 \times 250}$$

$$= 0.0036 \text{ v or } 3.6 \text{ mv}$$

and is in phase in this case with the applied voltage. If the detector is an essentially voltage-operated device, such as a vacuum tube amplifier with very high input impedance, the input to the detector is almost identically the same 3.6 mv after the detector is connected. We need know only the detector response for an input voltage of this amount in order to complete the problem.

If the detector is of low impedance we must continue the Thevenin solution. The impedance of the circuit as viewed from B-D with the voltage set equal to zero and the power source assumed to have negligible impedance is

$$Z_0 = \frac{1001 \times 1500}{2501} + \frac{(100 + j628)(150 + j942)}{250 + j1570} \tag{30}$$

$$= 600 + j377$$

The complete equivalent circuit for the computation of detector current and voltage is shown in Fig. 9-16. We can now compute the current through the

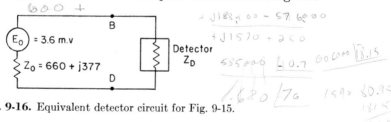

Fig. 9-16. Equivalent detector circuit for Fig. 9-15.

detector or the voltage across it for any specified value of detector impedance.

9-16. Classification of Bridge Circuits

In terms of the notation of Fig. 9-1, the constants of arm 4, considered as the unknown, may be found by the equation

$$Z_4 = \frac{Z_2 Z_3}{Z_1} \tag{31}$$

or, expressing each impedance by R and X

$$R_4 + jX_4 = \frac{(R_2 + jX_2)(R_3 + jX_3)}{R_1 + jX_1} \tag{32}$$

CASE I. If $R_2/X_2 = R_1/X_1$ (that is $\theta_2 = \theta_1$), eq. (32) becomes

$$R_4 + jX_4 = K(R_3 + jX_3) \tag{33}$$

where $$K = \frac{\mathbf{Z}_2}{\mathbf{Z}_1} = \text{real number}$$

In this case R_4 is determined by R_3, and X_4 by X_3 ($R_4 = KR_3$, $X_4 = KX_3$). This is a desirable situation because the resistive and reactive adjustments are independent of each other.

CASE II. If $\theta_2 = \theta_1 \pm 90°$, or $R_2/X_2 = -X_1/R_1$ the ratio $\mathbf{Z}_2/\mathbf{Z}_1$ is a pure imaginary, so (32) becomes

$$R_4 + jX_4 = \pm jK(R_3 + jX_3) \tag{34}$$

The two adjustments are independent, as above, but crossed over, since R_4 is measured by X_3 and X_4 by R_3.

Bridge arrangements coming under Cases I or II have been called, arbitrarily, "ratio-arm" bridges. These bridges give either a pure real or a pure imaginary number for the ratio $\mathbf{Z}_2/\mathbf{Z}_1$. The name is unimportant, and the idea unimportant except, as shown above, that under these conditions the two adjustments can be put in one arm (*adjacent* to the unknown) and be independent of each other in securing bridge balance.

Equation (32) may be changed by representing arm 1 by an admittance

$$R_4 + jX_4 = (R_2 + jX_2)(R_3 + jX_3)(G_1 + jB_1) \tag{35}$$

We can have different cases depending upon the $\mathbf{Z}_2\mathbf{Z}_3$ product in (35). The type of product may be expressed by the complex angle $\theta_2 + \theta_3$, and while any angle is possible in general, the following cases are of particular interest.

CASE III. $\theta_2 + \theta_3 = 0$

The product is a real number, and (35) becomes

$$R_4 + jX_4 = +K'(G_1 + jB_1) \tag{36}$$

R_4 and X_4 are balanced by admittance components of like phase, both in the arm *opposite* to arm 4.

CASE IV. $\theta_2 + \theta_3 = \pm 90°$

The product is a pure imaginary

$$R_4 + jX_4 = \pm jK'(G_1 + jB_1) \tag{37}$$

R_4 is measured by B_1 and X_4 by G_1. The two adjustments are independent and in the arm opposite the unknown.

An additional case might be given with $\theta_2 + \theta_3 = 180°$, but this requires a bridge made up entirely of pure reactances, and hence is of rather limited interest.

Bridges of Cases (III) and (IV), in which the product of the impedances of opposite arms is either a real number or a pure imaginary are called "product-arm" bridges. With them the two components of the unknown impedance are measured independently by the two admittance components of the opposite arm.

This classification of bridges into ratio-arm and product-arm types is of interest in showing the conditions for independence of balance adjustments. It does not necessarily follow that bridges outside these categories are unsuitable for use. The classification is also open to the objection that it places Owen and Maxwell bridges in different classes, although the Owen bridge may be regarded justifiably as a minor modification of the Maxwell circuit.

9-17. Balance Convergence of Bridge Circuits*

The discussion of the preceding section indicates the desirability in bridge balancing of having the two adjustments in the same arm, with the other arms giving either a pure real or a pure imaginary multiplying factor, so that the adjustments become independent of each other. It is not always feasible to make this combination in practice, as it may call for a variable type of element that is not available. In the case of capacitors, variable standards are not guaranteed to the same degree of accuracy as fixed standards. Therefore, accuracy requirements frequently dictate the use of fixed capacitors, and the adjustment is secured by a variable resistor in an adjacent arm. For these reasons a workable combination frequently differs from the ideal configuration. Some combinations, though not ideal, give good balancing conditions while others give troublesome interaction between controls. It is desirable that we be able to set up a criterion to judge bridge behavior in this respect.

The open-circuit voltage across the detector terminals of the bridge (see Fig. 9-1) is given by

$$E_0 = E\left(\frac{Z_1}{Z_1 + Z_3} - \frac{Z_2}{Z_2 + Z_4}\right) \tag{38}$$

$$= E\frac{Z_1 Z_4 - Z_2 Z_3}{(Z_1 + Z_3)(Z_2 + Z_4)} \tag{39}$$

For comparatively small variations near the balance point the numerator is the important element, as the denominator changes quite slowly. We can

---

*Reference 1, pp. 299 and 573.

simplify the study by confining our attention to the numerator and considering the denominator constant. The solution on this basis holds good only in the vicinity of balance, but this is adequate; the first rough steps in balancing can generally be achieved without too much trouble. For our simplifying assumption we may write, representing the relatively constant parts by the letter F,

$$E_0 = F(Z_1 Z_4 - Z_2 Z_3) \tag{40}$$

If we use $d = E_0/F$ to represent a quantity proportional to the detector response *near balance*

$$d = Z_1 Z_4 - Z_2 Z_3 \tag{41}$$

In terms of R and X components

$$d = (R_1 + jX_1)(R_4 + jX_4) - (R_2 + jX_2)(R_3 + jX_3) \tag{42}$$

Thus the locus of the d vector is linear in terms of any one resistance or reactance considered as the variable. (Note: *The detector response may be expressed equally well in terms of admittances, and this form will be more convenient for bridges with arms consisting of parallel elements.*)

In many places in electrical circuit work we draw diagrams of the locus of a current or a voltage phasor as a resistance or a reactance is varied. We find always that the result is a circle as, for example, the circle diagrams for induction motors and transmission lines. The same thing is true of an impedance bridge.* The locus of E_0, taken over a wide range, is a circle. In our approximation we are considering that the portion of the circle near the balance point can be approximated satisfactorily by a straight line.

We are using E_0 as a measure of the detector response. This is definitely true for a vacuum-tube type of detector. For a lower impedance device, such as telephones, the current depends on the circuit impedance as well as on E_0, but again, near balance the impedance changes slowly; hence we may still use (40) or (41) as a measure of detector response.

The inductance bridge of Fig. 9-17 will be used as an illustration. For this bridge

$$d = R_1(R_4 + jX_4) - R_2(R_3 + jX_3) \tag{43}$$

Suppose we investigate the detector response near balance as R_1 is varied. In this case d is the combination of two phasor quantities, one constant, and the other dependent on R_1. The equation may be easier to follow if we write it as follows

$$d = R_1(a + jb) + V_0 \tag{44}$$

where V_0 corresponds to the complex second term in (43). Equation (44) represents a straight line of slope $b/a = X_4/R_4$ laid off from the terminus of

---

*References 10 and 11.

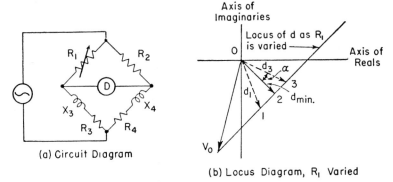

(a) Circuit Diagram

(b) Locus Diagram, R_1 Varied

Fig. 9-17. Study of detector response for an inductance comparison bridge.

V_0. The detector response is represented by a phasor from the origin to the point on the locus corresponding to the value chosen for R_1. By varying R_1, we can slide the end of the response phasor along the locus line.

The correct adjustment for minimum response is point 2 on the diagram, where the line 0-2 is normal to the locus. This is the *best* we can do with R_1 at present, until we improve the other component of balance. Before going to the other adjustment, we should mention that it is difficult to arrive *exactly* at point 2 when judging balance by the magnitude of d, either on a meter or in a telephone headset. We see that the magnitude of d changes very slowly in the vicinity of point 2, so there is a region of doubt between points 1 and 3. The length of d at any point may be written as $|d_{min}|/\cos \alpha$. For $\alpha = 20°$, $\cos \alpha = 0.943$, so an error of 20° gives an increase of 6 per cent in the detector response and this is about the limit of detection in telephones, though an amplifier and sensitive meter can work to smaller limits.

The next step in balancing the circuit is to adjust R_3. From (43) we see that the multiplier of R_3 is the real number $(-R_2)$, so the locus of d for variation of R_3 is a line of *zero slope*. The process of balancing is indicated in Fig. 9-18, wherein it is assumed that we do not reach the ideal balance point each time, but may overshoot or undershoot within the region of doubt, as indicated in Fig. 9-17(b). (It can be seen in Fig. 9-18 that some overshooting may actually help to reach balance faster.)

Convergence toward the true balance point 0 proceeds in a series of zig-zag steps. The speed with which convergence takes place depends on the angle β (as marked in Fig. 9-18), between the directions of the two locus lines. The best condition is an angle of 90°, which agrees with the discussion earlier in this chapter. The 90° relationship could be secured for Fig. 9-17(a) if we vary R_3 and X_3, which gives ideal convergence but depends on availability of a suitable variable standard for X_3.

As a final element in the discussion of the bridge circuit of Fig. 9-17(a), we shall consider the effect of the phase angle of the unknown. Figure 9-19

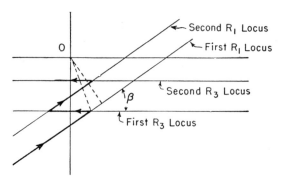

Fig. 9-18. Convergence of balance for circuit of Fig. 9-17, R_1 and R_3 varied.

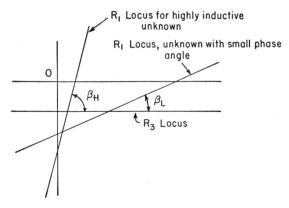

Fig. 9-19. Effect of phase angle of unknown on balance conditions in Fig. 9-18.

indicates the slope of the R_1 locus in case the unknown is highly inductive (phase angle near 90°), and in the reverse case of a low inductive angle near 0°. The diagram shows a large angle, β_H, between the R_3 and R_1 loci in case the unknown has a large phase angle, and in this case adjustments of R_1 and R_3 would operate quite satisfactorily to secure balance. If the unknown has a small phase angle, however, as with β_L in the figure, the convergence for variations of R_1 and R_3 is very poor, and we should use some other bridge arrangement.

Satisfactory convergence to a bridge balance depends on a combination of variables that gives a fairly large angle β between their loci of variation. An angle of 90° is ideal. Angles below 45° produce slow convergence Finally, an angle β less than the angle of doubt, α, might cause the operator to fumble around almost indefinitely and possibly **never** achieve a good balance.

9-18. Phasor Diagrams of Impedance Bridges

Phasor diagrams may be used to illustrate the phase relationships and the balancing processes in an impedance bridge. A diagram may be used in some cases to compute the magnitude of one of the components, but in general the complex algebra is more convenient for numerical work of this sort.

Building up a phasor diagram for a bridge follows the same process of combination of phasors at their correct angular relationships as followed in other phasor diagrams. There are only a few details in which a bridge may be regarded as being at all special and these will be illustrated by the circuit of Fig. 9-20(a), which will be used as a simple introduction to the subject. This bridge is an inductance comparison arrangement in which R_2 and R_3 have been selected as the components to be varied in securing balance. Resistance and reactance values in ohms are shown on the figure in order to make the problem definite. Resistances and reactances are shown by separate circuit symbols for convenience, although at least part of the resistances would in practice reside within the inductance coils. R_3 represents the total resistance of arm DA, which is assumed to be low enough in the first place to require the addition of a variable resistor to produce balance. (The components are arranged in the order shown merely for convenience in adding their voltages.)

With the oscillator connected from A to C the upper half of the bridge is purely resistive, so the phasors for the voltages AB and BC are directly in line and add up to the impressed voltage $\boldsymbol{E}_{AC}$. Adjustment of R_2 varies the relative lengths of the two phasors and moves the dividing point, B, either to left or right, but always along the horizontal line, as from B_1 to B_2. The intermediate point B, shown on the diagram, is the one that gives balance with the stated values of X_3 and X_4.

The current in branches AD and DC lags the applied voltage because of the inductive effect. The component phasors AE and ED add at 90° to give AD, and likewise DF and FC to give DC. Balance is secured in this bridge when the two triangles have equal angles, so that point D is on a straight line between A and C. If the diagram for the upper half of the figure were to be moved bodily and superposed on the lower diagram so that A falls on A and C on C, then B and D would coincide. This is the condition for "balance," that there be zero difference of potential between B and D. The two halves of the figure were shown separately in Fig. 9-20 to avoid the confusion that often results in the first approach to a seemingly complicated complete diagram.

The process of balancing the bridge becomes, in terms of the diagram, a matter of maneuvering B and D into coincidence. This is an area, or two-dimensional job, as B and D must be in the same position both in "latitude"

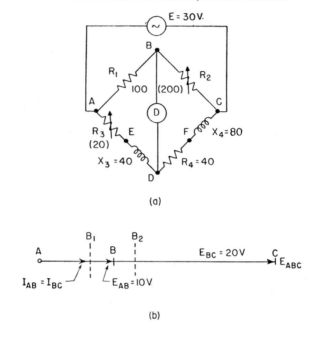

(a)

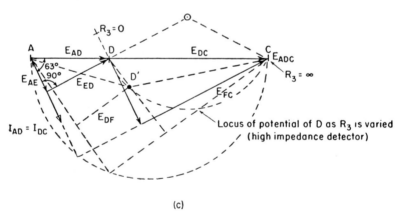

(b)

(c)

Fig. 9-20. Phasor diagram of an inductance comparison bridge. (a) Circuit diagram, inductance comparison bridge. (b) Diagram for upper half of bridge, balanced conditions. (c) Diagram of lower half of bridge, balanced conditions.

and in "longitude." This is the reason, of course, why two elements in a bridge must be variable, and elements must be varied that are capable of giving motion in the two directions. As shown above, adjustment of R_2 moves B in a horizontal direction. Conditions in the lower half are a little more complicated, in that a change of R_3 changes not only the magnitude but also the phase angle of the phasors. A consideraby unbalanced case (for *too large* a value of R_3) is shown by the dotted lines, moving D to D'. The distance from D to D' represents, to the established voltage scale, the voltage which appears across the detector, assuming it to be a high-impedance device that draws negligible current.

The path that D' follows in case of unbalance is a circular arc. At the point D where the arc crosses the horizontal line it is parallel to E_{AE}, and has an angle with the horizontal equal to $\tan^{-1} X_4/R_4$. The adjustment of D, accordingly, becomes more nearly on a vertical line for a high value of Q of the bridge arms. (A value of $Q = 2$ was used in drawing Fig. 9-20 in order to give angles of convenient size.) Proof that D' follows a circular arc will not be presented here. A discussion of locus problems and circuit inversions may be found in several texts on a-c circuits and requires more space than is available at this point.

REFERENCES

1. Hague, B., *Alternating Current Bridge Methods*, London: Sir Isaac Pitman and Sons, Ltd., 1946.

2. Cone, D. I., "Bridge Methods for Alternating-Current Measurements," *Trans. A.I.E.E.*, **39**, 1743–62 (1920).

3. Ferguson, J. G., "Classification of Bridge Methods of Measuring Impedance," *Trans. A.I.E.E.*, **52**, 861 (1933).

4. Seeley, Walter J., "A Brief Summary of Bridge Networks," *Elec. Eng.*, 108 (March 1940). (Contains bibliography.)

5. Owen, D., "A Bridge for the Measurement of Self-Inductance," *Proc. Phys. Soc. London*, **27**, 39–55 (1915).

6. Ferguson, J. G., "Measurement of Inductance by the Shielded Owen Bridge," *Bell System Tech. J.*, **6**, 375–386 (1927).

7. Laws, Frank A., *Electrical Measurements*, New York: McGraw-Hill Book Co., 1938.

8. Slonczewski, T., "Measuring Inductance with a Resistor," *Bell Labs. Record*, **12**, 3, 77–80 (1933).

9. Wilhelm, H. T., "Self-Contained Bridge for Measuring Both Inductive and Capacitive Impedances," *Bell Labs. Record*, **12**, 181 (Feb. 1934).

10. Seletzky, A. C., "Cross Potential of a Four-Arm Network," *Trans. A.I.E.E.*, **53**, 1734–40 (1934).

11. Seletzky, A. C. and Anderson, J. R., "Cross Current of a Five-Arm Network," *Trans. A.I.E.E.*, **53**, 1004–09 (1934).

12. Fortescue, C. L., "Measurement of Very Small Inductances," *J. Sci. Instruments*, **10**, 10, 301–5 (Oct. 1933).

PROBLEMS

NOTE 1: *The problems are stated with respect to bridge terminals lettered A, B, C, and D consecutively around the bridge.*

NOTE 2: *The term "picofarad," abbreviated pf, is frequently used to denote 10^{-12} farad; that is 1 pf = 1 $\mu\mu$f. The symbols pf and $\mu\mu$f may be used interchangeably. Small fractions of a micromicrofarad are encountered in some work (see pars. 12-10, 12-11, 13-8); in these cases the abbreviation μpf is more convenient than $\mu\mu\mu$f.*

9-1. A bridge is balanced with the following constants: arm AB, $R = 3000$ ohms in parallel with $C = 0.053$ μf; BC, $R = 1500$ ohms in series with $C = 0.53$ μf; CD, unknown; DA, $C = 0.265$ μf. The frequency is 1000 cycles per second. Find the constants of arm CD.

9-2. One arm of a 1000-cycle bridge consists of a resistance of 1000 ohms in series with a capacitance of 0.2 μf. Find the constants of a parallel combination equivalent to the original arm. Is there any restriction on the equivalence of these two arrangements?

9-3. A resistance of 500 ohms is in series with an inductance of 0.2 h. Find the values of a pure R in parallel with a pure L to be equivalent to the original circuit at 1000 cycles per second.

9-4. A coil has an inductance of 0.1 h and a series resistance of 80 ohms when measured at a frequency of 60 cycles per second. Find the values of pure L and pure R in parallel that are equivalent to the coil at the given frequency.

9-5. A bridge is balanced, with $f = 1000$ cycles, and the following constants: AB, 0.199 μf pure capacitance; BC, 500 ohms pure resistance; CD, the unknown; DA, $R = 417$ ohms in parallel with $C = 0.111$ μf. Find the R and L or C constants of the unknown, considered as a series circuit.

9-6. A 1000-cycle bridge has the following constants: arm AB, $R = 1000$ in parallel with $C = 0.5$ μf; BC, $R = 1000$ in series with $C = 0.5$ μf; CD, $L = 40$ mh in series with $R = 400$. Find the constants of arm DA to balance the bridge. The result is to be expressed as a pure R in parallel with a pure L or C.

9-7. An a-c bridge has in arm AB a pure capacitance of 0.2 μf; arm BC a pure resistance of 500 ohms; arm CD, $R = 50$ ohms in series with $L = 0.1$ henry; arm DA, $C = 0.4$ μf in series with a variable resistance R_3. $\omega = 2\pi f = 5000$.

(a) Find the value of R_3 for balance.

(b) Can complete balance be secured by adjustment of R_3? If not, specify the position and value of an adjustable resistance to complete the balance. Decade resistors are the only additional elements that may be used.

9-8. An a-c bridge has in arm AB a resistance R_1; BC, resistance R_2; CD, a series combination of R_x and L_x, paralleled by variable capacitance C; DA, a resistance R_3. Find the equations for the balanced condition for R_x, L_x, and Q_x in terms of the other bridge constants.

9-9. A bridge is made of the following parts: arm AB, capacitance $= C_1$; BC, resistance $= R_2$; CD, an unknown impedance $Z_4 = R_4 \pm jX_4$; DA, R_3 in parallel with C_3.

(a) Derive the balance equations for this bridge for Z_4 as the unknown.

(b) For which type of reactance in arm 4 can balance be attained?

(c) Express Q_4 in terms of the known constants.

9-10. Arm BC, $R = 1000$ ohms; CD, $L = 0.15$ h in series with $R = 100$ ohms; DA, pure R, to be determined; AB, $C = 0.2$ μf and a resistance.

(a) The resistance in arm AB is in parallel with the capacitor (Maxwell bridge). Find its value, and also the resistance of DA to balance the bridge. $f = 1000$ cycles per second.

(b) The resistance in arm AB is in series with the capacitor (Hay bridge). Find the constants as in part (a).

9-11. A balanced 1000-cycle bridge has the following constants: arm AB, $C = 0.5$ μf; BC, $R = 400$ ohms; CD, a coil having $L = 0.1$ h, $R = 100$ ohms.

(a) Find the constants of arm DA.

(b) Repeat, if the coil resistance is 20 ohms. How may the bridge be changed to avoid the need for so large a capacitor in arm DA?

9-12. The bridge shown in the diagram consists of fixed elements except for the adjustable resistance R_3. $\omega = 2\pi f = 10,000$.

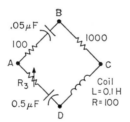

Prob. 9-12.

(a) Find the resistance of R_3 for the closest approach to bridge balance.

(b) Is the bridge completely balanced by this adjustment of R_3? If not, show where a decade *resistance box* can be added to improve the balance, and compute the resistance setting needed for exact balance.

9-13. A balanced 1000-cycle bridge has constants: arm AB, $R = 1000$ ohms in parallel with $C = 0.053$ μf; BC, $R = 1500$ ohms in series with $C = 0.53$ μf; CD, the unknown; DA, pure $C = 0.265$ μf. Find the R and L or C constants of the unknown, (a) as a series circuit; (b) as a parallel circuit.

9-14. A balanced bridge has the following constants: AB, $R = 500$ ohms; BC, $R = 1000$ ohms; CD, the unknown; DA, pure $C = 0.2$ μf. A voltage of 15 v at 1000 cycles is impressed from A to C.

(a) Find the constants of the unknown.

(b) The 1000-ohm resistor is changed to 1001 ohms. Find the voltage across the detector (high impedance).

(c) Repeat (b), with the generator and detector interchanged.

9-15. An "unknown" inductance of 0.22 h and 20 ohms resistance is measured by comparison with a fixed standard inductance of 0.1 h and 40 ohms resistance. They are connected in a bridge marked A-B-C-D consecutively around the corners so that X is in CD, S in DA. AB is a pure resistance of 750 ohms, BC is a pure resistance of amount to be determined. An oscillator giving 15 v at 1000 cycles per second is connected A to C. A vacuum-tube detector is connected B to D.

(a) Find the resistance of BC and show any necessary (and practical) additions to give both resistive and inductive balance.

(b) After balance is secured, the 750-ohm arm is changed to 751 ohms. What voltage appears across the detector input?

9-16. An unbalanced bridge has the following constants: Arm AB, $R = 2000$ ohms in parallel with $C = 0.2\,\mu f$; BC, $R = 1200$ ohms; CD, $L = 0.6$ h in series with $R = 500$ ohms; DA, $R = 2500$ ohms. The oscillator gives 15 v and is connected A to C. $\omega = 2\pi f = 5000$.

(a) For what changed constants in arm CD would the bridge be in balance?

(b) For the original bridge constants as stated above find the voltage across a vacuum-tube detector connected from B to D.

9-17. A bridge, balanced at 1000 cycles, has pure resistance ratio arms, A to B 1500 ohms, B to C 1000 ohms. The unknown is connected C to D. The arm DA has a 0.1 μf standard capacitor of negligible resistance, to which is added a series resistance of 10 ohms to give balance. The oscillator, connected B to D, gives 10 v. The detector is a vacuum-tube voltmeter.

(a) Find the constants of the unknown.

(b) Find the detector voltage for an increase of 10 ohms in arm BC.

9-18. A bridge is used at a frequency of 60 cycles per second to measure an "unknown" capacitance of 500 pf, with phase-defect angle of $0°12'$, by comparison with a standard capacitance of 200 pf with negligible loss. The standard is in arm AB, the unknown in BC. Arm CD is a resistance of 5000 ohms. Point D is grounded.

(a) Find the constants of arm AD for balance. (parallel components)

(b) What percentage error in the value of the capacitance or the dissipation factor of the unknown would be caused by the presence (not known to the operator) of a stray capacitance of 100 pf from point A to ground during the balancing process for the above "unknown" in arm BC?

9-19. An a-c bridge marked A-B-C-D consecutively around the corners has in arm AB a standard capacitor of 0.0001 μf with negligible losses; Arm BC, the unknown; Arm CD, pure $R = 10,000$ ohms; Arm DA, $R = 12,500$ ohms paralleled by a variable capacitor. Balance is secured with the variable set at 120 pf. The frequency is such that $\omega = 2\pi f = 10,000$. Point D is grounded.

(a) Compute the constants of the unknown as series components, and find its dissipation factor.

(b) Suppose now that an unknown stray capacitance of 50 pf had existed across arm CD during the balancing process for the unknown of part (a). Find the modification of bridge constants needed to secure balance, and the apparent value that would have been obtained for the dissipation factor of the unknown arm BC.

9-20. (a) Is this bridge circuit in complete balance? If not, specify two ways in which it can be made to balance, for X as given. (Give numerical values of any additions.)

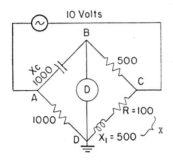

Prob. 9-20.

(b) For the bridge as shown, find the voltage across the detector terminals, assuming the detector to have very high impedance.

9-21. A balanced bridge has the following constants: AB, pure $C = 0.01$ μf; BC, the unknown; CD, $R = 2500$ ohms; DA, $R = 5000$ ohms in parallel with $C = 0.001$ μf. A voltage of 20 v at 1000 cycles is impressed from B to D.

(a) Find the impedance and power factor of the unknown.

(b) If the capacitor is removed from arm DA, what voltage appears across the detector terminals? (The detector is a vacuum-tube voltmeter.)

9-22. An oscillator giving five volts at $\omega = 2\pi f = 10,000$ is connected to the bridge shown in the diagram. The detector is a vacuum-tube device of very high input impedance.

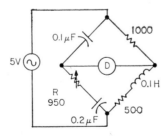

Prob. 9-22.

(a) Find the voltage indicated by the detector.

(b) If one millivolt is the smallest increment that can be read with assurance on the detector, to approximately what minimum increment can the variable resistor be balanced?

9-23. A balanced a-c bridge, marked A-B-C-D around its corners, has the following constants: AB, $R = 200$ in series with $C = 0.125\ \mu f$; BC, $R = 500$; CD, unknown; DA, $R = 2600$ ohms. An oscillator giving 5 v (rms) at $\omega = 2\pi f = 5000$ radians per sec. is connected from B to D.

(a) Find the constants (R and L or C) of the unknown.

(b) If the detector (high impedance) requires one millivolt for a perceptible signal, find the minimum unbalance (in per cent) of the reactive component of the unknown that can be detected, for the above conditions.

9-24. An a-c bridge has the following constants: arm AB, $C = 0.2\ \mu f$; BC, $C = 0.04\ \mu f$ in series with $R = 2000$ ohms; CD, $R = 8000$; DA, $R = 2000$. An oscillator giving 20 v (rms) and having negligible internal impedance is connected from A to C. $\omega = 2\pi f = 5000$. Find the current flowing in a detector having impedance $1350 + j4850$ at the operating frequency.

9-25. An a-c bridge marked A-B-C-D consecutively around the corners has the following impedance values: AB, pure $X_c = 2000$ ohms; BC, pure $R = 500$ ohms; CD, $R = 500$ in series with $X_L = 2000$ ohms. The detector is connected from A to C, and the oscillator B to D.

(a) Find the constants of arm DA to produce balance.

(b) The resistance of BC is changed to 1000 ohms, with all other values as in part (a). Find the current that flows through a detector of $1840 + j5480$ ohms if the oscillator is impressing 10 v on the bridge. The oscillator impedance is negligible

9-26. An a-c bridge has the following constants: arm A-B, a resistance of 3000 ohms in series with an inductance of 0.4 h; B-C, a resistance of 4500 ohms in series with an inductance of 0.6 h; C-D, pure resistance 3590 ohms; D-A, pure resistance 2410 ohms. An oscillator maintains an emf of 15 v (rms) at $\omega = 2\pi f = 10,000$ radians per second between terminals A and C. The detector is a telephone headset with impedance $1258 + j3600$ at this frequency. Find the current that flows in the detector for the above conditions.

9-27. A balanced a-c bridge has the following constants; AB, $R = 1000$ ohms; BC, $R = 2000$ ohms; CD, a standard capacitor of $0.01\ \mu f$, guaranteed to ± 0.1 per cent, paralleled by a variable air capacitor of $1000\ \mu\mu f$ range, on which any reading is guaranteed to ± 0.5 per cent; DA, a capacitor of $0.02\ \mu f$ nominal value guaranteed to ± 5 per cent. The resistance values are guaranteed to ± 0.1 per cent. Balance is secured with the variable set at $600\ \mu\mu f$.

(a) The standard is removed, and an unknown capacitor is connected in its place. Balance is obtained with the variable set at $200\ \mu\mu f$. Find the upper and lower limits of the unknown, as based on the guarantees.

(b) Suppose all capacitors in the circuit to have negligible losses except C_x, which has a dissipation factor of 0.01. Show an addition to the bridge to give complete balance when C_x is in the circuit. $\omega = 2\pi f = 10,000$.

9-28. A standard inductance of 1 mh has an accuracy of ± 0.10 per cent and a Q of 50 ± 0.25 at $\omega = 10^5$. With the standard inductance placed in the CD arm of the bridge, balance is obtained with the values: $R_{AB} = 1000$, $R_{BC} = 1000$, $R_{DA} = 2.00$ in series with $L_{DA} = 1$ mh. When the standard inductance is removed and replaced by an unknown inductance in the CD arm, balance is obtained by adjusting R_{BC} to 990 and R_{DA} to 1.00.

(a) What are the values of L and Q of the unknown?

(b) If the R_{BC} resistor at initial balance consists of 9-100-ohm and 10-10-ohm resistors, each accurate to ± 0.10 per cent, what is the accuracy of the determination of L_x if one of the 10-ohm resistors is removed for the unknown balance, the rest of the resistors remaining the same?

Assume the sensitivity of the detector is ample to detect the balance point with no additional uncertainty.

9-29. An a-c bridge operating at 1000 cycles per second, has in arm AB a decade resistance box guaranteed to ± 0.1 per cent; BC, a pure resistance of 3000 ohms ± 0.5 per cent; CD, a variable air capacitor of 1000 $\mu\mu f$ range, set at 508 $\mu\mu f$, paralleled by a standard capacitor of 2012 $\mu\mu f$ guaranteed to ± 0.05 per cent; DA, a capacitor rated 1800 $\mu\mu f$ ± 2 per cent. Point D is grounded. The oscillator is connected from A to C, detector B to D. Any reading on the air capacitor is guaranteed to ± 0.1 per cent. The decade box is adjusted to bring the detector indication to zero. The standard capacitor is replaced by an unknown, and balance is reestablished by changing the air capacitor to 788 $\mu\mu f$.

(a) Find the capacitance of the unknown.

(b) Within what accuracy limits can you guarantee this result?

(c) If C_x has a dissipation factor of 0.008, while all other capacitors in the bridge circuit have negligible losses, what *practical* addition can be made to the bridge to give complete balance when C_x is in the circuit? (Give numerical value of any addition.)

9-30. An unknown inductor is measured on the bridge (shown in the figure) in comparison with a standard inductor guaranteed as follows at 1000 cps:

$$L_s = 0.10060 \text{ henry} \pm 0.05\%, \qquad Q_s = 22 \pm 1.0\%$$

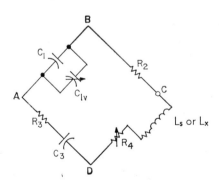

Prob. 9-30.

The bridge elements have guaranteed limits of error as follows:

$$C_1 = 0.1 \ \mu f \pm 1\% \qquad R_2 = 500 \text{ ohms} \pm 0.1\%$$
$$C_{1v} = 0.1\% \qquad R_3 = 2016 \text{ ohms} \pm 0.1\%$$
$$C_3 = 0.25 \ \mu f \pm 1\% \qquad R_4 = \pm 0.1\%$$

$$f = 1000 \text{ cps}$$

1st Balance (L_s): $C_{1v} = 435 \ \mu\mu f,$ $R_4 = 172$ ohms
2nd Balance (L_x): $C_{1v} = 785 \ \mu\mu f,$ $R_4 = 179$ ohms

Find L and Q of the unknown, and the limit of their errors by the guarantees·

9-31. A Schering bridge is set up to measure an unknown capacitance. Draw the circuit diagram. Consideration is to be given to the effect of residuals in the bridge members upon the values determined for the unknown. For each part below, state the effect on the apparent value of the capacitance or power factor of the unknown and give a supporting reason (not a long equation).

(a) What is the effect of losses in the standard capacitor?
(b) Of losses in the capacitor used for power factor balance?
(c) Of capacitance in the resistor paralleling the P.F. capacitor?
(d) Of capacitance in the other resistance arm?

9-32. An a-c bridge consists of a pure capacitance C_1 in arm AB, a pure resistance R_2 in BC, an inductive impedance $R_4 + j\omega L_4$ in CD, and a series combination of resistance R_3 and capacitance C_3 in DA. Impedance CD has a phase angle of 65°. Suppose that balance is to be secured by adjustments of C_1 and C_3. Draw a diagram showing the loci of the detector response near balance for variations of C_1 and C_3, and determine the angle between the loci.

9-33. A bridge marked A-B-C-D consecutively around the corners has the following components: arm AB, pure $C = 0.05 \ \mu f$; BC, $R = 1000$ ohms in series with $C = 0.05 \ \mu f$; CD, a variable resistance, R_4; DA, $R = 500$ ohms in parallel with a variable capacitance, C_3. $\omega = 2\pi f = 10,000$. Balance is obtained by alternate adjustments of R_4 and C_3.

Draw the locus of the detector voltage near balance for variations of R_4. For variations of C_3. Find the angle in degrees between these loci. (The bridge arms may be treated as impedances or as admittances, as preferred.)

9-34. A bridge has in arm AB a resistance R_1 in series with $C_1 = 0.1 \ \mu f$; arm BC, $R_2 = 1000$ ohms; arm CD, $R_4 = 100$ ohms, $L_4 = 0.2$ henry; arm DA, a resistance R_3. $\omega = 2\pi f = 5000$. An oscillator is connected from A to C, and a vacuum-tube detector from B to D.

(a) Find R_1 and R_3 for balance.
(b) Balance is secured by alternate adjustments of R_3 and R_1. Show the direction of the locus of detector voltage near balance for each adjustment, and find the angle between the loci.

9-35. This bridge may be used to measure an inductive unknown in terms of parallel components, L_4 and R_4. Construct the loci of the detector voltage phasor for variations of R_1 and R_2 near balance, and specify the angle between the loci in terms of the circuit components. Impedance or admittance analysis may be used, as seems most convenient.

9-36. An a-c bridge marked A-B-C-D consecutively around the corners has in arm A-B a resistance of 1500 ohms in parallel with a capacitance of 0.05 μf; arm B-C, a pure resistance of 1010 ohms; arm C-D, a coil having a resistance of 790 ohms

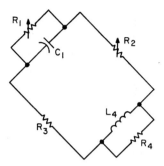

Prob. 9-35.

and an inductance of 0.06 henry; arm D-A, a pure resistance of 1200 ohms. An oscillator giving 10 v (rms) at $\omega = 2\pi f = 10,000$ radians per second is connected from B to D. The detector is a vacuum-tube voltmeter of very high input impedance.

(a) Find the detector voltage for the constants as given.

(b) Balance is secured by alternate adjustments of resistance R_1 in arm A-B, and resistance R_2 in arm B-C. (Other constants as given above.) Construct a diagram showing a locus of the detector voltage near balance for each of these adjustments, and find the angle between the loci. Consider whether analysis in terms of admittances or impedances will be simpler.

9-37. A balanced bridge circuit has $R = 3000$ ohms in arm AB, unknown R and C in series in BC, $R = 700$ ohms in CD, and $R = 900$ ohms in series with $L = 0.2$ henry in arm DA. An oscillator giving 20 v at $\omega = 2\pi f = 6000$ is impressed from A to C. Draw a phasor diagram for this bridge, and from it solve for the unknown R and C.

CHAPTER 10

Additional Measuring Circuits

10-1. Inductance Bridges — Anderson; Stroud and Oates

The Anderson and the Stroud and Oates bridges are essentially the same, merely with oscillator and detector interchanged, as will be shown below. They are modifications of the Maxwell inductance bridge of chap. 9.

(a) The Stroud and Oates bridge, shown in Fig. 10-1, appears at first glance to consist of six arms, in addition to detector and oscillator. Closer study indicates that the bridge is equivalent to the Maxwell inductance bridge, studied in chap. 9. The relationship may be shown by taking the three resistors, R_1, R_3, and R_5 of Fig. 10-1 and converting to an equivalent delta-connected set by the well-known wye-delta transformation equations. The resulting bridge, shown in Fig. 10-2, is seen to be the same as the

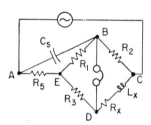

Fig. 10-1. Stroud and Oates inductance bridge.

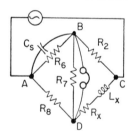

Fig. 10-2. Stroud and Oates bridge, after wye-delta transformation.

Maxwell bridge of Fig. 9-4, except that a redundant member, R_7, is added in parallel with the detector. R_7 serves no useful purpose and merely reduces to some extent the sensitivity of the detector.

The equations for the Stroud and Oates bridge may be derived by writing equations for the various loops in Fig. 10-1. However, the equations may be obtained more simply by the Maxwell equivalent given in Fig. 10-2,

$$L_x = C_s R_2 R_8 \tag{1}$$

Replacement of R_8 by its value from the wye-delta transformation gives

$$L_x = C_s R_2 \left(R_3 + R_5 + \frac{R_3 R_5}{R_1} \right) \tag{2}$$

Similarly
$$R_x = \frac{R_2 R_8}{R_6} = \frac{R_2 R_3}{R_1} \tag{3}$$

(b) The Anderson bridge is usually drawn as shown in Fig. 10-3. If the circuit is redrawn by placing R_5 and C_s to the left of point A, with the oscillator directly across from B to D, the bridge of Fig. 10-3 becomes the same as Fig. 10-1, except that the detector and oscillator are interchanged. It is axiomatic for bridge circuits in general that detector and oscillator may be interchanged at will, and a different name is usually not given for this change. The bridges of Fig. 10-1 and Fig.

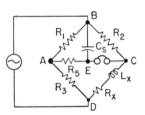

Fig. 10-3. Anderson bridge.

10-3, however, are exceptions and are usually known by the names attached to them here. The equations of the Anderson bridge are as given in (2) and (3) above.

The first characteristic feature noted in a comparison of the Anderson bridge with its prototype, the Maxwell, is the greater complexity. The Anderson bridge has more parts and is more complicated to set up and manipulate, and the balance equation is longer. The added junction point increases the difficulty of establishing suitable shielding for the bridge components. Any combination of resistors, R_1, R_3, R_5 in the Anderson bridge can be given exact delta equivalents, so it would seem at first sight that the Maxwell bridge can do anything the Anderson bridge does, but more simply. However, a study of convergence conditions, made as in par. 17 of chap. 9, indicates an advantage for the Anderson form in the case of extremely low-Q coils. The Maxwell and Hay bridges are satisfactory for all but a few special cases and are much easier to use.

If the bridge is set up for best conditions, and adjustment is made by R_5 and a resistor in series in arm X, the adjustments become independent of each other. This is a marked superiority over the sliding-balance condition met with extremely low-Q inductors on the Maxwell bridge. The Anderson bridge has been used in measurement of inductances from fairly large values down to the small residual inductances of decade boxes.

The following proportions give the greatest sensitivity, $R_1 = R_3 = \frac{1}{2}R_x$, $R_2 = R_x$, $L/C = 2R_x{}^2$. In case R_5 is very large, the sensitivity is better in the Stroud and Oates form, which then permits an increase in the applied voltage.

10-2. Inductance of Iron-Cored Coils with D-C*

The magnetic flux linking an iron-cored coil is not in direct proportion to the current flowing in the coil, but varies in a manner usually indicated by a magnetization curve. The flux usually increases fairly rapidly when the current build-up process is first begun. However, the flux increment corresponding to a particular size of current increment becomes smaller with continual increase of current as the core approaches a condition of magnetic "saturation." An induced voltage in a coil depends on a change of flux and hence becomes smaller for a given current change as the current is increased.

The basic expression for induced voltage in terms of a changing flux is usually replaced, for purposes of circuit study, by the "coefficient of self-inductance," multiplied by the rate of change of current. The new concept must be examined with care when applied to the case of an iron-cored coil. In terms of a flux change, the induced voltage is

$$e_i = -n\frac{d\varphi}{dt} \qquad \text{volts} \quad (4)$$

where φ = flux in webers, and

n = number of turns.

In terms of inductance, $\qquad e_i = -L\dfrac{di}{dt} \qquad \text{volts} \quad (5)$

where L = coefficient of self-inductance in henries. Comparison of the two expressions gives

$$L = n\frac{d\varphi}{di} \qquad \text{henries} \quad (6)$$

which helps to explain the nature of this quantity. Inductance may be represented by the slope of the φ-i curve, or magnetization curve, of a reactor. For an air-core coil there is a linear relationship between flux and current and, accordingly, L is a constant. An iron-cored coil does not have a linear flux-current relationship, and L has different values, depending on

---

*Reference 4; reference 1, p. 583.

the portion of the magnetization curve being used and on the manner of defining L. That is, for some purposes we may be interested in the average inductance, represented by the slope of a line from the origin to a particular point on the φ-i curve. For other purposes we are concerned with small variations in the neighborhood of a point on the curve (for very small variations, L equals n times the slope of the curve at the point in question). It is thus evident that the coefficient of self-inductance is ambiguous for iron-cored coils unless the conditions are carefully defined.

Many iron-cored coils are used as filter reactors in rectifier circuits and in other applications in which a combination of a d-c component and a superposed a-c are encountered. The inductance of the reactor to the a-c in such a case may vary over a wide range, depending on the magnitude of the d-c term. An air gap in the magnetic core tends to straighten the φ-i curve, and make L more nearly constant. An iron-cored coil without an air gap gives an extreme variation of L if the d-c component becomes high enough to produce a considerable degree of saturation.

The magnitude of the a-c component also has an effect on the apparent inductance of the reactor, though usually not to so great an extent as the d-c component in the common type of reactor application. To make inductance measurements on an iron-cored coil definite, the data should include the amount of d-c present, and the frequency and magnitude of the a-c.

The nonlinear character of the flux-current curve enters in the measurement problem in another way, by producing distortion of the current and voltage waveforms. That is, a sinusoidal voltage does not produce a sinusoidal current, or vice-versa, in an iron-cored coil. Even though the applied bridge voltage is sinusoidal, the detector voltage is distorted and hence may be analyzed by Fourier methods into the fundamental frequency component plus harmonics. The harmonics cause difficulties in the balancing process. If the detector consists of a vacuum-tube amplifier, a tuned circuit can be incorporated in it to pass the fundamental frequency and suppress the harmonics. The situation is a little more difficult when a telephone headset is used, but a person can train himself to listen for the null point of the fundamental frequency and disregard the harmonics. Due to the difficulties of defining inductance in the first place, and of determining balance in the second place, we cannot expect to measure an iron-cored coil with the same precision as an air-core coil. Fortunately, in most uses of the iron-cored coils, as in filter chokes, a low degree of accuracy is sufficient.

Many bridge circuits can be modified to permit the simultaneous application of d-c and a-c to the reactor. One thing that must be kept in mind in arranging the circuit is the fact that the amount of d-c specified for the reactor test must also pass through one of the bridge arms, and care must be taken not to overheat the precision resistor. The usual decade resistance box has a safe limit of the order of 15 to 20 ma for steps on the thousands dial and about 50 to 70 ma for the hundreds dial. Small reactors may be tested in a bridge made up of decade boxes if the values given here are adequate for the desired tests. If not, special resistors must be made up for the purpose. A bridge for the measurement of iron-cored coils over an adequate range of d-c is special in construction, though quite standard with respect to a-c bridge theory.

One of the most useful and convenient bridges for this purpose is shown in Fig. 10-4, as devised by H. T. Wilhelm of the Bell Telephone Labora-

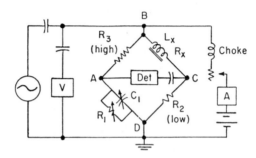

Fig. 10-4. Bridge for measurement of inductance with d-c present. (Maxwell bridge shown. Hay bridge may be used.)

tories.* A Maxwell bridge is shown here, but the Hay arrangement might be used instead, as the measured coils generally have high enough Q to permit using simplified Hay formulas. The Owen arrangement might also be used, except for possible difficulties in securing sufficiently high impedance in arm AB for the purpose of this bridge.

The a-c power supply and voltmeter are shown to the left, with condensers to block d-c from them. The circuit on the right, consisting of the d-c supply, ammeter or milliammeter, and rheostat provide adjustable amounts of d-c. The choke is used to minimize a-c flow through this branch. The meter reads the total d-c to the bridge, but this is practically the current to the unknown, as the branch B-C-D has much lower resistance than B-A-D, as discussed below. Hay or Owen circuits would provide a blocking condenser in the left branch, but the difference is rather unimportant.

The resistance arm CD must be able to carry the same direct current as the unknown without overheating and is made low in resistance to keep the required battery voltage to a moderate value. To offset the low resistance in this arm, the opposite arm is made high in resistance. Balance is secured by variations of C_1 and R_1 in order that R_2 and R_3 may be fixed resistors

*Reference 4.

(possibly with the choice of two or three values for range-changing purposes). R_2 may be of the order of 10 to 100 ohms, R_3 possibly 50,000 to 100,000 ohms. Such a ratio is not generally considered desirable, from considerations of accuracy, but has counterbalancing advantages in this case.

Inductance may be measured at one d-c value if desired, but more generally determinations are made over a range of values, and a curve is prepared showing the variation of inductance with change of the d-c. This information is frequently desired for the chokes used as smoothing reactors in circuits supplying d-c from rectifiers.

10-3. Other Incremental Measurements

The measurement of incremental inductance was presented in the preceding paragraph. Nonlinear capacitors and resistors are also encountered in scientific work, so it may be desired to determine their properties by bridge measurements. Nonlinear capacitors, in particular, have found numerous applications of late, made possible by the discovery of materials which have extraordinarily high initial dielectric constants, with marked decrease at higher levels of excitation, much as the magnetic permeability of steel decreases with "saturation."

Incremental effects may be studied by applying to the measuring bridge a steady d-c with a small superposed a-c. This may be done on several capacitance bridges, but the Schering circuit, used for the great majority of capacitance measurements, is easily applied to incremental determinations, and has excellent properties for the purpose. The circuit may be arranged as in Fig. 10-5, with a "biasing" direct voltage in series with an oscillator, both

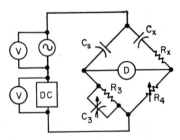

Fig. 10-5. Schering capacitance bridge with biasing voltage.

voltages being measured by appropriate voltmeters. The bridge balance equations are as derived in chap 9. No direct current (except leakage) flows with the power sources connected to the diagonal as shown in the figure, because of the blocking effect of the capacitors.

Nonlinear characteristics are met in the ceramic-type dielectrics (barium titanate, etc.) which are used for tuning purposes, for dielectric amplifiers, and for modulators. Another case in which a biasing voltage is needed is in the measurement of electrolytic capacitors. These capacitors (of unilateral type) are used in rectifier filter circuits, in which the capacitor functions with a combination of steady voltage and superposed ripple — it is therefore necessary to measure them in the same way. This problem is so important commercially that special bridges are made for the purpose, and other bridges have provisions for introducing the biasing voltage.

10-4. Resonance Bridge

Fig. 10-6. Resonance bridge.

The equations for this bridge may be obtained from Fig. 10-6 by inspection. Since arms 1, 2, and 3 are resistances, balance can be obtained only if arm 4 is purely resistive in over-all effect, that is, if

$$X_L = X_C$$

or

$$\omega L = \frac{1}{\omega C} \tag{7}$$

Resistance balance requires that

$$R_4 = \frac{R_2 R_3}{R_1} \tag{8}$$

This bridge has several applications. It may be used to determine the L and R of a coil if C and the frequency are known, by solution of (7) and (8). This is an accurate method of measuring the resistance of a coil, since with the reactance balanced out by the capacitor the resistance is measured as the sole element in the balance. This bridge is particularly appropriate for measurement of L and R of a coil at radio frequencies.

The resonance bridge may be used to determine frequency if L and C are known. From (7)

$$f = \frac{1}{2\pi\sqrt{LC}} \tag{9}$$

Another form of frequency bridge, using resistance and capacitance, will be presented in the next section.

10-5. R-C Frequency Bridge — Wien Bridge

The Wien bridge was used to a great extent in the past for the measurement of capacitance. Such measurements are now made most generally by the Schering circuit. The Wien bridge is presented here primarily as a frequency-measuring device; however, eqs. (13) and (16) may be solved for C_4 and R_4 if it is desired to use the bridge for capacitance determinations.

This bridge has a series R-C group in one arm, and a parallel R-C in an adjacent arm. A series R-C can be made equivalent to a parallel R-C at a particular frequency only. The bridge of Fig. 10-7 can, accordingly be expected to be sensitive to frequency; that is, a particular set of constants can give balance at *one* frequency only. Therefore, it can be used to determine frequency and, in fact, has its chief application for this purpose.

The impedances of arms 3 and 4 are

$$Z_3 = R_3 - j\frac{1}{\omega C_3} = \frac{\omega C_3 R_3 - j1}{\omega C_3} \qquad (10)$$

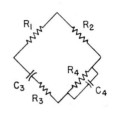

$$Z_4 = \frac{-j\dfrac{R_4}{\omega C_4}}{R_4 - j\dfrac{1}{\omega C_4}} = \frac{-jR_4}{\omega C_4 R_4 - j1} \qquad (11)$$

Fig. 10-7.   Wien bridge.

By equating $Z_1 Z_4 = Z_2 Z_3$, we obtain

$$-jR_1 R_4 \omega C_3 = R_2(\omega C_3 R_3 - j1)(\omega C_4 R_4 - j1) \qquad (12)$$

The real terms in the expansion give

$$\omega^2 R_2 R_3 R_4 C_3 C_4 - R_2 = 0 \qquad (13)$$

Therefore

$$\omega^2 = \frac{1}{R_3 R_4 C_3 C_4} \qquad (14)$$

or

$$f = \frac{1}{2\pi\sqrt{R_3 R_4 C_3 C_4}} \qquad (15)$$

The j terms in (12) give

$$R_1 R_4 C_3 = R_2 R_3 C_3 + R_4 C_4 R_2 \qquad (16)$$

Division by $R_2 R_4 C_3$ gives

$$\frac{R_1}{R_2} = \frac{R_3}{R_4} + \frac{C_4}{C_3} \qquad (17)$$

Some special combinations of constants are helpful in obtaining balance and in making computations.

Special Case No. 1
Make

$$R_1 = R_2$$
$$C_4 = \tfrac{1}{2}C_3$$
$$R_4 = 2R_3$$

Therefore

$$f = \frac{1}{2\pi R_3 C_3} \qquad (18)$$

Special Case No. 2
Make

$$R_1 = 2R_2$$
$$C_3 = C_4$$
$$R_3 = R_4$$

Therefore

$$f = \frac{1}{2\pi R_3 C_3} \qquad (19)$$

The second condition is convenient if it is desired to gang-control arms 3 and 4, making $C_3 = C_4$ and using equal variable resistors R_3 and R_4 that are controlled by a common shaft (or fixed resistors and variable capacitors). In this way we can secure a frequency-determining bridge balanced by a single control. If desired, the dial may be calibrated directly in terms of frequency.

Balancing is difficult unless the applied voltage is of pure wave form, due to the frequency-sensitive characteristics of the bridge. When the bridge is balanced for the fundamental frequency in the power supply, it is not balanced for harmonics, so the harmonics prevent a low null value if an amplifier-and-meter is used as detector, and tend to mask the balance point in ear phones.

As a result of its frequency characteristics the Wien circuit has another application, namely, as the frequency-determining element in some oscillators working in the audio and supersonic range.

10-6. Z-Y Bridge

The Z-Y bridge is called by its makers, the General Radio Company, a "universal" bridge, because it is capable of measuring any impedance from zero to infinity. The problems inherent in making measurements over a very wide range have been solved by measuring low values, up to 1000 ohms, as impedances, while higher values are measured as admittances. Actually, a small overlap has been provided, in that the impedance scale reads from zero to 1050 ohms, and the bridge components have been so selected that the same scale may be used for admittances in micromhos. The dial used for reactance is zero-center, so that both inductive and capacitive reactances may be measured without a change in connections; the same is true of the dial used for susceptances for admittance measurements. The resistance and conductance dials are also zero-center, so that it is actually possible to measure complex impedances in any of the four quadrants. Negative real components are, of course, not encountered with passive circuit elements; however, this capability of the bridge may be useful with certain amplifier or servo feedback elements.

The bridge is based upon a conventional resistance-capacitance arrangement, as shown in Fig. 10-8(a), but has a number of interesting features to provide the wide range of measurement, to give convenient adjustment, and to make provision for a zero balance before the unknown is introduced. The balance equation for the bridge may be written as,

$$Z_p = \frac{R_B(G_A + j\omega C_A)}{j\omega C_N} \tag{20}$$

which may be separated into,

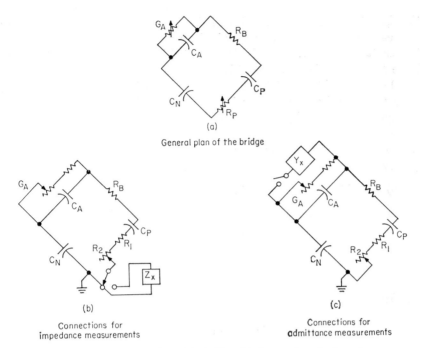

Fig. 10-8. (a) General plan of the bridge (b) Connections for impedance measurements (c) Connections for admittance measurements.

$$R_p = \frac{R_B C_A}{C_N} \qquad (21)$$

$$X_p = -\frac{R_B G_A}{\omega C_N} \qquad (22)$$

An unknown impedance is measured by placing it in series in the P arm, while an unknown admittance is connected in parallel with the A arm. One set of binding posts is provided for the unknown, and the change in position is accomplished by switching. The switch also has positions to remove the unknown from the circuit for the initial balance.

(a) *Impedance measurements.*

The switch serves, as shown in the diagram, first to bypass the unknown while the initial balance is made, and then to insert it in series in the P arm. From eq. (21), R_p is constant for the entire P arm, since it is related to three fixed quantities. Accordingly, R_2 must be decreased when the unknown is introduced, and the amount of the decrease is a measure of R_x. The initial balance is made with the main R_x dial set on zero, so the reading of this dial on final balance is the value of R_x. As shown by (22), G_A is the variable element used to balance X_x. G_A starts from a middle value with the dial set on zero, and hence either inductive or capacitive reactances may be measured, and the dial reading indicates both the amount and type of the reactance present. The dial is calibrated in ohms, and is direct reading if the frequency of measurement is the same as the frequency of calibration (1000 cps), provided that the Frequency Switch is in its middle position. Note that the factor ω occurs in the denominator of (3), and hence a multiplying factor, f_0/f, must be applied to the dial reading if the operating frequency, f, differs from the base, or calibration frequency, f_0. The range of reactance measurements may be extended in some cases by the Frequency Switch.

(b) *Admittance measurements.*

An unknown admittance is measured, as shown in the diagram, by placing it in parallel with arm A. The balance equations in revised form are, as obtained from (22)

$$G_A = \frac{C_N}{C_p R_B} \qquad (23)$$

and from (21),

$$B_A = \omega C_A = \frac{\omega C_N R_p}{R_B} \qquad (24)$$

Note in (23) that the total G_A is constant, and hence G_A (dial) must be decreased as G_x is added, and the G_x dial, starting from zero at the initial balance, gives a direct reading of the amount of conductance present in the unknown. From (24) we see that the susceptance is measured by R_p, accompanied by an ω factor. Readings on this dial, which may be either inductive or capacitive, must be multiplied by f/f_0 to obtain the susceptance in micromhos at the operating frequency.

We see that the G_A control that measures G for admittance measurements is the same that balances X in the measurement of impedance. This dial is marked, accordingly, "X or G." Similarly, the other dial is marked "R or B," and the dials carry markings for inductive or capacitive components, so the operator can tell at a glance the meaning of the adjustments.

The bridge is designed for the audio-frequency range from 20 to 20,000 cps. It is rated at an accuracy of approximately one per cent. Its chief feature is easy operation over a wide range of impedance values. A combination of initial and final balances is used to eliminate the effect of some of the bridge residuals.

10-7. Maxwell Commutated D-C Bridge for Capacitance

The Maxwell method of using commutated direct current in measuring capacitance is of considerable general interest, although it is not used in many laboratories on account of the equipment required. It has value as an absolute method of measurement, permitting determination of capacitance in terms of resistance and time (frequency). It has been used extensively at the Bureau of Standards.

The circuit consists of three resistors and a contact scheme for alternately charging and discharging the capacitance. A rotating commutator can be used, but the vibrating type is preferred. Figure 10-9 indicates

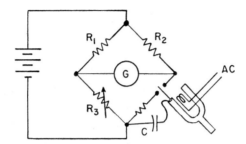

Fig. 10-9. Commutated d-c method for measurement of capacitance.

schematically a contact operated by a tuning fork which is driven electrically at its natural frequency. The detector is a d-c galvanometer which should have a period long in comparison with the period of the make-and-break, so that it averages the pulsating current flowing through it. The bridge is balanced to a null of the galvanometer by adjustment of one of the three bridge resistors.

The derivation of the exact balance equation is a rather long process, and not essential to our present purpose. It may be found in several references.* An approximate derivation may be made very easily. If a capacitor of capacitance C is charged to a voltage E, the charge stored in it is

$$Q = CE$$

If the charging process takes place in a time of $1/f$ sec, the *average* current is

$$I_{ave} = \frac{Q}{1/f} = fCE \tag{25}$$

If the charging process is repeated regularly at a frequency f, the arm of the bridge containing C has an apparent resistance given by E/I_{ave} or

$$R_{apparent} = \frac{1}{fC} \tag{26}$$

If we use this value of resistance in the usual Wheatstone bridge formula

---

*Reference 6, p. 121; reference 7, p. 375.

$$R_1 \frac{1}{fC} = R_2 R_3$$

we obtain

$$C = \frac{R_1}{fR_2 R_3} \tag{27}$$

where f is the frequency of the vibrating contact.

This derivation has neglected several things, particularly the resistances in the galvanometer and battery. The exact formula has the terms of (27) and a lengthy correction factor. Equation (27), however, is good to about one part in one thousand for a properly proportioned bridge.

The accuracy attained by this method at the Bureau of Standards is of the order of a few parts in 100,000. Curtis[†] states that under favorable conditions the uncertainty in a single measurement of a 0.25-μf air capacitor will not exceed 1 or 2 parts in a million. The method is not easily used in the ordinary laboratory as it requires the special contactor which must be driven at an accurately known frequency since this is a factor in the result.

10-8. Special Measuring Circuits

Numerous special measuring circuits have been devised to meet particular measuring needs in the most convenient manner. We shall study several of these devices, which — though not strictly bridge circuits — deserve to be included on account of their usefulness in certain types of measurement for which they either supplement or supplant bridge methods. The particular instruments or circuits which we shall consider are

    (1) Q-meter.

    (2) Bridged-T and parallel-T circuits.

    (3) Z-angle meter.

10-9. Q-meter

The Q-meter is an exceedingly useful instrument for measuring the characteristics of coils and capacitors at radio frequencies. The device is based on the fact that the ratio of the voltage across the coil or the capacitor of a series resonant circuit to the applied voltage is equal to the Q of the coil (if other resistances in the circuit are negligible). If a fixed voltage is applied to the circuit, a voltmeter across the capacitor can be calibrated to read Q directly.

The relationships for a series resonant circuit may be summarized by the following equations and the symbols of Fig. 10-10.

---

†Reference 6, p. 235.

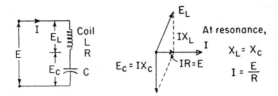

Fig. 10-10. Series resonant circuit.

For the resonant condition

$$\frac{E_c}{E} = \frac{IX_c}{IR} = \frac{IX_L}{IR}$$

Therefore

$$\frac{E_c}{E} = \frac{\omega L}{R} = Q \tag{28}$$

The basic Q-meter circuit is shown in Fig. 10-11. The current from the oscillator is adjusted to a fixed value I for all measurements, and this

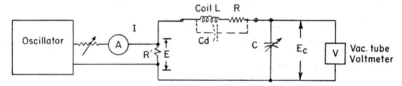

Fig. 10-11. Basic Q-meter circuit.

current flowing through the shunt R' introduces a fixed value of voltage, E, into the oscillating circuit. R' (0.04 ohm) is usually low in comparison with other resistances in the circuit, so practically all of the meter current I can be considered as flowing through it. Also the resistance it introduces into the oscillating circuit is negligible. The voltage E_c is read by a vacuum-tube voltmeter which has a direct-reading Q scale. The procedure in using the instrument is to connect the coil to the test terminals, tune the oscillator to the desired frequency, adjust C for resonance (as indicated by the greatest deflection for the E_c meter), adjust I to the fixed value, and read the E_c meter. This meter is calibrated to read Q directly if I is set to the unity index mark. The inductance of the coil can be computed from the known values of frequency and tuning capacitance C.

A coil has distributed capacitance between its parts, forming in effect a capacitive shunt across it. The capacitance has an effect on the impedance and phase angle of the coil, and this effect is dependent on the frequency. At any one frequency we can solve for the equivalent or effective values, L_e and R_e, which represent the L and R of an imaginary unshunted coil (no distributed capacitance) to give the same effect at the terminals. The Q-meter reads these effective values, which are, in general, the desired quantities. It is easy, however, to make determinations of L_e and R_e at several frequencies, and by computation or graph to solve for the distributed capacitance and the zero-frequency inductance (the "geometrical inductance").

A capacitor can be measured by connecting it in series with a coil to the test terminals and making two sets of readings, one with the unknown capacitor in circuit and the other with the capacitor shorted out. The constants of the capacitor can be computed from the two sets of readings. An alternative is to resonate a coil by means of the tuning capacitor, add the unknown capacitor in parallel with the coil, and retune. The change of reading of the tuning capacitor gives the capacitance of the unknown.

The choice of series or parallel connections depends on the magnitude of the unknown; lower values are measured in series, and higher values in parallel. General equations may be derived for either case, for application to any impedance, inductive or capacitive. The equations for the series connection are expressed most simply in series constants, and similarly parallel constants are used for the parallel connection.

(a) *Series connection, Fig. 10-12(a).*

If the circuit is resonated at a value C_1, with a Q-reading of Q_1 before introducing the unknown, then

$$\frac{1}{\omega C_1} = \omega L, \tag{29}$$

and
$$Q_1 = \frac{\omega L}{R} = \frac{1}{\omega C_1 R} \tag{30}$$

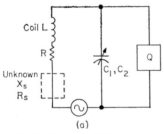
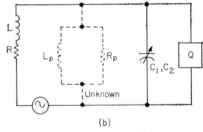

(a)
Series Connection Of Unknown

(b)
Parallel Connection Of Unknown

Fig. 10-12. Measurement of an unknown impedance in a Q-meter circuit. (C_1 = capacitance for first balance; C_2, second balance after connecting unknown.)

If an unknown with series components X_s (inductive or capacitive) and R_s is introduced in series, the new readings are C_2 and Q_2, giving for the unknown,

$$X_s = \frac{1}{\omega C_2} - \omega L = \frac{1}{\omega C_2} - \frac{1}{\omega C_1}, \tag{31}$$

or,

$$X_s = \frac{C_1 - C_2}{\omega C_1 C_2} \quad \begin{array}{l} \text{(inductive if } C_1 > C_2) \\ \text{(capacitive if } C_1 < C_2) \end{array} \tag{32}$$

The resistive component may be found in terms of reactance and Q,

$$R_s = \frac{X_2}{Q_2} - \frac{X_1}{Q_1} = \frac{1}{\omega C_2 Q_2} - \frac{1}{\omega C_1 Q_1} \tag{33}$$

$$= \frac{C_1 Q_1 - C_2 Q_2}{\omega C_1 C_2 Q_1 Q_2} \tag{34}$$

Therefore,

$$Q_x = \frac{X_s}{R_s} = \frac{(C_1 - C_2)Q_1 Q_2}{C_1 Q_1 - C_2 Q_2} \tag{35}$$

(b) *Parallel connection, Fig. 10-12(b).*

The characteristics of an unknown connected in parallel are expressed most simply in terms of parallel X_p and R_p. After the preliminary balance, as before, the addition of X_p requires an increase in tuning capacitance (if X is inductive) to a new value C_2, which may be regarded practically as C_1 resonating L as in the preliminary balance, and $(C_2 - C_1)$ resonating the unknown X_p, or,

$$X_p = \frac{1}{\omega(C_2 - C_1)} \tag{36}$$

To find R_p, we take the total conductance and subtract the conductance of the resonating coil,

$$\frac{1}{R_p} = \frac{\omega C_1}{Q_2} - \frac{R}{R^2 + \omega^2 L^2} \tag{37}$$

$$= \frac{\omega C_1}{Q_2} - \frac{1}{R} \cdot \frac{1}{1 + \frac{\omega^2 L^2}{R^2}}$$

$$= \frac{\omega C_1}{Q_2} - \frac{1}{R Q_1^2} \tag{38}$$

(In the last step, $1 + Q_1^2$ was replaced by Q_1^2).
After simplification and use of (30),

$$R_p = \frac{Q_1 Q_2}{\omega C_1(Q_1 - Q_2)} \tag{39}$$

Therefore,

$$Q_x = \frac{R_p}{X_p} = \frac{(C_2 - C_1)Q_1 Q_2}{C_1(Q_1 - Q_2)} \tag{40}$$

10-10. Bridged-T and Parallel-T Networks*

The bridged-T network consists of four impedances connected as shown in Fig. 10-13. It is possible to select combinations of impedances such that the detector connected from c to d gives a null indication. This circuit has the valuable feature that one side of the generator and one side of the detector are common, and this point may be grounded. The possibility of grounding both generator and detector is particularly desirable at radio frequencies; there stray capacitances make difficult the design of a satisfactory bridge circuit, one of the chief difficulties being in the construction of a shielded transformer of sufficiently low capacitance for connecting the oscillator or detector to the bridge. The transformer difficulty is removed in the T circuit by the possibility of grounding one side each of oscillator and detector.

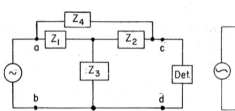

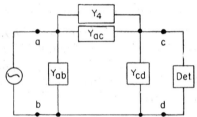

Fig. 10-13. Bridged-T network.   Fig. 10-14. Equivalent circuit derived from Fig. 10-11 by a wye-delta transformation.

The circuit may be analyzed most easily by a wye-delta transformation of the impedances Z_1, Z_2, and Z_3 to the form shown in Fig. 10-14. The detector response will be zero either if the impedance $c\text{-}d$ is zero or if the combined impedance $a\text{-}c$ is infinite. The former is not generally possible, hence the latter possibility is the one we shall investigate. It should be noted that the component impedances in the equivalent circuit may not be physically realizable, but they may nevertheless be used correctly for purposes of analysis.

Symbols have been placed in Fig. 10-14 in terms of admittances as being more convenient for this analysis. The admittances Y_4 and Y_{ac} are directly additive in the complex form, and the condition of infinite impedance a to c can be secured by the equivalent idea of equating the total admittance between these points to zero. The wye-delta transformation may be stated equally well in terms of admittances as of impedances. As this form is not so generally familiar, the formulas are given for convenience in Appendix II.

---

*Reference 8.

Figure 10-15 shows a circuit of this type that may be used for measuring a coil in terms of parallel inductance and resistance components. The equations for the detector null are

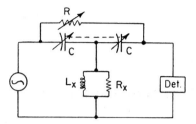

$$L_x = \frac{1}{2\omega^2 C} \qquad (41a)$$

and

$$R_x = \frac{1}{\omega^2 C^2 R} \qquad (41b)$$

Fig. 10-15. Bridged-T network for measurement of inductance.

The formula assumes that the two capacitors are of equal capacitance. This is not essential, as formulas may be derived using different values, but equal capacitances give simple relationships and may be secured conveniently by gang control. Note that the capacitance control balances the inductive component of the unknown, and the resistance control balances for the R_x (parallel) component. (The equations are not proved here, as a similar analysis is carried out below for the twin-T network.)

There are other possible circuit combinations that may be used in the bridged-T network. The one above serves as illustration of the idea, so we shall now turn our attention to the parallel-T.

The parallel-T network consists of two T networks in parallel between the oscillator and the detector. It has some of the properties of the bridged-T type and may be analyzed in the same way. The parallel-T has some advantages in circuit arrangement and balancing.

Figure 10-16 shows a circuit which forms the basis of the 821-A Twin-T Measuring Circuit formerly produced by the General Radio Company.

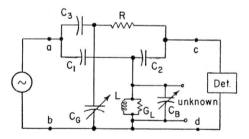

Fig. 10-16. Parallel-T measuring network.

This circuit incorporates the idea of an initial balance before connecting the unknown and a second balance after connecting it, so that the effect of stray circuit capacitances is eliminated. The constants of the unknown are found by the *difference* of the readings of C_G and C_B for the initial and final settings. The unknown is determined in terms of parallel conductance and susceptance components. The equations are

$$G_x = \frac{\omega^2 C_1 C_2 R}{C_3} \Delta C_G \tag{42}$$

$$C_x = \Delta C_B \qquad \text{(if capacitive)} \tag{43}$$

$$B_x = \omega \Delta C_B \qquad \text{(if inductive)} \tag{44}$$

The instrument is direct reading for capacitive unknowns, but requires computation from the susceptance found in (44) in case of inductive unknowns (ΔC_B is of opposite sign in the two cases).

The analysis of the circuit may be performed by wye-delta transformations. Figure 10-17 shows the transformations for the two T's separately

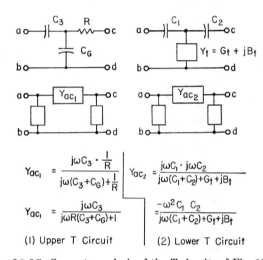

$$Y_{ac_1} = \frac{j\omega C_3 \cdot \frac{1}{R}}{j\omega(C_3 + C_G) + \frac{1}{R}} \qquad Y_{ac_2} = \frac{j\omega C_1 \cdot j\omega C_2}{j\omega(C_1 + C_2) + G_t + jB_t}$$

$$Y_{ac_1} = \frac{j\omega C_3}{j\omega R(C_3 + C_G) + 1} \qquad = \frac{-\omega^2 C_1 C_2}{j\omega(C_1 + C_2) + G_t + jB_t}$$

(1) Upper T Circuit (2) Lower T Circuit

Fig. 10-17. Separate analysis of the T circuits of Fig. 10-16.

for simplicity. Then the admittances a-c for the two parts can be added, and the total is equated to zero to derive the conditions for a null indication. The admittance in the upright of the lower T is indicated indefinitely by $G_t + jB_t$ so that we may solve for the characteristics needed at this place for balance.

For a null indication in the circuit of Fig. 10-16

$$\mathbf{Y}_{ac1} + \mathbf{Y}_{ac2} = 0$$

or $$\mathbf{Y}_{ac1} = -\mathbf{Y}_{ac2} \tag{45}$$

The values of ($\mathbf{Y}_{ac1}$) and ($\mathbf{Y}_{ac2}$) from Fig. 10-17 are substituted in (45), and the terms cross-multiplied to eliminate the fractions. Then the real and j terms can be separated, as in bridge equations, with the following results

$$G_t = \frac{\omega^2 C_1 C_2 R(C_3 + C_G)}{C_3} \tag{46}$$

$$B_t = -\omega\left(C_1 + C_2 + \frac{C_1 C_2}{C_3}\right) \tag{47}$$

Note, from (47), that the total susceptance B_t must be negative, — that is, inductive. This is the reason for the coil indicated by L in Fig. 10-16. The susceptance of L must be greater than the total capacitive susceptance in this branch by an amount that satisfies (47). The *total* B_t is the same for all balances. The fixed part consists of L and C_B, and the values are so chosen that balance can be attained before an unknown is added. If a capacitive unknown is connected, C_B must be decreased by an amount equal to C_x, which explains eq. (43). If, instead, an inductive unknown is connected, C_B must be increased, and B_x is computed from (44).

In eq. (46) for G_t, all terms are constant except the variable capacitance C_G. An initial value of C_G is needed to balance G_L. (G_L represents the small and unavoidable loss component of the coil L.) The change of C_G when the unknown is added, multiplied by $\omega^2 C_1 C_2 R/C_3$ as in (42) and (46), thus gives the conductance component of the unknown.

10-11. Z-Angle Meter

The Z-Angle Meter, as the name implies, measures an unknown directly in impedance and phase angle, rather than in inductance, capacitance, and resistance, as in the usual bridge. This device is useful for determining the constants of filters, loudspeakers, and other communication equipment. The meter is a development of the old idea of measuring an impedance by placing a variable resistor in series with it and adjusting the resistor until a voltmeter gives equal readings across resistor and unknown. The magnitude of the unknown impedance then is known to be the same as the resistance of the resistor and may be read directly if the resistor dial is calibrated.

The process of adjustment is tedious if a single voltmeter is used and switched back and forth from Z_x to R, as might be done in a laboratory setup. Also, an ordinary a-c voltmeter would be unsuitable because of the change in conditions produced by connecting it across one part of the circuit. The Z-Angle Meter overcomes these difficulties by using a double vacuum-tube voltmeter connected so that the indicating instrument reads the difference of the magnitudes of the two voltages. A zero indication on the instrument indicates the correct adjustment of R to make the two voltages equal, provided that the two sides of the vacuum-tube circuit have been correctly adjusted for equal response. After the two voltages have been adjusted to equality of magnitude, their difference can be used as a measure of the phase difference between them.

Figure 10-18 indicates the basic circuit. With the switch in "Bal" position, the variable resistance R is adjusted to bring the meter reading to

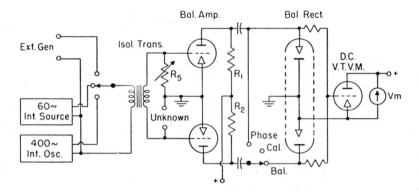

Fig. 10-18. Basic circuit of Z-angle meter. (*Courtesy of Technology Instrument Corp.*)

the null point (center scale), which thus indicates equality of the voltages across R and the unknown, as a result of the balanced-circuit construction. Note that in this position the voltages are rectified before the difference is taken, hence the comparison depends on magnitudes only without effect because of phase. With the switch at "Cal" the applied voltage is set to the standard value, then the switch is turned to "Phase," and the phase angle is read from the calibrated degree scale. In this position the outputs from the two halves of the amplifier are placed in parallel *before* rectification; therefore the voltage impressed on the detector is proportional to the vector sum of the output voltages. Due to the reversed grid-cathode relationships at the input to the amplifier, the output voltage depends on the phasor *difference* of E_z and E_R and so can be used conveniently as a measure of their phase difference.

The sign of the phase angle is obtained by depressing the button marked "Reactance Check" and noting the direction of the resultant meter deflection. An upward deflection indicates an inductive reactance, and a downward deflection indicates a capacitive reactance.

The Z-angle meter made by the Technology Instrument Corporation has a main dial calibrated in impedance in four ranges, 0–100, 0–1000, 0–10,000, and 0–100,000 ohms, selected by a range switch. The meter serves both as null instrument and phase-angle indicator as mentioned in the paragraph above. Power is supplied internally by either a 400-cycle oscillator, a 60-cycle connection, or by an external oscillator. Measurements may be made to 20,000 cycles to an accuracy within 1 per cent and to somewhat higher frequencies with increased error. Readings may be made very rapidly and conveniently with this instrument.

REFERENCES

1. Hague, B., *Alternating Current Bridge Methods*, London: Sir Isaac Pitman and Sons, Ltd., 1946.

2. Rosa, E. B., and Grover, F. W., "Measurement of Inductance by Anderson's Method," *Bull. Bur. Stds.*, **1**, 291 (1904).

3. Stroud, W., and Oates, J. H., "On the Application of Alternating Currents to the Calibration of Capacity Boxes, and to the Comparison of Capacities and Inductances, *Phil. Mag.*, 6th ser., **6**, 707–720 (1903).

4. Wilhelm, H. T., "Measuring Inductance of Coils with Superimposed Direct Current," *Bell Labs. Record*, **19**, 4, 131 (Dec. 1935).

5. Young, C. H., "A Five-Megacycle Impedance Bridge," *Bell Labs. Record*, **15**, 8, 261–65 (April 1937).

6. Curtis, H. L., *Electrical Measurements*, New York: McGraw-Hill Book Co. Inc., (1937).

7. Laws, Frank A., *Electrical Measurements*, New York: McGraw-Hill Book Co. Inc., 1938.

8. Tuttle, W. N., "Bridged-T and Parallel-T Null Circuits for Measurements at Radio Frequencies," *Proc. I.R.E.*, **28**, 23–29 (1940).

9. Hastings, A. E., "Analysis of a Resistance-Capacitance Parallel-T Network and Applications," *Proc. I.R.E.*, **34**, 126 (March 1946).

PROBLEMS

10-1. A bridge has the following constants: arm AB, $R = 800$ ohms in parallel with $C = 0.159$ μf; BC, $R = 1000$ ohms; CD, $R = 500$ ohms; DA, $C = 0.636$ μf in series with an unknown resistance. Find the frequency for which this bridge is in balance and the value of the resistance in DA to produce balance.

10-2. A bridge has in arm AB, $R = 800$ ohms in parallel with $C = 0.5$ μf; BC, $R = 400$ ohms in series with $C = 1.0$ μf; CD, $R = 1000$ ohms; DA, pure R of unknown value.
(a) Find the frequency for which this bridge is in balance.
(b) Find the resistance required in DA to produce the balance.

10-3. Prove eqs. (41a) and (41b) for the bridged-T network of par. 10.

10-4. A bridged-T has a capacitance C in each arm of the T and a resistance R in the upright. Find the type of impedance in the "bridge" member for which a null can be secured, and solve for its constants (in series components) in terms of R and C.

10-5. A bridged-T network consists of a pure inductance L in each arm of the T, and a resistance R in the upright. Find the R and L or C constants of the impedance (series constants) bridged across the top of the T so that there will be zero transmission between the input and the output terminals.

10-6. An a-c bridge has pure resistances in three arms: R_1 in AB, R_2 in BC, and R_3 in DA. Arm CD consists of a coil with series components L and R, paralleled by a variable capacitance C. Derive the balance equations for this bridge, to measure the constants of the coil in terms of the other components. Express Q of the coil in terms of bridge-arm constants at balance.

10-7. The bridge is the same as in problem 10-6, except that the inductive member is to be treated as a parallel combination of pure inductance L and a parallel pure resistance R. Derive the balance equations for determining L, R, and Q.

10-8. The horizontal members of a T-circuit consist of two equal pure inductances L. The vertical member, Y_x, consists of a parallel combination of a resistance R_x and an inductance L_x or capacitance C_x. Bridging the T is a series combination, R and C. Determine the relationships for Y_x for zero transmission through the network from input terminals to output terminals. (Either L_x or C_x is possible, depending upon the other constants. Find both expressions.)

10-9. Determine the constants of X (series components, R and L or C) for zero transmission through the network from the oscillator to the detector.

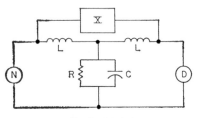

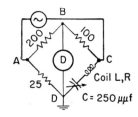

Prob. 10-9.   Prob. 10-10.

10-10. The bridge is in balance, with an applied voltage of 1-Mc frequency.

(a) Find the R and L of the coil.

(b) If a 200-$\mu\mu$f capacitor is added across arm AB, what value of C is now needed for balance?

(c) If the 200-$\mu\mu$f is added across AD instead of AB, find C for balanced conditions.

10-11. A parallel-T circuit has a resistance R in each arm and a capacitance $2C$ in the upright of one T. The other T has a capacitance C in each arm and a resistance $\frac{1}{2}R$ in the upright. Find the frequency for which the parallel-T gives zero transmission from input to output terminals. (This circuit is called an R-C filter. It suppresses a narrow band of frequencies. If used in the degenerative feedback loop of an amplifier, it permits amplification of the critical frequency, and suppresses others. See Reference 9.)

CHAPTER 11

Measurement of Mutual Inductance

11-1. Uses of Mutual Inductance in Bridge Circuits

Mutual inductance has been used in a great number of bridges for a variety of purposes. Known mutual inductances are used in some circuits for the measurement of unknown mutual inductances. Variable standard mutual inductances have been used as components in bridges for the measurement of self-inductance, capacitance, and frequency. Many such bridges are found in the literature, but some of the circuits represent minor modifications of other bridges; that is, changes to achieve greater accuracy or convenience for a particular type of measurement, or a special range of unknown.

We shall study only a small number of these circuits. Applications to capacitance and self-inductance determinations are of secondary interest because methods using a capacitance standard are generally more convenient and accurate for these measurements than a mutual inductor. They are also cheaper, more compact, and more generally available. The mutual inductance circuits may have particular merit in some cases, but this enters the field of special research rather than of general measurements, so we shall not attempt to go into much detail. Much reference material is available on the subject.

The derivation of the balance equations requires a different method when we have mutual coupling between the arms. It is necessary to write equations for the voltages around the loops and then to make solution for the unknown quantities.

11-2. Series-opposition Circuit*

This is not a bridge circuit, but it is a simple method of measuring mutual inductance if a variable mutual inductor of suitable range is available; hence it deserves mention at this point.

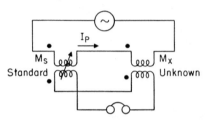

The two primary windings are connected in series to the oscillator and the two secondaries in series with the detector with one secondary voltage bucking the other. The dots on the inductors in Fig. 11-1 are relative polarity marks and are placed on the drawing in this case to empha-size the cross-over in connections to secure bucking action.

Fig. 11-1. Series-opposition circuit for measurement of mutual inductance.

The standard inductor is varied until the detector gives zero indication. For this setting

$$M_x = M_s \tag{1}$$

This is true, since, in the complex notation, the induced secondary voltage in either inductor equals $j\omega M \mathbf{I}_P$, and $\mathbf{I}_P$ is the same for both. Therefore the voltages can cancel only if the two mutual inductances are equal.

This method has had some modifications, by resistance shunts or otherwise, to permit use with fixed inductors, or where one inductance is larger than the other. Since the modified methods are subject to some errors, bridge methods seem more desirable if the method of Fig. 11-1 is not directly applicable.

11-3. Mutual Inductance Measured by Self-Inductance

If the terminals of the two coils whose mutual inductance is to be measured are available for series connection, measurement may be made by one of the self-inductance bridges considered in chap. 9. If connection is made so that the magnetic fields of the two coils are additive, the effective inductance of the two coils in series is

$$L_{e1} = L_1 + L_2 + 2M \tag{2}$$

If the connections to one coil are reversed

$$L_{e2} = L_1 + L_2 - 2M \tag{3}$$

By subtraction of (3) from (2)

---

*Reference 1, p. 414.

$$M = \tfrac{1}{4}(L_{e1} - L_{e2}) \tag{4}$$

The mutual inductance is thus obtained as one-fourth the difference of the self-inductances measured with series-additive and series-subtractive connections. This method is desirable only for fairly high coupling between the coils, otherwise the accuracy is poor due to the subtraction of nearly equal values in (4).

11-4. Heaviside Mutual Inductance Bridge*

The bridge of Fig. 11-2 may be used to measure mutual inductance in terms of a known self-inductance and known resistances. The same bridge, slightly modified, was used by A. Campbell to measure a self-inductance in terms of a known mutual inductance.

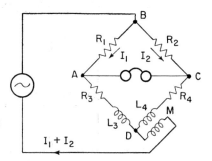

Fig. 11-2. Heaviside mutual inductance bridge.

The equations for balance may be derived by writing equations for the voltages in the two sides of the circuit. Zero response in the detector requires that

$$I_1 R_1 = I_2 R_2 \tag{5}$$

Also, the sum of the voltages from B to A to D must equal the sum B to C to D since these are parallel branches. This gives the equation

$$I_1(R_1 + R_3 + j\omega L_3) = I_2(R_2 + R_4 + j\omega L_4) + (I_1 + I_2)j\omega M \tag{6}$$

Substitution in (6) of one current in terms of the other from (5) gives the equation

$$R_2 R_3 + j\omega R_2(L_3 - M) = R_1 R_4 + j\omega R_1(L_4 + M) \tag{7}$$

By separation of real and j terms we have the resistance-balance requirement

$$R_2 R_3 = R_1 R_4 \tag{8}$$

and the equation for M

$$M = \frac{R_2 L_3 - R_1 L_4}{R_1 + R_2} \tag{9}$$

It should be noted that L_4, the self-inductance of one winding of M, must be known before M can be measured by this method.

---

*Reference 1, pp. 434, 437.

11-5. Carey Foster Bridge; Heydweiller Bridge*

This bridge was used ballistically by G. Carey Foster in 1887 and modified for a-c use by A. Heydweiller in 1894. Both names are met in connection with the bridge, which has been put to the two opposite uses of measuring capacitance in terms of mutual inductance and of measuring mutual inductance in terms of capacitance. The name Carey Foster is usually attached to the bridge for determination of capacitance, and Heydweiller for the reverse use.

Figure 11-3 shows the bridge with the addition of a Wagner earth connection. As drawn here, we have the unusual feature of one arm with zero

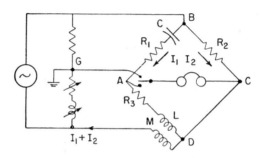

Fig. 11-3. Carey Foster bridge with Wagner earth circuit.

impedance. Diagrams may be found which look somewhat different, since with *C-D* shorted the other branches can be skewed around into different shape.

For zero response of the detector

$$I_1\left(R_1 - j\,\frac{1}{\omega C}\right) = I_2 R_2 \tag{10}$$

and for zero voltage *A-D* (since $E_{CD} = 0$)

$$I_1(R_3 + j\omega L) - (I_1 + I_2)j\omega M = 0 \tag{11}$$

The negative sign is taken for the mutual term, since it can be seen from the equation that negative coupling is needed for balance. Solution of these equations gives

$$M = R_2 R_3 C \tag{12}$$

The other condition derived from (11) and (12) is

$$L = \frac{M(R_1 + R_2)}{R_2} = CR_3(R_1 + R_2) \tag{13}$$

---

*Reference 1, p. 457; reference 2; reference 3, p. 6.

If the bridge is used for the measurement of capacitance, eqs. (12) and (13) may be rewritten,

$$C = \frac{M}{R_2 R_3} \tag{14}$$

$$R_1 = \frac{R_2(L - M)}{M} \tag{15}$$

M represents the *magnitude* of the mutual inductance, since it has already been assumed to be a negative quantity. In the measurement of capacitance, R_1 is not a separate unit but represents the equivalent series resistance of the capacitor, which may thus be determined from the constants of the bridge.

11-6. Campbell's Bridge for Comparison of Mutual Inductances*

The circuit is indicated in Fig. 11-4. There are two steps required in the balancing process,

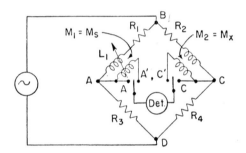

Fig. 11-4. Campbell's bridge for comparison of mutual inductances.

(1) Detector connected from A to C. This is a simple inductance-comparison bridge, which may be balanced by adjustment of R_3 (or R_4) and R_1 (or R_2). The requirement is,

$$\frac{R_3}{R_4} = \frac{R_1}{R_2} = \frac{L_1}{L_2} \tag{16}$$

(2) Detector connected from A' to C''. Keep adjustments as in (1). Adjust M_1 for balance. Then

$$\frac{M_1}{M_2} = \frac{R_3}{R_4} \tag{17}$$

*Reference 1, p. 422.

REFERENCES

1. Hague, B., *Alternating Current Bridge Methods*, London: Sir Isaac Pitman and Sons, Ltd., 1946.

2. Foster, G. Carey, "Note on a Method of Determining Coefficients of Mutual Induction," *Proc. Phys. Soc.*, **8**, 137–146 (1887).

3. "Some Methods of Measuring Inductance, Capacitance and Resistance," *Cambridge Inst. Co.* (1930).

CHAPTER 12

Shielding; Shielded Bridges

12-1. Coupling between Circuit Elements

We have assumed in our derivation of basic circuit relationships that a bridge consists of lumped impedance units connected only by the wires that we place in the circuit as connectors. This assumption works fairly well if the frequency is low, if the component impedances are not too high, and if the accuracy requirements are not too great. Actually, there are stray couplings from element to element and from element to ground, and these stray effects modify the conditions for bridge balance.

We speak of electric field and magnetic field couplings as separate entities, although they cannot strictly be considered as existing independently of each other. However, at low and moderate frequencies the two effects are sufficiently independent that we may consider each type separately. Accordingly, we refer to electromagnetic shielding and electrostatic shielding.

A magnetic field is produced by a current flowing in a wire, and more strongly by a coil of wire. For the small currents encountered in most of the bridge circuits the magnetic effect of a single turn is negligible, but the field due to a coil may be quite important. Particularly when there are two coils in the circuit, it is important to control the magnetic coupling between them. This can be done by large spacing, by the orientation of the coils (axes at 90°), by the shape of the coils, and by shielding. A particularly good shape for this purpose is the toroidal coil, for it has practically zero

319

external field. Coils with closed cores of magnetic material usually have small external fields and are easily shielded. Among the variable inductors, the Brooks inductometer (chap. 13) has a rather limited external field, due to two neighboring sets of coils connected for opposite directions of magnetic field.

Magnetic shielding may be accomplished by one or more shells of high-permeability magnetic material surrounding the part to be shielded. Multiple shells can be made to give very complete shielding. Alternating magnetic fields may also be screened by interposing highly conducting metal sheets such as copper or aluminum in the path of the magnetic flux. The eddy currents induced in the sheet give a counter mmf that tends to cancel the magnetic field beyond the sheet. This form of shielding becomes increasingly effective at higher frequencies. The shield cans used around coils in radio sets are examples of this type of shielding.

The major problem in most bridges is control of electric fields, that is, of capacitance effects between bridge components, and from them to ground. These effects can be minimized by wide separation of the parts, but while it is desirable to keep parts with large voltage difference well separated, it is not practicable to depend on this method entirely because it makes the bridge too bulky and also gives long leads (with inductance and capacitance) to join the parts. The effective way to control capacitances is to enclose bridge components in conducting shields connected to place the capacitance where it does no harm.

12-2. Shielding of Circuit Elements

Figure 12-1(a) and (d) indicate capacitive effects from resistors to their

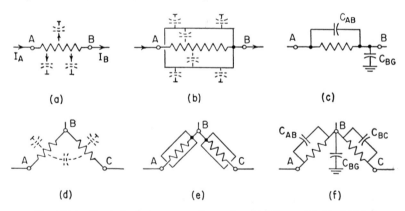

(a) (b) (c)

(d) (e) (f)

Fig. 12-1. Control of stray capacitance by shielding. (a) Unshielded resistor. (b) Resistor with shield. (c) Equivalent circuit of (b). (d) Capacitive coupling between resistors. (e) Resistors with shields. (f) Equivalent circuit of (e).

surroundings and between resistors. Some of these effects could be tolerated if they were constant, but in an unshielded bridge they change as pieces of equipment are moved in the vicinity of the bridge or as the operator changes his position with respect to the bridge. With capacitances present the current is different in different parts of one resistor, so the relationship $E = IR$ no longer has definite meaning, as I is indefinite. The uncertainties can be removed by enclosing the resistor in a metal shield connected to one end of the resistor, as shown in (b). There is capacitance from the resistor to the shield, but this is definite in amount and is independent of any other objects placed near the resistor. There is capacitance from the shield to ground (unless point B is at ground potential) but there is no capacitance from A to ground, except for effects beyond the limit of the shield. We say that the capacitance of the resistor to ground is concentrated at terminal B. Sketch (c) shows an equivalent circuit for the resistor. If both resistors of (d) have shields connected at the common point, as shown in (e), we have an effect for each as in (c), but no capacitance directly from one resistor to the other. The entire capacitance to ground is concentrated at terminal B.

The effect of shielding a resistor is to make its resistance definite. The shunt capacitance across it is increased somewhat in amount, but is made definite. Capacitance from the resistor to its surroundings is concentrated at the *terminals*, instead of being spread along the resistor as in the unshielded case. Similar results are obtained with shields on a capacitor and also with an inductor which has properly formed coils and shields.

Control of the position of the capacitance effects makes it possible to eliminate some errors that would otherwise be introduced into bridge measurements. If a capacitance can be placed across a *diagonal* of a bridge, it causes no error in the balance conditions but merely shunts the oscillator or detector, as the case may be. The shunt might be open to objection if it were very large, but the small amount found on most bridges neither overloads the generator nor decreases the sensitivity of the detector to a noticeable extent.

12-3. Double-shielding in Bridge Circuits*

In the preceding section it was shown that the resistance value of a resistor can be made definite and its capacitance to ground concentrated at one terminal by means of a shield. In this section we shall study further how a great number of the capacitances in a bridge circuit can be arranged to avoid an effect on the balance.

---

*Reference 3.

Figure 12-2 shows a bridge with shielded ratio arms AB and BC which are surrounded by an outer shield which is connected to terminal D. By this means the capacitance from the inner shield to the outer is in effect from B to D; that is, it simply shunts the detector, and does not affect bridge balance. The winding-to-shield capacitance of the detector transformer is also in shunt from B to D.

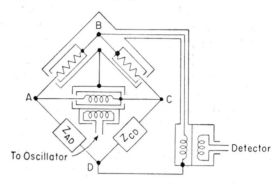

Fig. 12-2. Bridge with double-shielded ratio arms.

The input transformer secondary is shown with a shield connected to C. The winding-to-shield capacitance is thus placed from A to C, which simply shunts the supply and does not affect balance. The transformer has an outer shield connected to D to make capacitances definite. The inter-shield capacitance is thus from C to D, shunting Z_{CD}; it does enter into the balance relationships, and is the one capacitance not balanced out by this shielding arrangement. Its effect can be balanced, however, by an adjustable capacitance across the AD arm. This is easily arranged for a bridge with fixed one-to-one ratio arms.

Nothing was said above about grounding the bridge. For some measurements point D may be grounded directly, and for other types of measurement the bridge may be grounded by a Wagner earth connection.

The Campbell-Shackelton shielded ratio box used in the L&N shielded capacitance bridge consists substantially of this circuit, with the addition of a Wagner ground. There is one difference, in that the input transformer has a split shield, half connected to A and half to C. In this way the inter-shield capacitance is divided equally from A to D and C to D.

12-4. Leeds and Northrup Shielded Capacitance and Conductance Bridge

The L&N shielded capacitance bridge is primarily an equal-arm comparison bridge, with shielding and ground circuit added, so that it may be used for precision measurements of small capacitors and samples of insulating materials. The ratio box is as discussed in the preceding section.

The ground circuit consists of coarse and fine resistance adjustments and a double-stator capacitor, as shown in Fig. 12-3(b). The circuit is completed by a precision air capacitor C_1, a six-dial resistance box R_2, and a balancing unit made up of R_3 and C_3.

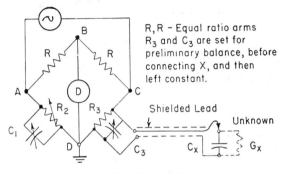

R, R – Equal ratio arms
R_3 and C_3 are set for
preliminary balance, before
connecting X, and then
left constant.

(a) Basic Circuit, Shielding Omitted

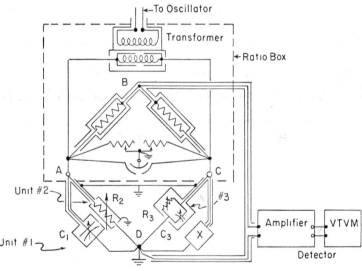

(b) Complete Circuit Diagram, Showing Shielding

Fig. 12-3. Circuit of the L&N shielded capacitance and conductance bridge.

The general procedure is to make a preliminary balance before connecting the unknown, using the ground-circuit adjustments (C_3 if necessary), with C_1 set at a low value and R_2 and R_3 at convenient and equal values. The unknown is connected, and a second balance is made by C_1 and R_2, other controls being kept constant. The leads to X should be in

place for the preliminary measurement, with the high-side lead disconnected at X. In this way the capacitance of the leads becomes a part of the
initial balance and does not enter into the value of X. The constants of X
are expressed in terms of the parallel components of the known arm. The
change of C_1 is equal to C_x since the ratio arms are equal. That is, if we use
plain letters to denote the initial value and primes for the second balance,

$$C_x = C_1' - C_1$$

R_2' equals the parallel combination of R_3 and R_x, and since R_2 equalled R_3
initially, we may write

$$R_2' = \frac{R_2 R_x}{R_2 + R_x}$$

Solution for R_x gives the value,

$$R_x = \frac{R_2 R_2'}{R_2 - R_2'} \qquad \text{ohms}$$

or, as a conductance,

$$G_x = \frac{R_2 - R_2'}{R_2 R_2'} \qquad \text{mhos}$$

Figure 12-4 shows a picture of the bridge, with the ratio box in the
center, decade resistance R_2 at the left, capacitor C_1 at the right front,
and balancing unit $C_3 R_3$ at the right rear. Connecting leads are run inside
metal tubes, for shielding. The bridge has a range up to $1100\mu\mu f$ on C_1,

Fig. 12-4. L&N shielded capacitance and conductance bridge. (*Courtesy of Leeds & Northrup Co.*)

guaranteed to $\pm 1\mu\mu f$, and may be used up to 50,000 cycles. In addition to the type of measurement described above, with D grounded, it can be used with D ungrounded for measurement of three-terminal capacitances, for substitution measurements or for measurements of inductance within a certain range.

The shielded ratio-arm construction and the grounding circuit would not be practical if the ratio arms were variable, because the capacitance effects would be different for each adjustment and could not be balanced out with certainty. The fixed and equal-ratio-arm construction used in this bridge keeps the stray capacitances fixed in amount and permits them to be included in the initial balance so that they do not produce error in the measurement of the unknown. On this account, small capacitances can be measured accurately.

12-5. Three-terminal Capacitances*

If both terminals of a capacitor are insulated from ground we have the condition shown in Fig. 12-5(a). Both sets of plates have capacitance to

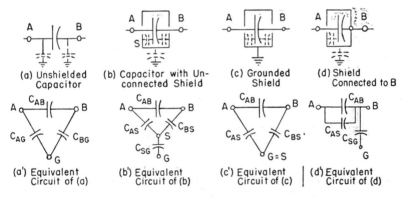

Fig. 12-5. Examples of three-terminal capacitors.

ground, as indicated in (a) by dotted lines. The equivalent circuit is shown in (a)'. We have in reality a delta connection of three capacitors, of which two are the ground capacitances. With large capacitors the ground capacitances may possibly be small enough in comparison to be ignored. When measuring small capacitances, such as a sample of insulating material or the interelectrode capacitance of vacuum tubes, the effects to the third terminal may be large in comparison and even larger than the direct capacitance whose value is desired. Conditions may be modified by shield-

---

*Ref. 5, p. 343.

ing or grounding, but in many cases we have the three-terminal effect indicated above.

With a shield added as in (b), the equivalent circuit is as shown in (b)'. If the capacitances are known, the equivalent circuit may be changed to the form of (a)' between terminals A, B, and G by means of a wye-delta transformation.

If the shield is grounded as in (c), the equivalent circuit is the same as (a)', with the added fact that the capacitances of A and B to ground are now definite.

If the shield is connected to B, as in (d), the capacitance from B to shield is short-circuited, and the capacitance from A to shield is added to C_{AB}. If point B may be grounded, the capacitance from shield to ground is short-circuited, and we have a two-terminal capacitor, with the value $C_{AB} + C_{AS}$.

Figure 12-6(a) shows a method used in testing samples of insulating material in sheet form. A is a circle of metal foil, usually attached to the

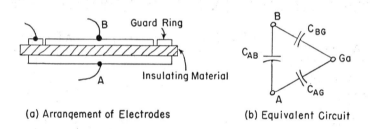

(a) Arrangement of Electrodes   (b) Equivalent Circuit

Fig. 12-6. Guard-ring electrodes for testing insulating material.

insulating material by rolling it on with a thin coating of petrolatum between them. B is a smaller foil circle, outside of which is a guard-ring electrode with a small gap between it and B. The purpose of the guard ring is to give a uniform rectilinear field between A and B free from uncertainties due to fringing of the electric flux at the edge. Sketch (b) shows the equivalent circuit, which is seen to be a three terminal capacitance. In use it is the direct capacitance A to B that is desired. The bridge should be arranged to give the guard ring the same potential as B, but not to measure its capacitance to A.

It is necessary in measuring capacitances, particularly small values, to remember the presence of the additional capacitances and to arrange the method so that no error is produced by them. In general, it is the "direct capacitance" A to B that is desired.

Figure 12-7 shows a method of measuring a direct capacitance on an ordinary grounded-point bridge. The subscript system is simplified here; C_x represents the desired direct capacitance, and C_a and C_b the capacitances to the third terminal. In this circuit C_b is directly in parallel with the detector and, hence, does not affect balance. C_a is across arm BC and affects the dissipation (loss) balance by requiring an addition to C_1 for balance, but does not affect the capacitance balance of the bridge unless it has appreciable losses. The circuit shown here is the Schering circuit of chap. 9, par. 12.

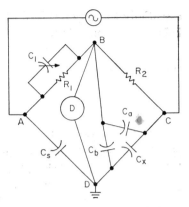

Fig. 12-7. Measurement of three-terminal capacitance on a grounded-point bridge.

A somewhat different method may be used on a bridge equipped with a Wagner earth circuit. This is shown on a comparison bridge in Fig. 12-8

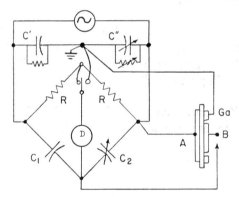

Fig. 12-8. Measurement of three-terminal capacitance on Wagner earth bridge. *Note:* The conductance component of C_{AB} may be balanced by capacitances across the R's or by resistances across C_1 and C_2. (See Figs. 12-3 and 12-13.)

with particular reference to the guard-ring electrode of Fig. 12-6, but it may be used with any three-terminal capacitance. C_2 is used for preliminary balance and for a second balance after the unknown is added, so that C_{AB} is the difference of initial and final values of C_2. Note that with balance of both main bridge and ground circuit, B and Ga are at the same potential, which is one result we wish to attain. Also, the capacitance from A to Ga is part of the ground circuit, in parallel with C'', and hence does not enter into the measured results. The L&N bridge of par. 4 may be used in essentially the way outlined above, but with resistors across C_1 and C_2 to balance the conductance component of the unknown.

(See par. 6 for an additional method of measuring a three-terminal capacitance.)

12-6. Ground Circuits and Guard Circuits

The Wagner earth is the most important and best known of the ground circuits. It was discussed in chap. 9, so its features will be mentioned very briefly here. Its use brings the detector to ground potential, hence capacitances from the detector corners of the bridge to ground are short-circuited. Also, capacitances from generator corners of the bridge to ground are in effect part of the ground circuit, and so do not affect balance. In order to get rid of effects within the power source it is necessary that the capacitances in it be concentrated at its terminals; this result is obtained by the use of a shielded input transformer.

The Wagner earth connection can thus remove disturbing effects due to all stray capacitances on a simple bridge if the stray capacitances are concentrated at the terminals. On special bridges, such as the Schering bridge when used to measure exceedingly small capacitances, the situation is more complex because there are special guard rings and guard circuits that must be maintained at the correct potential. The guard circuits have some points of similarity to the Wagner circuit, but have different detailed application and purposes.

A guard circuit is shown on a capacitance bridge in Fig. 12-9, used to give independent measurement of one capacitance of a three-terminal

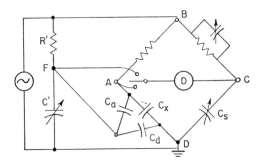

Fig. 12-9. Guard circuit on a capacitance bridge.

capacitance. With balance of both circuits, point F is at the same potential as A, so C_a has no effect. C_d is in parallel with C', and hence does not enter into the main bridge balance.

12-7. Shielded Schering Bridge*

The Schering bridge is a useful and versatile circuit and, accordingly, deserves more attention than we could give it in chap. 9. At that point we studied only the basic circuit and did not go into matters of shielding or details of adjustments.

It may be interesting to note that the circuit was devised by Phillips Thomas of the Westinghouse company and described by him in U. S. Patent 1,166,159, filed in 1913, and issued in 1915. The early development was mainly in Germany, however, by Professor Schering and others, so that the bridge now universally carries Schering's name. It has been developed in various forms for a wide variety of purposes. Most precision capacitance bridges, particularly those for the measurement of insulators, samples of insulating materials, and insulating oils, are of this type.

The bridge, arranged much as in chap. 9, is used for many low-voltage laboratory measurements. Shields are generally added to the resistors, capacitors, and transformers for the reasons discussed in par. 2 of this chapter. A selection of ratio-arm values may be provided to extend the range of the bridge.

Figure 12-10 shows the General Radio type 716-C capacitance bridge. It employs the Schering circuit, a complete system of shielding and multiple ratios. The variable standard capacitor has a capacitance from 100 to 1100 $\mu\mu$f, which, with ratios of 1, 10, 100, and 1000, covers values of the unknown up to 1.1 μf. The loss-adjustment dial is calibrated directly in dissipation factor, reading to 0.06 on the main dial, with ten additional steps of 0.05 on a parallel condenser. Switching is provided so that the bridge can be used for measurements by the substitution method, in which case it covers a range up to 1000 $\mu\mu$f with the internal standard, or higher with an external standard. The bridge has a frequency range from 30c to 300 kc. Power may be applied to either diagonal of the bridge, as desired. The circuit diagram is shown in Fig. 12-11. A modified form of this bridge is available for operation at 1 Mcs.

Figure 12-12 shows the circuit of an L&N bridge used for measurements of insulators, insulating oil, and the like, at high voltages. The standard air capacitor used with this bridge has a capacitance of 100 $\mu\mu$f and is available in two models: one for 10 kv and the other for 25 kv. The circuit is a modification of the Schering bridge. A guard circuit is provided so that the corners A and C and the detector are at ground potential. The bridge can be used to measure samples ranging from 40 $\mu\mu$f to 0.02 μf, and has a useful power-factor range from 0.0001 to 0.70. The C_4 dials are made direct reading in dissipation factor at 60 cycles by the value of 2652.5 ohms

---

*Reference 5, pp. 353–367, p. 578; references 6, 7, 8, 9, and 10.

Fig. 12-10. General Radio type 716-C capacitance bridge. (*Courtesy of General Radio Co.*)

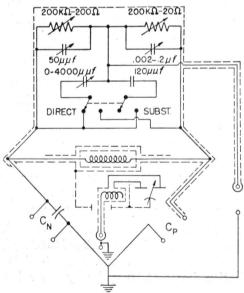

Fig. 12-11. Circuit diagram of the type 716-C capacitance bridge. (*Courtesy of General Radio Co.*)

selected for R_4; this makes $\omega R_4 = 10^6$, so $D = C_4$ in microfarads. (For small angles, power factor $\simeq D = \omega C_4 R_4$.)

The circuit has safety gaps to prevent high voltage on the control panel in the event of breakdown of the test specimen. The detector consists of an amplifier and a vibration galvanometer.

12-8. Conjugate Schering Bridge

The conjugate Schering bridge has the detector connected from the junction of the two capacitance arms to the junction of the resistance arms, which is the conjugate of the arrangement shown in Figs. 9-11 and 12-12.

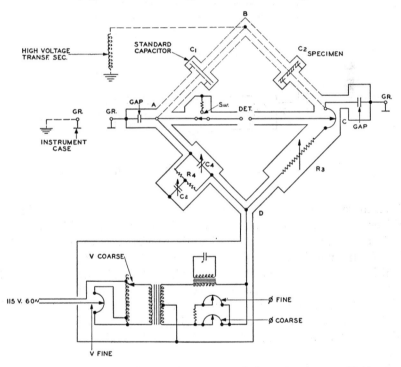

Fig. 12-12. Schematic diagram of modified Schering bridge circuit. (*Courtesy of Leeds & Northrup Co.*)

(The conjugate bridge is used ordinarily in the Type 716-C bridge of par. 7, except when using high voltage at 60 cps.) The conjugate arrangement has advantages of detector sensitivity and grounding when measuring very small samples for which the impedance is high.

Figure 12-13 is a diagram of a conjugate bridge with simple shielding.*

---

*Reference 10.

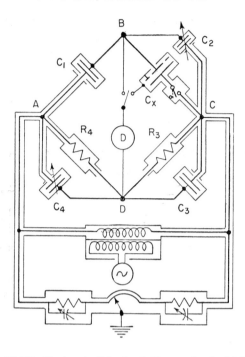

Fig. 12-13. Conjugate Schering bridge — simple shielding.

Much more elaborate shielding and grounding schemes are used in some bridges, but we do not have space to go into greater detail here.

As shown here, R_3 and R_4 are identical resistors of the order of 10,000 ohms for frequencies below 10 kc, and about 1000 ohms at 1 Mc. Measurement is made by the substitution method, so C_2 must have a range greater than the capacitance of the unknown and requires accurate calibration. C_4 balances the loss component of C_x, and C_3 is needed to insure a positive reading of C_4 when C_2 alone is in the circuit. C_1 is a fixed condenser for balancing. The usual grounding circuit and shielded input transformer are used.

The sequence in measurements is as follows:

(1) C_x in place and its switch closed. Balance the bridge by successive adjustments of C_2, C_4, and the Wagner ground. Call the values C_2' and C_4'.

(2) Turn the switch of C_x to the shield. Repeat the balancing process with C_2, C_4, and Wagner ground. Call these values C_2 and C_4. Then the capacitance of the unknown is

$$C_x = C_2 - C_2' \qquad (1)$$

The dissipation factor of X is

$$D_x = \frac{C_2}{C_2 - C'_2} \times \frac{\omega R_4(C'_4 - C_4)}{10^{12}} \tag{2}$$

if the capacitances are in $\mu\mu$f.

12-9. General Radio Type 1606-A Radio-frequency Bridge

The Type 1606-A bridge is an interesting example of the use of shielding to control capacitance between bridge components. The shielding is of multiple nature to maintain constant electric field configuration for the adjustable elements.

A schematic diagram of the circuit is shown in Fig. 12-14. It is, in general, of the Schering type, but with modifications to adapt it to induc-

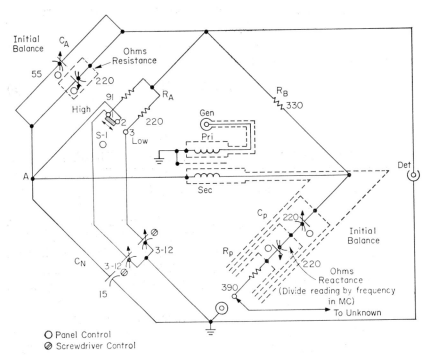

O Panel Control
Ø Screwdriver Control

Fig. 12-14. Schematic diagram of the Type 1606-A radio-frequency bridge. (*Courtesy of the General Radio Co.*)

tance and capacitance measurements at radio frequencies. One feature needed for accuracy of measurement at high frequency is double balancing; that is, a first balance before inserting the unknown, and another after it is connected. Separate dials are provided in this bridge, so that the main dials may be set on zero (or other desired point) while the initial balance is made by separate controls which are then locked during measurement of the unknown.

The balance equations, in terms of the lettering in the figure, are

$$R_P = R_B \frac{C_A}{C_N} \tag{3}$$

$$\frac{1}{j\omega C_P} = \frac{R_B}{R_A} \frac{1}{j\omega C_N} \tag{4}$$

The bridge is adjusted by making a first balance with the "Unknown" terminals short-circuited, using dials C_A' and C_P'. Then with the unknown in circuit the reactive balance is made on C_P and the resistive balance on C_A. The series-substitution method means that the total reactance of arm P is the same as before (and always capacitive), since the reactance adjustment is made in arm P, for which the reactance is determined by the three fixed quantities, R_A, R_B, and C_N, as shown by (4). The change of resistance of P upon insertion of the unknown is measured by the change of C_A. The constants of the unknown are given by

$$R_x = R_B \frac{C_{A2} - C_{A1}}{C_N} \tag{5}$$

$$X_x = \frac{1}{\omega}\left(\frac{1}{C_{P2}} - \frac{1}{C_{P1}}\right) \tag{6}$$

where subscripts 1 and 2 refer to the initial and final balance readings, respectively, on the main dials. The reactance dial is calibrated in ohms at a frequency of 1 Mc, and must be corrected by the proportion of 1 Mc to the actual frequency.

The resistance and reactance balances are independent of each other, and each depends on the setting of a variable air capacitor, which is the best arrangement for high frequencies. In this way, fixed resistors can be used for R_A and R_B. Decade resistors have much higher residual inductances and capacitances, which makes them unsuitable for high-frequency work. Two values of R_A are available, to give different starting points on C_A depending on whether an inductive or capacitive impedance is being measured. The resistor unit in arm P is needed to permit initial balancing, since there is from (3) a minimum value of R_P corresponding to the fixed values of R_B and C_N and the minimum capacitance in arm A.

The shielding of arm P is particularly interesting, since it consists of four shields. This number is needed because of the presence, of (1) the resistor unit, (2) the final-balance capacitor C_P, (3) the initial-balance capacitor C_P', and (4) the input transformer and the capacitance of the entire arm to ground. Figure 12-15 explains the shield functions by indicating the way in which they control and regularize the capacitive effects. The subscripts denote the shield numbers, marked in order from

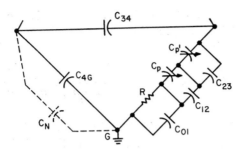

Fig. 12-15. Representation of shield capacitances of Fig. 12-14.

the inner shield to the outer. Shield 1 surrounds R, so that any internal capacitance to the shield is a constant shunt on R. Intershield capacitance C_{12} becomes a constant part of C_P, and isolates C_P from the surroundings. Capacitance C_{23} is a part of C_P' in the same way. C_{34} is in effect across the horizontal diagonal of the bridge, since the shield of the input transformer is connected to the left end of the winding; this capacitance, accordingly, does not affect balance conditions on the bridge. Finally, the capacitance from shield 4 to ground (the enclosing bridge cabinet) becomes a part of C_N, effectively in parallel with the capacitor units in this arm, and does no harm, since it is constant in amount.

12-10. Coil Measurements at Radio Frequencies

A coil possesses capacitance from each turn to each other turn. This complex of distributed capacitances may be represented as a single capacitance bridging the entire coil, for frequencies not too close to the "self-resonant" frequency where L and C effects resonate. The simplified circuit is indicated in Fig. 12-16.

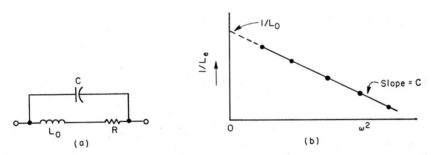

Fig. 12-16. Determination of the distributed capacitance of a coil.

The distributed capacitance becomes of great importance for coils used at high frequencies, and so it is frequently desired to evaluate this parameter. A measurement at a single frequency gives only the effective impedance at that frequency, and does not permit separation of L and C effects. The separation may be accomplished by measurements at two or more frequencies — preferably several, to permit averaging and assessment of regularity. The coil may be measured at a number of frequencies on a radio-frequency bridge, and the "apparent," or "effective," inductance determined for each reading.

The relationship may be derived to show the effect of frequency upon the effective inductance of the coil. Neglect of the resistive effect (usually permissible) permits the equation for the effective inductance to be put in the form.

$$L_e = \frac{L_0}{1 - \omega^2 L_0 C} \tag{7}$$

This may be inverted to the following,

$$\frac{1}{L_e} = \frac{1}{L_0} - \omega^2 C \tag{8}$$

That is, the plot of reciprocal-L against ω^2 can be expected to yield a straight line, as indicated in the figure.

The procedure, in accordance with the above theory, is to plot $1/L_e$ against ω^2 for each observed frequency. Regularity of the data may be judged by how well the points agree with a straight-line interpretation. It is desirable to have at least five points to permit averaging.

If the graph results in a satisfactory straight line, the coil constants may be found as follows:

L_0 = zero-frequency or "geometric" inductance — from the intercept on the zero-frequency axis.

C = slope of the curve = $\Delta(1/L_e) \div \Delta\omega^2$ (for any convenient and accurately read increments).

This method makes possible a quite definite determination of capacitance from the *slope* of the line, which is accurately established if the data is regular, as it can be for a careful set of measurements. The inductance L_0 depends on the intercept on the axis, but this can be well-defined, as the extrapolation is not excessive. Since the abscissa is ω^2, a good spread of points is obtained if the frequencies used cover a range of $\frac{2}{1}$ or $\frac{3}{1}$. In another scheme sometimes used, the capacitance required to resonate the coil is plotted against $(1/\omega^2)$ as abscissa; in this scheme the distributed capacitance appears as a small negative intercept after a rather long extrapolation — the accuracy is poor.

12-11. Bell Laboratories Bridge for Small Direct Capacitances

The measurement of very small capacitance becomes of great importance in some applications. One case of particular interest is the determination of the capacitance between the electrodes of vacuum tubes, a matter that is important in their development and testing. The capacitance between the plate and control grid of a screen-grid tube may be so small a fraction of a micromicrofarad that it cannot be measured with accuracy on an ordinary bridge. The bridge presented in this section was developed at the Bell Telephone Laboratories by C. H. Young for measurements of these very small capacitances.* Such capacitors are necessarily treated as three-terminal circuits, as each of the active electrodes may have much greater capacitance to the third terminal (usually ground) than it has to the other active terminal. The bridge must be designed to measure the small direct capacitance in the presence of much larger capacitances to the third terminal. Some of the bridges studied in earlier chapters may be used for such measurements, but are not satisfactory for very small values in the presence of large ground capacitances. The bridge to be studied here has been designed to give accuracy in measurements of this type and not to require the balancing of auxiliary guard or Wagner ground circuits.

Several interesting features of the bridge are shown in the diagram of Figure 12-17, which indicates the basic circuit without the shielding. The first item is the use of inductive ratio arms of a special type, with close inductive coupling between the two arms. This arrangement, which has been used on some other bridges,† has the merit of eliminating the effect of some of the ground capacitances on the balance conditions. The two arms are constructed, in the present bridge, on a toroidal core, with the AB and BC windings consisting of parallel wires wound at close spacing to minimize leakage inductance between them. An electrostatic screen is placed over these windings, and then the primary coil is wound over the screen. The windings A-B-C form the secondary of a transformer, with B as the center tap. Now, if the impedance of the transformer windings is low (low resistance and low leakage reactance) a load may be placed on one half of the secondary without having appreciable effect on the equality of the voltages of the two halves of the winding. As applied to the bridge, this arrangement permits capacitance to be connected across either AB or BC without entering into the balance relationships — which would not be true of resistance ratio arms. The unknown is shown in Figure 12-17 as C_x, with capacitances C' and C'' to the third terminal. C' is in reality connected from D to B, where it shunts the detector, but has no effect on balance.

---

*References 13 and 14.

†Reference 5, p. 421.

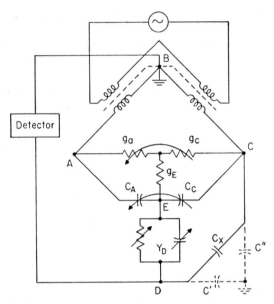

Fig. 12-17. Circuit of Young bridge for measurement of small direct capacitances.

C'' forms a capacitive load on the BC half of the transformer secondary, and so, as discussed above, has no effect on bridge measurements. Both C' and C'' are thus eliminated from the measurements without recourse to a Wagner ground circuit.

Another interesting feature of the bridge is the double-wye circuit shown in Figure 12-17. The combination of g_A and g_C, used to balance the conductance component of the unknown, is so arranged that one conductance is decreased as the other is increased, so that the total is kept constant for all settings. C_A and C_C are likewise arranged for variation so that the total is constant. The circuit is most easily analyzed by two wye-delta transformations of the admittances in this circuit (see Appendix II). The order of procedure and the nomenclature are shown in Figure 12-18, where (a) is the original circuit, (b) the circuit after the first transformation and (c) the circuit after the second transformation and with the direct admittance of the unknown added to help indicate the conditions for balance. The two new conductances, G_A and G_C, obtained from the first transformation, have the following values:

$$G_A = \frac{g_A g_E}{g_A + g_C + g_E} \tag{9}$$

$$G_C = \frac{g_C g_E}{g_A + g_C + g_E} \tag{10}$$

G_{AC} is not evaluated, as it is across the diagonal of the bridge and, hence,

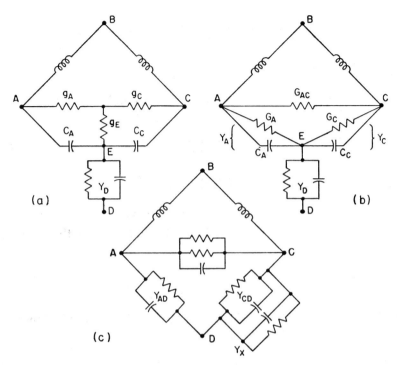

Fig. 12-18. Diagrams to accompany the development of formulas for the bridge of Fig. 12-17.   (a) Original circuit.   (b) after first wye-delta transformation.   (c) After second wye-delta transformation.

does not affect balance. The sum $(G_A + G_C)$ may be seen from (9) and (10) to be constant since $(g_A + g_C)$ is constant.

The sum of the admittances Y_A and Y_C in diagram (b) is

$$Y_A + Y_C = (G_A + G_C) + j\omega(C_A + C_C) \tag{11}$$

The sum is constant, and has a constant phase angle. The admittance Y_D is adjusted to have the same phase angle, or

$$\frac{\omega C_D}{G_D} = \frac{\omega(C_A + C_C)}{G_A + G_C} \tag{12}$$

With branch D adjusted in accordance with (10), Y_D may be written in the following way:

$$Y_D = b(Y_A + Y_C) \tag{13}$$

where b represents a real number.

The second wye-delta transformation may be made to place the circuit in the form of Figure 12-18(c). The values of Y_{AD} and Y_{CD} are obtained as follows:

$$Y_{AD} = \frac{Y_A Y_D}{Y_A + Y_C + Y_D} \tag{14}$$

$$Y_{CD} = \frac{Y_C Y_D}{Y_A + Y_C + Y_D} \tag{15}$$

If substitution for Y_D is made by (13), and Y_A and Y_C are expressed in terms of their components, (14) and (15) may be changed to the following:

$$Y_{AD} = (G_A + j\omega C_A) \frac{b}{b+1} \tag{16}$$

$$Y_{CD} = (G_C + j\omega C_C) \frac{b}{b+1} \tag{17}$$

If the unknown is added to arm CD as indicated and adjustment is made for balance, the following relationship must hold true, since the ratio arms are equal:

$$Y_{AD} = Y_{CD} + Y_X$$

or $$Y_X = Y_{AD} - Y_{CD} \tag{18}$$

Substitution of (16) and (17) in (18), and separation of conductance and susceptance components yields the relationships

$$C_X = (C_A - C_C) \frac{b}{b+1} \tag{19}$$

$$G_X = (G_A - G_C) \frac{b}{b+1} \tag{20}$$

$$= (g_A - g_C) \frac{g_E}{g_A + g_C + g_E} \frac{b}{b+1} \tag{21}$$

Both C_X and G_X are affected by the multiplier $b/(b+1)$, where b depends on the value of Y_D as given in (13). This fact makes possible the construction of a bridge with several decade multipliers by provision of appropriate values of Y_D, which may be connected into the circuit by a switching device. G_x is modified in addition by the conductance fraction shown in (21), but this multiplier may be taken into account in the calibration of the bridge.

This bridge as produced commercially has a capacitance standard consisting of one decade of 10-μμf steps and a 10-μμf variable air capacitor, giving a total of 110 μμf adjustable to 0.1 μμf. The conductance adjustment has a range of 10 micromhos, readable to 0.5 micromho. Multipliers of 1, 0.1, 0.01, 0.001 and 0.0001 are provided. On the lowest multiplier the bridge has a range of 0.011 μμf, adjustable to 0.00001 μμf. The bridge has a built-in detector and an oscillator operating at 465 kc. The standards of low capacitance, described in par. 8 of chap. 13 could very well be used to check the calibration of this bridge on the low ranges.

12-12. NBS Bridge for Small Capacitances*

Another form of transformer bridge has been developed as a result of work originating at the National Standards Laboratory of Australia, with further development at the National Bureau of Standards in Washington. This bridge is intended primarily for highly accurate measurements in the establishment of capacitance standards, and in the comparison of capacitors for standardizing work. The range of the bridge extends from 1 μf to a least count of 1 μpf (1 pf = 1 picofarad = 1 $\mu\mu$f).

The transformer, the central component of the bridge, is wound on a toroidal core of high-permeability material with great care given to the shielding between primary and secondary, and to the distribution of windings around the core. Transformers of both 1 : 1 and 10 : 1 ratio have been constructed, both accurate in ratio to better than one part in 10^6. The scheme of the bridge (less the conductance-balancing circuit) is shown in Fig. 12-19. The capacitances from the terminals to the shield of C_S

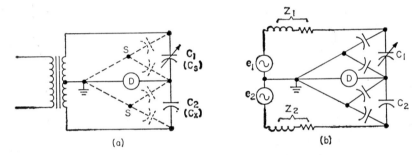

Fig. 12-19. Bridge for low-capacitance measurements. (a) Schematic of the transformer bridge (b) Equivalent circuit.

and C_X may be seen to shunt either the detector or the transformer secondary; the secondary windings are designed for low impedance to minimize the error caused by the loading effect of these terminal capacitances. The transformer is designed for operation primarily at a frequency of 1 kc, but may be used to at least 10 kc.

Decade capacitors are provided to cover the range from 1 μf to smallest increments of 1 μpf. The range from 1 μf to 0.001 μf makes use of silver-mica capacitors. Guarded and adjustable three-terminal capacitors were constructed for the lower ranges, using concentric cylindrical construction, with units of 100 pf, 10 pf and 1 pf in decade mountings. Below 1 pf an inductive divider is used in conjunction with a fixed capacitor for each decade.

---

*References 15, 16.

A form of capacitor that carries promise as an absolute standard, and which would be used with the bridge discussed here, is described in par. 13-8. An absolute standard means one that can be assigned a value entirely on the basis of theoretical relations and measured dimensions. The requirements on a suitable form of standard are very stringent, as the national laboratories have as a goal the establishment of standards accurate within a few parts per million. The form of standard capacitor based upon Lampard's theorem seems suitable for the purpose, but for reasonable size has small capacitance, of the order of 1 pf, and so requires accurate measurement of small capacitances, as by the bridge discussed in this section.

REFERENCES

1. Ferguson, J. G., "Shielding in High Frequency Measurements," *Trans. A.I.E.E.*, **48**, 1286 (Oct. 1929).

2. Shackelton, W. J. and J. G. Ferguson, "High-Frequency Measurement of Communication Apparatus," *Trans. A.I.E.E.*, **46**, 519–27 (1927).

3. Behr, L. and A. J. Williams, Jr., "The Campbell-Shackelton Shielded Ratio Box," *Proc. I.R.E.*, **20**, 969 (June 1932).

4. Laws, Frank A., *Electrical Measurements*, New York: McGraw-Hill Book Co. Inc., 1938.

5. Hague, B., *Alternating Current Bridge Methods*, London: Sir Isaac Pitman & Sons, Ltd., 1946.

6. "Modified Schering Bridge," Leeds and Northrup Co., *Catalog E-54* (1945).

7. Kouwenhoven, W. B. and A. Baros "A High-sensitivity Power-Factor Bridge," *Trans. A.I.E.E.*, **51**, 202–210 (1932).

8. Balsbaugh, J. C. and P. H. Moon, "A Bridge for Precision Power-Factor Measurements on Small Oil Samples," *Trans. A.I.E.E.*, **52**, 528–34 (1933).

9. Balsbaugh, J. C., Kenny, N. D., and Herzenberg, A., "The M.I.T. Power-Factor Bridge and Oil Cell," *Trans. A.I.E.E.*, **54**, 272–81 (1935).

10. "Test Methods for A-C Capacitance, Dielectric Constant and Loss Characteristics of Electrical Insulating Materials," *A.S.T.M.*, D 150–54T.

11. Kouwenhoven, W. B. and Leo J. Berberich, "A Standard of Low Power Factor," *Trans. A.I.E.E.*, **52**, 521–27 (1933).

12. Balsbaugh, J. C., Howell, A. H., and Dotson, J. V., "Generalized Bridge Network for Dielectric Measurements," *Trans. A.I.E.E.*, **59**, 950–6 (1940).

13. Young, C. H., "Measuring Inter-Electrode Capacitances," Bell Labs Record, **24**, No. 12, 433–38 (Dec. 1946).

14. Young, Clarence H., U.S. Patents 2,309,490 and 2,326,274.

15. Thompson, A. M., "The Precise Measurement of Small Capacitances," *Trans. I.R.E.*, **I-7**, p. 245 (Dec. 1958).

16. McGregor, M. C., J. F. Hersh, R. D. Cutkosky, F. K. Harris, F. R. Kotter, "New Apparatus at the National Bureau of Standards for Absolute Capacitance Measurement," *Trans. I.R.E.*, **I-7**, p. 253 (Dec. 1958).

PROBLEMS

12-1. The bridge is balanced at a frequency of 1000 cycles per second.

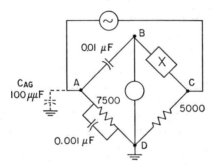

Prob. 12-1.

(a) Find the approximate R, C constants of the unknown, neglecting the stray capacitance C_{AG}.

(b) Find the true value of X, considering C_{AG}. Does C_{AG} have greater effect on R or on C of unknown?

12-2. An a-c bridge is marked A-B-C-D consecutively around the corners and has the following constants; AB, pure capacitance 0.01 μf; BC, pure resistance 2000 ohms; CD, unknown; DA, capacitance 0.02 μf in series with resistance of 8000 ohms. The bridge is balanced at a frequency such that $\omega = 2\pi f = 50,000$.

(a) Find the unknown for the bridge constants as stated above.

(b) There is a stray capacitance of 100 $\mu\mu f$ across arm DA, in addition to the constants above. Find the true value of the unknown.

12-3. A balanced a-c bridge has in arm AB a pure capacitance $C = 0.01$ μf; arm BC, the unknown; arm CD, pure $R = 5000$ ohms; arm DA, pure $R = 10,000$ ohms in parallel with $C = 200$ $\mu\mu f$. Point D is grounded. $\omega = 2\pi f = 10,000$ radians per second.

(a) Find the constants of the unknown as a series combination of R and L or C.

(b) How much stray capacitance from point C to ground would cause a 10 per cent error in the result for R_X? All other constants as above.

12-4. A balanced a-c bridge has the following constants: arm AB, a resistance of 5000 ohms in parallel with a capacitance of 0.1 μf; arm BC, resistance 5000 ohms; arm CD, the unknown; arm DA, resistance 4000 ohms. There is a stray capacitance of 100 $\mu\mu$f from each corner of the bridge to ground. $\omega = 2\pi f = 10,000$. Terminal D is grounded. Find the constants of the unknown in series components (R and L or C).

12-5. An a-c bridge marked A-B-C-D around the corners has in arm AB a resistance of 5000 ohms; BC, a resistance of 10,000 ohms; CD, a capacitance of 0.002 μf. $\omega = 2\pi f = 50,000$. There is a stray capacitance of 100 $\mu\mu$f from each corner of the bridge to ground. Find the constants in arm DA (series constants) to give balance if point B is grounded.

12-6. A balanced a-c bridge has the following constants: arm AB, $R = 2500$ ohms in series with $C = 0.4$ μf; BC, unknown; CD, $R = 8000$ ohms; DA, $C = 0.01$ μf. Point D is grounded. There are stray capacitances of 0.0002 μf each from A, B, and C to ground. $\omega = 2\pi f = 10,000$.

(a) Find the constants of the unknown, neglecting the stray capacitances.

(b) Repeat, including the stray capacitances.

12-7. A capacitance bridge has a variable resistance A to B; 1000 ohms pure resistance B to C; unknown capacitance C to D; standard capacitance of 0.001 μf D to A. The unknown has a direct capacitance between terminals of 0.002 μf, and stray capacitances of 0.0001 μf from each terminal to ground. $\omega = 2\pi f = 10,000$.

(a) If bridge terminal A is grounded, find R_{AB} for balance on the actual bridge. Find the apparent C_X that would be derived from the observed bridge-arm constants, if the presence of the stray capacitances were not known.

(b) Same, except that B is grounded. Show how to complete the balancing of the bridge, giving numerical values of any additions.

(c) Same, except that C is grounded.

12-8. In the sketch, (1), (2), (3), and (4) represent stray capacitances of parts of the circuit. Explain briefly for each whether it causes error in balance conditions of the bridge.

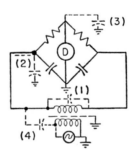

Prob. 12-8.

12-9. An a-c bridge is balanced with the following constants, arm AB, $C = 0.1$ μf: BC, unknown; CD, $R = 10,000$; DA, $R = 5000$ in parallel with $C = 0.001$ μf. Point D is grounded. There is a stray capacitance of 200 $\mu\mu$f from C to ground. $\omega = 2\pi f = 10,000$.

(a) Find the constants of the unknown as a series circuit (R and L or C), neglecting the stray capacitance. Find the dissipation factor of the unknown.

(b) Repeat, including the stray capacitance.

12-10. An a-c bridge is balanced with the following constants, arm AB, $R = 20$ in series with $C = 0.5\ \mu\text{f}$; BC, $R = 5000$; CD, unknown; DA, $R = 10,000$. There are stray capacitances of 100 $\mu\mu\text{f}$ from A to ground and 100 $\mu\mu\text{f}$ from C to ground. Point D is grounded, $\omega = 2\pi\text{f} = 10,000$.

(a) Find the apparent constants (R and L or C, in series) of the unknown, neglecting stray capacitances.

(b) Find the true constants of the unknown, considering stray capacitances.

12-11. A Maxwell inductance bridge, balanced at 1000 cycles, has constants: arm AB, $R = 3970$ ohms in parallel with $C = 0.2\ \mu\text{f}$; BC, $R = 5000$ ohms; CD, the unknown; DA, $R = 10,000$ ohms. The oscillator is connected from A to C, the detector from B to D, and D is grounded. There is stray capacitance of 200 $\mu\mu\text{f}$ from A to ground and 200 $\mu\mu\text{f}$ from C to ground.

(a) Find the apparent constants of the unknown, ignoring the stray capacitances.

(b) Find the constants of the unknown, considering the stray capacitances.

CHAPTER 13

Characteristics of Bridge Components

13-1. Construction of Precision Resistors for A-C Use*

Resistors hold an important place in many types of a-c measurements. We have seen their use as an essential part in all a-c bridges, and also in other types of measuring equipment. It would be ideal for these applications if a resistor could be constructed that would have the effect only of pure resistance of the desired amount at all frequencies. All electrical circuits, however, possess some inductive and capacitive effects, and even the best precision resistors are not exempt. It is the constant attempt to construct resistors for measurements use with the lowest possible residual inductance and capacitance.

Inductance is the circuit constant that denotes the presence of magnetic flux linking the circuit when current is caused to flow through the circuit members. The amount of flux linking the circuit can be kept to a low value by keeping the "going" and "return" sides of the circuit as close together as possible. This may be accomplished in a resistor by using the "bifilar" construction in which the resistance wire is formed into a long loop or "hairpin," with the two sides as close together as possible; that is, as close as the insulation on the wire permits. Figure 13-1(a) shows such a loop in the straight form, and (b) represents the loop wound around a circular cylinder to give a more compact construction.

---

*Reference 1; pp. 86–127; reference 2, pp. 122–150.

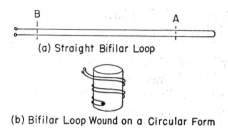

(a) Straight Bifilar Loop

(b) Bifilar Loop Wound on a Circular Form

Fig. 13-1. Bifilar construction of resistors.

The bifilar type of construction is successful in reducing the inductance to a low value. However, it has the undesirable feature of giving a large capacitive effect if the loop is long. Capacitance is the term that indicates the presence of charges on conductors that have a difference of potential between them. This effect is large in the bifilar winding, because of the closeness of the two sides of the circuit; however, the relative importance depends also on the length and resistance of the loop. The capacitive effect is small in the neighborhood of A in Fig. 13-1(a), since the difference of potential between wires is small there. The effect at B, per unit length, is much greater than at A, because of the greater difference of potential. It helps to divide a long loop into two or more shorter loops placed in series, but mounted at a small distance from each other; this construction has been used in the past even for rather high resistances, but has been generally superseded by other types of winding. The bifilar form of construction is used for $\frac{1}{10}$ ohm and 1-ohm (per step) decades. Other forms are used generally for the higher decades in resistance boxes intended for a-c use.

A simple but effective construction is to wind the resistance wire in a single winding on a thin card made of mica or other insulating material. Inductance is fairly low due to the presence of currents in opposite directions, separated only by the thickness of the card. Capacitive effects are small since the starting and finishing ends of the winding are at opposite ends of the card, and the difference of potential between adjacent wires is due to the IR drop in only one turn of wire. This winding may be considered a solenoid which has been flattened to a sheet, thereby reducing the cross-sectional area and the enclosed magnetic flux. The effectiveness for alternating current use depends upon reduction to a very thin form.

The card form of construction is sometimes used with two windings in parallel, wound in opposite directions around the card, with one winding spaced between the turns of the other. This form, called the Ayrton-Perry winding, gives better cancellation of magnetic fields than the single winding. It is more expensive to construct, and requires either more wire, or else finer wire, because of the two paths in parallel. This becomes an objection in the high-resistance units that must use very fine wire, at best.

Many other types of winding have been devised with the purpose of keeping residual inductance and capacitance to low values. One form, originated by Curtis and Grover, is indicated in Fig. 13-2 as wound on a slotted ceramic cylinder (a slotted flat card may be used instead). In this end view the second turn is drawn outside the first so that current directions can be marked on them. From the starting terminal A the wire is taken around the form clockwise for one turn, then across the center, and around one turn counter-clockwise, and so on. This construction causes adjacent wires to carry current in opposite directions, hence magnetic effects are small. Capacitive effects also are small since the potential difference between adjacent wires is the IR drop of one turn. The electrical characteristics of the winding are thus excellent. However, the winding job is a slow and tedious hand process, not adapted to machine methods. The card windings discussed above have good characteristics and are well suited to machine production.

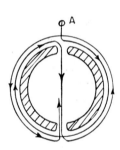

Fig. 13-2. Curtis and Grover winding.

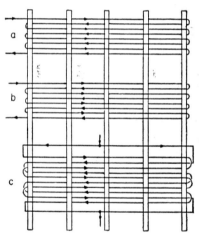

Fig. 13-3. Woven-fabric resistors. (*Courtesy of Leeds & Northrup Co.*)

Another form of winding that is very good for high resistances is the woven resistor. This is in the form of a woven strip, or ribbon, in which the resistance wire forms the weft (transverse) and strong silk or cotton threads the warp. Inductance is low because oppositely directed currents flow in adjacent wires. Capacitance is small because the IR drop is low between adjacent wires, and the starting and finishing ends of the resistor are far apart. Figure 13-3(a) shows the ordinary type of weaving. This gives good results, but can be improved, since adjacent wires are not quite parallel due to the crisscross effect as the wires go over and under the lengthwise threads. Figure 13-3(b) shows an improved weaving produced by the Leeds and Northrup company, which causes adjacent pairs of wires to lie close together. Very good characteristics are obtained by this method. Diagram (c) shows a weaving consisting of two wires in parallel, similar to the Ayrton-Perry resistor mentioned above. There are always two wires side by side carrying currents in opposite directions, so the inductance is low. This weaving is particularly useful for the lower-valued resistors. The woven fabric is adaptable to a wide range of resistance values by the simple expedient of cutting the ribbon to the required length.

13-2. Frequency Errors of Resistors

A resistor may be represented to a first approximation by the circuit of Fig. 13-4, which shows inductance in series with the resistance and this combination shunted by a capacitance. R represents the resistance as measured with d-c. (Skin-effect is negligible at the lower frequencies for fine wire of high-resistance material, and even at radio frequencies causes minor error compared with other effects, except for low resistance units.)

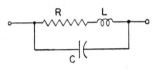

Fig. 13-4. Equivalent circuit of a resistor.

The equivalent impedance of the circuit may be found by the complex notation as follows:

$$Z_{equiv} = \frac{(R + j\omega L)\left(-j\dfrac{1}{\omega C}\right)}{R + j\left(\omega L - \dfrac{1}{\omega C}\right)} \qquad (1)$$

$$= \frac{\omega L - jR}{\omega CR - j(1 - \omega^2 LC)} \qquad (2)$$

If the fraction is rationalized in the usual way and simplified, the expression becomes

$$Z_{equiv} = \frac{R + j\omega[L(1 - \omega^2 LC) - CR^2]}{(\omega CR)^2 + (1 - \omega^2 LC)^2} \qquad (3)$$

The second term in the denominator when expanded is

$$1 - 2\omega^2 LC + \omega^4 L^2 C^2$$

The last term may be neglected as the square of a very small number, since for any measurement case $\omega^2 LC \ll 1$. If we incorporate this fact in the denominator and split Z into real and j terms we have

$$R_{equiv} = \frac{R}{1 - \omega^2 C(2L - CR^2)} \qquad (4)$$

$$X_{\text{equiv}} = \frac{\omega[L(1 - \omega^2 LC) - CR^2]}{1 - \omega^2 C(2L - CR^2)} \tag{5}$$

Since X is always comparatively small, we may drop the $\omega^2 LC$ in the numerator of (5) in comparison with unity. Then X has the value

$$X_{\text{equiv}} = \frac{\omega(L - CR^2)}{1 - \omega^2 C(2L - CR^2)} \tag{6a}$$

$$\simeq \omega(L - CR^2) \tag{6b}$$

The phase angle of the resistor is of great interest, because it is an important item in bridge-balance conditions. If we call the phase angle φ, we may write from (4) and (6a)

$$\tan \varphi = \frac{X_{\text{equiv}}}{R_{\text{equiv}}} = \omega\left(\frac{L}{R} - CR\right) \tag{7}$$

The characteristics of resistors are sometimes expressed in terms of a "time constant." For a simple L and R circuit, L/R is called the time constant, and for a C and R circuit the time constant is CR. It may be noted that eq. (7) contains terms of this sort. As another approach to the matter, reference is sometimes made to the "equivalent inductance" of a resistor as,

$$\text{equivalent inductance} = L' = (L - CR^2) \tag{8}$$

from (6b). This is a way of saying that the over-all inductive effect of a resistor is reduced by the shunt capacitance. In these terms, the time constant is

$$\text{time constant} = \frac{L'}{R} = \frac{L}{R} - CR \tag{9}$$

The time constant of a resistance box may be either positive or negative, depending on which of the terms, L/R or CR, predominates. The low-valued decades (1 or 10 ohms) generally have a positive time constant, which means that they display a net inductive effect. The capacitance usually predominates for the 1000-ohm decade, giving a negative time constant. The 100-ohm decade is likely to have a time constant nearly zero. For a good resistance box, the time constants of the decades are usually only a few units times 10^{-8} sec, certainly less than 10^{-7} sec, except for the 1- or $\frac{1}{10}$-ohm decades.

Figure 13-5 shows time-constant curves that are representative of the results that may be attained in a decade box of good construction, using woven resistors in the higher decades. The time constants are all less than 10^{-7} sec except for the 1-ohm decade. With constants of this magnitude, the resistance error and the phase angle are reasonably small up to 50,000 cycles per second, or somewhat higher.

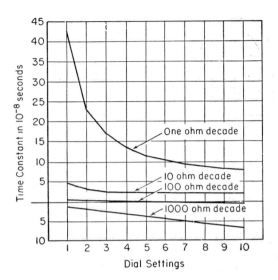

Fig. 13-5. Time constants of a decade box.

Now, returning to eq. (4), we shall investigate the effect of frequency on the magnitude of the effective resistance. Equation (4) may be changed by long division or by a binomial expansion (chap. 3) to the form

$$R_{equiv} = R[1 + \omega^2 C(2L - CR^2) + ...]$$

where the terms of higher power are dropped as being negligibly small. R_{equiv} is the a-c resistance, while R is the d-c value. Therefore the frequency error is

$$\Delta R = R_{equiv} - R = \omega^2 CR(2L - CR^2)$$

The fractional error is

$$\frac{\Delta R}{R} = \omega^2 CR \left(2\frac{L}{R} - CR \right) \tag{10}$$

Note that the frequency error in the magnitude of the resistance *may be either positive or negative*, depending on the constants, and hence may be zero. That is

$$\frac{\Delta R}{R} = 0 \quad \text{if} \quad 2\frac{L}{R} = CR \tag{11}$$

and this expression is *independent of frequency*.

Resistors show large errors at high frequencies, even with the best construction for the purpose. Analysis in terms of lumped constants, as carried out above, is not adequate at the very high frequencies, but must be restated to recognize the distributed effects. Figure 13-6(a) indicates a

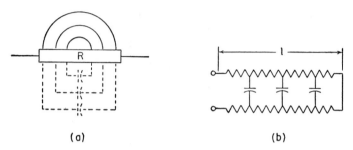

Fig. 13-6. Diagram showing the effect of distributed capacitance in a high-frequency resistor. (a) Resistor, rod type: above, electric field lines; below, equivalent capacitances. (b) Equivalent circuit of the resistor, drawn as a short-circuited transmission line.

resistor of the rod type, showing in the upper side the electric field lines originating on the charges present because of the difference of potential between the ends. The lower side shows the same thing, but in terms of the capacitances that would be used in an equivalent circuit to represent this condition. The circuit is redrawn in (b), with the resistor doubled back at the center for convenience. This representation is equivalent to a transmission line short-circuited at the load end; hence an analysis will be made utilizing transmission-line equations. The important terms in this case are the resistance and capacitance, with small inductive effects which will be dropped for simplicity.

The input impedance of a short-circuited line is

$$Z = Z_0 \tanh Pl$$

where P = propagation constant

$$= \sqrt{yz} \qquad \text{in general}$$

$$= \sqrt{j\omega cr} \qquad \text{in this case}$$

$$= \sqrt{\omega cr}\underline{/45°}$$

(Small letters will be used for the constants per unit length, and capitals for the total for the line.) Also

$$Z_0 = \text{characteristic impedance of the line}$$

$$= \sqrt{\frac{z}{y}} \qquad \text{in general}$$

$$= \sqrt{\frac{r}{j\omega c}} \qquad \text{in this case}$$

$$= \sqrt{\frac{r}{\omega c}}\underline{/-45°}$$

With these values substituted the impedance becomes

$$\mathbf{Z} = \sqrt{\frac{r}{\omega c}} \underline{/-45°} \tanh \sqrt{\omega cr} \, \underline{l/45°}$$

$$= rl \frac{\tanh \theta}{\theta}$$

$$= R \frac{\tanh \theta}{\theta},$$

where $\theta = \sqrt{\omega cr} \, \underline{l/45°} = \sqrt{\omega CR} \underline{/45°}$

At low frequencies, θ is small, the ratio $\tanh \theta/\theta$ is unity, and $\mathbf{Z} = R$. At high frequencies, for the constants as given, $\tanh \theta/\theta$ decreases in value. Eventually, $\tanh \theta$ approaches one as a limit, and θ continues to increase with increase of frequency. For any one type and proportion of resistor, θ becomes a function of the frequency and resistance R. Accordingly if the relative resistance R_{ac}/R_{dc} is plotted against $f \cdot R_{dc}$, the points for different resistance values should fall on the same curve. For the highest-grade metallized resistor the decrease of resistance is 1 per cent when $f \cdot R$ (in megacycles and megohms, respectively) is 10 and at even smaller values for wire-wound or lower-grade metallized resistors.

Another factor that contributes to the reduction of resistance at high frequency is the dielectric loss in the shunt capacitance. The loss increases with frequency, which, in terms of parallel R-C, requires a lower value of R at higher frequency. The dielectric loss and the distributed capacitance in combination give a reduction of resistance with increase of frequency known as the "Boella effect." The original article by Boella, in Italian, was presented in 1934, with discussion later in references 10 and 11.

13-3. Design of Resistors for Minimum Error

It may be noted from (7) that the phase angle is zero if $L/R = CR$, and from (11) that the resistance error is zero if $2L/R = CR$. Thus it is not possible to secure both zero phase angle and zero resistance error from the same design. Conditions are highly satisfactory if the phase angle is zero, for this avoids the introduction of certain errors in bridge measurements. Also, if this condition is met, the remaining resistance error is usually small.

Precision resistors have been constructed with the idea of eliminating error by the balancing of the L and C constants to give the conditions outlined above. This may be done for a single resistor, as for example, a fixed arm of a bridge. (Low-valued R-F resistors have been made this way, good up to a frequency of 30 Mc with accuracy excellent for such a frequency value.) The idea of adjusting L and C for minimum error is scarcely applicable to a decade box. It is easier in decade-box construction to keep L and C residuals to the lowest practical values, in which case the characteristics are satisfactory up to 50 kc or so, and up into the radio-frequency range, except for the very high or very low resistances.

13-4. Characteristics of Decade Boxes

We have talked so far mainly of the residuals of a resistor unit as such and have said nothing of the effect of mounting a number of resistors to form a decade box, except for discussion of the characteristics shown in Fig. 13-5. When resistors are mounted together capacitance effects are increased, and there is further increase due to switches, metal panel, and shielding. The inductance is increased also due to the loop formed by connecting wires, switches, and resistors. The inductance and capacitance effects in the assembled decade box are quite appreciable, but cannot be stated as definite values as they depend on the type of construction.

Figure 13-7 shows an internal view of an L&N decade box, with shielded construction and a six-resistor type of switching arrangement. In this case

Fig. 13-7. L&N Type 4754 a-c resistance box.

the switch blades are stationary and the resistors are moved, thus keeping the circuit configuration practically constant. This company makes another type of decade box in which only one coil is active in each decade at one time and it is moved to the brushes, thus keeping absolutely constant circuit geometry for all dial settings.

Figure 13-8 shows an internal view of a General Radio decade box. The switch construction is clearly shown. The resistors are of the flat-card type, being unifilar for the 1000-ohm decade, and Ayrton-Perry for the

Fig. 13-8. General Radio type 1432 decade resistance box; internal view showing construction of switches and resistors. (*Courtesy of General Radio Co.*)

1, 10, and 100-ohm decades. There is an interesting table in the catalog of the General Radio Company showing the residual inductance and capacitance of their decade boxes.

The General Radio Company makes also a "compensated decade" construction having constant inductance on all dial settings. This is accomplished by a double set of switch contacts so that a copper coil is switched into circuit as a manganin coil is switched out. The constant inductance idea is useful in radio-frequency circuits where the inductance can be tuned out if it is constant in amount. The inductance of one of these compensated decade boxes stays constant for all switch settings within 0.1 μh.

13-5. Construction of Standard Inductors*

Inductors used for measurement purposes may be subdivided first into self-inductors and mutual inductors. Each class may be further classified as fixed or variable. The fixed inductors are of two types, absolute and secondary standards. The absolute standards are so constructed that their inductance may be computed from the dimensions to a high degree of accuracy. These coils usually have a single-layer winding on a large ceramic or pyrex glass form, carefully machined to a cylindrical shape, with a helical groove to space the wire. Standards of this sort are used only for absolute measurements at the national laboratories. Secondary standards are constructed in any convenient manner and are then calibrated by reference to an established standard of inductance or capacitance.

---

*Reference 1, pp. 127–161; reference 2, pp. 345–355.

An inductance standard should have the following properties:

(a) Permanence and constancy of inductance.

(b) Negligible effects from temperature and humidity.

(c) Largest possible ratio of L to R (Large time constant).

(d) Inductance should be independent of current strength. (No magnetic materials in the field).

(e) Effective inductance and resistance should be affected to a minimum by frequency. This may require stranded wire in some coils, but in all cases there should be no masses of metal subject to induction of eddy-currents. Also, the distributed capacitance should be kept to a low value.

(f) The inductor should desirably be unaffected by external magnetic fields and should produce a minimum interfering field of its own. A uniform toroidal coil is the only one meeting this requirement fully, but an astatic system of two adjacent coils with oppositely directed fields is often an acceptable substitute.

The requirement of permanence calls for care in the selection of the material of the coil form, but the selection is further limited by consideration of temperature and humidity effects. Coils have, in general, considerable change of inductance with temperature, due to changes of dimensions. An inductance standard thus presents difficulties of several sorts in design and difficulties of other kinds in use. Standard mutual inductors are needed in a number of circuits and cannot well be replaced by anything else. For many measurements a self-inductance standard can be replaced by a capacitance standard, with improvement in accuracy and ease of shielding and with reduction of necessary space.

Variable inductors are useful in many circuits, so we should make some study of their design. The inductance of a single coil depends on the number of turns and on the dimensions, neither of which lends itself to smooth incremental change. The principle adopted for most variable self-inductors is the use of two coils in series, mounted so that the mutual inductance between them may be varied. As is well known, the total series inductance is $L_1 + L_2 \pm 2M$. If $L_1 = L_2$, and the coils may be closely coupled, the total inductance may be varied from a small value to a maximum nearly four times the inductance of one coil. Several forms of construction have been used. In one type (the Aryton-Perry) the stationary coil is wound on the inside of part of a spherical surface. The movable coil is on a part-spherical surface of slightly smaller radius. The movable coil may be turned 180° on its lateral axis, from a position in which the coil axes coincide with subtractive effects, to a position in which the axes coincide with additive effects. A slight variation on this type has a cylindrical rather than spherical form, but is otherwise similar.

An improved variable inductor, or "inductometer," was devised by H. B. Brooks of the Bureau of Standards.* A photograph of this instrument as made by the Leeds and Northrup Company is shown in Fig. 13-9. The

Fig. 13-9. Brooks Inductometer. (*Courtesy of Leeds & Northrup Co.*)

middle disk has two link-shaped coils set in it, and the top and bottom plates have matching coils in their inner surfaces. A diagram of the construction is shown in Fig. 13-10. The top and bottom plates each have a coil with half the turns of the moving coil. Much study was given to the shape and proportion of the coils to produce a calibration curve that is very nearly linear. Figure 13-9 shows the index mark on the stationary member and the calibration marks on the movable part. The calibration of inductometers of this type is guaranteed to ±0.3 per cent of the maximum value. The two groups of coils are connected with opposite polarities; that is, when the flux direction is upward in the right-hand coils, it is downward on the left. One coil serves as the return flux path for the other, and the magnetic field does not spread out very greatly. The Brooks inductometer is a great improvement over the Ayrton-Perry in this respect, and also in compactness and time constant. The inductometer may be used as a

---

*Reference 3.

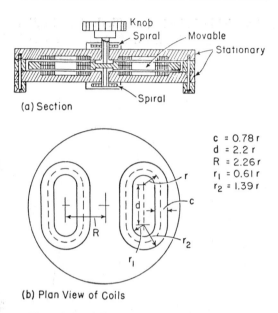

(a) Section

$$c = 0.78\,r$$
$$d = 2.2\,r$$
$$R = 2.26\,r$$
$$r_1 = 0.61\,r$$
$$r_2 = 1.39\,r$$

(b) Plan View of Coils

Fig. 13-10. Diagram of Brooks inductometer.

variable self-inductor if the fixed and movable coils are connected in series, and as a mutual inductor if the coils are in different circuits.

13-6. Frequency Errors of Inductors

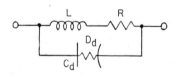

Fig. 13-11. Equivalent circuit an inductor.

An inductor may be represented approximately by the equivalent circuit of Fig. 13-11, for frequencies not too close to the self-resonant point. Inclusion of the loss component of the distributed capacitance, indicated in the figure, gives a more accurate representation of loss conditions at the frequencies, but is of minor importance in the lower part of the frequency range. Since the effective inductance of a coil changes tremendously as the self-resonant frequency is approached, it is not usable as a standard in that neighborhood; in fact, it should be used as a standard only to a small fraction of the resonant frequency.

The circuit of Fig. 13-11 is basically the same as Fig. 13-4 except for loss in the capacitance, and the equations are the same with this exception up to the point at which we begin to make approximations. The approximations are different in the case of a coil, as compared with a resistor, due to the very different ratio of L to R. For the present we shall omit D_d and use eqs. (4) and (5), with neglect of terms in CR^2 as compared with L. This gives, for $\omega^2 LC \ll 1$,

$$R_{\text{eqiuv}} = \frac{R}{1 - 2\omega^2 LC}$$

$$\simeq R(1 + 2\omega^2 LC) \tag{12}$$

$$X_{\text{equiv}} = \frac{\omega L(1 - \omega^2 LC)}{1 - 2\omega^2 LC} \tag{13a}$$

$$\simeq \omega L(1 + \omega^2 LC) \tag{13}$$

Therefore $$L_{\text{equiv}} \simeq L(1 + \omega^2 LC) \tag{14}$$

The terms in parentheses in (12) and (14) represent correction factors to take into account the effect of frequency. Note that the additive term is twice as great for R as for L. The correction factor may be quite appreciable for large multilayer coils, even at moderate frequencies, because of large capacitance. The distributed capacitance may be determined by measurements of L and R at two different frequencies, and then the correction factors can be computed for any other frequency. Another method, better in that it permits graphical averaging of several determinations, was presented in chap. 12 (par. 10); this method permits an accurate evaluation of the L and C_d elements. The equation used there is a slight modification of the present (13a) or (16).

A more complete picture of the behavior of the residuals of a coil may be obtained by beginning with the complete circuit of Fig. 13-11. D_d represents the dissipation factor of the insulating material that makes up the dielectric of the distributed capacitance. The circuit may be analyzed conveniently by means of the parallel circuit operators in the following way. For the complete circuit:

$$G = \frac{R}{R^2 + \omega^2 L^2} + \omega C_d D_d \simeq \frac{R}{\omega^2 L^2} + \omega C_d D_d \tag{15}$$

$$B = \frac{\omega L}{R^2 + \omega^2 L^2} - \omega C_d \simeq \frac{1}{\omega L} - \omega C_d \tag{16}$$

The latter forms are obtained by neglecting R^2 in comparison with $\omega^2 L^2$, which is a reasonable approximation if the coil has a fairly high value of Q. The condition for resonance may be obtained by setting B of (16) equal to zero, giving as the result

At resonance $$\omega_0 = \frac{1}{\sqrt{LC_d}} \tag{17}$$

$$f_0 = \frac{1}{2\pi\sqrt{LC_d}} \tag{18}$$

The dissipation factor of the entire circuit may be found by the quotient G/B from (15) and (16). When these terms are reduced to a common denominator and divided, we obtain

$$D = \frac{(R/\omega L) + \omega^2 L C_d D_d}{1 - \omega^2 L C_d} \tag{19}$$

The expression may be changed in form, and possibly given greater significance by the following substitutions. Let r represent the ratio of the operating frequency to the self-resonant frequency or

$$\omega^2 L C_d = \frac{\omega^2}{\omega_0^2} = r^2 \tag{20}$$

Also, as symbol for the component of the dissipation factor due to the d-c resistance of the coil winding

$$D_c = \frac{R}{\omega L} \tag{21}$$

With these substitutions in (19), the expression for D becomes

$$D = \frac{D_c + r^2 D_d}{1 - r^2} \tag{22}$$

There is an additional loss component D_e due to eddy currents (skin effect) in the wire of the winding. This loss may be minimized by the use of insulated stranded wire (Litzendraht), which is of particular value in coils for high-frequency operation.

The relationship of the various loss components is shown qualitatively in Fig. 13-12. D_c is the important component at low frequencies, and it

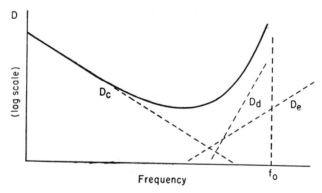

Fig. 13-12. Variation of the dissipation factor of an inductor with frequency.

decreases with frequency as shown in (21) and as illustrated in the figure. The factor $(1 - r^2)$ has little effect in the lower part of the range, but has great influence as the resonant frequency is approached. The contributions of D_d and D_e become important at the higher frequencies. The curve for the total dissipation factor comes to a minimum at a frequency considerably below the resonant point and then rises rather sharply.

Some inductors are "iron-cored," which means that they are wound on cores consisting of ferromagnetic material in some form. The cores may consist of laminations of electrical sheet steel or of some of the nickel iron alloys. They may be continuous or may have an air gap. The laminations must be made thin for high frequencies to limit the effect of eddy currents. For still better loss characteristics a "dust core" is used, in which the magnetic material is in powdered form with the particles insulated and held together by a binding material. The presence of a high-permeability core increases the inductance of a coil above the value it would have with an air core, or, expressed differently, a given inductance may be obtained from a smaller coil if an iron core is used. However, all magnetic materials show a change of permeability with change of the magnetic field strength, so the inductance of an iron-cored coil becomes a function of the current flowing in it. This fact tends to eliminate iron-cored coils as components of measuring circuits, although some dust-core coils are produced that may be used with moderate accuracy if care is taken in the amount of current passed through them. Toroidal coils on dust cores have the valuable features of compactness, zero external field, and a lower dissipation factor than an air-core coil.

The D_c component of the dissipation factor is much reduced by the presence of an iron core, owing to the smaller winding required for a given inductance. The reduction is partially offset by the introduction of new loss components due to hysteresis and eddy currents in the core. The distributed capacitance is increased somewhat by the conductivity of the core.

Analysis of the circuit of Fig. 13-11, begun in eqs. (15) and (16), may be completed by finding the values of the equivalent series inductance and resistance, to take into account the effect of D_d. The results are as follows:

$$L_{\text{equiv}} = \frac{1}{\omega}\frac{B}{Y^2} = \frac{L}{1 - r^2} \tag{23}$$

$$R_{\text{equiv}} = \frac{G}{Y^2} = \frac{R(1 + r^2 D_d/D_c)}{(1 - r^2)^2} \tag{24}$$

For $r^2 \ll 1$, eq. (23) reduces to the value obtained in (14), with no change due to D_d, and (24) becomes equal to (12) except for introduction of the term in D_d.

13-7. Air Capacitors*

Capacitors for measurements purposes are classified as to the dielectric material, whether air or a solid material. Air capacitors of the parallel plate type are very well known and used in many places, but only in small capacitance values. Solid-dielectric capacitors are more compact and for this reason are used almost entirely above about 0.001 μf and for smaller values also where adjustability is not required.

The parallel-plate variable air capacitor is an exceedingly useful tool for many measurements. At high frequencies it is the most desirable variable element, having better characteristics than variable resistors or variable inductors. In the modern form it is direct-reading, accurate, and stable. Figure 13-13 shows external and internal views of a General Radio Com-

Fig. 13-13. General Radio type 722-D precision air capacitor. Left, external view. Right, internal view. (*Courtesy of General Radio Co.*)

pany Type 722 Precision Capacitor having capacitance values from 100 to 1100 μμf, which are guaranteed within ±1 μμf. A correction chart is supplied giving corrections at 100 μμf intervals, by which the error may be still further reduced. There is a low-capacitance section, with a range from 25 to 110 μμf, guaranteed to ±0.2 μμf. This capacitor has a rugged cast frame, precision ball bearing, and an accurately cut worm drive. The insulators of steatite (quartz available for special purposes) are placed to avoid high field gradients, thus achieving very low dielectric losses. The frame, plates, shaft, spacer rods, and spacers are all made of aluminum or aluminum alloys so that all parts have the same coefficient of linear expansion. The capacitor is entirely shielded by an aluminum panel and metal-lined box. It is produced in both two-terminal and three-terminal models.

*Reference 1, pp. 162–198; reference 2, pp. 355–375; reference 4.

Fixed capacitors have been used in many forms in measurement work, ranging from the vacuum type to air capacitors under several atmospheres pressure in high voltage work, but usually at atmospheric pressure. In the latter form they frequently serve as the standard in Schering bridges for measurement of the characteristics of insulators and insulating bushings. The Bureau of Standards has a bank of ten 0.01-μf air capacitors which are used as secondary standards for the measurement of working standards sent there for calibration. Air capacitors have extremely low losses, usually assumed zero, if the dielectric which supports the insulated electrode is good material and placed in a region of low field gradient. There are small loss components caused by the insulating supports, the resistance in the conducting members, and the dirt or other contamination on the surface of the electrodes. It has been found, for example, that moisture absorbed in the oxide layer on the surface of aluminum plates has an effect on the dissipation factor. The losses, however, are smaller than for any other form of standard.

Air capacitors seem at first consideration to be well suited as primary electrical standards. It has been found difficult, however, to secure the extremely high stability needed for this purpose. Even carefully treated metal parts do a small amount of warping with time, so the long-time stability is not so high as desired in a primary standard.

13-8. Air Capacitors of Low Range

Standards of small capacitance became important some years ago, largely through need in the measurements of the characteristics of vacuum tubes. A development at the National Bureau of Standards provided fixed, continuously variable, and decade capacitors reaching several decades below the 1 $\mu\mu$f level.* More recently a new approach has been opened by a group at the National Standards Laboratory of Australia, with further work at the U.S. National Bureau of Standards. This work has resulted in capacitors of calculable value, useful both as standards in measurements, and possibly with adequate accuracy to form the basis of determination of the units on an absolute basis. We shall give some attention to both of these developments.

In the first group we shall consider the interesting standards of very low capacitance that were developed at the National Bureau of Standards. They are useful in particular as standards in bridges for the measurement of the interelectrode capacitance of vacuum tubes, in which the values range downward from a few micromicrofarads in plain triodes to a small fraction of a micromicrofarad between some electrodes of a shield-grid tube.

---

*Reference 12.

Low-capacitance standards are necessarily *three-terminal* capacitors, because the capacitance of each electrode to ground or shield may be much greater than the direct capacitance between the measured electrodes; definite and precise values can therefore be assigned only on a three-electrode basis. The system must, of course, be completely shielded so that the ground capacitances are entirely definite in value and cause no discrepancy in the direct capacitance. Down to about 0.1 μμf the ordinary guard-ring construction may be used, but below this value the center "island" becomes very small or else the spacing between the plates is large. In this case wide guard rings are needed to control field conditions in the measured part so that the capacitance may be computed. The development at the Bureau modifies the construction to a "guard-well" form, so called

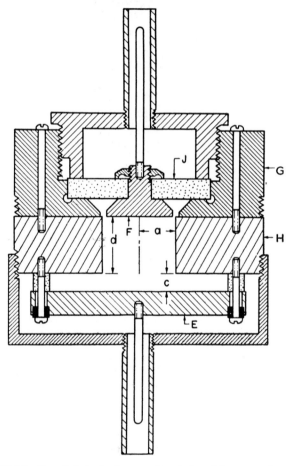

Fig. 13-14. Guarded-electrode capacitor designed for a very low value of capacitance. (*Courtesy of National Bureau of Standards.*)

because the insulated electrode is placed at the bottom of a "well," as shown in Fig. 13-14.* The insulated electrode F is supported by the Pyrex-glass collar J, and the surfaces of F and G are ground and lapped to an optically flat surface. The high-voltage electrode E is supported by insulators at a small distance from the ring H, which is carefully made to have flat and parallel faces. The capacitance from F to E can be controlled by the thickness of H, since an increase of H causes more of the electric flux from E to fringe to the inner surface of H and less to reach the bottom, F. Formulas have been derived that permit the accurate computation of the capacitance of this configuration from the dimensions shown as "a", "d," and "c" on the figure. Units of convenient size have been constructed down to 0.001 $\mu\mu$f. The solidity and permanence of construction in addition to the possibility of accurate computation of capacitance seem to qualify them well for service as primary standards. It is believed that a capacitance of 0.1 $\mu\mu$f can be attained to an accuracy of 0.1 per cent and 0.0001 $\mu\mu$f to an accuracy of 2 per cent. A variable capacitor and a decade capacitor in this same range of values are described in the reference.

Another form of air capacitor of small capacitance depends on Lampard's theorem, a recently discovered theorem in electrostatics.† We may first define the geometry to which the theorem applies. Suppose a hollow cylinder of conducting material to have a right-section which has one axis of symmetry $x' - x$, as in Fig. 13-15. The cylinder is divided

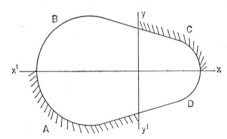

Fig. 13-15. Cross-section of cylinder, to illustrate capacitance theorem.

by slits on $x' - x$ and $y' - y$, parallel to the longitudinal axis, to give the four segments A, B, C, D. The theorem states that the cross-capacitance between opposite segments, as A and C, for connection of all segments as a three-terminal capacitor equals‡

$$C = \frac{\log_e 2}{4\pi^2} \text{ e.s.u. per cm of axial length}$$

---

*Reference 12.
†Reference 13.
‡Reduced to numerical values,
$$C = 0.017558 \text{ e.s.u./cm} = 0.019535 \ \mu\mu\text{f/cm}$$

It is interesting to note that the result does not depend on crosswise dimensions or on the shape used, provided only that the symmetry condition is satisfied. The cross section may consist of straight lines or circles, or other forms, with no restriction other than symmetry. The fact that the axial length is the only dimension that must be determined accurately makes this type of capacitor appear attractive as an absolute standard, for which capacitance can be determined from the dimensions. Small departures from symmetry are found not to be serious, as the mean of the two cross-capacitances (*A-C* and *B-D*) eliminates the error to within a second-order effect. (Guard sections are used at the ends to eliminate fringing from the measured dimension.)

One departure of practice from theory is in the division of the cylindrical surface into segments. Theory assumes that the divisions are of infinitesimal width, whereas a slit of finite width is required in practice. However, the division may be placed in a re-entrant corner, as in Fig. 13-16(a), giving

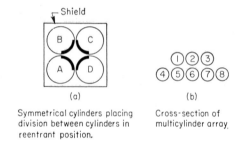

(a)

Symmetrical cylinders placing division between cylinders in reentrant position.

(b)

Cross-section of multicylinder array.

Fig. 13-16. Practical electrode arrangements to utilize the capacitance theorem.

negligible effect of the slit on the cross capacitance. Another arrangement is indicated in (b), which shows the cross section of a multi-cylinder array. Bars 2 and 6 have insulated end (guard) sections, while the others are plain bars. A capacitive determination can be made for the direct capacitance of the central section of bar 2 to 5 and 7, with the other bars connected to the third terminal (ground). (This gives a capacitance twice the formula value, above.) A second balance, using bar 6 to 1 and 3, gives a check for symmetry.

The capacitance is small, as expressed by the formula. For a 10-inch guarded section, the capacitance is practically $1 \mu\mu f = 1$ pf. Special bridges must be used for measurements in this range. New developments in bridges for this purpose have been described in several papers.*

---

*References 14, 15.

13-9. Solid-dielectric Capacitors

Compared with an air capacitor solid-dielectric capacitors are used in the larger values because of their compactness, which is secured by the higher dielectric constant and the thinner layer of dielectric. Several dielectric materials are used in making capacitors, the most common of which are impregnated paper, mica, and polystyrene. Paper is used for many common by-pass and power-factor correction capacitors, but is not much used in measurement work (except as dictated by the high cost of large mica capacitors) because of a relatively high dissipation factor. Mica is most generally employed in high-grade capacitors because of a combination of good qualities. The following summary gives an idea of the characteristics of the materials, necessarily in a somewhat approximate way. The values for paper depend not only on the type of paper, but even more largely on the impregnating material used. Mica is a naturally occurring substance and varies considerably between different samples.

It should be pointed out in a discussion of precision capacitors that the term "dielectric constant" is something of a misnomer, for many materials show changes of this "constant" with frequency. The causes are not completely understood, but they are known to be related to matters of molecular structure. There is an effect due to interfacial polarization, generally at low frequency, and an effect due to dipole polarization at very high frequency. Some of the better dielectrics, such as quartz, polystyrene, and mica show very little change of constant with frequency, at least in the usual electrical range; quartz is practically constant from 100 c to 100 Mc. Some of the phenolic materials on the other hand, show very large changes.

(a) *Impregnated paper.* ϵ_r (relative dielectric constant = 2.4–3.5, D = 0.005–0.02.

(b) *Mica.* ϵ_r = 5.7, D = 0.0001–0.0005 at 1 kc. Temperature coefficient of capacitance = +30 ppm/°C (ppm = parts per million.) $\Delta C/C$ = change in capacitance due to interfacial polarization = 0.1 per cent at 1 kc, and increases by factor of 10 for decrease of frequency of 3 decades. D changes in same proportion. K and D considerably affected by moisture.

(c) *Polystyrene.* ϵ_r = 2.4, D = 0.0003, not affected by frequency. Temperature coefficient of capacitance = −140 ppm/°C. Slightly affected by moisture.

Polystyrene has an advantage over mica in constancy of characteristics over a wide frequency range, but is subject to a much larger temperature effect. There is little choice with respect to the dissipation factor. Mica continues, as in the past, to be the material used for most precision capacitors.

Paper capacitors are made by winding together on a winding machine long strips of metal foil and treated paper. In some capacitors, contact is made to the metal strips only at the ends, but this gives an objectionable inductance for any use except at very low frequencies. A better construction is to allow the two metal strips to extend beyond the paper at opposite ends of the roll and then, upon completion of the winding, to crimp the protruding metal together, thus shorting all turns of one strip together. This reduces both resistance and inductance components of the winding. Mica capacitors are made in the older method by stacking alternate sheets of mica and strips of metal foil, allowing the foil strips to extend alternately to right and left of the mica, so that they can be connected together to form the terminal connections. The mica-and-foil stack is placed between metal end plates with screws to hold the assembly under considerable pressure. The unit is heated to drive out moisture, and then is dipped in wax to prevent the re-entry of moisture. The assembly of a capacitor of this type is entirely a hand job, which explains the high cost. Another form of construction is the silvered-mica capacitor, in which films of silver are formed directly on the surfaces of the mica. This may be accomplished by spraying the surfaces with a suitable silver solution after masking the areas required for margins. The mica sheets are fired at a high temperature for a short time so that a tight bond is formed between the silver and mica. The laminations are then stacked, with lead foil strips as connectors, and held by electrodes that are crimped over the ends. The silvered-mica capacitor is reported to have a lower dissipation factor and a smaller temperature coefficient than the clamped type described above.* Experience with the silver-mica type, though covering only a few years, indicates that it is much more stable with age than the foil type. Foil capacitors show, in general, an increase of capacitance with time, although there are great individual differences; the increase may be attributed to slow yielding of mica or foil under the clamping pressure, giving more intimate contact in microscopic surface irregularities. Observation by the author of a group of about twenty capacitors over periods of five to ten years showed an average rate of increase for foil construction of about 0.03 per cent per year, with a range from about zero to 0.06. Some capacitors appear to approach a stable value after several years, while others continue to drift. The magnitude of change, and the uncertainty regarding it, require reevaluation at fairly short intervals. Silver-mica units appear to have little or no aging effect, on the basis of present experience with them.

Figure 13-17 shows a standard capacitor made by the General Radio Company. It is of silver-mica construction, and is mounted in a cast aluminum case, and sealed to exclude moisture. The capacitor terminals are connected to two insulated binding posts, with a third grounded to the case, so the capacitor can be used as two-terminal or three-terminal as

---

*Reference 4.

Fig. 13-17. General Radio Type 1409 standard capacitor. (*Courtesy of General Radio Co.*)

needed. The capacitance is adjusted within ±0.05 per cent of the nominal value, and a certificate states the result of comparison precise to ±0.01 per cent with a standard certified to ±0.03 per cent by the National Bureau of Standards. The temperature coefficient of capacitance is stated to be 35 ± 10 ppm/°C (parts per million).

Mica capacitors are available in decade mounting, beginning usually with 0.001 μf steps. Below 1 μf there is a choice of mica or paper dielectric, accompanied by a considerable price differential, but above 1 μf paper is generally used on the basis of cost. It should be noted that even high-grade decade capacitors are usually not guaranteed to better than 1 per cent except for some very high-priced boxes, whereas fixed capacitors may be obtained to 0.1 per cent or better. For this reason fixed standards are generally used where accuracy is an important consideration.

13-10. Frequency Errors of Capacitors

A capacitor fails to present a perfect and constant capacitance at all frequencies due to several effects. There are I^2R losses in the plates and connecting wires, interfacial polarization and power loss in the dielectric, and inductance in the connections. At high frequencies the resistances of plates and connecting wires are increased by skin effect. The major effects

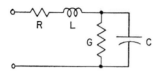

Fig. 13-18. Equivalent circuit of a capacitor.

are represented by the circuit of Fig. 13-18, which shows R and L for the connecting leads in series with a pure capacitance C paralleled by a conductance G to represent the losses in the dielectric.

At low frequencies the effects of R and L are negligible, and the capacitor acts as a pure capacitance C in parallel with a conductance G, or in series with a resistance $R_s = G/(\omega C)^2$. The inductance becomes increasingly important as the frequency is raised, eventually producing a condition of resonance between L and C. At frequencies below resonance L partially neutralizes the effect of C, giving an apparent, or effective, terminal capacitance expressed approximately by the equation

$$C_e = \frac{C}{1 - \omega^2 LC} \tag{25}$$

As an illustration of the possible error in using an air capacitor at high frequency, suppose that $L = 0.05$ μh, $C = 1000$ $\mu\mu$f and $f = 10$ Mc. In this case $X_L = 3.14$ ohms and $X_C = 15.9$ ohms, so L causes an error of 20 per cent in the capacitive effect. This indicates the care that must be taken to reduce the residuals to low values in high-frequency circuits.

The circuit of Fig. 13-18 is very nearly equivalent (for $G^2 \ll \omega^2 C^2$) to a pure capacitance C_e in series with a resistance

$$R_s = R + \frac{G}{(\omega C)^2} \tag{26}$$

Or else, C_e may be placed in parallel with a conductance

$$G_P = G + R(\omega C)^2 \tag{27}$$

In either case, the total dissipation factor is

$$D = \omega CR + \frac{G}{\omega C} \tag{28}$$

Curves typical of the variation of capacitance and dissipation factor of a solid-dielectric capacitor (0.1 μf, mica) are shown in Fig. 13-19. The downward slope of the $\Delta C/C$ curve for low frequencies is caused by interfacial polarization in the dielectric, the upward slope at high frequencies is caused by the series inductance. The slope of the D curve at low frequencies is characteristic of the dielectric material. G has a small constant component and a dielectric component that varies approximately as the first power of the frequency, modified by polarization conditions. Both $\Delta C/C$ and D curves descend to a minimum and then rise sharply, the capacitance approximately as the square of the frequency (eq. 25), and D as the $\frac{3}{2}$ power,

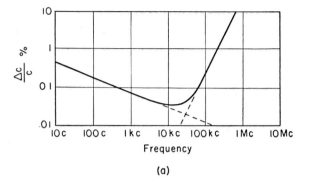

(a)

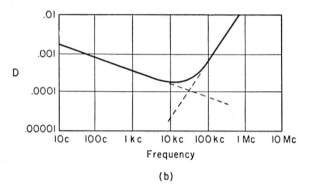

(b)

Fig. 13-19. Frequency characteristics of a 0.1 μf mica capacitor. (a) Change of capacitance as a function of frequency (changes based on the value C would have if there were no interfacial polarization or series inductance). (b) Dissipation factor as a function of frequency. (*Courtesy General Radio Co.*)

due to the first term in (28). (R increases as the $\frac{1}{2}$ power of f at high frequency, by skin effect.)

13-11. The Effect of Residuals in Bridge Circuits

The residuals that are present in a bridge circuit affect the balancing process and require a different setting of the adjustable elements than if the residuals were not present. An error is introduced into the computed value of the unknown if the marked constants of the standards are used in the computations, and the residuals are neither balanced-out in the bridge nor introduced into the computations.

The residual effects enter into the circuit in two ways; first, they are present within the standards themselves, as studied in the foregoing sections; second, they occur because of coupling between the elements of the circuit. The inter-element coupling may be of electric or magnetic form, either of which may be controlled by appropriate shielding. The form of most frequent concern is the capacitance of the various parts of the circuit to ground. If the capacitances to ground are concentrated at the four corners of the bridge (as they can be placed by shielding), their effect on bridge balance may be eliminated by the use of a Wagner ground connection. An alternative, on a solidly grounded bridge, is to measure the capacitances and include them in the bridge computations. Information regarding the residuals within the standards is frequently essential in accurate measurements. It may sometimes be found in satisfactory form in the equipment-maker's catalogue, but if this is not available or sufficiently definite, the data should be obtained under the actual operating conditions by measurement. A very good discussion of methods of correction for residuals and of making the needed measurements may be found in reference 9.

A study of bridge equations is needed as an aid in making corrections and as a means of estimating the amount of the error introduced by the residuals. The equations derived in chap. 9 took into account only the main, or intentional, constants, so it is necessary to repeat the process with the addition of the important residuals. The representation of the elements will be simplified as compared with the studies made in the foregoing sections, otherwise the equations become very unwieldy. It will be found more convenient in many cases to introduce the residuals in terms of the Q or D of the bridge arm, rather than by adding R, C, or L symbols for the extra components. The choice of Q or D is a matter of convenience to simplify the circuit expressions; however, it is usually best to choose the symbol that is small for a small amount of the residual. For example, a small stray capacitance shunting a resistor gives it a small value of Q. $(Q = B/G = \omega CR)$ whereas D is merely reduced from infinity to an inconveniently large number by C. Therefore, Q is the better quantity to use in this case. Similarly, D is the better term to represent the addition of a small resistive effect to a capacitor.

Bridges will be studied that use resistors and capacitors as standards, as this covers most of the bridges and the most useful bridges, which were considered in previous chapters. A large part of the desirability of resistors and capacitors as standards is the fact that their residuals are much smaller than those of inductors. A resistor will be represented as R paralleled by a small C, as this is the net effect, at least in the higher ranges of R. For this arrangement, $Q = \omega CR$ and will be given a positive sign in the following derivations. (The inductive effect may predominate for low-valued resistors, in which case Q may be given a negative sign in the formulas obtained here.) In other words, we show a resistor by values of R and Q

that represent it correctly at the given frequency, so that we do not need to work with R, L and C symbols on the diagram and in the equations. The losses in a capacitor may be represented by either a series R or a parallel G, and we may use either method, depending on whether the circuit requires a series or parallel combination with other elements. For a capacitance with series resistance, $D = \omega C R$.

It becomes important in many cases to distinguish between *first order* and *second order* effects of residuals. Thus, for a bridge arm consisting of a resistor paralleled by a small value of C, the *first order* effect of C is to introduce a phase angle in the impedance of this arm. This is frequently the important part of the action of C. The change in the *magnitude* of resistance due to C is very small — a second order effect that is often negligible.

(a) *Schering bridge.*

This bridge is shown in Fig. 13-20 with a series residual R_1 in arm 1 and parallel capacitances across arms 3 and 4. The symbol C_3 will be used to represent the total of stray and lumped capacitance of arm 3. In arm 1, when a good capacitor is used, R_1 is so small that it has negligible effect on the *magnitude* of the impedance, so $\mathbf{Z}_1$ may be written

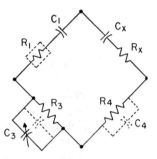

Fig. 13-20. Residuals in a capacitance bridge (Schering circuit).

$$\mathbf{Z}_1 = \frac{1}{\omega C_1} \underline{/-90° + \tan^{-1} D_1}$$

and $$D_1 = \omega C_1 R_1 \qquad (29)$$

The quantity D is very small, so the tangent and the angle in radian measure are practically identical. Therefore $\mathbf{Z}_1$ may be changed to

$$\mathbf{Z}_1 = -j\frac{1}{\omega C_1} \underline{/D_1}$$

Arms 3 and 4 have impedances practically equal to their resistance, with small phase angles given by the Q's

$$\mathbf{Z}_3 = R_3 \underline{/-Q_3} \quad \text{where} \quad Q_3 = \omega C_3 R_3$$

$$\mathbf{Z}_4 = R_4 \underline{/-Q_4} \quad \text{where} \quad Q_4 = \omega C_4 R_4$$

The values above may be substituted in the equation,

$$\mathbf{Z}_x = \frac{\mathbf{Z}_1 \mathbf{Z}_4}{\mathbf{Z}_3},$$

giving $$\mathbf{Z}_x = -j\frac{R_4}{\omega C_1 R_3} \underline{/D_1 + Q_3 - Q_4} \qquad (30)$$

For the small angle the cosine is practically one, and the sine equals the angle

$$\mathbf{Z}_x = R_x - j\frac{1}{\omega C_x} = -j\frac{R_4}{\omega C_1 R_3}[1 + j(D_1 + Q_3 - Q_4)]$$

Multiplication and separation of real and j parts gives for the j parts

$$C_x = \frac{C_1 R_3}{R_4} \tag{31}$$

which is the result as obtained in chap. 9 without the residuals. For the real parts

$$R_x = \frac{R_4(D_1 + Q_3 - Q_4)}{\omega C_1 R_3} \tag{32}$$

The dissipation factor of X is $\omega C_x R_x$, or

$$D_x = Q_3 - Q_4 + D_1 \tag{33}$$

The dissipation factor of the unknown is supposed to be read on the dial of the C_3 capacitor. However, as shown by the equation, the value of D_x should be modified by the Q_4 and D_1 terms. These quantities are small, but so is D_x for a good capacitor, so the corrections cannot be ignored. This is a case in which the residuals are of very great importance. Even an initial balance or substitution method does not remove all the uncertainty for D, because it gives the difference for the two conditions, but not the absolute value of either.

(b) *Maxwell bridge.*

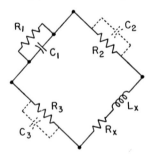

Figure 13-21 indicates stray capacitances in arms 2 and 3. (Stray capacitance in R_1 may be combined with C_1 and conductance of C_1 with R_1, so that the symbols on the diagram stand for the total effects.) The constants of the arms may be expressed

Fig. 13-21. Residuals in a Maxwell inductance bridge.

$$Y_1 = \frac{1}{R_1} + j\omega C_1 \qquad Q_1 = \omega C_1 R_1$$

$$Z_2 = R_2\underline{/-Q_2} \qquad Q_2 = \omega C_2 R_2$$

$$Z_3 = R_3\underline{/-Q_3} \qquad Q_3 = \omega C_3 R_3$$

Combination of these quantities in the equation

$$R_x + j\omega L_x = Z_2 Z_3 Y_1$$

gives the results

$$L_x = C_1 R_2 R_3 \tag{34}$$

which is independent of the residuals, except for the addition to C_1 by the residual capacitance of R_1. Q_x, however, is found to have the value

$$Q_x = \frac{Q_1}{1 + Q_1(Q_2 + Q_3)} \tag{35}$$

If the computation of Q_x is made neglecting Q_2 and Q_3, then $Q_x = Q_1$ and the denominator of the fraction is assumed to be unity. Actually, the denominator should be greater than unity, so the true Q_x is less than the value that is computed without the residuals. Neglect of the residuals can cause serious error in the value of Q_x, particularly for high values of Q.

(c) *Hay bridge.*

Results for the Hay bridge parallel the Maxwell very closely. Figure 13-22 shows residual capacitances across arms 2 and 3. A similar capacitance could be shown across R_1, but is not generally necessary, because R_1 is small and hence little affected by a shunting capacitance for the cases for which the Hay circuit is usually employed. A similar configuration is encountered in the Owen bridge, but with larger value of resistance, and is treated there more exactly.

The constants of arm 1 may be expressed in the following way

$$\mathbf{Z}_1 = R_1 - j\frac{1}{\omega C_1}$$

$$= \frac{D_1 - j1}{\omega C_1}, \tag{36}$$

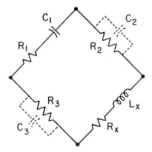

Fig. 13-22. Residuals in a Hay inductance bridge.

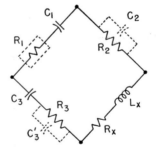

Fig. 13-23. Residuals in an Owen inductance bridge.

where $D_1 = \omega C_1 R_1$. $\mathbf{Z}_2$ and $\mathbf{Z}_3$ are the same as for the Maxwell bridge. Solution of the balance equation gives the following results

$$L_x = \frac{C_1 R_2 R_3}{1 + D_1^2} \tag{37}$$

$$Q_x = \frac{1}{D_1 + Q_2 + Q_3} \tag{38}$$

$$D_x = D_1 + Q_2 + Q_3 \tag{39}$$

Corrections for Q_2 and Q_3 may be of great importance for high-Q coils. For example, suppose that Q_2 and Q_3 each have the quite possible value of 0.001. If $D_1 = 0.01$, the true value of Q_x is 83.3, but if Q_2 and Q_3 are neglected the computed result is 100, an error of 20 per cent. This shows the great importance of the residuals in measurements of high values of Q.

(d) *Owen bridge.*

Figure 13-23 shows residuals R_1, C_2, and C_3'. The reactance of arm 3 may be computed in the following way

$$X_3 = \frac{1}{\omega C_3} + \frac{\omega C_3'}{(1/R_3^2) + (\omega C_3')^2}$$

The last term in the denominator may be neglected, with the resulting simplification

$$X_3 = \frac{1}{\omega C_3} + \omega C_3' R_3^2$$

$$= \frac{1 + Q_p D_s}{\omega C_3} \tag{40}$$

where the following symbols are used for the series and parallel effects in arm 3

$$D_s = \omega C_3 R_3 = \frac{1}{Q_s} \tag{41}$$

$$Q_p = \omega C_3' R_3$$

With these abbreviations Z_3 becomes

$$Z_3 = R_3[1 - j(Q_s + Q_p)] \tag{42}$$

Solution of the equation for balance yields the following results

$$L_x = C_1 R_2 R_3 \tag{43}$$

$$Q_x = \frac{1}{Q_3 + Q_2 + D_1} \tag{44}$$

or

$$D_x = Q_3 + Q_2 + D_1 \tag{45}$$

The symbols have the following values, with parentheses inserted to indicate the residuals,

$$Q_3 = \frac{1}{\omega C_3 R_3} + \omega(C_3')R_3 = Q_s + Q_p$$

$$Q_2 = \omega(C_2)R_2$$

$$D_1 = \omega C_1 (R_1)$$

The results indicate, as in the Maxwell and Hay bridges, that the residuals have negligible effect on the inductive balance, but may cause serious error in Q for high-Q coils.

(e) *Resonance bridge.*

Capacitive residuals are shown across the three resistive arms in Fig. 13-24. This circuit is frequently used at radio frequencies for which the residuals may be fairly large, so the equations should be examined with considerable care.

The impedance for arm 1 may be obtained in the following way

$$Z_1 = \frac{(1/R_1) - j\omega C_1}{(1/R_1{}^2) + (\omega C_1)^2}$$

$$= R_1 \frac{1 - jQ_1}{1 + Q_1{}^2} \qquad (46)$$

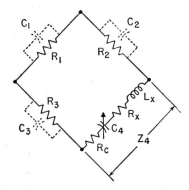

Fig. 13-24. Residuals in a resonance bridge.

where $\qquad Q_1 = \omega C_1 R_1$

If similar expressions are written for Z_2 and Z_3 and substituted in the balance equation, the following results are obtained:

$$R_4 \text{ (total)} = \frac{R_2 R_3}{R_1} \frac{1 + Q_3(Q_1 - Q_2) + Q_1 Q_2}{(1 + Q_2{}^2)(1 + Q_3{}^2)} \qquad (47)$$

$$L_x = \frac{1}{\omega^2 C_4} - \frac{R_2 R_3}{\omega R_1} \frac{(Q_3 + Q_2 - Q_1)}{(1 + Q_2{}^2)(1 + Q_3{}^2)} \qquad (48)$$

Second-power terms have been retained in (47), because this bridge is often used when a careful determination of coil resistance is desired, and this may be in a high-frequency circuit in which the Q's of the resistors may be fairly high. The expressions may be simplified in many cases if the Q's are small. The resonance bridge is particularly well suited to measurements of coil resistance, as the resistance balance is dominant once the L-C tuning is performed. The bridge should be set up with care, because the junction of coil and condenser is a point of high potential and, hence, subject to capacitive effects.

REFERENCES

1. Hague, B., *Alternating Current Bridge Methods*, London: Sir Isaac Pitman and Sons, Ltd., 1946.

2. Laws, Frank A., *Electrical Measurements*, New York: McGraw-Hill Book Co. Inc., 1938.

3. Brooks, H. B. and Weaver, F. C., "A Variable Self and Mutual Inductor," *Bull. Bur. Stds.*, **13**, 569–80 (1917).

4. Christopher, A. J. and Kater, J. A., "Mica Capacitors for Carrier Telephone Systems," *Trans. A.I.E.E.*, **65**, 670 (Oct. 1946).

5. Field, R. F., "Residual Impedances in the Precision Condenser," *Gen. Radio Exp.*, X, 5 (Oct. 1935).

6. Field, R. F., "Frequency Characteristics of Decade Condensers," *Gen. Radio Exp.*, XVII, 5 (Oct. 1942).

7. Arguimbau, L. B., "Losses in Audio-Frequency Coils," *Gen. Radio Exp.*, XI, 6 (Nov. 1936).

8. Sinclair, D. B., "Radio-Frequency Characteristics of Decade Resistors," *Gen. Radio Exp.*, XV, 6 (Dec. 1940).

9. "Impedance Bridges Assembled from Laboratory Parts," General Radio Co., Cambridge, Mass.

10. Editorial, "Behaviour of High Resistances at High Frequencies," *Wireless Engineer and Exp. Wireless*, **12**, 141–3, 291–5 (June 1935); 413–4 (Aug. 1935).

11. Puckle, O. S., "Behaviour of High Resistances at High Frequencies," *Wireless Engineer and Exp. Wireless*, **12**, 141, 303–9 (June 1935).

12. Moon, Charles, and Sparks, C. Mathilda, "Standards for Low Values of Direct Capacitance," *NBS J. Research*, **41**, 5, 497–507 (Nov. 1948).

13. Lampard, D. G., "A New Theorem in Electrostatics with Application to Calculable Standards of Capacitance," *Proc. Inst. Elec. Engrs.* (London), **104**, Part C, p. 271–280 (1957).

14. Thompson, A. M., "The Precise Measurement of Small Capacitances," *Trans. I.R.E.*, **I-7**, p. 245–253 (Dec. 1958).

15. McGregor, M. C., J. F. Hersh, R. D. Cutkosky, F. K. Harris, F. R. Kotter, "New Apparatus at the National Bureau of Standards for Absolute Capacitance Measurement," *Trans. I.R.E.*, **I-7**, p. 253–261 (Dec. 1958).

PROBLEMS

13-1. A coil has an inductance of 10 mh, a resistance of 50 ohms and a distributed capacitance of 50 $\mu\mu$f. Find the series resistance and inductance equivalent to this coil at a frequency of 10,000 cycles per second. Repeat for 100,000 cycles per second.

13-2. A 2000-ohm resistor has a series inductance of 6 μh and a shunt capacitance of 25 $\mu\mu$f.

(a) Find the current and its phase angle with the voltage if 10 v at 100,000 cycles per second are impressed on the resistor. Express this as a complex number in Cartesian and polar forms, using voltage as reference.

(b) Find the apparent series resistance and reactance at this frequency.

13-3. A 1000-ohm bifilar-wound resistor has an inductance of 2 μh and a shunt capacitance of 350 $\mu\mu$f. Find the apparent resistance and the phase angle at frequencies of 1000 and 10,000 cycles per second.

13-4. A precision 1000-ohm bridge resistor has an inductance of 5 μh and a distributed capacitance of 20 $\mu\mu$f. Find the phase angle of this resistor when used at a frequency of 50,000 cycles per second. Use no formulas.

13-5. A shielded resistor has a resistance of 1000 ohms when measured on a d-c bridge. It has an inductance of 5 microhenrys, a capacitance of 40 $\mu\mu$f from terminal A to the shield, and an equal value from terminal B to the shield. Find the phase angle of this resistor at 100,000 *cycles* per second if both terminal B and the shield are connected to ground. Use no formulas.

13-6. A coil has a resistance of 100 ohms and an inductance of 0.1 henry, when measured at a very low frequency. The distributed capacitance is 0.005 μf. Find the effective resistance and inductance of this coil at a *frequency* of 2000 *cycles* per second, without the use of formulas.

13-7. A coil has a resistance of 200 ohms and an inductance of 1.0 h when measured at very low frequency. The distributed capacitance is 200 $\mu\mu$f. Find the percentage change in effective inductance when this coil is used at a frequency of 1000 cycles per second.

13-8. A coil has an effective inductance of 0.102 henry when measured at a frequency of 1000 cycles per second and 0.110 h at 3000 cycles per second. Find the effective inductance at 2000 cycles per second.

13-9. The leads to the plates of a 100 $\mu\mu$f capacitor have a resistance of 0.5 ohm and an inductance of 0.3 microhenry. The dielectric losses at the operating frequency of 10 megacycles are equivalent to a leakage path of 50,000 ohms. Find the equivalent capacitance and the dissipation factor of the capacitor at this frequency.

CHAPTER 14

Bridge Accessories

14-1. Types of Detectors

There are many kinds of detectors available for a-c bridges. We shall list the more common ones and make a brief study of their characteristics. The following kinds will be considered:

(a) Galvanometers
   Vibration galvanometer
   Excited-field a-c galvanometer
(b) Telephone receivers
(c) Amplifiers, followed by a
   Galvanometer
   Telephone receiver
   Indicating instrument
   Tuning eye
   Cathode-ray oscilloscope
(d) Radio receiving sets

14-2. Galvanometers

(a) *Vibration galvanometer**.

The vibration galvanometer has a permanent-magnet field and a very narrow moving coil supported by a taut suspension. When alternating

---

*Reference 1, p. 261; reference 2, p. 465.

current is passed through the coil, it tends to deflect alternately in one and then in the other direction for the two halves of the cycle. The purpose of the narrow coil and taut suspension is to provide a system in which the natural frequency of vibration is the same as the frequency of the alternating current. The system is made tunable by changing the active length of the suspension for coarse adjustment and then its tension for fine adjustment. The scale shows simply a narrow vertical line of light when no current flows in the coil. When current of the resonant frequency is passed through the coil, the line of light widens out into a ribbon.

The use of mechanical resonance makes possible a high degree of sensitivity for the frequency for which the system is tuned, and this entails the additional feature of low sensitivity for frequencies remote from the resonant value. Thus if the bridge-supply voltage is of distorted wave form, the vibration galvanometer permits balancing for the fundamental component, since it gives very slight response for harmonics of the supply frequency. The frequency-selective characteristic is a particular merit of this instrument.

Vibration galvanometers may be constructed for frequencies up to several hundred cycles per second, but they mainly find application at power frequencies of 25 and 60 cycles per second. Amplifiers are sometimes used between the bridge and the galvanometer to increase the sensitivity. However, if an amplifier is to be used at all, it is simpler to add a tube or two and then use the indication of a rugged and cheap d-c milliammeter. The trend has been toward such arrangements in recent years, since stable, high-gain amplifiers have become available. The frequency-selective properties of the vibration galvanometer can be duplicated in an amplifier by the inclusion of tuned circuits.

(b) *Excited-field a-c galvanometer*[*].

An excited-field galvanometer has a field structure consisting of soft-iron laminations excited by a coil carrying current of the same frequency as used on the bridge. The moving coil produces a torque in the same direction for both halves of a cycle like an ordinary a-c wattmeter, since the field flux and the moving-coil current both reverse direction in the second half-cycle as compared with the first. The spot of light thus behaves in the same manner as on a d-c galvanometer.

Figure 14-1(a) indicates a sinusoidal current i_b flowing in the moving coil and a field flux of the same time phase. Under this condition the torque is always in the same direction and the galvanometer coil is deflected. However, if the flux should be at 90° time-phase from i_b, as in sketch (b), the average torque is zero and no deflection results. In this case of a 90°

---

*Reference 1, p.248.

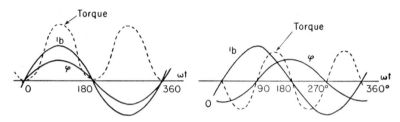

Fig. 14-1. Effect of phase relationships on galvanometer torque. (a) φ in phase with i_b. (b) $\varphi 90$ from i_b.

phase between the two circuits the behavior of torque and deflection are the same as in a wattmeter.

If the bridge is unbalanced and conditions are as in (b), the galvanometer gives no evidence of the unbalance. It is thus necessary to adjust the phase of φ to correspond with (a) in order to achieve maximum sensitivity. This is the function of the phase-shifter indicated in Fig. 14-2. The

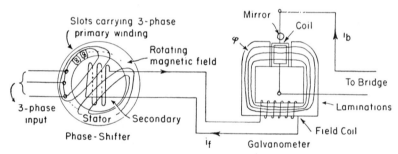

Fig. 14-2. Diagrammatic view of excited-field a-c galvanometer with phase shifter.

phase-shifter has a laminated-iron stator with a three-phase winding similar to a three-phase induction motor and receives power from a three-phase supply. The rotor laminations carry a winding which is used to excite the galvanometer field. The rotor is ordinarily at rest but may be set to any desired position by a handwheel and a fine adjustment of the worm-gear type. The handwheel has stops to permit exact 90° travel.

(NOTE: A two-phase stator winding may be used with a capacitor in one phase with connection to a single-phase supply. This is similar to a single-phase capacitor motor. If available, the three-phase input is usually given preference.)

As is well known in motor theory, the combination above results in a rotating magnetic field. This field cuts across the rotor windings and induces a voltage in them. The *phase* of this voltage depends on the *position* of the rotor coil. If the rotor is moved 90° in the direction of the rotating field, the phase of the rotor voltage is retarded 90° in time-phase. Adjustment of the rotor position thus gives a convenient means of controlling the phase of the galvanometer field current.

In balancing a bridge it is desirable to secure the conditions of Fig. 14-1(a) for maximum sensitivity. (The best way to do this is to adjust the phase-shifter for *zero* deflection of the galvanometer, then to shift 90°, because the zero point is better defined than the maximum.) However, the adjustment of the field should not be regarded as a drawback of this instrument but rather the reverse, for it gives a "phase-sensitive" detector. All bridges have double balance conditions, for resistance and for reactance. If the R and X controls are the ideal 90° apart, as discussed in chap. 9, this galvanometer permits independent adjustment for each component of the unbalance. For example, in Fig. 14-3, suppose we have a bridge unbalance

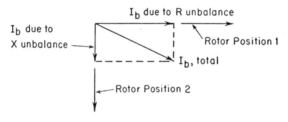

Fig. 14-3. Bridge adjustment for galvanometer of Fig. 14-2.

as the result of incorrect R and X adjustments. If the phase-shifter rotor is in position 1, the galvanometer has maximum sensitivity for the R unbalance and is *not affected by X unbalance*. That is, R can be set at its correct, final position in spite of X unbalance. Then we put the phase shifter to position 2 and balance X. This is a faster method than approaching balance by successive approximations as we must do with an amplitude-responsive detector. Actually, we should check back from position 2 to 1 because the phase shifter may not be exactly in the correct position, but the second adjustment should be quite small.

Galvanometers of the excited-field type are confined practically to power-line frequency owing to the power requirements of the phase-shifter and galvanometer field. (We shall discuss later an electronic circuit that achieves the same result but in a more flexible manner.) These galvanometers are not often applied to impedance bridges because of both frequency limitations and cost of equipment, but they can be used in this way. We shall meet galvanometers of this sort in the next chapter in a study of testing methods for instrument transformers.

While the a-c galvanometer as a null instrument in a-c circuits appears to be a direct counterpart of a D'Arsonval galvanometer in a d-c circuit, nevertheless there is a distinct difference in behavior. The coil of the d-c galvanometer responds to an unbalance of the bridge to which it is connected, but does not introduce a voltage (except motionally). For the a-c case, however, a voltage is induced in the moving coil by transformer action by the field flux except when the coil lies parallel with the field of the electromagnet. Consequently, the current through the moving coil of the a-c galvanometer (and likewise its torque) has two components, one due to the bridge unbalance, and the other due to the voltage induced by the field magnet. The torque produced by the induced current may be much larger than the component originating in the bridge unbalance. Accordingly, the use of the a-c galvanometer as a null detector is not fully effective unless the torque of the induced current is nullified. (If the circuit of the moving coil is inductive, the torque of the induced current opposes any deflection, and thus reduces the sensitivity of the galvanometer to the unbalance of the bridge.)

The torque of the induced current depends on its phase relationship to the field flux. The induced voltage is given by $n\,d\phi/dt$, and so has a 90° phase relationship to the flux; if the current is brought into phase with the voltage by the addition of series inductance or capacitance (as required by the bridge circuit to which it is connected) to give a purely resistive effect around the galvanometer coil circuit, then the interfering torque is zero, and the effective sensitivity of the galvanometer is increased.* The tuning of the coil circuit, revised each time the bridge constants are changed, is a considerable bother, and counts against the a-c galvanometer for many bridge applications.

A small steel strip, adjustable in position near the moving coil, is sometimes mounted on an a-c galvanometer. Adjustment of the strip modifies the field near the moving coil, and can be used to bring the electric (field-energized) zero into agreement with the mechanical (rest) zero.

14-3. Telephones†

Telephones are the most common and one of the most sensitive detectors used in the audio-frequency range. Considering their simplicity and low cost, they do a remarkable job. Used directly without an amplifier they permit balancing of many bridges to the limit of accuracy of the components. The two chief objections to telephones as detectors are the difficulty of using them in noisy places and the fatigue that they cause the operator if used for long periods. Many bridge measurements are made at 1000 cycles, which is in the region of greatest sensitivity for the human ear and of good response for phones. A telephone receiver is likely to have a

---

*Reference 7.
†Reference 1, p. 251.

resonant point somewhere between 700 and 1200 cycles, and accordingly may be expected to have high sensitivity at 1000.

Various designs and resistance values are met in telephone headsets, hence it is not possible to be very definite. However, it may be interesting to have some sample values to indicate the order of magnitude. The following table shows values of L, Q, and the minimum voltage for a perceptible signal as measured on two headsets among several picked at random in the laboratory.

CHARACTERISTICS OF TELEPHONE HEADSETS AT 1000 CPS

		Measured			Computed		
Phones	R_{DC}	L	Q	E_{min}, mv	X	R	I_{min}, μa
A	1500	3.45	3.2	1	21600	6800	0.04
B	120	0.09	2.0	0.13	570	280	0.21

The data are approximate, for even two headsets of the same model may give considerably different results. In particular, the minimum current for an audible signal is a very uncertain sort of value. It depends not only on the phones, but also on the observer and on the noise conditions in the room at the time the test is made. Accordingly, it is clear that the above values should be regarded as giving an order of magnitude only. The current required for audibility is seen to be less than 1 μa.

14-4. Detectors Using Amplifiers

Vacuum-tube amplifiers are such common articles that it does not seem necessary to discuss them or to go into detail regarding circuits. Bridge amplifiers are of fairly standard design, except for a few features. For one thing, the frequency response may be deliberately peaked at the bridge-operating frequency in order to discriminate against harmonics in the bridge supply. Sometimes the peaking circuit is a separate plug-in unit so that it may be changed easily. Another necessity in bridge amplifiers is good shielding as they customarily operate at a very low input-signal level. Adequate shielding is particularly important for a bridge and amplifier working at 60 cycles with surrounding 60 cycle circuits in the laboratory.

The use of an amplifier with a galvanometer or telephones needs no further comment as we have discussed these instruments. Amplifiers for this purpose do not need high gain as these devices are quite sensitive in themselves. More gain is needed, but the detector is more rugged when a d-c meter is used instead of a galvanometer.

One detector of the amplifier-meter type is the General Radio Type 1231-B. It is entirely self-contained, operating from batteries inside the case (a line-operated power supply is optional). It may be used as an amplifier with linear input-output relationship and a gain up to 80 db. Also, by pressing the "Null" button a circuit is brought into play that gives the panel meter an approximately logarithmic response. This is very convenient as a large unbalance does not throw the meter off-scale, and the sensitivity becomes much greater as the bridge-balance point is approached. For this reason the operator does not need to make frequent changes of the gain control; in fact, once a rough balance is made, the gain control can be left at the maximum setting. Provision is made to plug in a filter to give peaking at 60, 400, or 1000 cycles as desired.

Another convenient and sensitive arrangement is a high-gain amplifier, about 80 db, followed by a multirange vacuum-tube voltmeter. The voltmeter can be easily switched to a lower scale as balance is approached. As 80 db represents a voltage ratio of 10,000, the 0.1-v scale corresponds to 10 μv from the bridge for full-scale reading. Such sensitivity can be used only if the circuit is exceptionally well shielded to guard against stray 60 cycle pick-up. For large bridge unbalances it is desirable to reduce the amplifier gain (or bridge input voltage) rather than to depend on a high voltmeter scale, as the distortion in the amplifier and the narrowness of the null make detection of the balance point difficult if the amplifier is driven beyond the limits of linear response.

A *tuning-eye tube* may be used instead of a meter at the output of an amplifier. One advantage is greater ruggedness, but the visibility is not so good and the meter is preferable for accurate tuning.

A *cathode-ray oscilloscope* may be used (following an amplifier) in two ways. If the horizontal sweep is linear with time, the pattern on the screen is a number of cycles of the bridge-output voltage. The size of the waves can be observed when adjustments are made for the bridge balance. By another method, the horizontal sweep can be supplied by the bridge oscillator. The pattern on the screen by this procedure gives information regarding the kind, as well as the amount, of the unbalance. If sinusoidal wave forms are assumed and if the bridge unbalance has a component at 90° phase angle from the oscillator voltage, an ellipse is produced. On the other hand, if the bridge and oscillator voltages are in phase, an inclined straight line results. If both unbalances are present, we have an ellipse with inclined axis. It is thus possible to judge the two bridge adjustments separately — the one to close the ellipse to a straight line and the other to make the axis of the ellipse horizontal. The two bridge adjustments are not necessarily at 0° and 90° from the oscillator voltage, but a phase-shift circuit can be included in the amplifier to bring them to this relationship.

Another way to secure phase-sensitive indication is by means of vacuum-tube circuits. One type is indicated in Fig. 14-4(a), but there are several variations of this idea. As shown, the amplified bridge voltage is impressed on a phase-shift bridge consisting of two condensers and two resistors. It is possible to take from this circuit two voltages 90° apart, either of which may be selected by a switch and impressed between the grids and cathodes of the double-triode output tube. The plates of this tube are supplied with voltage at bridge frequency but 180° apart because of the transformer connection, hence the two halves of the tube conduct alternately, that is, each half conducts when its plate is positive. The indicating meter receives pulses in opposite directions in alternate half-cycles but cannot follow rapid variations because of its inertia; the resulting reading equals the average value of the current flowing in the instrument coil. The pulses are equal if the circuit is symmetrical and the signal voltage on the grids is zero; in this case the meter pointer stays at zero (at the center of the scale). If the grid potential is different during the two halves of the cycle, the plate-current pulses are no longer equal and the meter pointer deflects. Accordingly, if the bridge-output voltage has a component, as selected by the switch S, in phase with the plate voltage, the meter deflects in one direction or the other, and the direction of deflection is an indication of whether this particular bridge adjustment should be increased or decreased to secure balance. S can then be thrown to the 90° voltage and the circuit can be balanced for this component. The meter thus indicates the direction of adjustment required for balance, and the phase-shift circuit separates the R and X adjustments once it is set to the correct position.

The phase-shift bridge of Fig. 14-4(a) gives the choice of two voltages, either of which may be controlled in phase. The constants may be selected so that one voltage gives the maximum detector sensitivity for the resistance adjustment on the bridge and the other for the reactance adjustment. The circuit has the merit of permitting a large change of phase angle without changing the magnitude of the voltage but is somewhat inconvenient to use because it requires separate phase adjustment in each circuit. There are many modifications that may be made in designing circuits of the general type being discussed here. The arrangement of Fig. 14-4(b) separates the functions of continuous phase adjustment and the 90° shift. Switch S makes selection of two voltages that are 90° apart and that are equal in magnitude if the condenser reactance equals the resistance. The adjustable resistance R may be used to shift the phase of both voltages, so that one adjustment automatically brings both voltages to the desired positions. The capacitor shown in dotted lines across the transformer terminals is useful in improving the wave form and sensitivity; it also increases the impedance of the transformer if tuning is made for the bridge frequency.

A general remark can be made about phase-sensitive detectors, including the last circuit and the oscilloscope arrangement. They require setting of a phase-shift circuit to correspond with the R and X adjustments

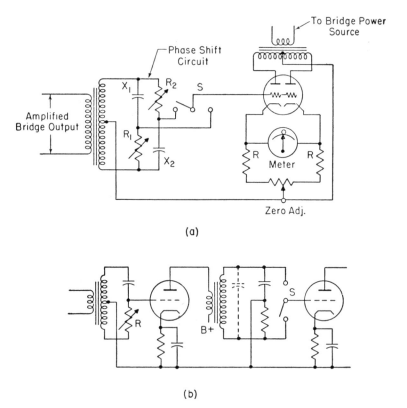

(a)

(b)

Fig. 14-4. Circuit components for a phase-sensitive detector. (a) Circuit of a phase-sensitive detector using a balanced output tube. (b) Separate continuous-shift and 90°-shift circuits for a phase-sensitive detector.

on the bridge, and this setting requires trials and a certain amount of time. This may be worthwhile when several measurements are to be made on essentially the same circuit. Otherwise, a plain amplitude device, such as telephones, or an amplifier with a meter is probably faster and easier to use.

14-5. Radio Receiving Set

A well-shielded radio receiving set is a satisfactory detector for radio-frequency measurements. A communications-type receiver is particularly well suited for this work because it is both sensitive and well shielded. If the generator supplying the r-f bridge is modulated with an audio-frequency tone, this tone is heard in the phones at the receiver output if the bridge is unbalanced. If the generator gives an unmodulated output, the oscillator in the receiver can be adjusted to give a beat-frequency note for balancing.

14-6. Frequency Classification of Detectors

The following list classifies detectors according to the frequency range in which they are particularly useful. The dividing lines between the frequency bands are generally rather indefinite. For example, the combination of the ear and telephone receivers is insensitive at 60 cycles and, again, in the upper range above 4000 to 5000 cycles, but it is difficult to set fixed limits.

(a) 25 to 200 cycles
    Galvanometer
    Amplifier with meter, oscilloscope, etc.
(b) 200 to 4000 cycles
    Telephones
    Amplifier with phones, meter, oscilloscope, etc.
(c) 4000 cycles to lower radio frequencies
    Amplifier with meter, etc.
(d) Radio frequencies
    Radio receiving set

14-7. Bridge Power Supplies

Many different types of power supply have been used at various times to furnish the small amount of alternating current needed for an impedance bridge. We shall discuss a few of the more common types without attempting to make the list complete. The author has even employed a d-c generator with condensers to block the d-c, using the commutator ripple as the bridge supply. This is not recommended for general use but serves the purpose if no other equipment is available.

(a) *Microphone hummers and tuning forks.*

An ordinary buzzer produces alternating current of ragged wave form. If a microphone button is used instead of a make-and-break, the wave form is improved, though still far from sinusoidal. A tuning fork may be used with microphone button and coil drive or with a coil and vacuum-tube drive. The tuning fork gives good frequency stability, particularly if it is temperature-controlled. The wave form may be improved by filters in the output circuit. Tuning forks have been superseded by vacuum-tube oscillators for some purposes but still have applications on account of the frequency stability. Microphone hummers are used in some bridges as a compact source of bridge tone.

(b) *Vacuum-tube oscillators.*

Since an enormous variety is available in this field, only a few will be discussed. Beat-frequency oscillators have been used for many purposes and have the advantage of covering a wide frequency range on a single dial. There are many types of oscillators with *L-C* circuits as the frequency-determining element. In recent years the *R-C* type of oscillator has come into use quite widely because of certain advantages in construction and tuning characteristics. There are two general types in common use: those depending on phase-shift effects and those using a Wien bridge type of circuit.

Figure 14-5 shows the circuit of an oscillator which depends on the phase shift of three *R-C* circuits to give the 180° phase difference needed between

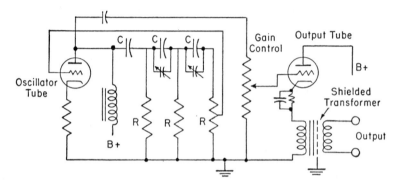

Fig. 14-5. Circuit of an *R-C* phase-shift oscillator.

plate and grid voltages to maintain oscillations.* The amount of phase shift in an *R-C* circuit depends on the frequency, thus there is only one frequency for which this circuit gives a 180° shift, and this is the frequency of oscillation. The oscillator shown here is a fixed-frequency device (1000 cycles) with trimmer adjustments for setting the exact value. However, it

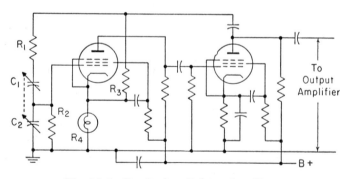

Fig. 14-6. Circuit of an *R-C* tuned oscillator.

*Reference 4.

can easily be made variable by means of variable condensers, and the resistances may be switched to other values for range-changing purposes. The output transformer is connected as a cathode follower, which results in good wave form over a wide range of load impedances.

Figure 14-6 represents the frequency-determining portion of an oscillator operating on a variation of the Wien bridge circuit.* In a straight Wien bridge oscillator a constant regeneration is supplied bucked by degeneration from the R-C circuit, so that oscillation is prevented for all frequencies except the one for which the bridge is in balance. The oscillator of Fig. 14-6 bears some resemblance to the Wien bridge with respect to tuning but utilizes the R-C circuit in a different manner. The oscillator consists of a two-stage amplifier having both positive and negative feedback loops with the R-C element in the positive loop. The negative feedback is given by R_3 and R_4, with a lamp bulb used for R_4 to stabilize the circuit by providing increased degeneration in case the amplitude of oscillation tends to increase. The frequency of R-C oscillators depends on the inverse first power of the variable capacitance, hence a full decade of change can be covered on the dial. This is in contrast to L-C oscillators, which involve the inverse square root of C, thus giving not much more than a 3:1 frequency change for the usual variable condenser.

14-8. Bridge Transformers

Shielded transformers are needed between the power supply and the bridge and also between the bridge and the detector. A shield is placed between primary and secondary windings and between each winding and the core. A shield consists of a thin sheet of copper wrapped around the coil. (The ends must, of course, be separated by insulation where they overlap in order to avoid forming a short-circuited turn.) The shielding actually increases some of the capacitances but permits control of their position as, for example, across opposite corners of the bridge so that the balance is not affected. It is important to keep the direct capacitance between the primary and secondary windings of the input transformer to a very low value; otherwise the coupling between windings introduces a part of the generator voltage across an arm of the bridge and thus produces erroneous results.

Figure 14-7 shows a shielded transformer made by the General Radio Company. It has separate shields around primary and secondary with spacers between primary and secondary shields to minimize intershield capacitance. The coil leads are brought out by shielded conductors. The direct capacitance from primary to secondary is held to about 0.3 $\mu\mu$f.

---

*Reference 3, p. 511; reference 5.

(a)

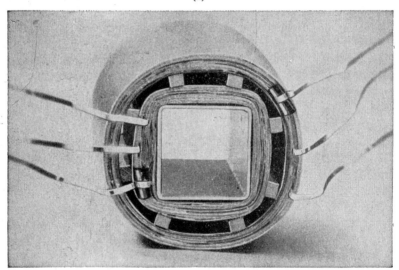

(b)

Fig. 14-7. General Radio type 578 shielded transformer. (a) Assembled models. (b) View showing coil construction. (*Courtesy of General Radio Co.*)

REFERENCES

1. Hague, B., *Alternating Current Bridge Methods*, London: Sir Isaac Pitman and Sons, Ltd., 1946.

2. Laws, Frank A., *Electrical Measurements*, New York: McGraw-Hill Book Co. Inc., 1938.

3. Cruft Laboratory Staff, *Electronic Circuits and Tubes*, New York: McGraw-Hill Book Co. Inc., 1947.

4. Ginzton, E. L., and L. M. Hollingsworth, "Phase-Shift Oscillators," *Proc. I.R.E.*, **29**, 43 (1941).

5. Terman, F. E., R. R. Buss, W. R. Hewlett, and F. C. Cahill, "Some Applications of Negative Feedback with Particular Reference to Laboratory Equipment," *Proc. I.R.E.*, **27**, 649–55 (1939).

6. Lamson, Horatio W., "Electronic Null Detectors for Use With Impedance Bridges," *Trans. A.I.E.E.*, **66**, 535–40 (1947).

7. Higgins, T. J., and W. Kneen, "Basic Theory and Experimental Verification of the A-C Galvanometer," *Trans. A.I.E.E.*, p. 235 (July 1954).

CHAPTER 15

Instrument Transformers

15-1. Use of Instrument Transformers

Transformers are used in a-c systems for the measurement of the basic quantities current, voltage, and power and for the connection of instruments that indicate other quantities or conditions, such as power factor, frequency, and synchronism. Transformers are used also to connect overcurrent, undervoltage, and various other relays to a power system. In all of these applications the transformer serves a measuring function and thus requires consideration for both its accuracy characteristics and the means of making experimental determination of the accuracy. Transformers used in connection with instruments for measurement functions are referred to as instrument transformers. The type used for measurement of current is called an "instrument current transformer" or, more generally, by the shortened form "current transformer" or simply by "C.T." on diagrams. Transformers for voltage measurement are similarly "potential transformers" or "P.T." for short.

There are two principal reasons for the use of transformers for making measurements in a-c circuits: first, to multiply the range of an instrument, and second, to insulate an instrument from a high-voltage line. The extension of range is better accomplished for an a-c ammeter by a transformer than by a shunt (as used for d-c ammeters). In the first place, it is more difficult to achieve accuracy with a shunt on alternating current, since the division of current between meter and shunt depends upon the reactances as well as the resistances of the two paths. Even more important, the transformer produces practically the same instrument reading regardless of the constants of the instrument or, in fact, the number of instruments connected in the circuit within reasonable limits. The flexibility of arrangement is thus much greater if a transformer is used. The transformer method is better than a series resistor for a voltmeter on a high-voltage line with respect to economy of power, safety, and flexibility.

The fact that the meters in the secondary circuit of an instrument transformer are insulated electrically from the primary winding is of exceedingly great importance in high-voltage systems. Large a-c generators usually operate at voltages of 12,000 to 20,000 v and sometimes higher. Transmission-line voltages range from this order up to the present top value of about 350,000 v. It is necessary to have a measure of current and voltage in these generators and lines, both for metering purposes and for the operation of protective relays and other equipment. It would be out of the question to bring the high-voltage lines directly to switchboard instruments for the higher voltage systems, and even for potentials of a few thousand volts it would be difficult to insulate the equipment to provide safety for the operating personnel. With the use of instrument transformers the situation is changed completely, for only the secondary leads from the instrument transformers are brought to the switchboard, and only low voltages exist between the wires and from the wires to ground. Instrument transformers are so important for both insulating and range-adjusting purposes that it is difficult to imagine operation of an a-c power system without them.

15-2. General Phasor Diagram for a Power Transformer

It may be helpful to review transformer theory in general before taking up the study of instrument transformers, since their action must be based on the same fundamental laws as other transformers. There are differences in detail as the result of the special operating conditions, which are different also for current and for potential measurements.

Figure 15-1(a) is the usual schematic diagram showing the primary and

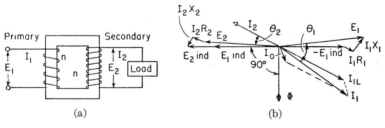

Fig. 15-1. Diagram for a power transformer. (a) Schematic diagram of transformer. (b) Phasor diagram.

secondary coils and the common magnetic core. Figure 15-1(b) shows the complete phasor diagram, illustrating current and voltage relationships. This diagram may be built most easily by beginning with the phasor to represent Φ, the magnetic flux common to both windings. There are induced voltages in both windings at 90° from Φ caused by the change of flux through the windings. The terminal voltage E_2 on the secondary side differs from the induced voltage by the I_2R_2 and I_2X_2 voltage drops in the secondary, as shown. The angle θ_2 between E_2 and I_2 is the phase angle of the load. The phasor diagram was oriented as shown in Fig. 15-1 in order to place the input voltage and current phasors in the general neighborhood of the positive x axis. While the position of the diagram as a whole is not of great importance, the present orientation seems logical as it brings the input quantities to the position found on many other phasor diagrams.

Before the load was connected to the secondary, the flux in the core was of such amount that the back voltage in the primary was nearly equal to the applied voltage and differed from it only by that small amount needed to make the no-load exciting current I_0 flow in the primary coil. When the load is connected and I_2 flows in the secondary, the balance between applied voltage and back voltage is disturbed because of the demagnetizing effect of I_2. This allows greater current to flow in the primary until balance is again established. The core flux and back voltage are only slightly less for the loaded condition than for the unloaded case in the usual power transformer as a result of the small impedance drops in the windings. Accordingly, balance is attained under load with a current in the primary which we may consider as made up of a *load component* of such magnitude that $n_1I_{1\,\text{load}} = n_2I_2$, which offsets the demagnetizing effect of I_2, and the *excitation component* I_0, which is nearly the same as at no-load.

The applied voltage on the primary side is made up of a voltage equal but opposite to the primary induced voltage, plus I_1R_1 and I_1X_1 voltages for the primary winding. The angle θ_1 between E_1 and I_1 differs from θ_2 because of the voltage drops in the two windings and the primary current component I_0.

15-3. Current Transformer — Ratio of Currents

Instrument current transformers are used in the measurement of the current flowing in a line wire of an a-c system. The primary winding is connected in series in the line wire, and the secondary is closed through the metering circuit, which may include an ammeter and other devices such as the current coil of a wattmeter or a watt-hour meter. The connection is indicated in Fig. 15-2. It should be understood that the figure is only schematic with respect to the arrangement of the transformer windings.

Actually, the primary and secondary are wound on the same side of the core in order to keep the leakage flux between windings to a low value.

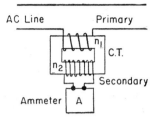

Fig. 15-2. Current transformer connected in a single-phase line.

The operation of a current transformer differs from a power transformer in two respects. First, the operation represents nearly a short-circuit condition, in that the secondary load ("burden" is the correct term for an instrument transformer) is of very low impedance. Second, the current in the secondary winding is determined by the *primary current* and *not* by the secondary circuit impedance (within practical limits). The primary current is determined by the load on the a-c system, and the secondary current is related to the primary current by approximately the inverse of the turn ratio. The secondary burden can be varied over a considerable range with only a minor effect on the secondary current, which is a valuable feature of the transformer in measurement work.

The actual ratio of currents may be found to a better approximation than the turn-ratio value mentioned above by a study of the relationships of Fig. 15-3. The currents in this diagram are multiplied by the turns in

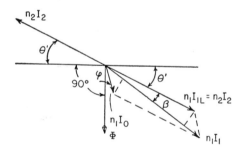

Fig. 15-3. Ampere-turn diagram for a current transformer.

order to represent the ampere-turns of primary and secondary. This brings primary and secondary phasors to the same general size, since the $n_1 I_1$ load component equals $n_2 I_2$. $n_1 I_0$ represents the exciting mmf needed to set up the flux through the magnetic reluctance of the core apart from the demagnetizing effect of one winding on the other.

An ideal current transformer would give a secondary current in an exact and fixed ratio to the primary current and exactly 180° from it on the phasor diagram. Actual transformers depart from the ideal because of the presence of the I_0 component of the primary current. The nature of the departure may be found by a study of the diagram of Fig. 15-3.

By projection of phasor lengths on n_1I_1

$$n_1I_1 = n_2I_2 \cos \beta + n_1I_0 \cos (90° - \theta' - \varphi - \beta)$$

$$\frac{I_1}{I_2} = \frac{n_2}{n_1} \cos \beta + \frac{I_0}{I_2} \sin (\theta' + \varphi + \beta) \tag{1}$$

Since β is usually a very small angle (of the order of a degree or less), the ratio of currents has the approximate value

$$\frac{I_1}{I_2} \simeq \frac{n_2}{n_1} + \frac{I_0}{I_2} \sin (\theta' + \varphi)$$

$$\frac{I_1}{I_2} \simeq \frac{n_2}{n_1} \left[1 + \frac{I_0}{I_1} \sin (\theta' + \varphi) \right] \tag{2}$$

The turn ratio n_2/n_1 may be regarded as the approximate ratio of the transformer. The second term in the bracket is a correction which gives the effect on the current ratio of the exciting component of the primary current. The correction term depends on the ratio of I_0 to I_1, hence it is desirable to have I_0 as small as possible. This requires that a high-permeability core material be used and that the flux density be kept low. The factor $\sin (\theta' + \varphi)$ depends on the type of burden, being small for resistive burdens and greater for inductive burdens. The angle φ, which is the phase angle between the flux and the "equivalent sine wave" of exciting current, depends on the core material; it is small for materials having small magnetic losses.

The angle β may be evaluated from Fig. 15-3 as follows:

$$\tan \beta = \frac{n_1I_0 \sin (90° - \theta' - \varphi - \beta)}{n_2I_2 \cos \beta} \tag{3}$$

The right side of eq. (3) may be simplified by neglect of minor effects of the angle β and still give results to a satisfactory degree of approximation. Since β is small, $\cos \beta \simeq 1$, and (3) becomes

$$\tan \beta \simeq \frac{n_1I_0 \cos (\theta' + \varphi)}{n_2I_2} \tag{4}$$

For small angles the tangent is practically equal to the angle in radians. Also, n_2I_2 in (4) may be replaced with sufficient exactness by n_1I_1, which gives as a final result

$$\beta \simeq \frac{I_0}{I_1} \cos (\theta' + \varphi) \quad \text{radians} \tag{5}$$

Equation (2) shows that the actual ratio of currents departs from the turn ratio by an amount depending on the ratio I_0/I_1 and on the phase angle of the burden. The ratio I_0/I_1 changes with change of I_1 because of nonlinearity in the magnetic behavior of the core material; it also depends on the impedance of the secondary burden. The current ratio I_1/I_2 accordingly varies slightly with the magnitudes of I_1 and the burden. The nominal current ratio is frequently used in measurements and calculations not requiring great accuracy, but, as indicated in the foregoing discussion, the true ratio cannot always be the same fixed value. Accordingly, we speak of the "ratio error" of a current transformer as the difference between the true ratio and the nominal ratio. The "ratio correction factor," or RCF is the factor by which the nominal ratio must be multiplied to give the true ratio.

$$\text{Ratio Correction Factor} = \text{RCF} = \frac{\text{True Ratio}}{\text{Marked Ratio}} \tag{6}$$

Equation (5) gives the angle by which I_2 on the phasor diagram fails to be directly opposite to I_1. A 180° phase difference of I_1 and I_2 with respect to the direction of flow around the core can be taken into account in a circuit by the interchange of a pair of leads; so the matter of real concern is the angle β which marks the departure from an exact 180° relationship. The angle β is called the "phase-angle error" of the transformer. The relationships of Fig. 15-3 can be described by the statement, "I_2 reversed *leads I_1 by the angle β.*"

The ratio and phase-angle errors must be taken into account in precise measurements. The ratio error is the only concern in case the current transformer is used for an ammeter or other purely current-operated device. However, for power measurements correctness of phase angle is also important, and the phase-angle correction must be made if the change of angle has an appreciable effect on the power measurements. The application of this matter to both potential and current transformers will be considered at greater length in a later section.

15-4. Experimental Methods for Determination of Current Transformer Accuracy

It is important to measure the accuracy of current transformers not only as an original check on the correctness of factory production, but also to determine the ratio and phase-angle errors for various values of secondary burden. There are several methods of measuring the characteristics of current transformers. Two of the most common methods will be studied here.

(a) *Silsbee comparison method.*

In this method an unknown transformer is calibrated by comparison with a calibrated transformer of the same nominal ratio. It is necessary to know the characteristics of the standard transformer for the burden placed upon it by the test set. The method is fast and convenient for routine tests and involves a minimum of equipment.

(b) *Primary and secondary shunt method.*

In this method the secondary current is compared with the primary current by balancing voltage drops across shunts in the primary and secondary circuits. This is a fundamental method that does not depend on the presence of a previously calibrated transformer. We shall study both methods in the following sections.

15-5. Silsbee Comparison Method*

The Silsbee method determines the characteristics of an unknown transformer by comparing it with a standard transformer of the same nominal ratio. The method is convenient and fast and well adapted to routine tests that must be conducted in large numbers in a power system. The equipment is not particularly bulky and therefore may easily be used for field testing, that is, testing transformers in place if they cannot be moved easily to the laboratory. The requirements of the standard are not severe because the only necessity is that its calibration be known accurately for the burden imposed upon it by the testing set. There are two forms of the Silsbee test: one is a deflection method and the other is a null method.

(a) *Deflection method.*†

The two transformers are connected with their primaries in series to the power supply and their secondaries in series, as shown in Fig. 15-4(a). "B" is the burden, usually adjustable in amount, in the secondary of the tested transformer. An ammeter is included in the circuit of the standard so that the current may be set to any desired value.

It is essential that the transformers be connected with the correct polarity in order to establish the current flow path *around the loop* formed by the two secondaries. In this way the current in the middle member of the circuit is the *difference* of the two currents and hence gives a convenient measure of the difference of characteristics of the two transformers. If X and S are identical and have equal burdens, the current in the middle member is zero. The small difference-current, ΔI in the figure, is measured by the wattmeter W, which has field excitation from a phase-shifter (chap. 14, par. 2). The wattmeter is used rather than an ammeter so that ΔI may be determined in terms of components in phase and in quadrature with I_s.

Figure 15-4(b) shows phasor relationships of I_s and I_x to the common primary current I_p. The difference of I_s and I_x is very greatly exaggerated to make ΔI visible because with good transformers ΔI is likely to be a

---

*Reference 7.
†Reference 16, p. 577.

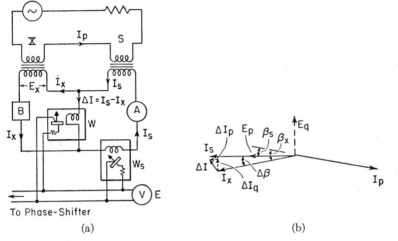

Fig. 15-4. Silsbee current-transformer testing: deflection method. (a) Circuit diagram. (b) Vector diagram.

fraction of a per cent of I_s and $\Delta\beta$ to be less than 1°. ΔI, which is the vector difference of I_s and I_x, is shown divided into components ΔI_p in phase with I_s and ΔI_q in quadrature. ΔI_p is a measure of the difference in magnitudes of I_s and I_x, and ΔI_q of the angle $\Delta\beta$.

If the letter N is used to represent the complex ratio of primary current to secondary current, we may write

$$N_x = \frac{I_p}{I_x}$$

$$N_s = \frac{I_p}{I_s}$$

Therefore the ratios are related as follows:

$$N_x = N_s \frac{I_s}{I_x} = N_s \left(1 + \frac{\Delta I}{I_x}\right) \tag{7}$$

since $$I_s = I_x + \Delta I$$

The correction term in (7) may be expressed with little change in value as a fraction of I_s, or

$$N_x = N_s \left(1 + \frac{\Delta I}{I_s}\right) \quad \text{phasor ratio} \tag{8}$$

The phase-shifter is adjusted so that W_s reads zero, which indicates that the voltage applied to the wattmeter potential circuits is in quadrature with I_s. The reading of W therefore has the value

$$W_q = \Delta I_q E$$

or
$$\Delta I_q = \frac{W_q}{E} \tag{9}$$

The voltage is now shifted 90° by the phase-shifter, and the new reading of W gives

$$\Delta I_p = \frac{W_p}{E} \tag{10}$$

In this condition W_s gives a maximum reading which may be used to compute I_s, as $I_s = W_s/E$. This may be a more accurate value of I_s than given by the ammeter, particularly for small currents.

We may now write for the magnitude of the ratio

$$\frac{N_x}{N_s} = \left(1 + \frac{W_p}{EI_s}\right) \tag{11}$$

Similarly for the angle

$$\Delta\beta = \beta_x - \beta_s = \tan^{-1}\frac{\Delta I_q}{I_s} \simeq \frac{W_q}{EI_s} \quad \text{radians} \tag{12}$$

W must be a sensitive instrument. Its current coil may be designed for a small current (about 0.25 amp) for tests of transformers designed for the usual secondary current of 5 amp. Care must, of course, be taken to avoid connecting X into the circuit with wrong polarity. A good feature of this method of testing is the fact that all equipment in the secondary circuit may be used to test transformers of any ratio, provided only that they are built for a 5-amp secondary current. It is, of course, necessary to provide a standard transformer of each ratio value at which tests are to be conducted.

If an equation of voltages is written around the secondary loop of the unknown transformer, we have

$$E_x = I_x Z_B - \Delta I Z_W$$

from which the total burden on X may be obtained as

$$Z_x \text{ total} = \frac{E_x}{I_x} = Z_B - \frac{\Delta I}{I_x} Z_W \tag{13}$$

(Z_W = impedance of wattmeter W current coil.)

An equation around the S circuit gives similar results except that the ΔI term is positive. This study shows the desirability of keeping the detector impedance low so that the impedance is not, in effect, transferred from one transformer to the other, thus causing the actual burdens to differ from the apparent values.

A special type of wattmeter should be used for the detector, W. In the usual wattmeter there is mutual inductance between the current and potential coils, which means a voltage induced in one circuit by a change of current in the other. This causes no appreciable error in ordinary usage, but in the present application a voltage induced in the ΔI circuit could cause a large error because the other voltages in the circuit are small. The error may be avoided by a special type of wattmeter that has a calibrated torsion-head mounting for the control spring. In taking readings, the pointer is brought back to the zero position, and the reading is taken from the torsion head. The zero position corresponds to 90° between the axes of the fixed and movable coils, which gives zero mutual inductance between them.

(b) *Null method.*

The null method has some points of resemblance to the deflection method but adds a slide wire, variable resistor, and a mutual inductor so that the detector may be brought to a null indication. The detector is an a-c galvanometer with its field supplied by a phase-shifter (chap. 14, par. 2). The slide wire, shown as a-b-c in Fig. 15-5, is of low resistance (about 0.2 ohm) and has a center tap to which the galvanometer is connected. The slider may be placed at any point from a to c as required for balance. The symbol r_2 represents the resistance from the slider to b, and r_1 is the total resistance in the ΔI path from d to e, including the secondary winding of the inductometer. It

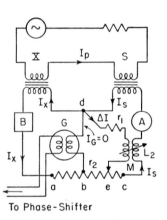

Fig. 15-5. Diagram for Silsbee current-transformer testing set: Null method.

is possible to find settings of r_2 and M so that the galvanometer deflection is zero for two positions of the phase-shifter approximately 90° apart. For this condition points d and b must be at the same potential, and we may write from d to b.

$$\Delta I(r_1 + j\omega L_2) + j\omega M I_s - I_x r_2 = 0 \qquad (14)$$

Substitution of $\Delta I = I_s - I_x$, and collection of terms gives

$$\frac{I_s}{I_x} = \frac{r_1 + r_2 + j\omega L_2}{r_1 + j\omega(L_2 + M)} \qquad (15)$$

This ratio, as shown in (7), gives the relative current ratios of the two transformers. We can separate magnitude and phase angle by converting (15) to the polar form

$$\frac{N_x}{N_s} = \frac{\sqrt{(r_1 + r_2)^2 + \omega^2 L_2{}^2}\Big/\tan^{-1}\dfrac{\omega L_2}{r_1 + r_2}}{\sqrt{r_1{}^2 + \omega^2(L_2 + M)^2}\Big/\tan^{-1}\dfrac{\omega(L_2 + M)}{r_1}} \tag{16}$$

The reactances are very small in comparison with the resistances, since they are needed only because of the small phase difference between I_s and I_x. The magnitude of the ratio is therefore very closely

$$\frac{N_x}{N_s} \simeq \frac{r_1 + r_2}{r_1} = 1 + \frac{r_2}{r_1} \tag{17}$$

The angle $\Delta\beta$ may be obtained by the difference of denominator angle and numerator angle in (16), since $I_x/I_s = N_s/N_x$, which is the reciprocal of (16). Since the angles are so small, the angles in radian measure are practically equal to the tangents

$$\Delta\beta = \beta_x - \beta_s = \frac{\omega(L_2 + M)}{r_1} - \frac{\omega L_2}{r_1 + r_2}$$

$$= \frac{\omega M}{r_1} + \frac{\omega L_2 r_2}{r_1{}^2} \tag{18}$$

r_2 has been dropped from the denominator of the second term in (18), since (17) indicates that it is very small in comparison with r_1.

If there is doubt regarding the sign of $\Delta\beta$, the burden on X can be increased a small amount. If M must be increased for the new balance, then β_x is greater than β_s. In some cases balance will occur with the slider to the left of point b in Fig. 15-5. Equations (17) and (18) may still be used if r_2 is given a negative sign, which may easily be proved by the complete derivation for this case following the method outlined above.

15-6. Primary and Secondary Shunt Method of Testing Current Transformers

The circuit for the primary and secondary shunt method is shown in Fig. 15-6. The primary shunt is of fixed value, for convenience, and is

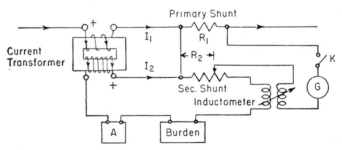

Fig. 15-6. Primary and secondary shunt method for testing current transformers.

placed in the primary line in series with the primary winding of the transformer. The secondary circuit of the transformer includes the secondary shunt, an inductometer, an ammeter, and a variable burden. The galvanometer circuit includes the two shunts and one winding of the inductometer. The detector must be an a-c type of galvanometer, preferably the excited-field form with phase-shifter (see chap. 14, par. 2), so that the R_2 and M adjustments may be made independently of each other.

The main idea of the method is to balance the IR drop of one shunt against the IR drop of the other (relative instantaneous polarities are shown in the figure). If there were no phase-angle error, the drop I_1R_1 could be balanced by an equal voltage I_2R_2. As a result of the phase difference, there is a component that must be balanced by a quadrature voltage supplied by the mutual inductance M. The galvanometer deflection is brought to zero by alternate adjustment of R_2 and M.

The voltage around the galvanometer circuit at balance is zero

$$I_1R_1 - I_2R_2 - j\omega M I_2 = 0 \tag{19}$$

Therefore, the ratio of currents, expressed in the complex notation, is

$$\frac{I_1}{I_2} = \frac{R_2 + j\omega M}{R_1} \tag{20}$$

$$= \frac{R_2}{R_1}\sqrt{1 + \frac{\omega^2 M^2}{R_2^2}} \bigg/ \tan^{-1}\frac{\omega M}{R_2} \quad \text{phasor ratio} \tag{21}$$

The magnitude of the ratio is

$$\frac{I_1}{I_2} = \frac{R_2}{R_1}\sqrt{1 + \frac{\omega^2 M^2}{R_2^2}} \tag{22}$$

Since $\omega M/R_2 \ll 1$, this may be written

$$\frac{I_1}{I_2} \simeq \frac{R_2}{R_1}\left[1 + \frac{1}{2}\frac{\omega^2 M^2}{R_2^2}\right] \tag{23}$$

Frequently the last term in (23) is negligible and the following is adequate:

$$\frac{I_1}{I_2} \simeq \frac{R_2}{R_1} \tag{24}$$

The phase-angle error of the transformer is the angle of the ratio, as given in (21). Since the angle is always very small, tan β may be replaced by β in the radian measure, or,

$$\beta = \frac{\omega M}{R_2} \quad \text{radians} \tag{25}$$

or

$$\beta = 3438\frac{\omega M}{R_2} \quad \text{minutes} \tag{26}$$

15-7. Standard Burdens for Current Transformer Testing

If a current transformer is always to be used with the same burden, it would be logical to make test measurements with that particular burden. However, for general purposes, such as publishing catalogue data about transformers, and for comparisons of different transformers, it is desirable to have standard burdens which will be used by all manufacturers. Table 4 gives the standards established by the American Standards Association. As indicated in the table, 5 amp is the usual standard

TABLE 4

STANDARD BURDENS FOR CURRENT TRANSFORMERS
ASA Standard Burdens
(Standard C 57.13, 1954)

Designation	Burden Characteristics		Sec. Burden at 60 cps, 5 amp		
	R, ohms	L, mh	Z	VA	PF
B-0.1	0.09	0.116	0.1	2.5	0.9
B-0.2	0.18	0.232	0.2	5.0	0.9
B-0.5	0.45	0.580	0.5	12.5	0.9
B-1	0.5	2.3	1.0	25	0.5
B-2	1.0	4.6	2.0	50	0.5
B-4	2.0	9.2	4.0	100	0.5
B-8	4.0	18.4	8.0	200	0.5

rating for the secondary current. Standard 5-amp instruments are used in the case of ammeters and of wattmeter current coils. Switchboard ammeters are calibrated in terms of the actual primary current, but the meters actually carry a maximum which is normally 5 amp. Proposals have been made to adopt a lower value as standard in order to facilitate running leads to greater distances from transformer to switchboard, but there are objections to the higher voltages involved and no general change has been made.

15-8. Reduction of Current-transformer Error

The discussion of par. 3 brought out the errors that are produced in the ratio and phase angle of a current transformer owing to the presence of the excitation component of the primary current. Improvement of the accuracy, then, depends upon minimizing this component or nullifying in some way its effect in introducing errors. The simplest and most obvious methods of improving the accuracy are related to the idea of keeping the excitation current as small as possible. Various methods are considered briefly in the following paragraphs.

(a) *Low flux density.*

The excitation component may be made small by keeping the magnetic-flux density low. This may be done by the use of a core of large cross section, or of a large number of turns, or by a combination of the two. For this reason current transformers are designed for much lower flux densities than the usual power transformers. Another matter to consider in this connection is the fact that current transformers used for relaying service are frequently required to have fair accuracy at currents many times the rated current (10 to 20 times), in order that relaying action will be correct in the event of a short-circuit on the system, particularly when differential relaying schemes are used.

(b) *High-permeability core material.*

The excitation component may be made small by the use of high-permeability core material. Some special materials are even better than the highest grade of silicon steel with respect to permeability, particularly at low flux densities. Permalloy is very good at low flux densities, but the saturation density is lower than for other magnetic materials, which is one feature that limits its use in current transformers. Hipernik (50 per cent Fe, 50 per cent Ni) has high permeability at low inductions and also a reasonably high saturation density. It is used frequently for current transformers.

(c) *Modification of turn ratio.*

The accuracy of a current transformer may be improved, at least with respect to ratio, by a modification of the ratio of turns. Instead of using the turns in exact inverse of the desired current ratio, a change of a turn or a few turns may be made in the secondary winding. (The change is made in the secondary because the primary generally has such a small number of turns that a one-turn change would be too great.) For the usual type of load, as shown in Fig. 15-3, the actual secondary current is less than the turn-ratio value as a result of the presence of I_0. Correction can be made by a small reduction of secondary turns, but this correction can be exact only for a particular value of current and burden impedance. The transformer may be marked as being "compensated" for this condition.

(d) *Use of shunts.*

If the secondary current is too great, it may be reduced by a shunt placed across the primary or secondary. This method makes an exact correction only for a particular value and type of burden. It also reduces the phase-angle error.

(e) *Wilson compensation method.*[*]

A compensated type of design, originated by M. S. Wilson of the

---

[*]See reference 5.

General Electric Company, is shown in Fig. 15-7. A few turns of wire are passed through a hole in the core and connected in series with the secondary winding. A short-circuited turn is placed around one portion of the core to improve the phase-angle relationships.

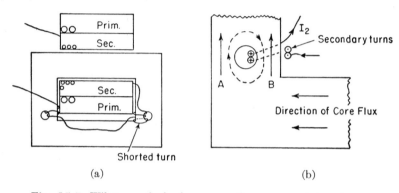

(a)   (b)

Fig. 15-7. Wilson method of compensating a current transformer. (a) Diagram of the transformer. (b) Detail of one corner, showing relative flux directions.

The auxiliary secondary turns are connected to magnetize in the same direction around the core as the main secondary winding and thus their effect opposes the flux set up by the primary. The auxiliary turns tend to set up a circulating flux around the hole as indicated by the dotted line in (b). The flux effects are additive in section A of the core and subtractive in B. At low densities, the core flux in effect passes through A, with reversed flux through B. However, as the densities increase (larger I_1 and I_2), section A tends to saturate (or, at least, is beyond the point of maximum permeability), so the flux in this section increases in less than linear proportion to the current. The action is equivalent to transferring a portion of the core flux to B, where it passes through the auxiliary secondary turns and gives the effect of an increased number of active secondary turns. An increase of secondary turns for large currents means a reduction of I_2, as compared with an uncompensated transformer, and an increase of RCF. This is the action needed to flatten the RCF curve (see Fig. 15-8). The RCF curve can be raised or lowered as a whole by adjustment of the number of turns in the main section of the secondary winding, so the important thing in the compensation is to make the curve as flat as possible over a large range of secondary current values.

The shorted turn around part of the core makes the flux in that part lag in phase behind the main flux, just as "shading coils" are used in small a-c motors. The small lag effect on I_2 is the action needed to bring it more nearly in line with I_1 and hence decrease the phase-angle error.

(f) *Two-stage design.*

The two-stage design utilizes a second current transformer to correct the error in secondary current of the first transformer. The method is applicable in general only to a watt-hour meter because a second coil is needed in the meter to carry the error-correcting current, unless an auxiliary transformer is used.

(g) *Wound-core construction.*

An improvement in magnetic characteristics may be effected by the use of the wound-core construction that has been used for some time in distribution transformers. By special treatment of the steel and by using it to carry flux always in the direction of the grain (as produced by rolling the sheet metal) the magnetic properties are improved. This improvement may be utilized in current transformers to decrease the ratio and phase-angle errors.

15-9. Operation of Current Transformers

The ammeters, wattmeter current coils, relay coils, and so forth, connected in the secondary of a current transformer, are placed in series; hence, the same current passes through each instrument. With more instruments in circuit a higher voltage is required to make the current flow and this, in turn, requires more flux in the core and a greater magnetizing component of the primary current. The result is an increase in the ratio and phase-angle errors of the transformer. It should be noted in this connection that an increase of the burden impedance of a current transformer means an increase of the volt-ampere burden. This is opposite to the case for a potential transformer, which is a fact that occasionally causes confusion. The differentiation in the two cases comes from the difference in operating conditions; in the current transformer the current depends on the primary current and is nearly independent of the burden impedance, which thus produces a change in the voltage. In the potential transformer the voltage is fixed and the total secondary current changes with change of burden.

The phasor diagram of Fig. 15-3 may be used to analyze the effect of the impedance and type of burden on the operating characteristics of a current transformer. If we consider I_1 to be fixed and increase the burden impedance, the result is an increase in the $n_1 I_0$ component because of the need for greater voltage and, accordingly, greater core flux. As $n_1 I_0$ increases, $n_2 I_2$ decreases slightly and β increases. The ratio I_1/I_2 and the RCF, which is the true ratio divided by the nominal ratio, become greater. (It should be remembered that Fig. 15-3 is qualitative only and that I_0 actually is very small — of the order of 1 per cent of I_1. It is exaggerated in the drawing merely to enable the reader to see it.)

We can investigate also the effect of changing the power factor of the burden* for a fixed impedance. If the two burdens considered have the same magnitude of impedance, the n_1I_0 requirements are the same and the difference enters in the size of the angle θ'. For a unity-power-factor burden, θ' is small and the angle between n_1I_1 and n_1I_0 is large. Under this condition the n_1I_0 component has a large effect on the phase-angle error β but a rather small effect on the magnitude of n_2I_2 and hence on the RCF. For a burden with considerable inductive effect, on the other hand, θ' is larger and the angle between n_1I_0 and n_1I_1 is small. This produces a large effect on the RCF but introduces little phase-angle error. For a burden such that n_1I_1 is directly in line with n_1I_0, the effect of n_1I_0 on the RCF is at a maximum and β is zero.

We have considered so far the effect of changing the impedance of the burden for a fixed value of primary current. Now, to complete the picture, we shall study what happens with a fixed burden when the current changes. If the magnetic behavior of steel were linear, we would expect n_1I_0 to vary in direct proportion as n_1I_1 varies, which would make β and the RCF constant for variations of current. However, core materials do not show linear behavior, even in the range of low flux densities used in current transformers. Beginning at zero induction, the permeability increases at first, reaches a maximum, and then decreases. Accordingly, the effect of the n_1I_0 component is larger at low values, then decreases relatively to n_1I_1 for larger values of I_1 because of the increased permeability. The magnetizing component again increases relatively to n_1I_1 if the core density passes the value for maximum permeability. Curves for current-transformer operation are usually plotted with I_2 as abscissa and RCF and β as the ordinates. Curves of this sort are shown in Fig. 15-8.

The qualitative curves of Fig. 15-8 will serve to summarize our study of current-transformer behavior. The RCF values of curve A show the general decrease with increased current that we mentioned in the previous paragraph. In this case the turn ratio has been altered to give correct ratio for 3 amp, or the transformer is "compensated for 3 amp" and this amount of burden. The RCF is greater for curve B, which is for a greater burden but still at unity power factor. Curve C is for the same burden impedance as B, except at 0.5 power factor, and shows greater values of RCF. The

---

*The angle θ' shown on Fig. 15-3 is the phase angle for the entire secondary circuit, including the resistance and reactance of the transformer secondary as well as the burden. The transformer quantities are small but not necessarily negligible in comparison with the burden, so θ' may differ appreciably from θ_2, which is the phase angle of the burden.

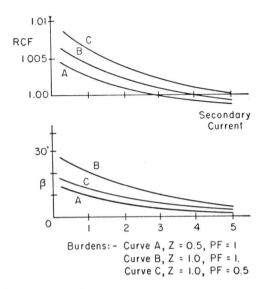

Fig. 15-8. Qualitative characteristic curves for a current transformer.

curve of phase-angle error in C is close to A because it tends to be higher, owing to greater burden impedance, and lower, as the result of the greater phase-angle of the burden. Curve B has larger values, owing to the greater burden impedance.

15-10. Precautions in the Use of Current Transformers

Current transformers are always used with the secondary circuit *closed* through ammeters, wattmeter current coils, or relay coils. A current transformer should *never* have its secondary *open* while the primary is carrying current. Failure to observe this rule may lead to serious consequences both to the operating personnel and to the transformer.

In normal operations, as indicated in Fig. 15-3, the secondary has demagnetizing ampere-turns only slightly less than the primary ampere-turns, and the flux density in the core is consequently low. If, however, the secondary circuit is opened, the demagnetizing effect of the secondary disappears, and flux is built up in consequence of the *entire* primary ampere-turns and not to the small $n_1 I_0$ component, as before. The core flux increases many-fold as a result. (Except in low-voltage circuits, the primary current is not affected appreciably by the happenings in the current-transformer circuit.)

There are two serious objections to the increased core flux that results from the open secondary circuit. For one thing, the large flux induces a high voltage in the secondary winding, and this voltage is frequently so great that it constitutes a serious danger to operators of the equipment. The hazard is particularly great since the current transformer is normally a low-voltage device and usually not treated with the caution that a high-voltage circuit deserves.

The transformer may be injured by the open-circuit condition. Insulation break-down is possible, though modern transformers are designed to withstand the voltage. The iron loss is increased very greatly, but the heating is not a matter of much concern, except possibly for a long period. However, even if the open-circuit condition is only momentary, the accuracy of the transformer may be seriously impaired by the residual magnetism remaining in the core in case the primary circuit is broken or the secondary circuit re-established at an instant when the flux has a large value. The residual effect causes the transformer to operate at a displaced position on the magnetization curve, which affects the permeability and, hence, the calibration. A transformer so treated must be first demagnetized and then recalibrated.

Current transformers are usually provided with a secondary short-circuiting switch, which should be closed before the secondary is opened for the removal of a meter or other circuit changes. In this way the dangers of open-circuit operation are avoided.

15-11. Types of Current Transformers

Current transformers are made in many kinds of mounting, depending on the type of use and the voltage of the circuit. We may list the following main divisions, each of which has several subdivisions.

(a) Portable type
(b) Indoor type
     Wound type
     Bar type and window type
(c) Outdoor type
(d) Bushing type

Figure 15-15(a) shows a precision portable current transformer made by the General Electric Company. It gives a choice of ratios, obtained by changing links on the top plate. Figure 15-15(b) is a Weston transformer of a different type. It gives the choice of four ratios by means of binding posts and additional ranges by placing the primary conductor through the opening in the case. One conductor through the opening gives a current ratio of 800/5, two turns 400/5, and so on. Figure 15-17(a) shows a Westinghouse current transformer for use on a 115-kv system. One bushing is used for incoming and outgoing primary leads because the difference of potential between them is small. Figure 15-19(a) illustrates the porcelain-clad construction, in which the core-and-coil unit is mounted in a shallow cast-iron base and extends up into the porcelain housing, which is thus both housing and high-voltage bushing.

Figure 15-18 shows two types of indoor current transformers for low-voltage circuits. The picture in (a) is of a wound-primary type, in which the primary winding consists of a few turns of heavy conductor to whose projecting ends the primary cable or bus-bar is bolted. The secondary terminals are under the sealed cover just above the X_1 mark, which indicates secondary polarity. The transformer in (b) includes the laminated core and secondary winding but no primary winding as such. The primary consists of the bus-bar, which is passed through the opening in the insulating sleeve.

The bushing-type transformer is similar in idea to the bar type in the respect that core and secondary are mounted around the single primary conductor. It has a circular core that carries the secondary winding, forming a unit which may be installed in the high-voltage bushing of a circuit-breaker or power transformer. The primary "winding" is the conductor in the bushing. The design is thus based on a one-turn primary, so that the primary ampere-turns must be the same in number as the primary amperes. The accuracy is accordingly poor for low-current designs that are often needed in high-voltage circuits. The bushing type is frequently used, however, if the accuracy is at all adequate, as its cost is very much less than that of a separate conventional current transformer built for high voltage. Various things have been tried to improve the accuracy of the bushing-type transformer, including the superposition of an alternating current of higher frequency than the power current.

15-12. Operation of Potential Transformers

Instrument potential transformers are used to operate voltmeters, the potential coils of wattmeters, the potential coils of relays, and others, from high-voltage lines. For all these purposes it is important that the secondary voltage be an accurately known fraction of the primary voltage, and (for some purposes) primary and secondary voltage should be in phase. There is no essential difference in basic theory between a potential transformer used for metering service and a regular power transformer. The general phasor diagram of Fig. 15-1 for a power transformer serves equally well, qualitatively, for a potential transformer. The two chief differences in a measuring transformer are, first, the attention to accuracy in the voltage ratio, and second, the minimizing of voltage drops in the windings to avoid phase shift and ratio-error effects. Small voltage drops are secured by design for small leakage reactance and the use of a large copper conductor. In other words, the loading on a potential transformer is limited by accuracy considerations, while in a power transformer the load limitation is on a heating basis. Actually, potential transformers are able to carry loads on a thermal basis many times their rated loads for measurements purposes (from 2 to 3 times for low-voltage potential transformers; up to 30 or more times for some high-voltage transformers).

Two phasor diagrams will be given to show the kind of phase-angle effects that occur for different types of burdens. The same notation is used as in Fig. 15-1. Figure 15-9(a) is for a unity-power-factor burden, and

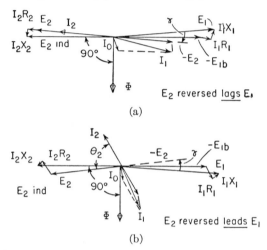

Fig. 15-9. Effect of power factor on phasor relationships of a potential transformer. (a) Unity power factor burden. (b) 50 per cent power factor burden.

(b) for approximately 0.5-power-factor inductive burden. The voltage drops are, of course, very much magnified to make them visible. We can decide from the diagrams that E_2 reversed lags E_1 for a resistive load, and leads for an inductive load. Also, the ratio E_1/E_2 is greatest when the phase-angle of the burden is the same as the phase-angle of the transformer-impedance triangle. As another element in the analysis, if we increase the volt-ampere burden (and therefore the current) the voltage drops and the RCF increase in a linear manner.

From the diagrams and the discussion we can make an analysis of the form of the RCF and phase-angle-error curves. If we assume first that the turn ratio is the same as the desired voltage ratio, we have curve 1 of Fig. 15-10. Even at light load we have some voltage drop, largely as the result of I_0, so the RCF is slightly greater than 1 and increases with increase of burden. If the secondary turns are increased slightly or the primary turns decreased, we have the condition of curve 2, i.e., an RCF slightly less than 1 at no-load and 1 at an intermediate load. The 0.8-power-factor curve starts

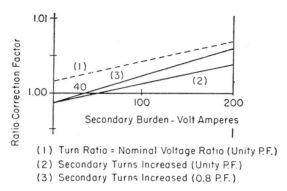

(1) Turn Ratio = Nominal Voltage Ratio (Unity P.F.)
(2) Secondary Turns Increased (Unity P.F.)
(3) Secondary Turns Increased (0.8 P.F.)

Fig. 15-10. Ratio-correction factor curves of a potential transformer.

from the same point, but has a somewhat greater slope. If this curve has an RCF of 1.00 at 40 va, we say that the transformer is "compensated for 40 va, 0.8 power factor."

The variation of phase-angle error is illustrated in Fig. 15-11. We saw above that inductive loads cause E_2-reversed to lead E_1 and resistive loads

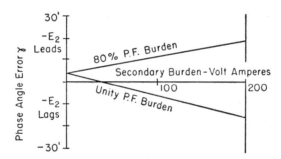

Fig. 15-11. Phase-angle-error curves of a potential transformer.

to lag. The no-load point is the same for either type of load and is on the inductive side because of the nature of I_0. Accordingly, the curves for inductive loads have forms somewhat as in Fig. 15-11. The question of the convention regarding plus and minus signs for γ may be raised at this point. Figure 15-11 follows the same convention used with current transformers of calling the angle plus when the secondary quantity reversed leads the primary. However, the meaning of plus and minus is indicated on the graph to avoid possible confusion. Both this convention and the opposite one may be found in use, but it seems simpler and more logical to follow the same convention for both current transformers and potential transformers. The present ASA standards* utilize this notation on the accuracy curves.

---

*Reference 2.

15-13. Standards for Potential Transformers

The standard secondary voltage for a potential transformer is either 115 or 120 v, depending on which gives an even multiple of the high-side voltage. For example, the following combinations are found: 4600/115, 4800/120, 115,000/115. There have been efforts to reduce the great number of voltage standards in use in various power systems and to concentrate on certain "preferred" values. On this basis, for instance, 4600 would be superseded by 4800. On the preferred list 120 volts is the secondary standard for primary voltage up to 24,000 and 115 volts is adopted above this value.

In power rating, 200 va is the most general value at 60 cycles. Some light-burden transformers for the lower voltages are rated 50 va. For higher voltages, 24,000 volts and above, a rating of 500 va is general. Actually, a power of 500 va, or even 200 va, is not needed for most metering purposes, but the design is based on considerations other than load capacity. The 200-va transformer is usually compensated for 40 va, as this represents a reasonable operating burden.

The following burdens have been adopted as standards for testing potential transformers, from "American Standards for Testing Transformers, Regulators, and Reactors," ASA, C 57.13.

TABLE 5

STANDARD BURDENS FOR POTENTIAL TRANSFORMERS
(ASA Standard C 57.13 — 1954)

Designation of Burden	Secondary Volt-Amperes	Burden Power Factor
W	12.5	0.10
X	25	0.70
Y	75	0.85
Z	200	0.85

15-14. Testing of Potential Transformers

Potential transformers may be tested by a comparison method similar to the Silsbee method for current transformers, discussed in par. 5. The high-voltage windings of the unknown and standard transformers are excited in parallel from a high-voltage supply transformer. Calibrated resistors are connected across the secondaries to form voltage dividers that permit comparison of the unknown transformer with the standard. It is necessary to have a calibrated standard transformer of the same nominal ratio as the unknown.

Potential transformers may be calibrated by another method that is somewhat analogous to the current-transformer test studied in par. 6. In the potential-transformer case a voltage divider is connected across the high-voltage supply, and the secondary voltage is balanced against a portion of the divider. Figure 15-12 shows a regulating transformer,

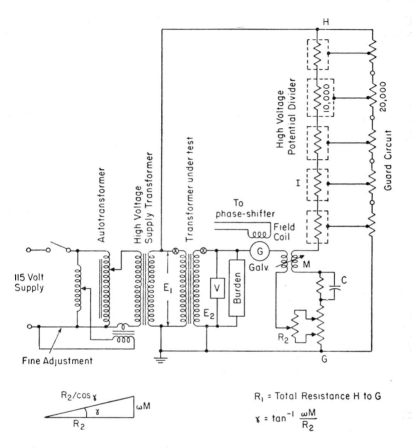

Fig. 15-12. Circuit for determination of potential transformer characteristics.

high-voltage-supply transformer, and the voltage-divider system with connections to the potential transformer that is to be tested. The divider is made of sections, each designed for 1000 v and consisting of an accurate 10,000-ohm resistor in an oil-filled metal container. Any desired number of these units may be used, depending on the potential transformer to be tested. The lower unit in Fig. 15-12 is less than 10,000 ohms by the amount in M and the voltage divider below it. The mutual inductor M is used to balance the quadrature component arising from the phase difference between E_2 and E_1. The capacitor C is bridged across part of the resistance to neutralize the self-inductance of the primary of M, so that the potential divider as a whole is purely resistive.

The metal containers for the resistors of the voltage-divider system are insulated from ground and connected to resistors of the guard circuit, which thus holds each container at the midpotential of the resistor it encloses. The guard system is essential for accurate results, as otherwise capacitive effects to ground would cause the current in the upper resistors to differ from that in the lower ones, and the true voltage ratio would not be the same as the known resistance ratio. The circuit is adjusted by variations of R_2 and M to give a null indication on the galvanometer. The equations for the balanced condition may be derived in the following way, using the symbols on Fig. 15-12,

$$E_1 = IR_1 \tag{27}$$

$$E_2 = IR_2 + j\omega M I \tag{28}$$

$$\frac{E_1}{E_2} = \frac{R_1}{R_2 + j\omega M} \tag{29}$$

Therefore the magnitude of the ratio is

$$\frac{E_1}{E_2} = \left|\frac{E_1}{E_2}\right| = \frac{R_1}{R_2} \cos \gamma \tag{30}$$

where γ is defined by the small triangle in Fig. 15-12. It may be seen from (29) that γ is the phase angle of the ratio of E_1 to E_2 and, therefore, the phase-angle error of the transformer. This angle may be defined from the same triangle. Therefore,

$$\text{Phase-angle Error} = \gamma = \tan^{-1} \frac{\omega M}{R_2} \tag{31}$$

or

$$\gamma \simeq \frac{\omega M}{R_2} \qquad \text{radians} \tag{32}$$

$$\simeq 3438 \frac{\omega M}{R_2} \quad \text{minutes} \tag{33}$$

The angle γ is ordinarily a small angle (less than 1°), so its cosine is practically unity. With this change, (30) becomes

$$\frac{E_1}{E_2} = \frac{R_1}{R_2} \tag{34}$$

Equations (34) and (32) or (33) are the usual relationships for this testing equipment.

The method studied above is useful for low- and medium-voltage transformers. The construction of the shielded resistor and the power loss in it become objections at high voltages, for which other arrangements are more desirable. A capacitive voltage-divider may be used, with appropriate circuits for comparing the secondary voltage with a known fraction of the primary voltage.* A related idea is the capacitance bridge used to determine the ratio and phase angle of high-voltage potential transformers.†

15-15. Use of Instrument Transformer Corrections

We have seen that both current and potential transformers are subject to errors in ratio and phase angle. The errors in good modern transformers are small and may be ignored for many purposes. However, they must be considered in precision work; also in some power measurements they may have a large effect. Voltmeters and ammeters are affected by ratio errors only, but wattmeters are influenced in addition by phase-angle errors. Corrections can be made for these effects if test information is available about the instrument transformers and their burdens.

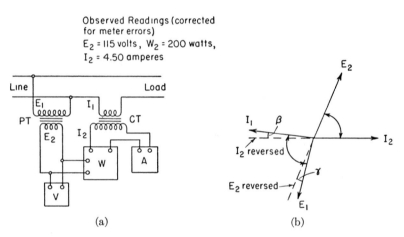

Fig. 15-13. Application of instrument transformer corrections. (a) Circuit. (b) Phase relationships.

Probably the best way to illustrate instrument transformer corrections is by means of a sample problem. To start with let us take the circuit and data of Fig. 15-13, which shows a single-phase line with voltmeter, ammeter, and wattmeter connected to the circuit by instrument transformers. We shall assume that any instrument errors were corrected before the data were entered in the figure. Also the following corrections will be assumed to have been found from the instrument transformer calibrations, taking into account the burdens in the circuit.

---

*Reference 10.
†Reference 11.

Current transformer: Nominal ratio 25/5

$RCF = 0.997$
$\beta = 15'$ (I_2-reversed leads I_1)

Potential transformer: Nominal ratio 11,500/115

$RCF = 0.995$
$\gamma = -25'$ (E_2-reversed lags E_1)

Nominal values, without corrections, are

$E = 11,500$ v
$I = 22.5$ amp
$W = 200 \times (25/5) \times (100/1) = 100,000$ watts

With corrections for RCF's

$E_1 = 11,500 \times 0.995 = 11,440$ v
$I_1 = 22.5 \times 0.997 = 22.43$ amp
$W_1 = 100,000 \times 0.995 \times 0.997 = 99,200$ watts

Now, to make the phase-angle correction for the wattmeter, we compute first the apparent phase-angle as found on the secondary side of the transformers.

$$\cos \theta_a = \frac{200}{115 \times 4.5} = 0.3865$$

$$\theta_a = 67°16'$$

The directions of the phase-angle corrections are shown in Fig. 15-13(b), where I_2-reversed leads I_1, as in the data and E_2-reversed lags E_1. Corrections of this sort may be made on a mechanical basis for routine computations, but a sketch as used here is the best way of being sure of signs for an occasional case. From the sketch and data the true phase-angle is

$$\theta_1 = 67°16' + 15' + 25' = 67°56'$$

$$\cos \theta_1 = 0.3751$$

Therefore, the true power is

$$W_1 = 99,200 \times \frac{0.3751}{0.3865} = 96,280 \text{ watts}$$

(There should be a correction of 1' or so for the angle between the voltage and current in the potential circuit of the wattmeter, but this has been ignored here.)

In this case the total correction from the nominal value of 100,000 watts to the true value of 96,280 watts is 3.7 per cent. The greater portion of the correction is caused by the phase-angle errors. This is not presented as an "average" case, but it does show possible effects. If the load phase-angle had been small, the change in cosine due to β and γ would have been very much less. The phase-angle errors of the instrument transformers have negligible effect for purely resistive loads, but become extremely important as the load phase angle approaches 90°.

Corrections of the same sort may be made for three-phase circuits. The usual two-wattmeter connection is shown in Fig. 15-14, where the current and potential transformers are omitted for simplicity. We shall assume a

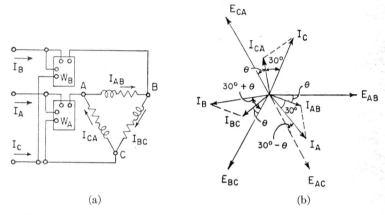

(a)	(b)

Fig. 15-14. Two-wattmeter measurement of power as applied to a balanced inductive load.

(a) General Electric type P-3 transformer. (*Courtesy of General Electric Co.*)

(b) Weston model 461 transformer. (*Courtesy of Daystrom, Inc., Weston Instruments Division.*)

Fig. 15-15. Portable current transformer.

balanced delta-connected inductive load of small phase angle and phase sequence *A-B-C*. The reading of Wattmeter *A* depends on I_A, E_{AC}, and the cosine of the angle between them, which has a value of $(30° - \theta)$ for the balanced case. Similarly, the angle for Wattmeter *B* is $(30° + \theta)$. If we install instrument transformers in a circuit similar to Fig. 15-14, we must remember the $(30° - \theta)$ and $(30° + \theta)$ angles in making the phase-angle corrections for the instrument transformers. Otherwise, the corrections are made as in the single-phase example above. Instrument transformer connections for a three-phase, three-wire circuit are illustrated in Fig. 15-16(b), which includes two voltmeters, two ammeters, and two separate watt elements (they may be combined in one "polyphase wattmeter"). The symbol "R" denotes the coil of an overcurrent relay.

Fig. 15-16(a). Outdoor metering set, consisting of two current transformers and two potential transformers, as designed for voltages up to 69 kv. (*Courtesy of Westinghouse Electric Corp.*)

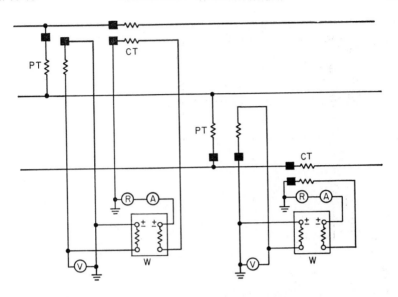

Fig. 15-16(b). Diagram of Fig. 15-16(a).

15-16. Types of Potential Transformers

Figure 15-17(b) shows a cutaway view of a potential transformer for operation on a 69 kv system. It is oil-filled, and uses oil-filled bushings, which minimizes the over-all height of the unit. The tank is a close fit on the core and coils, thus reducing the required volume of oil. Figure 15-17(a) is a companion current transformer for 69 kv systems. The leads from the two ends of the primary winding are brought up through the same insulator, since there is only a small voltage between them, thus saving the expense of another high-voltage insulator.

Figure 15-16(a) illustrates a single-unit outdoor metering set designed for the metering of a 69 kv three-phase, three-wire circuit. It contains two current transformers and two potential transformers, which may be connected to the instruments and relays as indicated in part (b) of the figure. (Note the polarity markings on the transformer terminals in the diagram.)

Figure 15-19(b) shows a 69-kv potential transformer of compact design. The volume of insulating liquid has been greatly reduced in this type. The potential transformers of Fig. 15-17 and 15-19 are of two-bushing construction, which is necessary when neither side of the line is at ground potential. Some potential transformers, connected from line to neutral of grounded-neutral systems, have only one high-voltage bushing.

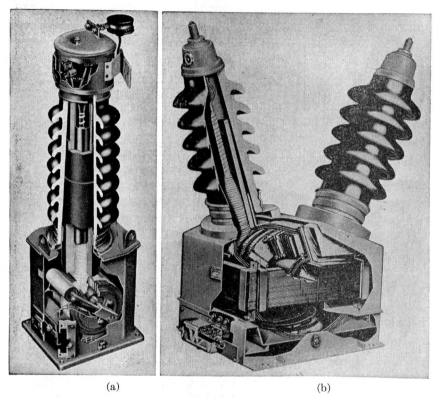

(a)   (b)

Fig. 15-17. Cutaway views of instrument transformers of 69 kv class, showing use of oil-filled bushings and low-volume tanks. (a) Current transformer. (b) Potential transformer of two-bushing type. (*Courtesy of Westinghouse Electric Corp.*)

Fig. 15-18(a). Indoor current transformer: Butyl-molded with wound primary, 10 to 800 amperes, for 5,000-v circuit.

Fig. 15-18(b). Indoor current transformer: Butyl-molded bar-primary 200 to 800 amperes for 600-v circuit. *(Fig. 15-18 courtesy of General Electric Co).*

(a)   (b)

Fig. 15-19. Instrument transformers for 69,000-v circuit: (a) Porcelain-clad current transformer, 200/400 to 5 amp, 25/125 cps. (b) Potential transformer, 69,000 to 115 v, 50/60 cps, 500 va. *(Courtesy of General Electric Co.)*

REFERENCES

1. Laws, Frank A., *Electrical Measurements*, New York: McGraw-Hill Book Co. Inc., 1938.

2. *American Standards for Transformers, Regulators and Reactors*, A.S.A., C 57.13 (1954).

3. Brooks, H. B., and F. C. Holtz, "The Two-Stage Current Transformer," *Trans. A.I.E.E.*, **41**, 383 (1922).

4. Boyajian, A., and W. F. Skeats, "Bushing-Type Current Transformers for Metering," *Trans. A.I.E.E.*, **48**, 949 (1929).

5. Wilson, M. S., "A New High-Accuracy Current Transformer," *Trans. A.I.E.E.*, **48**, 783 (1929).

6. Camilli, G., and E. S. Townsend, "New High-Voltage Potential Transformers," *Gen. Elec. Rev.*, (Aug. 1946), p. 8.

7. Silsbee, F. B., "A Method for Testing Current Transformers," *NBS, Sc. Paper* 309 (Nov. 1917).

8. Silsbee, F. B., "Methods for Testing Current Transformers," *Trans. A.I.E.E.*, **43**, 282 (1924). (Contains bibliography.)

9. Spooner, T., "Current Transformers with Nickel-Iron Cores," *Trans. A.I.E.E.*, **45**, 701–7 (1926).

10. Hague, B., *Instrument Transformers*, London: Sir Isaac Pitman and Sons, Ltd., 1936.

11. Bousman, H. W., and R. L. Ten Broeck, "A Capacitance Bridge for Determining the Ratio and Phase Angle of Potential Transformers," *Trans. A.I.E.E.*, **62**, 541 (1943).

12. Park, J. H., "Effect of Wave Form on the Performance of Current Transformers," *J. Research N.B.S.*, **19**, 517 (1937).

13. Silsbee, F. B., R. L. Smith, N. L. Forman, and J. H. Park, "Equipment for Testing Current Transformers," *J. Research N.B.S.*, **11**, 93–122 (1933).

14. Camilli, G., "A Survey of Bushing-type Current Transformers for Metering Purposes," *A.I.E.E. Tech. Paper* 50–63 (1950). (Contains bibliography.)

15. Silsbee, Francis B., and Francis M. Defandorf, "A Transformer Method for Measuring High Alternating Voltages and its Comparison with an Absolute Electrometer," *J. Research N.B.S.*, **20**, 317–36 (March, 1938).

16. Harris, Forest K., *Electrical Measurements*, New York: John Wiley & Sons, Inc., 1952.

PROBLEMS

15-1. (a) Draw two good qualitative phasor diagrams for an instrument potential transformer for the following burdens: (1) a very small burden at unity power factor; (2) a large burden at unity power factor.

(b) Indicate the effect of the amount of burden on the RCF and phase-angle error, as shown by (1) and (2).

15-2. Draw two good qualitative phasor diagrams for a potential transformer with 0.5 power factor inductive burden, one diagram for a very small burden, the other for a large burden. Indicate the effect of the amount of burden on the RCF and phase-angle error.

15-3. Draw two good qualitative phasor diagrams for an instrument current transformer, one showing a unity-power-factor burden, the other a 0.5-power-factor lagging burden. Indicate the effect of the power factor on the ratio correction factor and phase-angle error.

15-4. If a current transformer is "compensated for 2.5 amp" and a certain burden, will the meter reading at 1 amp be lower or higher than it should be (with the same burden)? Why? At 5 amp? Sketch a typical RCF vs. I_2 curve.

15-5. An instrument current transformer with 20 to 1 turns ratio is carrying 100 amp in its primary coil. For the given burden the primary magnetizing ampere-turns are 2 per cent of the total primary ampere-turns and may be considered to be in phase with the flux. Neglect the secondary leakage reactance.

(a) Draw the phasor diagram and find the actual current ratio and the phase-angle error, if the burden is purely resistive. Find the RCF.

(b) Would the RCF and the phase-angle error be greater or smaller for a highly inductive burden, other elements being the same? Explain by phasor diagrams.

15-6. An instrument transformer with nominal 2000 to 5 amp ratio has 1 primary turn and 398 secondary turns. When carrying a primary current of 1200 amp and a particular secondary burden, the magnetizing component is 1.0 per cent of the primary current.

(a) Draw the phasor diagram and compute the ratio correction factor and phase-angle error, if the burden is purely resistive. Assume the magnetizing current to be in phase with the core flux and neglect secondary reactance.

(b) Same, except with an inductive burden of 0.5 power factor.

15-7. A current transformer with nominal 1000 to 5 amp ratio has 1 primary turn and 199 secondary turns. When carrying a primary current of 1000 amp with a particular secondary burden of 0.8 power factor (inductive), the magnetizing component is 1.0 per cent of the primary current and 20° ahead of the core flux in time phase. Find the RCF and phase-angle error for the given burden, neglecting secondary leakage reactance. Draw the phasor diagram and make computations from it (no formulas).

15-8. An instrument current transformer with a nominal ratio of 600/5 amps has 2 primary turns and 237 secondary turns. For the burden in use, the secondary circuit has a phase angle of 55°. A magnetizing force of 10 amp-turns, 15° from the flux, is required when the secondary current is a full rated value with the particular burden used in this test. Draw the phasor diagram, and from it compute the ratio correction factor and the phase angle error (no formulas).

15-9. A current transformer with a name-plate rating of 1000 to 5 amps has a one-turn bar primary. With full rated current in the primary and a particular inductive burden the RCF is 0.9950 and the phase angle error is 20 minutes. Assume that the excitation component of current leads the core flux by 15°, and the phase angle of the entire secondary circuit (burden plus secondary windings) is 45°. Draw a phasor diagram, mark it clearly, and from it solve for the number of secondary turns (no formulas).

15-10. An instrument current transformer with nominal ratio of 600/5 amps has 2 primary turns and 238 secondary turns. The complete secondary circuit has a phase angle arc cos 0.8. The primary magnetizing component, leading the flux by a phase angle of 20°, is 18 amp-turns when the secondary current is at rated value with the given burden.

Draw a phasor diagram for this case, and mark it clearly. From the diagram compute the ratio correction factor and the phase-angle error (no formulas).

15-11. An instrument current transformer with nominal ratio of 600/5 amps has two turns in the primary winding. For the burden in use, the complete secondary circuit has a phase angle of 45°. A magnetizing force of 15 amp-turns, leading the flux by a phase angle of 20°, is required when the primary current is at full rated value, with the burden used on this transformer.

(a) Draw the phasor diagram, and from it compute the number of turns to be used in the secondary winding in order that the RCF will be as nearly unity as possible. (Use no formulas.)

(b) Compute the phase-angle error (in minutes) for the condition of (a).

15-12. A current transformer with a one-turn primary and a 1000/5 amp nominal ratio has an RCF of 1.0036 when the secondary current is 2.5 amps, and the burden is such that the phase angle of the secondary circuit is 60°. The core magnetizing component is 5 ampere-turns, which may be assumed to be in phase with the core flux.

Find the RCF and the phase angle error when the secondary current is 5 amps, for the same burden. The magnetizing component is now 8 ampere-turns. Use no formulas. Give a clearly marked phasor diagram.

15-13. Two transformers of the same nominal ratio, 500 to 5 amp, have their primaries connected in series, and their secondaries as in Fig. 15-4(a). With the current in the secondary of the standard transformer adjusted to rated value the current in the middle conductor is $\Delta I = 0.05\epsilon^{-j126.9°}$, expressed with respect to I_s as the reference. It is known that the standard has a ratio correction factor of 1.0015 and phase-angle error of +8′. Find RCF and β for the unknown transformer.

15-14. An instrument potential transformer of nominal 24,000/120 v ratio is tested by comparison with a calibrated transformer of the same ratio. The standard transformer is known to have an RCF of 0.9985 and a phase angle error of −12′

(E_2 reversed lags E_1). The difference voltage ($E_x - E_s$) is $0.50\underline{/216.9°}$ v with respect to E_s when the transformers are operating at rated voltage. Draw a phasor diagram of the voltages. Compute the RCF and phase angle error of X.

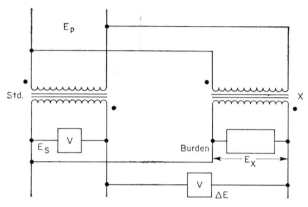

Prob: 15-14

15-15. An instrument potential transformer rated 2400 to 120 v has 10,000 turns in the primary coil and 500 in the secondary. The resistances and leakage reactances of the windings are: $R_1 = 150$ ohms, $X_1 = 400$, $R_2 = 0.5$, $X_2 = 0.8$. The excitation component of the primary current is 0.0075 amp, leading the core flux by 20°. The transformer is operating with an inductive burden of 24 v − a at a power factor of 0.8, and a terminal voltage of 120 v.

(a) Draw a good phasor diagram, and mark all quantities clearly.

(b) Find the primary current and terminal voltage, and the ratio correction factor and phase-angle error, for this condition of operation. Use the secondary terminal voltage as reference in your calculations, $E_2 = -120 + j0$.

15-16. An instrument potential transformer rated 4800 to 120 v has 15,000 turns in the primary winding and 375 turns in the secondary. With 4800 v applied to the primary, and the secondary terminals open-circuited, the primary current is 0.0075 amp, lagging the voltage by 73.7°. With a particular burden connected to the secondary, the primary current is 0.0180 amp, lagging the voltage by 53.1°. The resistances and leakage reactances of the windings are, $R_1 = 600$, $X_1 = 1000$, $R_2 = 0.4$, $X_2 = 0.6$ ohm.

(a) Draw a complete phasor diagram.

(b) Solve for the secondary current and terminal voltage, using the applied primary voltage $E_1 = 4800 + j0$ as reference.

(c) From these results compute the actual ratio, the RCF, and the phase angle error.

15-17. An instrument potential transformer rated 6900 to 115 v has 22,500 turns in the primary winding, and 375 turns in the secondary. With 6900 v applied to the primary, and the secondary terminals open-circuited, the primary current is 0.0050 amp, lagging the voltage by 73.7°. With a particular burden connected to the secondary, the primary current is 0.0125 amp, lagging the voltage by 53.1°. The resistances and leakage reactances of the windings are, $R_1 = 1200$, $X_1 = 2000$, $R_2 = 0.4$, $X_2 = 0.7$ ohm.

(a) Draw a complete phasor diagram.

(b) Solve for the secondary current and terminal voltage, using the applied primary voltage $E_1 = 6900 + j0$ as reference.

(c) From these results compute the actual ratio, the RCF, and the phase-angle error.

(d) If the RCF is to be brought close to unity for the above conditions by change of the number of primary turns, what number of turns, approximately, would you use?

15-18. Current, voltage, and power readings were taken on a single-phase inductive load by meters connected to the secondaries of instrument transformers. The meter readings (after correction of meter calibration errors) were $E = 115$ volts, $I = 4.24$ amperes, and power $= 320$ watts. Transformer data for this condition are:

	Potential Transformer	Current Transformer
Nominal ratio.........	6900/115	100/5
RCF................	0.994	1.002
Phase-angle error......	$-22'$ ($-E_2$ lags E_1)	$8'$ ($-I_2$ leads I_1)

Find the true values of load voltage, current, and power.

15-19. Voltmeter, ammeter, and wattmeter connected to a single-phase line by current and potential transformers give readings $E = 120$ v, $I = 4.0$ amp, power $= 240$ watts after correction of calibration errors. The current transformer has a nominal ratio of 500/5, an RCF of 0.998 and a phase-angle error of $15'$ (I_2 reversed leads I_1). The potential transformer has a nominal ratio of 20/1, an RCF of 1.005, and a phase-angle error of $-15'$ (E_2 reversed lags E_1).

(a) Find the true volts, amperes, and watts for the load, assuming it to be inductive.

(b) Same as (a), except that the load is capacitive.

15-20. Current, voltage, and power readings on a nominal 4800-v single-phase line are taken on instruments connected into the circuit by current and voltage instrument transformers. The readings (after corrections for errors in meter calibration) are, $I = 3.62$ amps, $E = 117.5$ v, power $= 120$ watts. The load is inductive. Data for the instrument transformers under these conditions are:

	Current Transformer	Potential Transformer
Nominal ratio.........	500/5	4800/120
RCF................	1.005	0.998
Phase-angle error......	$12'$	$-24'$
	I_2 reversed leads I_1	E_2 reversed lags E_1

(a) Find the true values of the line current, voltage, power and power factor.

(b) Find the percentage error in the calculated load power if the instrument-transformer corrections are ignored.

15-21. On a special test conducted on a nominal 12,000-v three-phase line, the power to a balanced inductive load is read by the two-wattmeter method (using two separate wattmeters), with current and potential transformers between the lines and the instruments. The readings of the instruments (after correction for any calibration errors of the instruments) are: amperes per line = 4.50, volts between lines = 121.0, total power = 283 watts. The instrument transformers have the following characteristics:

	Potential Transformer	Current Transformer
Nominal ratio.........	12,000/120 v	400/5 amp
RCF.................	0.993	0.997
Phase-angle error......	−24 min.	+15 min.
	(E₂ reversed lags E₁)	(I₂ reversed leads I₁)

(a) Find the reading of each wattmeter.

(b) Find the load volts, amperes, and kilowatts, using the nominal instrument transformer ratios, without corrections.

(c) Apply the corrections, and find the true values of the load volts, amperes and power. What percentage error in power is occasioned by neglect of the corrections?

15-22. Power to a balanced three-phase inductive load is measured by the two-wattmeter method, with current and potential transformers between the lines and the instruments. The secondary readings (after correction for errors of meter calibration) are: amperes per line = 4.50, volts between lines = 117.5, total power = 295 watts.

The instrument transformers have the following characteristics:

	Potential Transformer	Current Transformer
Nominal ratio.........	7200/120	500/5
RCF.................	0.994	0.991
Phase-angle error......	−24′	+18′
	(E₂ reversed lags E₁)	(I₂ reversed leads I₁)

(a) Find the reading of each of the two wattmeters.

(b) Apply the corrections, and find the true values of load volts, amperes per line, and power. What percentage error in power is occasioned by the neglect of corrections in this case?

15-23. A 3-phase delta load is measured by using instrument transformers and the two-wattmeter method. The load is balanced and operates at 4000 v line-to-line. The line currents are 173.2 amp each and the load requires 600 kw at a lagging power factor. The phase sequence is A-B-C, the current transformers are in lines A and B, and the potential transformers are across AC and BC. The nominal ratios and the corrections for the above load are as follows:

	Current Transformer	Potential Transformer
Nominal ratio.........	400/5	4000/100
RCF................	1.0008	0.9978
Phase-angle error......	47' (secondary leads)	5' (secondary lags)

The wattmeter ratings are 150 v, 5 amp, and 500 watts full scale.

(a) Sketch the circuit, showing all transformers and instruments. Show polarity marks.

(b) Find the reading of each voltmeter, ammeter, and wattmeter, neglecting transformer errors.

(c) Repeat (b), taking transformer errors into account.

15-24. A bank of high-voltage power factor correction capacitors operating on a nominal 12,000-v, 60-cycle, single-phase line is tested under approximately rated conditions by measurement of voltage, current, and power (sine wave forms are verified at the start). The instruments are, of course, connected to the circuit by means of instrument transformers. The readings of the instruments (after correction for any calibration errors of the instruments) are:

$$E = 118.5 \text{ v}, \quad I = 3.31 \text{ amp}, \quad \text{power} = 1.60 \text{ watts}$$

The instrument transformer data for the burdens imposed by the given instruments are:

	Potential Transformer	Current Transformer
Nominal ratio.........	12,000/120	20/5
RCF................	0.992	1.004
Phase-angle error......	−8'	+16'
	(E_2 reversed lags E_1)	(I_2 reversed leads I_1)

(a) Determine the true value of the dissipation factor of the capacitors, and the erroneous value that would be obtained if instrument transformer errors were neglected.

(b) Determine the capacitance of the capacitor bank.

CHAPTER 16

Magnetic Measurements

16-1. Types of Magnetic Measurements

A convenient initial classification of magnetic measurements is into d-c and a-c tests, since quite different methods and objectives are found generally in the two cases. The d-c branch may be further subdivided into measurements of field strength, flux, permeability, B-H curves, and hysteresis loops. The a-c measurements are concerned mainly with losses in magnetic materials under conditions of alternating magnetization.

Magnetic measurements are more difficult to make and essentially less accurate than electrical measurements for two main reasons. First, we do not measure flux as such, but only some effect produced by it, such as the voltage induced by a change of flux. As a second and greater difficulty, flux paths are not definite as are electrical circuits — we met some difficulty in precision a-c bridges in making the electrical circuits definite, but the magnetic situation is radically worse and not subject to the same control. As one result of the factors mentioned above, most magnetic measurements are by deflection, rather than null, methods, which is another reason for limited accuracy.

16-2. The Ballistic Galvanometer

The ballistic galvanometer differs from the galvanometers studied in chap. 5 in details of usage rather than in fundamental ideas. As a consequence of the conditions of use the design is modified, but the basic construction is the same as met before in the d'Arsonval type. In circuits we have encountered so far, the current through the galvanometer is steady as long as the key is depressed. In the ballistic case the current flow takes place in a short period of time. The coil receives a momentary impulse, which causes it to swing to one side and then to return to rest, either gradually or after several oscillations, depending on the damping. The deflection is read at the extreme point of the first throw.

The discharge through a ballistic galvanometer must occur in a short time, that is, before the coil has moved appreciably from its rest position, in order that the throw shall be directly proportional to the quantity of electrical discharge. Galvanometers for ballistic use are designed, accordingly, to have a long period of the order of 20 to 30 sec, compared with 4 to 6 sec for the ordinary type. The long period is secured by large inertia of the moving system and small restraining torque of the suspension. The inertia has been secured in some galvanometers by the addition of dead weight to the moving system. It is preferable to design the coil for greater width so that the added copper not only increases inertia but adds to the sensitivity as well.

Equations for the behavior of the ballistic galvanometer may be derived in the following way. The torque developed by the coil at any instant is

$$T_d = Bi \times 2Ln\frac{W}{2} = i(BLnW) = K_1i \tag{1}$$

where L, W, and n are, respectively, the length, width, and number of turns of the coil, and B is the air-gap flux density (see chap. 5). The torque of acceleration is

$$T_a = J \times \text{acceleration} = J\frac{d\omega}{dt} \tag{2}$$

where J is the moment of inertia of the coil about its axis and ω is the angular velocity. Now if the coil is close to its zero position during the time the discharge takes place, the torque of the suspension is practically zero, and if the damping torque is negligible in comparison with the driving torque during this time, we may equate the two values of torque derived above. Therefore, during the brief discharge period

$$J\frac{d\omega}{dt} = K_1i \tag{3}$$

By integration we obtain,

$$J\omega\Big]_{\omega=0}^{\omega=\omega_0} = K_1\int_{t=0}^{t=t_0} i\,dt \tag{4}$$

where the subscript zero refers to conditions at the end of the discharge period. The integral on the right side of (4) is the quantity of charge that has passed through the coil during this period. Therefore,

$$\omega_0 = \frac{K_1}{J} Q \tag{5}$$

This equation tells us that the velocity which the coil acquires from the impulse and with which it begins its swing is proportional to the quantity of charge that passed through it. This depends on (3), which, in turn, is true only if the entire discharge takes place before the coil has moved appreciably from its rest position. This is the reason for the requirement of a long period for the galvanometer swing.

The equation of motion of the coil follows the derivation in chap. 5, eq. (17), up to the point at which the arbitrary constants are evaluated. In the present application the initial conditions are

$$\theta = 0 \quad \text{and} \quad \frac{d\theta}{dt} = \omega_0 \quad \text{at } t = 0 \tag{6}$$

which takes the end of the discharge period as the time-origin for the swing. If the relationships in (6) are substituted in the equations for the under-damped case, we obtain

$$\theta = \omega_0 \epsilon^{-D\,t/2J} \frac{\sin \beta t}{\beta} \tag{7}$$

where, in general,

$$\beta = \sqrt{\frac{S}{J} - \frac{D^2}{4J^2}}$$

or

$$\beta = \sqrt{\frac{S}{J}} \tag{8}$$

if the damping is small. Notice in (7) that deflections are proportional to ω_0, which from (5) is proportional to Q. The deflection of the galvanometer may accordingly be used as a measure of the quantity of electricity dis-charged through it.

The amplitude of the first swing, θ_1, for the "undamped" case $(D \simeq 0)$ is from (7)

$$\theta_1 = \omega_0 \sqrt{\frac{J}{S}} \tag{9}$$

The ratio of successive swings (first to one side, then to the other side of zero) with damping present is found by the exponential multiplier in (7) for a time interval such that $\beta t = \pi$, or $t = \pi/\beta$. The ratio of successive swings is

$$r = \frac{\theta_1}{\theta_2} = \epsilon^{(\pi/2)(D/J\beta)} \tag{10}$$

The natural logarithm of this ratio is referred to as the "logarithmic decrement" of the galvanometer and has the value

$$\lambda = \ln\left(\frac{\theta_1}{\theta_2}\right) = \frac{\pi}{2}\frac{D}{J\beta} \tag{11}$$

The discussion in the present paragraph has been introduced mainly to explain this term, which will be encountered frequently in studies of ballistic galvanometers.

The third swing may be found in the same manner as in (10), except that $\beta t = 2\pi$. The exponent is twice as great, therefore

$$\frac{\theta_1}{\theta_3} = r^2$$

In general

$$\frac{\theta_1}{\theta_n} = r^{n-1}$$

Greater accuracy in evaluation of the damping ratio is secured by permitting several swings and finding r from the equation above or from

$$r = \sqrt[n-1]{\frac{\theta_1}{\theta_n}}$$

The case of critical damping is of particular interest to us here. If we take the equation for critical damping from chap. 5 and substitute the initial conditions from (6), we obtain

$$\theta = \omega_0 \epsilon^{-(D/2J)t}t \tag{12}$$

and

$$\omega = \frac{d\theta}{dt} = \omega_0 \epsilon^{(D/2J)t}\left(1 - \frac{D}{2J}t\right) \tag{13}$$

Maximum deflection is found for $\omega = 0$ or $t = 2J/D$ from (13). Substitute this value in (12) and call the deflection θ_1, then

$$\theta_1 = \omega_0 \frac{2J}{D}\frac{1}{\epsilon} \tag{14}$$

or

$$\theta_1 = \omega_9 \sqrt{\frac{J}{S}}\frac{1}{\epsilon} \tag{15}$$

since $D^2/4J^2 = S/J$ for the critical case.

It is interesting to compare (15) with (9). The deflection in the critically damped case is $1/\epsilon$, or 36.8 per cent, of the undamped deflection, but it is still a direct measure of ω_0 or Q. The loss of sensitivity is not serious, since conditions can usually be arranged to give ample deflection. The speed and convenience of working with the critically damped galvanometer is so much better than with the underdamped instrument that the sensitivity is a secondary consideration.

We may summarize the results of our study in the following equation of the charge passing through the galvanometer

$$Q = K_2\theta \tag{16}$$

The usual working units in (16) are

K_2 = galvanometer sensitivity in microcoulombs per millimeter deflection*

θ = deflection in millimeters

Q = charge in microcoulombs

In the theoretical study θ was in radians and Q in coulombs, but the units above are the ones in common use in laboratory work. The sensitivity K_2 has been shown in our discussion to depend on the damping, being $1/\epsilon$ as great with critical damping as with no damping. Curves may be found showing the relationship of sensitivity to damping in order to give an idea of the nature of the variation.† The sensitivity for a set of measurements, however, should be obtained by a calibration conducted under the actual conditions of use. Sensitivities given in the equipment makers' catalogues are for the undamped case.

16-3. Measurement of Magnetic Flux by Ballistic Galvanometer

Figure 16-1 represents a bar magnet surrounded by a pick-up coil or search coil that is connected in series with a galvanometer and variable resistor. The resistor may be used to control sensitivity, but preferably it should be set to give critical damping, and thereafter the sensitivity is controlled within satisfactory limits by the number of turns in the search

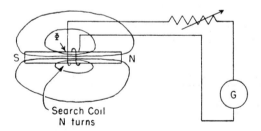

Fig. 16-1. Measurement of flux by a ballistic galvanometer.

---

*Note: Deflection for reflecting galvanometers is expressed in millimeters on a scale 1 meter from the galvanometer. For galvanometers with attached scales, deflection is expressed in scale divisions (regardless of size).

†Reference 7, 9.

coil. If the magnet is withdrawn quickly from the coil, an impulse of short duration is produced in the galvanometer, and this may be used as a measure of the flux. The voltage induced in the coil at any instant is

$$e = N \frac{d\varphi}{dt} \quad \text{volts} \tag{17}$$

if the flux is measured in webers and N is the number of turns in the search coil. If R represents the total resistance of the circuit, including search coil, galvanometer, and series resistor, the current that flows is

$$i = \frac{e}{R} = \frac{N}{R} \frac{d\varphi}{dt} \quad \text{amperes} \tag{18}$$

By integration with respect to time for the period of the discharge

$$\int_{t=0}^{t=t_0} i \, dt = \frac{N}{R} \int_{\phi=\Phi}^{\phi=0} d\phi \tag{19}$$

Therefore, the quantity of charge through the galvanometer (neglecting sign) is

$$Q = \frac{N\Phi}{R} \quad \text{coulombs} \tag{20}$$

The deflection of the galvanometer is

$$\theta_1 = \frac{Q}{K_2} = \frac{N\Phi}{K_2 R} \tag{21}$$

or

$$\Phi = \frac{K_2 R \theta_1}{N} \quad \text{webers} \tag{22}$$

Two points should be noted in this use of the ballistic galvanometer. First, the change of flux must be made in a time interval that is short in comparison with the period of the instrument. Second, the galvanometer is here used in a closed circuit and hence is subject to electromagnetic damping. The sensitivity factor, K_2, must be evaluated for the resistance used in the test measurements.

16-4. Calibration of a Ballistic Galvanometer

(a) *Capacitor method.*

A ballistic galvanometer may be calibrated by means of the circuit of Fig. 16-2. The resistor and K_2 are used to bring the galvanometer to rest quickly after a deflection. The condenser is charged in the upper position of K_1 and discharged by temporary contact in the lower position. The quantity of charge may be computed as $Q = CE$, so the constant K_2 is derived from the charge divided by the observed deflection. This, as shown, is the *undamped* sensitivity because the galvanometer circuit has infinite resistance. A shunt may be added, as indicated by the dotted lines (with K_2 removed), and a new calibration can be obtained for the shunted galvanometer. The shunt also gives damping and, if nearly equal to the critical value, obviates the need for K_2 to control the return swing. If the shunt is much below the critical value and is placed directly across the galvanometer, the action is sluggish. Damping conditions may be improved by a combination of shunt and series resistance.

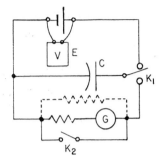

Fig. 16-2. Calibration of a ballistic galvanometer by a condenser.

This circuit has been used to measure an unknown capacitor by comparing deflections with a standard capacitor, but this arrangement is neither as convenient nor as accurate as an a-c impedance bridge for this purpose. Capacitor measurements and also the calibration described above are open to the objection that a capacitor with solid dielectric is not definite in d-c use, owing to the phenomenon of "absorbed charge." Such a capacitor absorbs charge over a considerable period of time when charging on d-c and, likewise, continues to give up charge over a considerable period on discharge. The term capacitance does not have a definite meaning under these conditions except with respect to a definite schedule of charge and discharge.

(b) *Calibration by a solenoid.*

The galvanometer may be calibrated for flux measurements by use of a test coil wound closely around (or else placed inside) the middle of a long solenoid, as shown in Fig. 16-3. For a solenoid of length many times the

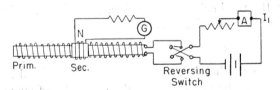

R = total resistance of Galvanometer circuit

Fig. 16-3. Calibration of a ballistic galvanometer by a solenoid.

diameter, the field near the center is practically uniform and may be computed accurately. We thus establish a calibration for flux measurements by means of a known flux, and the calibration and damping remain constant if we keep the same total resistance in the galvanometer circuit at all times.

The flux linking the search coil is given in the *MKS* system by

$$\Phi = 4\pi \times 10^{-7} N_1 I_1 A \quad \text{webers} \tag{23}$$

where

    N_1 = primary turns *per meter* length of solenoid
    I_1 = primary current, amperes
    A = area of cross section of test coil in square meters, if inside
        = area of cross section of solenoid, if test coil is closely wound outside

The circuit is operated by throwing the reversing switch from one position to the other. This gives a flux change twice as great as above, so by substitution in (20)

$$Q = \frac{8\pi N_1 I_1 A N_2}{R} \times 10^{-7} \quad \text{coulomb} \tag{24}$$

where

    N_2 = turns on the test coil
    R = resistance of the test coil and galvanometer circuit

The sensitivity factor K_2 can thus be established from Q and the observed deflection as in (16).

The calibration for flux measurements may be placed in convenient form, once K_2 is known. If we use the galvanometer for measurement of an unknown flux, we may write, from (22),

$$\Phi_x = \frac{\theta_{1x}(K_2 R)}{N_x} \tag{25}$$

where

    Φ_x = flux change of unknown
    θ_{1x} = deflection in millimeters
    N_x = number of turns used in the search coil for the unknown *provided* that the total galvanometer circuit resistance, R, is the same as when the calibration was made.

(c) *Calibration by mutual inductor.*

This is practically the same test as in (b) but expressed in a somewhat different way. If we have a known mutual inductor of convenient value,

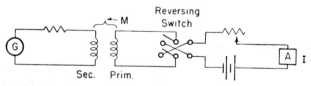

R = total resistance of Galvanometer circuit

Fig. 16-4. Calibration of a ballistic galvanometer by a mutual inductor.

we may connect it as shown in Fig. 16-4. The deflection θ_1 produced by the reversal of a known primary current I is observed.

For a changing primary current

$$e_2 = M \frac{di_1}{dt} \tag{26}$$

so the galvanometer current is

$$i_2 = \frac{M}{R} \frac{di_1}{dt} \tag{27}$$

By integration,

$$\int_{t=0}^{t=t_0} i_2 \, dt = \frac{M}{R} \int_{-I}^{+I} di_1 \tag{28}$$

Therefore

$$Q = \frac{2MI}{R} \quad \text{coulombs} \tag{29}$$

and

$$K_2 = \frac{Q}{\theta} = \frac{2MI}{R\theta_1} \tag{30}$$

16-5. Use of Shunts with Ballistic Galvanometers

A shunt may be used with a ballistic galvanometer, provided that proper calibration is made. The shunt has two effects, first in changing the current through the coil, and second, in increasing the damping.

It may appear that the charge from the search coil will not divide between the two paths in accordance with the resistances as a result of the inductance of the galvanometer coil and the irregular nature of the search coil current. That it does so divide may be shown from the equation written for the symbols of Fig. 16-5

$$i_s R_s = i_g R_g + L_g \frac{di_g}{dt} \tag{31}$$

By integration,

$$R_s \int_0^{t_0} i_s \, dt = R_g \int_0^{t_0} i_g \, dt + L_g \int_0^0 di_g$$

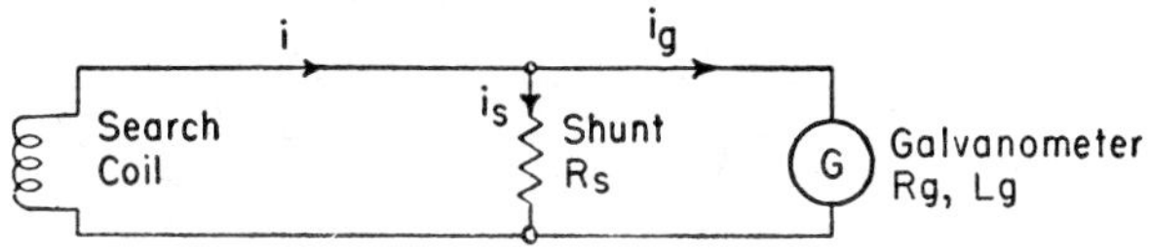

Fig. 16-5. Shunted ballistic galvanometer.

The last term is zero, since i_g is zero for both initial and final conditions. Therefore,

$$\frac{Q_s}{Q_g} = \frac{R_g}{R_s} \tag{32}$$

and L_g has no effect on the division of charge between the two paths. For the total charge from the search coil

$$\frac{Q_g}{Q_t} = \frac{R_s}{R_g + R_s} \tag{33}$$

The shunt not only provides a multiplying factor, as shown by (33), but also changes the resistance of the damping circuit. This is true even with an Ayrton shunt. Accordingly, the ballistic galvanometer should be calibrated with each shunt that is to be used, since the application of a shunting factor from (33) is not adequate by itself.

16-6. Fluxmeter

The fluxmeter is a special form of ballistic galvanometer in which the torque of the suspension is made negligibly small, and the electromagnetic damping is very heavy. It possesses advantages over the ballistic galvanometer for some kinds of magnetic measurements. Besides portability, it has the advantage that the flux change need not take place in a short time. The deflection obtained for a given flux change is nearly independent of the time taken in making the change.

The construction of the meter is indicated in Fig. 16-6. The coil is supported by a silk fiber from a spring support. The current connections are made by spirals having the minimum possible spring effect. The objective in the suspension and spirals is to make the restoring torque of the system as nearly zero as possible. The moving coil and the search coil are

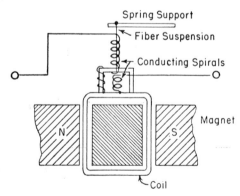

Fig. 16-6. Diagram of construction of a fluxmeter.

kept to relatively low resistance, so that the moving system is heavily overdamped by electromagnetic action. The meter coil follows flux changes very rapidly and is practically deadbeat in its action. After being deflected the coil stays almost stationary and moves extremely slowly toward the zero position; thus the fluxmeter is much easier to read than a ballistic galvanometer. A mechanical or electrical return must be used to bring the pointer back to zero. The fluxmeter is usually of the portable type, carrying a pointer and scale.

While the fluxmeter is essentially a ballistic galvanometer with low restoring torque and heavy electromagnetic damping, the equations we derived above for the galvanometer cannot be used for the fluxmeter because of the differences in operating conditions. The derivation for the ballistic galvanometer was based on the assumption that the discharge through the coil takes place before it moves appreciably from the rest position and that damping has negligible effect during this period. [See eq. (3).] This is not true for the fluxmeter because of the heavy damping action. With low circuit resistance a large accelerating torque acts on the meter coil whenever the search-coil voltage exceeds the back voltage caused by the motion of the meter coil. As a result, the meter coil responds rapidly and is well along in its movement while the flux change is still in progress, regardless of whether the change is fast or slow. The almost complete absence of a counteracting suspension torque permits the meter to add the effects of separate flux changes over a (short) period of time and gives independence of the over-all time, within reasonable limits. It is better, on account of these special characteristics, to begin directly with circuit equations for the fluxmeter than to try to adapt the ballistic-galvanometer equations to this purpose.

The voltage generated in the fluxmeter coil by its motion becomes the controlling element in this meter because of the low resistance of the moving system and of the absence of torque from the suspension. The coil voltage could be expressed as the change of flux linkages, but since the turns and magnet strength are constant it is more convenient for our purpose to relate voltage to coil velocity or

$$ e_m = K_1 \frac{d\theta}{dt} \tag{34} $$

where $K_1 = BLnW$ is the same coil constant met in equation (1).

The quantities needed in the analysis are indicated in Fig. 16-7. The voltage induced in the search coil is indicated by a small "generator" marked e_c and the back-voltage caused by the motion of the fluxmeter coil by another generator, e_m. In the electrical circuit

$$ e_c = iR + L\frac{di}{dt} + e_m \tag{35} $$

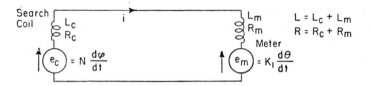

Fig. 16-7. Analysis of fluxmeter action.

If we insert the value of e_c in terms of flux change and e_m from (34) and solve for i,

$$i = \frac{N}{R}\frac{d\varphi}{dt} - \frac{K_1}{R}\frac{d\theta}{dt} - \frac{L}{R}\frac{di}{dt} \tag{36}$$

The equation of coil motion will be written as in chap. 5, with the term for suspension torque omitted. The net torque is equated to the amount produced by the current

$$J\frac{d^2\theta}{dt^2} + D\frac{d\theta}{dt} = K_1 i \tag{37}$$

where

J = the moment of inertia

D = the constant of friction and air damping

If i is substituted from (36) into (37) and if the terms are collected, we have

$$J\frac{d^2\theta}{dt^2} + \left(\frac{K_1^2}{R} + D\right)\frac{d\theta}{dt} = \frac{K_1 N}{R}\frac{d\varphi}{dt} - \frac{K_1 L}{R}\frac{di}{dt} \tag{38}$$

We shall now integrate all terms with respect to time. The first term drops out since it integrates to $J\omega$, and ω is zero for both initial and final conditions. The last term is also zero since i is zero at both limits. The result, with subscripts 1 and 2 to denote initial and final values, respectively, and after multiplication by R/K_1 is given by

$$\left(K_1 + \frac{DR}{K_1}\right)(\theta_2 - \theta_1) = N(\varphi_2 - \varphi_1) \tag{39}$$

where $R = R_c + R_m$.

The air damping effect and the resistance are both small, and accordingly the first term is nearly equal to K_1. Therefore, approximately,

$$K_1(\theta_2 - \theta_1) \simeq N(\varphi_2 - \varphi_1) \tag{40}$$

That is, the change of flux linkages of the search coil is proportional to the change of pointer-angle on the meter, so the meter may be calibrated in terms of flux.

Equation (40) tells an interesting physical fact. The quantity on the right is the change of flux linkages with the search coil, and the quantity on the left is the change of flux linkages with the meter coil. If the operator withdraws a magnet from the search coil, reducing the flux linkages with that part of the circuit, the meter coil moves in a direction to restore the flux linkages and thus to keep the total constant for the circuit. The fluxmeter does this within the limits set by the degree to which air damping and suspension torque are negligible.

We can see the effect of shunting the meter by a simplified form of derivation. We found above that the inductance terms did not enter in the result and hence we shall drop them at the start in the following study.

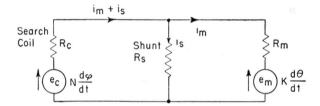

Fig. 16-8. Simplified circuit diagram for a shunted fluxmeter.

Accordingly, we shall use the simplified diagram of Fig. 16-8 for the shunted case. We may write the two circuit equations

$$N \frac{d\varphi}{dt} = (i_m + i_s)R_c + i_m R_m + K_1 \frac{d\theta}{dt} \tag{41}$$

$$i_s R_s = i_m R_m + K_1 \frac{d\theta}{dt} \tag{42}$$

By substitution for i_s in (41) of its value from (42) and collection of terms, we have

$$N \frac{d\varphi}{dt} = i_m \left(R_c + R_m + \frac{R_c R_m}{R_s} \right) + K_1 \frac{R_c + R_s}{R_s} \frac{d\theta}{dt} \tag{43}$$

If we solve for i_m from (43), substitute the value in (37), and solve as before, we obtain

$$\left[K_1 + \frac{D}{K_1} \left(\frac{R_c R_s}{R_c + R_s} + R_m \right) \right] (\theta_2 - \theta_1) = \frac{N R_s}{R_c + R_s} (\varphi_2 - \varphi_1) \tag{44}$$

and in the approximate form

$$(\theta_2 - \theta_1) = \frac{N}{K_1} \frac{R_s}{R_c + R_s} (\varphi_2 - \varphi_1) \tag{45}$$

Comparison of (40) and (45) shows that the main effect of the shunt is to apply a shunting factor $R_s/(R_c + R_s)$ to the reading. The resistance of the circuit, as shown in the correction term in (39) or (44), is not of extreme importance. However, for accurate work the total resistance should be brought to the same value used in calibration by the addition of series resistance, if necessary.

Figure 16-9 shows a multirange fluxmeter in convenient, portable form. A wide range of values can be covered by the combination of five multiplying ranges and the choice of the search coil to be used with it.

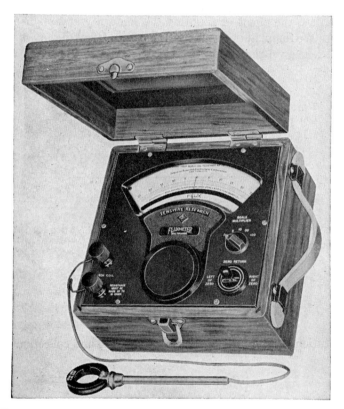

Fig. 16-9. Multirange portable fluxmeter. (*Courtesy Sensitive Research Instrument Corporation*)

16-7. Use of Ballistic Galvanometer and Fluxmeter

Flux measurements may be made in open- or closed-frame electromagnets if the current in the magnet coil can be either switched on and off or reversed. A search coil of a suitable number of turns must be wound around the magnet. If the flux in the field pole of a large motor or generator is being measured, even a one-turn search coil may give too much deflection, and it will be necessary to shunt the meter. The fluxmeter has the advantage in such measurements that the flux need not be changed in a short time — changes should not be made too rapidly, particularly in large machines.

The flux of a permanent magnet can be measured if the search coil can be moved on and off the magnet. It must be placed always in the same position with respect to the magnet to give consistent results.

The field strength between magnet poles can be measured if there is room to insert and remove the search coil. Another method, used in fields of considerable extent is a "flip coil," which is a coil mounted so that it can be rotated an exact 180° in a short period of time. The fluxmeter indication can then be interpreted in terms of field strength.

B-H curves of ring-shaped samples of magnetic material may be obtained by a ballistic galvanometer or fluxmeter in conjunction with search and magnetizing coils wound around the sample. Similar tests are run on straight-strip samples by means of various forms of "permeameter." Both of these subjects will be considered in the following sections.

16-8. Normal Magnetization Curve of a Ring Sample

A ring-shaped sample has one advantage for magnetic testing in that it gives a continuous piece of the material of constant cross section without joints or air-gap. It may be solid metal or may be built up of ring-shaped sheets of laminated material. The ring is prepared for the test by first winding on it a thin layer of insulating tape and then a search coil of a suitable number of turns, evenly distributed around the ring. Insulating tape is applied over the search coil, and then the magnetizing coil is wound on the ring. The magnetizing coil should be spaced uniformly around the ring and must consist of a suitable number of turns of wire of adequate current-carrying capacity to give the number of ampere-turns required for the tests.

The equipment may be connected as shown in Fig. 16-10. The magnetizing winding is connected to the power supply through a reversing switch, ammeter, and rheostat. (S_2 and R_2 will be used later for deter-

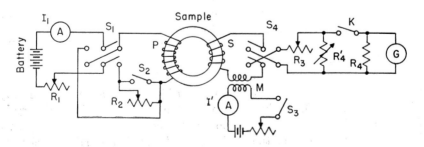

Fig. 16-10. Circuit for magnetic measurements on a ring sample.

mination of the hysteresis loop. S_2 will be kept closed for the present measurements.) G represents a ballistic galvanometer, to which resistors R_3 and R_4 are added to control sensitivity and damping. R_4 is set approximately to the critical damping resistance of the galvanometer in order to control the return swing. Sensitivity is adjusted ordinarily by R_3, but an additional shunt, R_4', can be connected as shown to decrease the sensitivity for the deflection direction without affecting the return swing. S_4 may be used to keep the galvanometer deflection always in the same direction. The circuit through S_3 and M is used for calibration of the galvanometer; calibration may thus be performed without disturbing the circuit and with exactly the same resistance that is used in the test measurements. (S_3 may be replaced by a reversing switch, if desired.)

For the circuit as described above, the galvanometer is overdamped in the deflection direction (K closed). Overdamping involves some decrease in sensitivity, but this can be made up by an increase in the search coil; in other respects overdamping is an advantage. For one thing, the deflection is less dependent on the time of discharge. This is a help because flux changes are slow in some systems, particularly in those involving solid magnetic materials. With overdamping, the light spot reaches maximum deflection in a shorter time and stays at the maximum point longer, making the deflection easier to read.

The sample must be demagnetized before readings are taken; this is accomplished by application of a high value of field strength which is carried through cyclic changes of gradually decreasing magnitude until zero is reached. Demagnetization may be performed in the circuit of Fig. 16-10 by closing S_1, increasing I_1 to a high value, making several reversals of S_1 (S_4 open), decreasing I_1 a small amount, making more reversals, and so on until zero current is reached. The reversals should be rather slow, say one per second, if the magnetic system consists of a large mass of solid iron. Laminated material may be demagnetized much faster, even using alternating current of the ordinary power frequency. In the a-c method, the magnetizing coil of the ring specimen is connected to a smoothly variable a-c supply before connection in Fig. 16-10; the current is increased to a value high enough to give saturation and then gradually reduced to zero.

After demagnetization, R_1 is set to give a low current for the first test reading, and S_1 is reversed several times to put the sample in a definite cyclic state. S_4 and K are closed, and readings of G are taken for several reversals. The flux change can be computed from the deflections and the calibration established by M (the number of turns in the search coil must, of course, be known). It should be noted that the observed flux change is from tip to tip of the hysteresis loop and hence must be divided by two in finding the ordinate of the B-H curve. B equals the half flux change divided by the cross-sectional area of the core.

H in the MKS system is measured in ampere-turns per meter. It can be computed for any value of I_1 if the number of turns of the test coil and the mean circumference of the test ring are known. The use of an average length is not strictly correct, but a computed correction is scarcely more reliable because of nonlinearity of core behavior. Error from this source is not important if the width of the ring is small in comparison with the mean radius. In the CGS system, H in oersteds equals $0.4\pi NI/\text{Length}$ (in centimeters), if I is in amperes. Similarly, the units for B are webers per square meter in MKS and gausses (maxwells per square centimeter) in CGS.

After one pair of B-H values is found, as described above, R_1 is changed to give a slightly greater value of I_1 and the process is repeated. Further readings are taken up to the desired limit. The readings must be taken with regular, uninterrupted increases of I_1. If at any time I_1 is increased accidentally above the value at which a reading is wanted, the ring must be demagnetized below this point and then increased to the desired value. The B-H curve plotted from the observed points taken by the method described in this section is the one going through the tips of loops of increasing size and is called the normal magnetization curve or normal induction curve. It is indicated by the dotted curve from the origin in Fig. 16-11.

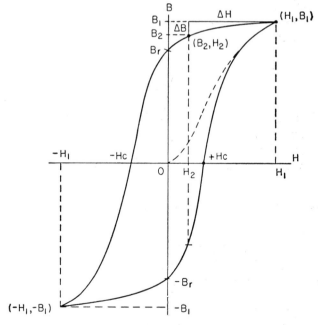

Fig. 16-11. Hysteresis loop of a steel sample.

A search coil always encloses some air space and insulation space in addition to the magnetic material that is being tested. It may be necessary in some cases to correct the observed value of flux to take into account the flux in the air space. The correction is negligible for a close-fitting search coil if the specimen material has high permeability, but may become important under other conditions. There is always some "air" space between the sheets of laminated material, the minimum amount being the layer of oxide on the surface of the sheets, with increase in some cases by a coating of insulating varnish. The "stacking factor" — the ratio of iron thickness to the total thickness of a stack of laminations — has a maximum of about 0.92 for tight stacking to a much lower value for thin laminations or varnish-coated laminations. In any case in which a correction for air flux is needed, it may be made by subtracting the correction term $(A_c - A_s)\mu_0 H / A_s$ from the apparent flux density, where A_c and A_s are the cross-sectional areas of the search coil and the specimen, respectively, and μ_0 is the permeability of free space ($4\pi \times 10^{-7}$ in the MKS system). (The "apparent" flux density is the flux indicated by the search coil divided by the cross-sectional area of the specimen.)

Ring samples receive limited use in spite of their advantage in being free of joints, largely because of the work and expense involved in preparing them. The process of winding the search and magnetizing coils is rather tedious unless a special winding device is available. Also, special stampings may be required for laminated materials if an annular form is adopted. Sheet materials from which laminations are made have properties along the grain produced by the rolling process that are different from the properties in the crosswise direction. A ring stamping gives an average of the properties in the different directions, which may at times be useful information, but generally a test for a single direction is preferred. Ring samples may also be formed by winding a long, narrow ribbon of the material into a ring form. This construction has the desirable quality of giving a sample with the same grain orientation for the entire length, and suffers only from the coil-winding problem. Straight strip samples may be tested in a permeameter, which will be described later.

16-9. Determination of a Hysteresis Loop

Data may be obtained for a hysteresis loop, shown in Fig. 16-11, by means of the circuit of Fig. 16-10. The current is set to the value corresponding to the desired value of maximum H. The switch S_1 is reversed several times so that the sample is in a definite cyclic condition. The point (H_1, B_1) at the top end of the loop becomes the reference from which other points are located. The sample can be returned to a cyclic condition at this point at any time by manipulation of S_1. To make directions easy to remember, we shall assume that the data are plotted so that the right-hand tip of the loop corresponds to the right-hand position of S_1.

The position of (H_1, B_1) is determined carefully by measurements as in par. 8. With conditions at (H_1, B_1), S_2 is opened, which introduces R_2 and decreases the magnetizing current. ΔB is obtained from the galvanometer deflection and calibration and H_2 from the new value of current. After this observation S_1 is thrown to the left, S_2 is closed, and S_1 is returned to the right, bringing the sample back to the starting point. Several observations can be made of the point (H_2, B_2), after which R_2 is changed and a new point found. This method may be used for any point from (H_1, B_1) to (O, B_r). The B_r-point may be found with S_2 closed simply by moving S_1 from the right to the open position. It must be remembered in finding ΔB values that the 2 factor does not enter as in the reversals from $+B$ to $-B$ in the previous section.

Points between B_r and $-B_1$ cannot be obtained by the present switching arrangement, but we can obtain points between $-B_r$ and $+B_1$ in the following way. With the material in the cyclic state (S_2 closed), S_1 is thrown to the left, bringing conditions to point $(-H_1, -B_1)$. S_2 is opened and then S_1 is thrown to the right. This changes H from $-H_1$ to $+H_2$ and gives a ΔB that is measured upward from $-B_1$. Any point from $-B_r$ to $+B_1$ can be obtained in this way by an appropriate setting of R_2. We now have the entire right half of the loop and can plot the left half by symmetry. (Air-flux corrections may be made, if needed, by the method described in par. 8.)

The H_c-points measure the "coercive force," which is the reversed value of H needed to "coerce" the material back to the unmagnetized state, which means $B = 0$. H_c cannot be read directly by the above method, but if enough points are found the loop can be drawn and H_c read from the graph.

Care must be taken throughout the measurements to keep the material in the cyclic state. If a mistake is made in the switching operations, S_2 should be closed and S_1 reversed several times to return the material to the correct condition.

It is sometimes desirable in measurements involving high values of H to determine the B_r-points, and then to measure other densities from the B_r-points as much as possible, making only momentary contact in the $H_{\max}$ positions. This serves to minimize heating in the magnetizing coil for the large values of current.

16-10. Permeameters

Straight strip or bar specimens are more convenient than ring specimens because they are easier to prepare and because they may be inserted in, and later removed from, stationary coil forms. However, it is harder to test them in such a way that the applied field is uniform in value and subject to accurate measurement.

A straight sample may be tested by placing it in a long solenoid. The flux may be measured by a search coil wound around the middle of the sample. H is more difficult to determine accurately, as the field of the solenoid is modified by the poles of the specimen. The effect of the specimen can be minimized by making its length great in comparison with the lateral dimension. Generally, this is not a practical arrangement, however, as it requires a very long and awkward solenoid for the testing of samples of moderate size. There are various forms of "permeameters" that have been devised to avoid these difficulties and to test straight samples under controllable field conditions. They consist generally of a fixed steel frame to which straight samples can be connected, but differ widely in the arrangement of magnetizing and pick-up (search) coils and in the means of guarding against leakage fluxes and mmf drops at the joints. An extensive study of the many forms of permeameters would take much more space than is available here, so we shall limit the discussion to two types.

(a) *Burrows permeameter.*

The magnetic circuit of the Burrows permeameter is made up of the test specimen, an auxiliary specimen of similar characteristics, and two connecting yokes. Each specimen has a magnetizing coil and search coils

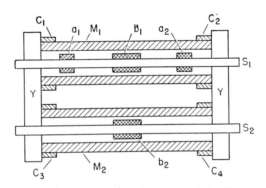

Fig. 16-12. Diagram of the magnetic circuit of the Burrows permeameter.

S_1 — Specimen to be tested.

S_2 — Specimen of similar characteristics.

Y,Y — Yokes.

M_1 — Magnetizing coil for test specimen.

M_2 — Magnetizing coil for auxiliary specimen.

b_1,b_2 — search coils (equal).

a_1,a_2 — search coils, each one-half of b_1.

C_1,C_2,C_3,C_4 — Compensating coils.

as indicated in Fig. 16-12. The purpose of the various parts can best be understood by a brief outline of the procedure followed in making adjustments preparatory to taking a reading.

The current in M_1 is set at the value computed from the desired value of H, the number of turns of M_1, and the length of the specimen between yokes. The search coils b_1 and b_2 are connected in series opposition to the ballistic galvanometer, and the current in M_2 is adjusted until the galvanometer shows no deflection when the currents in M_1 and M_2 are reversed. This means that S_1 and S_2 carry equal fluxes at their centers. Coil b_1 is now connected to the galvanometer in opposition to the combination of a_1 and a_2, which together have the same number of turns as b_1. The current through C_1, C_2, C_3, and C_4 is adjusted until no galvanometer deflection results when the currents in all the coils are reversed. The effect of C_1 and C_2, as set by this adjustment, is to give uniform flux for the entire length of S_1 if the sample is uniform in its properties. Coils C_1 and C_2 make this condition possible by supplying the mmf needed for the yokes and joints. The addition of the compensating coils may have disturbed the equality of b_1 and b_2, hence the preliminary adjustments are rechecked and modified if necessary. When the adjustments are all satisfactory, the flux in S_1 is read by connection of b_1 alone to the galvanometer, followed by observation of the ballistic deflection when the currents in all the coils are reversed simultaneously.

The Burrows method results in uniform conditions along the specimen, provided that the material of the specimen is itself uniform. The method is slow, as may be judged by the number of adjustments needed for each test point, as outlined above. Moreover, the description covered the steps needed in deriving data for the normal induction curve, which required only simple reversals from tip to tip of the loop. If it is desired to find points along the hysteresis loop, additional rheostats, switches, and adjustments are needed. Correct adjustment of the currents in M_1, M_2, and the compensating coils is secured when no deflection is given on the galvanometer successively for b_1 against b_2 or b_1 against $(a_1 + a_2)$ at the time when a change is made from the tip of the loop to the intermediate point. The change of the currents can be controlled by short-circuiting switches across rheostats in the three circuits. Gang-operated switches are needed for this purpose as for the reversing process.

The Burrows permeameter has been used for magnetic measurements at the Bureau of Standards and in other laboratories and has been accepted as a standard by the A.S.T.M.* for many years. It has been used for magnetic testing of many kinds and also for the calibration of specimens that serve as reference standards in some of the simpler test methods.

---

*Reference 10.

(b) *Fahy simplex permeameter.*

The Fahy permeameter has been widely used for magnetic measurements of bar and laminated materials and is accepted as a test method by the A.S.T.M. It is simple in operation and, except possibly for very uniform samples, gives accuracy equal to the Burrows. The average characteristics of nonuniform specimens are better obtained by the Fahy permeameter, as it inherently averages the *B-H* values for the length of the test member.

The permeameter, as shown in Fig. 16-13, has a single magnetizing coil which is wound around the cross bar of a heavy *U*-shaped frame made of

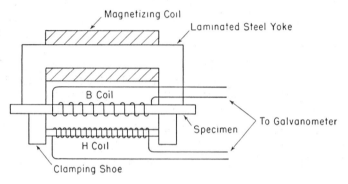

Fig. 16-13. Diagram of the magnetic circuit of the Fahy simplex permeameter.

laminated steel. The specimen, with a search coil wound around it, is clamped across the ends of the *U*. The *H*-coil is wound on a nonmetallic strip which is mounted as close as possible to the specimen, with termination on steel blocks at the ends. The same magnetic potential difference is thus applied to the *H*-coil and to the specimen. Readings of *H* are taken by a ballistic galvanometer in the same way as for the flux of the search coil around the sample. The *H*-coil must have a large number of turns in order to give a suitable galvanometer deflection.

Figure 16-14 shows the permeameter and the control box that is used with it. The control box contains switching arrangements, resistors for control of the galvanometer sensitivity, and an ammeter and mutual inductor for calibration. Separate sets of resistors are provided so that convenient calibration factors can be set up independently for the *B* and *H* circuits, giving constants such as 10 or 100, so that the galvanometer is practically direct reading. The permeameter as shown can be used for fields up to 300 oersteds, but an adapter is available that makes tests possible up to 2500 oersteds on short samples.

Compensating coils are not used at the ends of the specimen as in the Burrows device. The absence of compensating coils simplifies adjustments

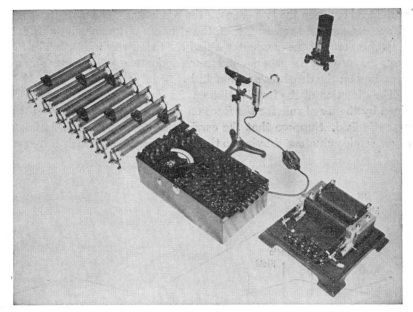

Fig. 16-14. Fahy Simplex Permeameter, with control box, galvanometer, and rheostats. (*Courtesy of Minneapolis-Honeywell Regulator Co., Rubicon Instruments.*)

and operating techniques and is not a serious source of error. With the yoke made of high-grade laminated steel of heavy cross section, the field applied to the specimen is uniform to a satisfactory degree without compensation, and the full-length B and H coils serve to average the variations.

The Fahy permeameter is simple in operation. With the current in the magnetizing coil set to a suitable value, the current is reversed with the galvanometer connected to the B-coil, and B is computed from the deflection. The galvanometer is then transferred to the H-coil, and readings are taken for this quantity. Data for a normal magnetization curve may thus be taken very rapidly. (If the sample is considerably smaller in cross section than the B-coil in which it is mounted, a correction is needed for the air flux, particularly at high H. This correction is made as discussed in par. 8.) A hysteresis loop may be determined equally well by application of the method discussed for the ring sample in par. 9. Some Fahy permeameters have been equipped with a flip coil for H for use in hysteresis-loop determinations.

16-11. Magnetic Measurements Using the Hall Effect*

If a strip of conducting material carries current in the presence of a transverse magnetic field, as indicated in Fig. 16-15, a difference of potential

---

*Reference 6.

is produced between the two edges of the conductor. The magnitude of the voltage depends on the current, on the strength of the magnetic field, and on the property of the conductor called the "Hall coefficient." The "Hall effect" is present in metals and semiconductors in varying amounts, depending on the densities and mobilities of the carriers.

The direction of the force action on the moving charges may be determined by the usual rule for the force on a current-carrying conductor in a magnetic field. Suppose that the carriers are electrons, as in metals and n-type semiconductors, moving to the left in Fig. 16-15, which corresponds

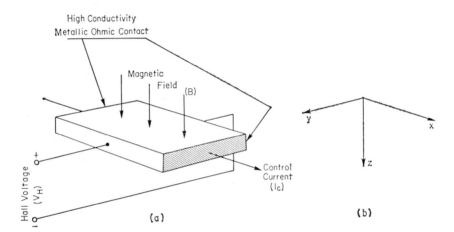

Fig. 16-15. Principle of the Hall effect. (a) Diagram of a Hall element. (b) Notation for the axes. (*Courtesy of the Westinghouse Electric Corp.*)

to a conventional current direction to the right. Then if the magnetic field is in the direction shown, the force on the electron stream is toward the rear face of the conductor, so that it becomes negatively charged, leaving the front face positive. Equilibrium is attained when the resulting electric field balances the deflecting force. The Hall coefficient may be written

$$R_H = \frac{E_y}{J_x B_z} \qquad (46)$$

where E_y is the electric field intensity, J is the current density, and the subscripts x, y, z, denote the directions on a set of standard orthogonal axes. (E is in the $-y$ direction for the polarities and directions of Fig. 16-15). R_H has the dimensions cm^3/coulomb, or m^3/coulomb, depending on which unit of length is adopted.

For interest in the voltage between electrodes, (46) may be changed to

$$E = R_H J B \tag{47}$$

If the conductor has a width w between electrodes, the voltage is

$$V = Ew = \frac{R_H J w t B}{t} = \frac{R_H I B}{t} \text{ volts} \tag{48}$$

where t is the thickness of the strip. (R_H, B, and t in consistent units. Multiply by 10^{-8} for gauss and centimeter units.)

Our concern in this chapter is with the measurement of magnetic fields, so we shall consider here the use of the Hall effect for this purpose. We may make a device for measuring steady fields by using a thin wafer of appropriate material through which a constant current is passed, and measuring the transverse voltage produced. The voltage becomes a measure of the magnetic field, provided that disturbing effects such as temperature are properly compensated. Materials show wide differences in R_H and its dependence on temperature. It is desirable, for measurement purposes, to choose a material with minimum sensitivity to temperature (see par. 17-28). Small voltage changes may be compensated by the use of thermally sensitive resistors in the current or voltage circuit. Reference may be made to chap. 17 for further consideration of Hall effects as applied to other types of measurement.

A Hall element, as viewed in (48), may be considered a multiplying device, giving an output proportional to the product of two input quantities. There are many possible applications. As shown in chap. 17, a watt-meter may be formed in this way, with a coil which gives B proportional to the load current, and a current proportional to load volts. Applications to electrical instrumentation are suggested, such as a device to measure the horsepower of an engine by the product of inputs proportional to torque and speed.

16-12. Other Methods of Measuring Magnetic Quantities

A bismuth conductor has the property of experiencing a change in electrical resistance if the conductor is placed in a transverse magnetic field. This effect has been used by several experimenters.* One or two bismuth spirals are placed in some sort of thin mounting that can be inserted in the magnetic field, as for example, between armature and poles of a d-c motor. The spiral can be made one arm of a Wheatstone bridge (or two spirals may be used in opposite arms). By calibration the resistance can be related to field strength. One difficulty is a rather low sensitivity to weak fields.

---

*References 4 and 5.

(a)

Fig. 16-16. (a) Gauss meter being used to measure field in gap of small magnet. (b) Cut-away view showing construction. (*Courtesy of General Electric Co.*)

(b)

Another, and more serious, trouble is the effect of temperature on the resistance of bismuth, so that large errors can result unless the temperature is carefully controlled or the effect compensated.

The General Electric Company makes a small and simple "gauss meter" that can be usefully employed in testing field strength in places such as the gap between the poles of a magnet. It consists of a tiny permanent magnet mounted at the end of a long, thin shaft which is supported and protected by a bronze tube. The assembly is used as a probe and has an outside diameter less than one-tenth of an inch. The head of the meter consists of a bearing, scale, pointer, and spring and has no electric or magnetic elements in it. The small probe magnet tends to align itself with any magnetic field around it. The person using the meter turns the head until the maximum indication is given on the scale; this gives the strength of the field. This meter is made possible by a new magnetic material called Silmanal, which has a very high coercive force (about ten times that of alnico). The gauss meter is shown in Fig. 16-16.

An a-c bridge method is sometimes used for measurement of permeability and core loss of iron samples, particularly in the audio-frequency range. If a coil with the iron sample as its core (closed core) is connected to one of the a-c inductance bridges and its inductance measured and the air-core value of inductance is then determined either by measurement or calculation, the ratio of the two values is the permeability of the core material. It should be noted that the permeability obtained in this way is an "effective" value, representing an averaged effect over a cycle of flux change and is not the permeability belonging to any one value of flux density. The permeability so measured may be related to the cyclic $B_{\max}$ (or B_{eff}) by computation from the frequency, sectional area of core, and the number of turns and voltage of the coil. This value is useful in a-c applications but should not be expected to check results from d-c tests. A bridge method is described in the A.S.T.M. Standards.*

16-13. A-C Measurements — Epstein Core-loss Test

(Note: Units will be stated here in the MKS system. However, as most references and test codes will be found to use the CGS units, the latter will be included in parentheses for convenience.)

Laminated steel is an essential component of many kinds of electrical equipment, motors, generators, transformers, and others. Since losses and efficiency are stressed for many reasons, knowledge of losses in sheet steel is a matter of great importance alike to manufacturer and user of the

---

*Reference 10, 11, 12, 13.

material. Measurements in the power-frequency range are usually made by the Epstein method.

Figure 16-17 illustrates the equipment needed to make this test. The chief item is the test frame, which provides the mounting for the samples

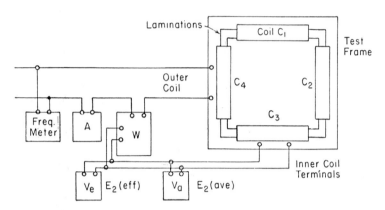

Fig. 16-17. Equipment for Epstein core-loss test.

and the magnetizing coils. The samples are straight strips (50 cm by 3 cm for the larger A.S.T.M. form or 28 cm by 3 cm for the 25 cm Epstein with double-lapped joints) arranged in four equal bundles and inserted in the test coils. There are windings on all four sides, each of which consists of two coils of an equal number of turns. The four inner coils are connected in series to supply the voltmeters and the potential coil of the wattmeter. The four outer coils are the magnetizing winding.

The inner coils constitute a search coil that measures the flux in the core. The voltage induced in the coil is, by the usual transformer equation,

$$E_2 = 4FN_2fB_mA \quad \text{volts} \tag{49}$$

where

F = form factor = 1.11 for a sine wave

N_2 = total series turns of secondary

f = frequency, cycles per second

B_m = maximum flux density in webers/m^2 (gausses)

A = cross-sectional area of steel in m^2(cm^2)

(Note: The factor 10^{-8} is added to (49) if B_m is in gausses and A in cm^2.)

The cross-sectional area is not measured as such but is computed from the mass of the sample and its length and density or

$$A = \frac{M}{4ld} \tag{50}$$

where

M = mass of sample in kilograms (grams)

l = length of one side in meters (centimeters)

d = density

   = 7500 kg/m³ for high-resistance silicon steel (7.5 g/cm³)

   = 7700 kg/m³ for low-resistance silicon steel (7.7 g/cm³)

By substitution of (50) in (49)

$$E_2 = \frac{4FN_2fB_mM}{4ld} \quad \text{volts} \tag{51}$$

Since all factors in (51) can be known and f is read on the frequency meter, this equation gives a direct relationship between E_2 and B_m.

The connection of the wattmeter potential coil to the inner set of coils on the samples serves to eliminate the I^2R in the magnetizing winding from the wattmeter reading. The E_2 voltage represents the back-voltage in the primary, since primary and secondary have the same number of turns. If the wattmeter were supplied with the input voltage, this would include not only the back-voltage but the IR (and IX) in the primary, and hence the wattmeter reading would include the I^2R in the exciting winding. For the connection shown in the figure the wattmeter reads only the power to the sample and to the wattmeter potential coil and the voltmeters, or

$$W_f = W - W_i \tag{52}$$

where

W_f = iron loss in the sample

W = wattmeter reading

W_i = instrument loss

$$= E_2^2 \left(\frac{1}{R_{ve}} + \frac{1}{R_{va}} + \frac{1}{R_w} \right) \tag{53}$$

R_{ve}, R_{va}, and R_w are the resistances of the effective-value voltmeter, average-value voltmeter, and wattmeter potential circuit, respectively.

Losses in iron are different for different wave forms even though the effective voltage value is the same. The loss depends rather on the maximum flux density, provided that there are no actual reversals of slope of the flux curve between negative and positive flux maxima. It thus becomes important for standardizing purposes to have the standard sinusoidal shape, which requires that power for these tests should be drawn from a "sine wave generator" specially built to give good wave form. The current required to set up sine-wave flux is of distorted form on account of the shape of the hysteresis loop, and hence the generator output voltage becomes distorted if the generator has appreciable internal impedance. It is important for the generator to have low internal impedance and ample kva capacity in order that the load current will have no serious effect on the wave form of the terminal voltage. Regulation of voltage should be effected by a generator field rheostat and adjustment made by steps, if necessary, by a transformer but *not* by a *rheostat* in the generator output circuit, as this would produce distortion of the voltage wave. Two voltmeters are used across the secondary, one a high-grade electrodynamometer instrument reading the effective value and the other a "flux voltmeter" that gives a reading proportional to the average value of the voltage loops. The average-value reading is a measure of the total flux change from negative maximum to positive maximum. The "average" meter is frequently calibrated to read the same as the "effective" meter on a sine-wave supply; the difference of the instrument readings in iron-loss testing thus gives directly a measure of the departure of the voltage wave form from the standard sinusoidal shape.

The losses in magnetic materials are sensitive to changes of temperature. Accordingly, the A.S.T.M. specifies that the tests for iron loss shall be made at a temperature of $25°C \pm 5°$. In addition, there are some stipulations regarding details of testing for which reference should be made to the standards.*

16-14. Treatment of Iron-loss Data

For some comparison purposes, the loss in a sample of electrical sheet steel is specified for the standard conditions $B_m = 1$ and 1.5 webers/m^2 (10,000 and 15,000 gausses), and $f = 60$ cycles per second. Generally, however, more complete information is desired. This may take the form of a set of measurements (Run 1) in which B_m is held constant at 1 weber/m^2 and the frequency is varied, and another set (Run 2) in which frequency is held at 60 cycles, and B_m is varied. Additional densities and frequencies may be used if a complete set of curves is desired.

The first step in the treatment of the data is to reduce the loss to a unit basis, either watts per kilogram, or watts (or ergs per second) per cubic centimeter. Values are frequently expressed in this country on a watts per pound basis.

Loss in steel with alternating magnetization is caused in part by hysteresis and in part by eddy currents. The total loss measurement may be separated by treatment of the data of Runs 1 and 2 indicated above. In the following material the symbol W will be used to represent the loss *per kilogram*. The relationships are expressed as follows:

$$W = W_h + W_e \tag{54}$$

*Reference 10.

where

W_h = hysteresis loss, watts per kilogram

$$= K_h f B_m{}^x \tag{55}$$

and

W_e = eddy-current loss, watts per kilogram

$$= K_e f^2 B_m{}^2 t^2 \tag{56}$$

In these equations f and B_m have the meanings that were previously given to them and

K_h = hysteresis coefficient
K_e = eddy-current coefficient
t = thickness of laminations in meters (centimeters)

If we insert (55) and (56) in (54) and divide by f, we have

$$\frac{W}{f} = (K_h B_m{}^x) + (K_e B_m{}^2 t^2)f \tag{57}$$

Therefore, if we take the data of Run 1, in which B is constant, the quantities in parentheses are constant, which means that we may expect a straight line if W/f is plotted against f. This is shown in Fig. 16-18. The

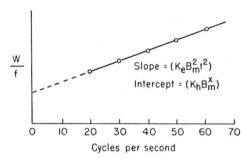

Fig. 16-18. Separation of hysteresis and eddy-current losses.

intercept on the vertical axis is $(K_h B_m{}^x)$ and the slope is $(K_e B_m{}^2 t^2)$. The readings should cover as wide a frequency range as possible (at least 30 to 60 cycles and preferably 20 to 70) in order to make the extrapolation of the curve definite. The slope is determined from the curve, and if t is known and $B_m = 1$ weber/m² (10,000 gausses), K_e may be found.

The data of Run 2 may be used to determine K_h and x in the hysteresis-loss equation. First reduce each test value in Run 2 to a watts/kilogram basis. Compute B_m and then the eddy-current loss from the K_e as found above. Subtraction of W_e from the total gives W_h. By logarithms of (55)

$$\log W_h = \log (K_h f) + x \log B_m \tag{58}$$

Plot $\log W_h$ against $\log B_m$ (which is most easily done on log-log paper). This, as indicated in Fig. 16-19, should give a straight line, since f is constant for this run. The slope of the curve gives the exponent x. (x is frequently specified as 1.6; however, it will be found to vary with different materials and different ranges of flux density.) Theoretically we can find $(K_h f)$ by extending the straight line back to the ordinate for which $\log B_m = 0$, or $B_m = 1$. Unfortunately, this is a very great extrapolation, so the graphical method is not satisfactory. Computation is probably better, even though it does not remove all uncertainty. Take one of the experimental points that lies accurately on the curve in Fig. 16-19, or better, read values of W_h and B_m for a point on the curve. Substitute these values and the known f and x in eq. (55) and solve for K_h.

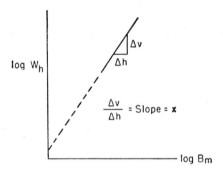

Fig. 16-19. Determination of hysteresis constants.

REFERENCES

1. Golding, E. W., *Electrical Measurements and Measuring Instruments*, London: Sir Isaac Pitman and Sons, Ltd., 1948.

2. Laws, Frank A., *Electrical Measurements*, New York: McGraw-Hill Book Co. Inc., 1938.

3. Spooner, Thomas, *Properties and Testing of Magnetic Materials*, New York: McGraw-Hill Book Co. Inc., 1927.

4. Smith, G. S., "A New Magnetic Flux Meter," *Trans. A.I.E.E.*, **56**, 441 (1937).

5. Smith, G. S., "Use of Bismuth-Bridge Fluxmeter for A-C Fields," *Trans. A.I.E.E.*, **58**, 52 (1939).

6. Pearson, G. L., "Magnetic Field-Strength Meter Employing the Hall Effect in Germanium," *Rev. Sci. Instruments*, **19**, No. 4, 263 (April 1948).

7. "Notes on Moving Coil Galvanometers," *Note Book ED(1)*, Philadelphia: Leeds and Northrup Co., 1943.

8. Bozorth, R.M., "Magnetism," *Rev. Mod. Phys.*, **19**, No. 1, 29 (Jan. 1947).

9. Harris, Forest K., *Electrical Measurements*, New York: John Wiley & Sons, Inc., 1952.

10. "Standard Methods of Testing Magnetic Materials," *A.S.T.M.*, A34–55.

11. "Standard Methods of Test for Normal Induction and Hysteresis of Magnetic Materials," *A.S.T.M.*, A341–55.

12. "Standard Methods of Test for Alternating Current Core Loss and Permeability of Magnetic Materials," *A.S.T.M.*, A343–54.

13. "Standard Methods of Test for Alternating Current Magnetic Properties of Laminated Core Samples," *A.S.T.M.*, A346–58.

CHAPTER 17

Electrical Indicating Instruments

17-1. What an Indicating Instrument Does

An electrical indicating instrument is used to permit determination of quantitative information regarding an electrical entity of some sort. Since electrical quantities are not, in general, directly visible, it is necessary to utilize some property of the quantity to develop a force that, in turn, produces an indication. Frequently a force is produced by the electromagnetic action between currents or between a current and the field of a permanent magnet, and the force is used to rotate a shaft against a spring force and to deflect a pointer on a calibrated scale. In some instruments measurement is made by the force action between electric charges, and in still others the heating effect of an electric current is used to give a measurable effect. The production of torque by electromagnetic action, however, is by far the most generally used method.

Instruments with a rotatable element require, with a few exceptions, the following components:

(a) Deflecting torque.
(b) Control torque.
(c) Damping torque.

17-2. Supporting the Moving Element

The force or torque developed by the moving element of an electrical instrument is necessarily small in order that the power consumption be kept to the lowest possible level, in order that the introduction of the instrument into a circuit may cause the minimum change in circuit conditions. As a consequence of the low force level available, however, the problem of supporting the moving element becomes of great importance. With the actuating force so small, the friction of bearings or pivots must be exceedingly minute or else the action will be erratic; the instrument pointer will stick at one point of the scale, then jump to another, and will generally give an erratic and unreliable performance.

Several types of support are used, depending upon the sensitivity required and the operating conditions to be met. Support may be of the following types:

(a) Suspension.

(b) Taut suspension.

(c) Pivot-and-jewel bearing (double).

(d) Unipivot.

The first two types are customarily used in instruments of the galvanometer class, but there is no sharp dividing line between galvanometers and indicating meters. Some sensitive wattmeters, electrometers, and electrostatic voltmeters use flexible suspensions. Type (a) consists of a fine, ribbon-shaped metal filament for the upper suspension and a coil of fine wire, giving negligible restraint, for the lower connection (in case a second connection is needed for current-carrying purposes). This type of suspension requires careful leveling of the instrument, so that the moving element hangs in its correct position. This construction is therefore not suited to field use and is employed only in those laboratory applications in which very great sensitivity is needed. A clamp to support the moving element during transit is always provided.

Type (b) has the flat-ribbon suspension both above and below the moving element, with the suspension kept under tension by some sort of spring arrangement. Exact leveling is not required if the moving element is properly balanced. The taut suspension has been used in many galvanometers of portable type, and in some indicating instruments. A double-filament form, used in Greibach instruments, provides up to four current connections, as well as restraining torque and support; other designs have a single strip at each end of the moving coil. In either case the coil deflection depends upon flexure, and not on a pivot moving in a bearing, and hence has the freedom from friction needed for good operation of a high-sensitivity movement. One form of taut ribbon suspension which has been developed for instrument use will be discussed in the following section.

Type (c) is the support commonly used in all kinds of indicating instruments and is of such great importance that it will be studied in detail in paragraph 17-4. In the Unipivot construction of Type (d), used in some milliammeters or microammeters, the coil is of circular shape, with the pivot extending inward, giving a single support at the center of the system. This configuration is not sensitive to small discrepancies in levelness of the instrument.

17-3. Taut Band Suspensions

Figure 17-1 illustrates a development of the taut band suspension, consisting of a short band at each end, kept under tension by strip springs.*

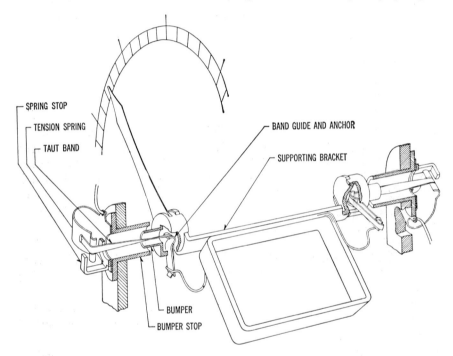

┌ SPRING STOP

┌ TENSION SPRING   ┌ BAND GUIDE AND ANCHOR

┌ TAUT BAND   ┌ SUPPORTING BRACKET

└ BUMPER

└ BUMPER STOP

Fig. 17-1. Diagrammatic view of an instrument movement using a taut band suspension. (*Courtesy of the Westinghouse Electric Corp.*)

The suspension shown here is for a d-c movement with a 250° scale, using a concentric magnet assembly. Galvanometers with taut suspensions had to be used with the coil axis vertical, because the suspensions were too flexible to give proper support in the horizontal position. In the new instrument design, however, the light-weight coil is supported by a rather wide strip under enough tension so that the axis may be placed in either vertical or horizontal position; in the horizontal position the sag is no greater than the side play in a pivot-and-jewel support.

---

*Reference 2.

The coil and mounting have stops that limit axial or sidewise motion in the event of severe shocks; this form of suspension is immune to treatment that would injure a pivot-and-jewel bearing. The amount of twist in a short ribbon (250° in a 0.4-in. length in the design shown) appears rather great at first sight. However, instruments of this design have been put through several million cycles from zero to full scale with no indication of fatigue or permanent set. Pivots and jewels would be badly worn for much shorter service.

One great feature of this flexure type of support is the entire absence of sliding friction in a bearing. Instruments with coils supported on pivots have been judged by the torque-to-weight ratio of the moving part, as a measure of performance in the presence of friction. This criterion has no significance for suspension movements, since there is no bearing friction. The absence of even the small friction of jewel bearings makes possible very sensitive meters of high performance, drawing very small power from the measured source.

17-4. Mechanical Features of Pivot-and-jewel Bearings

Since the double pivot-and-jewel arrangement is used in so many indicating instruments, and since indicating instruments of various sorts play such an essential part in so many branches of work, it is important that the construction and behavior of the bearings be understood. The bearings have an apparent simplicity that may cause their true importance to be overlooked. Also, they can be badly abused, and the accuracy of the metering function may be lost by unintelligent handling of the instrument.

Figure 17-2 shows a photomicrograph of a pivot. This pivot was made from a steel wire 0.015 in. in diameter, a fact that establishes a scale for the picture. In manufacture, the wire is first machined to a conical point and cut to length in a small lathe, after which the pivots are heat-treated to give

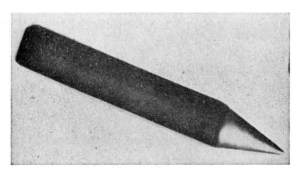

Fig. 17-2. Photomicrograph of an instrument pivot (diameter 0.015 in.). (*Courtesy of Daystrom, Inc., Weston Instruments Division.*)

them the desired degree of hardness. A large quantity of pivots is placed in a "tumbling barrel" together with lead shot and fine abrasive and tumbled for a period of time that controls the resulting radius of the pivot end. A pivot of this sort is pressed into the hollow shaft of the moving element. Figure 17-3 illustrates the general mounting arrangement for a d-c instrument.

Fig. 17.3. Enlarged view showing details of top bearing. The hardened steel pivot bears in the conical aperture of the sapphire jewel, which in turn is set in a finely threaded screw for adjustment. (*Courtesy of Daystrom, Inc., Weston Instruments Division.*)

The pivot radius used in an instrument depends on the type and weight of the moving element and upon the conditions of use. The radius varies from about 0.0005 to 0.002 in. for most meters. A large radius increases the load-carrying capacity of the pivot, but also increases the effective radius of the friction torque. Airplane instruments frequently use a radius of 0.005 in. on account of the heavy bearing loading caused by the vibration; however, the vibration in this case also serves to prevent sticking, which might ordinarily occur with so large a radius.

Bearings of precision instruments have customarily been made of sapphire — originally natural sapphire and later synthetic. The combination of steel pivot and sapphire gives the lowest value of friction. In recent years glass has come into use as jewel material for some instruments. Action in this direction was accelerated greatly during the war when simultaneously foreign sources of sapphire were cut off and an enormous increase took place in the number of instruments needed. Different kinds of glass and methods of production were tried, leading to the production of glass jewels that are satisfactory for many types of service. Glass has the advantage of being relatively cheap and also of being easily shaped. A small-diameter glass rod is heated and a forming tool is pressed into the end, producing the desired shape of cavity. Sapphire, on the other hand, must be ground to shape. The grinding process is rather slow and difficult because of both the hardness of the material and the nature of the grinding process. In this process the relative velocity between the tool and the jewel is zero at the center, just where the jewel shape is most important.

Figure 17-4(a) is an idealized picture of pivot and jewel, indicating their geometric shapes. The jewel has a somewhat larger angle than the pivot and a larger radius. In this idealized case there is only point contact between the two members, and this condition is of course impossible. The moving element, though light, does have finite weight and, accordingly, must have a finite area of support. The actual condition is more as shown in Fig. 17-4(b) in enlarged form. The sketch indicates deformations of both jewel and pivot, as must occur. (The sketch is qualitative only, with no promise as to the exact shape of the contact line.)

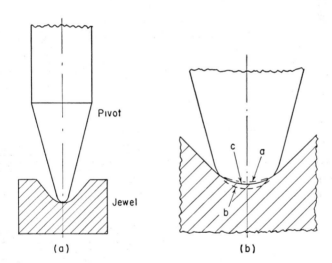

Fig. 17-4. Diagram of pivot-and-jewel bearing. (a) Showing geometric shapes. (b) Enlarged view, indicating deflection of surfaces under load: *a*, unloaded jewel surface; *b*, unloaded pivot surface; *c*, surface of contact.

If the pivot loading is moderate, the deformations shown in the figure are in the elastic range for both materials, and their chief effect is to produce a finite radius of action for frictional forces. The designer should proportion the bearing so that the stresses are in fact well within the elastic limit for static conditions. Conservative operating loads are important since the effect of any shock or rough handling of the instrument is a possible several-fold increase of the stress.

The high value of the pivot-point stress is not generally appreciated, probably because of the apparent small size and light weight of the moving element. Suppose we take as example a d-c moving-coil instrument of medium size, with a movement that weighs 1 gram (0.0022 lb). If we assume the circle of contact, as in Fig. 17-4(b), to have a diameter of 0.0003 in., the *average* stress over the area is about 30,000 lb/in.[2]! Even with a somewhat larger circle the stress at the center may reach a very high value. Moreover, if an instrument is subjected to a shock (such as by laying it down roughly on a table), the dynamic effects may cause the stress to be two or more times the static value. It is easy to see that rough handling of an instrument may cause the pivot to be deformed permanently or the jewel to be cracked. Such an accident increases the normal pivot friction enormously and gives unsatisfactory action until repairs are made. Jewels are sometimes spring-mounted to lessen the likelihood of damage from shock. The effect of impact loading on the combination of sapphire jewel and steel pivot is generally the deformation of the pivot, since sapphire is harder than steel. The jewel may, however, be cracked by the heavy load. It is important that these facts be kept in mind, to the end that instruments be handled carefully and not be subjected to avoidable shock.

Instrument bearings are used dry. Any lubricant, such as oil, would be of questionable value in the first place and would soon cause trouble by gumming and by collecting dust.

17-5. Orientation of the Moving-element Axis

If an instrument is used with the axis of its moving element in a vertical direction, as is done with a portable instrument when placed with the scale horizontal, the conditions are as pictured in Fig. 17-5(a). The entire load is carried on the lower bearing, the pivot rides centrally in the jewel, and the top bearing serves as a guide to keep the movement in alignment. The upper jewel cannot be tight on the pivot because a little play must be allowed to avoid undue friction and binding. The top end of the shaft may move slightly, but this causes no difficulty, as the motion is small and the small side force can produce little friction.

With the axis horizontal, as in most switchboard instruments, conditions are as shown in Fig. 17-5(b). The jewels cannot be so tight as to hold the moving element exactly in line and, moreover, there would be no supporting force if the axes were in line. Some play is needed, so the moving element takes a position with its axis below the axis of the jewels, as shown in exaggerated form in the sketch. The support is now on the *side* of the pivot and not on the end. If the moving element is given a deflecting torque, the general tendency is for the pivot to roll within the jewel. Figure 17-5(c) represents section *A-A* taken through the point of contact in (b). The small circle is the pivot and the larger circle is the section of the jewel. As

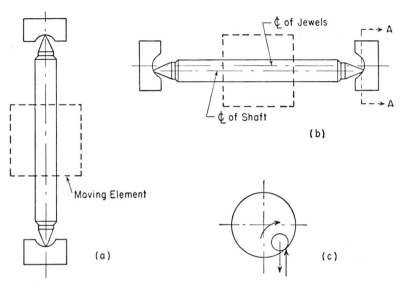

Fig. 17-5. Diagram showing effect of instrument position on bearing action. (a) Axis vertical. (b) Axis horizontal. (c) Enlarged section, *A-A.*

the moving element rotates, the pivot rolls up the jewel to a certain extent and then may, particularly with vibration, slide back to the bottom, thus introducing an element of uncertainty in the behavior. When the pivot is in the position shown in Fig. 17-5(c), there is a torque produced by the weight of the movement acting about the point of contact; this torque adds to the effect of the control springs and gives a change of indication. Electrical indicating instruments are, in general, less accurate when used with the axis horizontal rather than vertical.

17-6. Torque-weight Ratio

The torque-weight ratio is a constant that is of great interest to the designer of an electrical instrument, as it is to a considerable extent a measure of the merit of a design; at least, it sets a limit beneath which a good design should not go. A high ratio indicates, other things being equal, less trouble from bearing friction. Suppose, for example, that we take a d-c milliammeter, remove the control springs, and substitute more flexible springs. The meter is now more sensitive, that is, it gives full-scale deflection for a smaller current and torque than before. The weight is the same, so the torque-weight ratio is smaller. If we go too far in this direction, we have finally a very sensitive instrument; yet it is less reliable because friction effects are more prominent since other torques have been reduced. Adherence to a minimum torque-weight ratio guards against trouble of this sort.

The torque-weight ratio may be expressed in different units. One set that is frequently used is as follows:

$$\frac{\text{Torque in milligram-centimeters per } 100° \text{ deflection}}{\text{Weight in grams}}$$

The ratio covers a wide range for different types of meters; naturally it is held to higher levels for high-accuracy instruments than for those of low accuracy. One maker uses the criterion, derived from experience, that the torque-weight ratio (with units as above) should be equal to 80 divided by the accuracy rating of the instrument. That is, for a "$\frac{1}{2}$ per cent meter" the ratio should be 160, for "2 per cent meter" it should be 40, etc. This is not a hard-and-fast limit, since reasonably good instruments can be made with possibly half these values, but trouble would probably result with much further reduction.

17-7. Damping

Proper damping is an essential property of a good indicating instrument. If a voltmeter, for example, has very little damping and is suddenly connected to a line, the pointer swings far beyond its correct reading, then swings back too far in the opposite direction, and settles to rest only after many oscillations. This is time-consuming and annoying to the user and may practically prevent the taking of readings if the line voltage has frequent variations. Damping is second to accuracy as a necessary quality in an instrument, but not much inferior in importance.

Damping may be accomplished in several ways, which may be placed in two general classifications insofar as indicating instruments are concerned. These are air-vane damping and electromagnetic damping (of which there are several forms).

(a) *Air-vane damping.*

A light aluminum vane, mounted on the moving element, swings in a close-fitting stationary chamber. The amount of damping action depends to a considerable extent upon the clearance between the two members. (The clearance must be great enough, as a minimum, to prevent danger of actual rubbing, which would introduce serious friction.) This device may be considered analogous to an oil dashpot for damping, except that air is used as the damping fluid instead of oil. This type of damping is shown in Figs. 17-17, 17-18, 17-19, and 17-20.

(b) *Electromagnetic damping.*

Damping is secured in some d-c permanent-magnet moving-coil instruments by winding the coil on a light aluminum frame. Circulating currents are set up in the frame as the coil deflects, thus giving a retarding torque proportional to the velocity. In some cases the conducting frame is not used, but a resistor is connected across the coil terminals, giving a path for circulating currents and damping action as before. Further study of this action will appear in the section on d-c instruments.

Magnetic damping of another sort is illustrated in Fig. 17-6. An aluminum vane attached to the moving element passes between the poles of a permanent magnet, thus producing a damping torque proportional to the velocity.

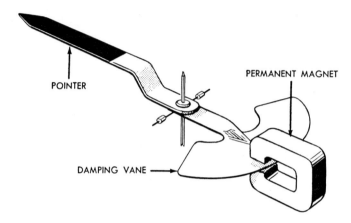

Fig. 17-6. Magnetic-damping mechanism. (*Courtesy of General Electric Company.*)

A mathematical study of damping was made in chap. 5 with respect to galvanometers, and the results may be applied directly to indicating instruments. It is shown there that damping may fall into three classes:

I. Underdamped or oscillating case.

II. Overdamped case. The pointer approaches the final position slowly. Sluggish action.

III. Critically damped case. The pointer comes to its final position in the minimum time that may be obtained without overshooting.

In the damping of an electrical instrument it is generally desired to produce a condition close to case III. However, there is a preference for damping slightly on the underdamped side of case III, so that the pointer overshoots its final position a small amount, then settles back to the reading without further noticeable oscillation. The slight overshoot gives assurance to the user that the instrument has not been injured as the result of rough handling.

17-8. Errors and Guarantees

Some general features with regard to the errors in electrical indicating instruments can be studied at this time, apart from special considerations that apply to individual types. A knowledge of the different kinds of errors that exist in a meter and how they vary along the scale is essential to the intelligent use of the device. Too many users record instrument readings without critical appraisal of their reliability, apparently assuming that because the pointer appears to indicate a particular value, that value must be correct. Neither the instruments nor those who use them are infallible.

Some errors affect the instrument indication about equally at all parts of the scale, while others increase in proportion to the reading. Figure 17-7

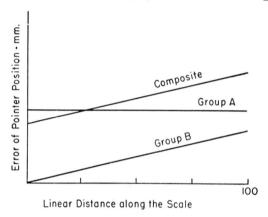

Fig. 17-7. Curves showing variation along the scale of various types of instrument error.

illustrates these conditions. The abscissa represents the scale of the instrument. Errors, plotted as ordinates to a linear scale, represent the actual distance the instrument pointer is from its true position. The errors are classified into two groups:

Group A

This group comprises those effects that tend to produce errors of the same magnitude at any point of the scale. It includes:

(1) *Scale error.* The scale markings may not be exactly in the correct place because of insufficient care during calibration or small variations by the draftsman in inking the scale markings. Low-priced instruments have mass-produced scales that do not account for the idiosyncrasies of individual instruments. Scale variations are equally probable at any point on the scale.

(2) *Zero error.* Failure of the user to adjust the pointer to zero before

taking readings causes an equal-distance error all along the scale. This matter should always be checked before taking readings.

(3) *Reading error.* The user does not always judge the reading exactly. Precision of reading becomes more difficult if the pointer has a broad tip. Interpolation errors may enter, particularly if the scale markings are too large or too small.

(4) *Parallax error.* This is a reading error caused by the observer not having his line of sight on the pointer exactly at right angles to the plane of the scale. High-grade testing instruments are equipped with a mirror beneath the scale and a knife-edged pointer to help eliminate this error.

(5) *Friction error.* The friction of the pivot in the jewel may cause the pointer to come to rest a short distance from its correct position. This error is small for a bearing in good condition, but may become very large if the pivot is badly worn or if it has been damaged by rough handling of the instrument. Friction is more serious for sensitive instruments designed for low operating torque. The user can minimize frictional effects by tapping *gently* on the case with his finger tip before taking a reading.

Group B

This group includes those effects that produce errors proportional to the pointer deflection.

(1) Incorrect resistance in voltmeters, ammeter shunts, potential circuit of wattmeters.

(2) Effect of temperature in changing the resistances mentioned in (1), even if they were correct at their initial adjustment.

(3) Effect of temperature on the characteristics of the control springs. An increase of temperature makes the springs more flexible in behavior.

(4) Effect of temperature on the strength of permanent magnets in d-c instruments. An increase of temperature causes a decrease of strength of a magnet.

(5) Effect of frequency in a-c instruments. A voltmeter, for example, has inductance in its operating coil, which means that the circuit as a whole cannot be purely resistive. The circuit impedance is greater for high than for low frequencies, thereby causing error proportionate to scale reading at a given frequency. Frequency errors enter in various ways in different instruments.

The total error experienced in using an instrument is made up of a number of the effects discussed above. It is not possible to make up a definite total curve, as the components enter to different extent in various cases. The "composite" curve of Fig. 17-7 illustrates the general effect that may be expected in combination of Groups A and B. Smaller errors may be anticipated at midscale than at full scale but not in proportion to the readings. If the instrument is a d-c voltmeter, the error in volts is proportional to the ordinate of the composite curve, because of the linearity of the scale. In an a-c voltmeter, however, a 1-v division is smaller in the lower part of the range, so the composite curve interpreted in volts would tip upwards at the lower part of the range.

The upshot of the above discussion is that the error at different parts of the scale is more nearly constant in actual *amount* than it is as a *percentage of the reading* being taken. Instrument makers recognize this fact and guarantee the accuracy of the instruments in percentage of *full-scale* reading. The abbreviated marking "Accuracy: 0.5 per cent" really means that the error at any point of the scale will not exceed 0.5 per cent of the *full-scale* reading. This fact, not always understood, has important consequences in the use of electrical instruments. The percentage accuracy may be very poor for a reading taken on the lower part of the scale. For example, the reading taken at 20 per cent of the scale on a "0.5 per cent" ammeter may be in error by 2.5 per cent of the observed value and still be in accord with the maker's guarantee. This emphasizes the importance of selecting an instrument, when possible, so that the reading comes well up on the scale.

D-C PERMANENT-MAGNET MOVING-COIL INSTRUMENTS (PMMC)

17-9. Principles; Types of Construction

The permanent-magnet moving-coil instrument, sometimes referred to as the d'Arsonval movement, is the type used most generally for d-c measurements. There are a few other kinds applied where simplicity and low cost are the first requisites, as in the charge indicator on the dash of an automobile, but they are not used where any degree of accuracy is required.

The action of the PMMC instrument depends on the force developed on a current-carrying conductor in a magnetic field. A multiturn coil of fine wire is pivoted to turn in the air gap of a permanent magnet. The torque produced by the coil is measured by the deflection it gives against the opposing torque of the "control springs" (which serve also to conduct current to and from the coil). The PMMC instrument is thus a type of "spring balance" which measures a current in terms of the force it produces in a constant magnetic field. Figure 17-8 is a phantom view of a standard type of PMMC movement, showing a number of features of construction. First, there is the permanent magnet of horseshoe form, with soft-iron pole pieces attached to it. In the center is a cylinder of soft iron to give a uniform radial field and to reduce the air portion of the magnetic flux path. The coil can be seen, also the pointer (made of aluminum tubing), and the front control spring. Figure 17-9 is a separate view of the moving coil, showing

Fig. 17-8. View showing construction features of a permanent-magnet moving-coil instrument. (*Courtesy of Daystrom, Inc., Weston Instruments Division.*)

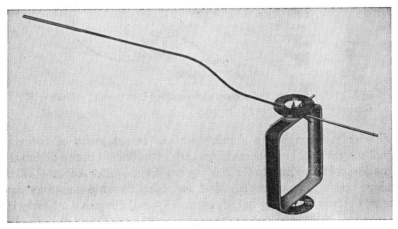

Fig. 17-9. Moving element of a permanent-magnet moving-coil instrument. (*Courtesy of Daystrom, Inc., Weston Instruments Division.*)

the frame, winding, control springs, and pointer. There are three balance arms attached to the coil, two in the plane of the coil and a larger one at

180° from the pointer. The balance weights consist of small springs that grip the balance arms and can be moved in or out as required to give balance. It is important to have the moving element in balance in order to assure that the position of the pointer does not change in case the instrument is placed at various angles in the vertical plane. The Y-shaped member is the zero-adjuster, which is attached to the fixed end of the front spring and which engages an eccentric pin through the instrument cover so that the zero can be adjusted from outside the case. The springs are made of phosphor bronze, which is a good spring material and not subject to magnetic forces as a steel spring would be.

Figure 17-10 shows a different magnet arrangement, although the coil is much the same as before. This design uses short, broad magnets and is

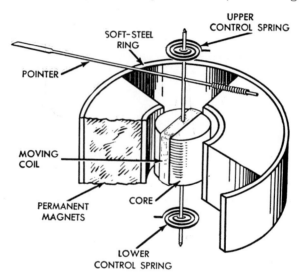

Fig. 17-10. Diagram of a concentric-magnet instrument. (*Courtesy of General Electric Company.*)

possible only through the development in recent years of permanent magnet materials of high coercive strength. One benefit of this design is the magnetic shielding given to the moving coil by the outer soft-steel ring. A standard instrument, unless shielded, is subject to error caused by strong external fields, particularly in the neighborhood of bus bars that carry large currents.

An exploded view of a concentric-scale instrument is shown in Fig. 17-11, which needs little comment. Note that the magnet is magnetized radially, that is, the entire inner surface of the ring-shaped magnet is one pole, and the soft-steel core is the other pole. Only the outer side of the moving coil is active in producing torque, as the inner side is inside the

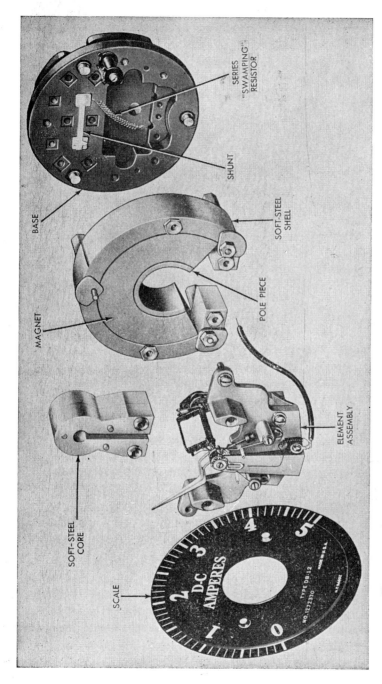

Fig. 17-11. Exploded view of a concentric-scale instrument. *(Courtesy of General Electric Company.)*

middle core and has no field (and also is on the axis). By this construction the coil travel is about 240°; thus a long scale is obtained in an instrument of small over-all dimensions.

Figure 17-12 shows an instrument in which the magnet is inside the moving coil, and the outside member is a soft-steel ring to complete the

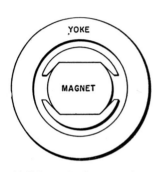

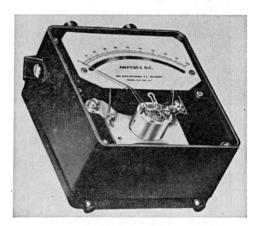

(a) Schematic diagram of the core-magnet magnetic system.

(b) Interior view of an ammeter, showing the core-magnet mechanism.

Fig. 17-12. Construction of a permanent-magnet moving-coil instrument with the magnet inside the moving coil. (*Courtesy of Daystrom, Inc., Weston Instruments Division.*)

magnetic circuit. This construction is very compact and light and gives such excellent shielding from external fields that a conductor carrying 1000 amperes 16 inches from the instrument causes no readable change of indication. A conventional instrument with horseshoe magnet has a much heavier magnet and requires a large and heavy steel enclosure to provide shielding. The internal magnet, or "core-magnet" design, becomes practicable only due to modern magnetic materials of high coercive force.

17-10. Torque Relationships

The torque of a d-c indicating instrument of the PMMC type is expressed by the same equation that was developed earlier for a galvanometer. The equation may be restated with the symbols of Fig. 5-2, as in eq. (2) of chap. 5,

$$T = BnWLi \qquad (1)$$

with the same remarks as to the units as given there. L is the axial length and W the width of the coil, n the number of turns, i the current, and B the flux density in the part of the air gap in which the coil is positioned.

The equation shows several things that are of interest from the design standpoint. If a movement is being designed for a large instrument with a long and heavy pointer, greater torque should be provided than for a small movement, both to overcome the increased friction and also to maintain a satisfactory speed of response. The greater torque is obtained, in general, by an increase in coil dimensions. For a small instrument with a short pointer (probably also a low-accuracy rating) a small coil is adequate.

The equation shows also that benefit is derived from an increase in the flux density. In early instruments a density of 1000 to 1400 gausses was considered very good and was attained only by the use of a long magnet. Modern magnetic materials have made possible densities of the order of 4000 gausses.

Adjustment of magnet strength is used as a control of uniformity in quantity production. The magnets are magnetized to a higher value than desired for the completed instrument. They are then put into a test jig with a standard instrument movement and treated with alternating current until indication comes down to the standard value. This treatment serves not only to bring all magnets to a uniform strength, but also to "stabilize" them by means of the partial demagnetization, improving their permanency and making them less subject to future demagnetization from stray effects.

The power required by the PMMC movement to give full-scale deflection is surprisingly small. For instruments of fairly standardized design the power may be about 25 microwatts for a small instrument and about 200 microwatts for a large switchboard type. These values may be reduced by a factor of 4 to 10 if improved magnetic materials are used. If the instrument is restricted to use in the horizontal position (axis vertical), a further division by four may be made, and satisfactory behavior is still obtained. Designers sometimes squeeze down still more for the sake of extreme sensitivity for special uses, but further reduction may give difficulties.

17-11. D-C Voltmeters

The basic PMMC instrument considered so far is in reality a milliammeter or microammeter. It can be converted into a voltmeter by the addition of a series resistor or into an ammeter by the use of a shunt. In the process of these conversions, however, some modifications are made in the interests of increased accuracy, and as the modifications are different for voltmeter and ammeter, the discussions of the two instruments will be given separately.

If we have a movement that requires 10 ma for full-scale deflection, we can make from it a voltmeter of 300-v range by connecting enough resistance in series so that the total is 30,000 ohms; or we can make a meter of 150-v range by making the total resistance 15,000 ohms. If in either case we divide the number of ohms by the scale range, that is, 30,000/300 or 15,000/150, we have as the quotient 100 ohms/v. This is often referred to as the sensitivity of the voltmeter. The current drawn at full-scale is the reciprocal of this number, or 0.01 amp.

In some circuits the current drawn by the voltmeter becomes an important matter. A current of 0.01 amp is insignificant if we are testing a large generator, but it may change very materially the thing that we are trying to measure if the circuit itself has high resistance as, for example, many vacuum-tube circuits. Such conditions can be helped by the use of a voltmeter of higher sensitivity. Many voltmeters are made with sensitivities of 1000 ohms/v and some higher (5000 to 20,000 ohms/v, or more). The series resistance is usually mounted inside the case for voltmeters of moderate ranges, that is, up to a few hundred volts. For higher voltages the resistor is mounted separately to avoid excessive heating inside the case and is called a multiplier.

The series resistor is made of a material of low temperature coefficient, usually manganin (see Table 2, chap. 4), to avoid errors due to changes of temperature. There are temperature effects in other parts of the instrument, however, that should be considered in designing the resistor. The springs become more flexible with heat by about 0.04 per cent per °C. This effect in itself would cause the pointer to read high at higher temperature. The magnets become weaker at higher temperature, which partially nullifies the effect of the springs. The net change for spring and magnet is about 0.013 to 0.02 per cent per °C in a direction causing an increase of indication. If the voltmeter circuit were made entirely of copper, the instrument would read low at high temperature, as copper increases in resistance about 0.4 per cent per °C. If manganin, which is hardly affected by temperature, is combined with copper in the proportion of 20/1 to 30/1, the total electrical circuit increases in resistance just enough to counteract the change of springs and magnet, and the over-all temperature error for the voltmeter is practically zero. The 20/1 requirement is easily met, except that it does place a limit on how low a range can be built. For example, suppose that we have a simple PMMC movement requiring 10 ma for full scale and having a coil resistance of 2 ohms. This gives a drop of 20 mv across the coil for full-scale deflection. If we were to try to use this instrument as a voltmeter of 20-mv range, the temperature error would be extremely bad, as we would be using an all-copper circuit. We could make a voltmeter with a 500-mv or even 250-mv range with little error, but below this the temperature effect would begin to be serious. Manufacturers of these instruments like to arrange the design so that the temperature error will not be more than half the guaranteed tolerance for a temperature change of

10°C. Instruments in the millivolt range may be compensated as shown later in Fig. 17-15.

The damping of a PMMC voltmeter is obtained by winding the coil on an aluminum frame. Damping caused by circulating currents in the coil windings of voltmeters is small, in general, because of the high resistance of the circuit. If the movement were built with higher sensitivity than needed and then shunted to the desired sensitivity, the shunt would play an important part in the damping action.

17-12. D-C Ammeters

A PMMC movement can be converted into an ammeter by the use of a shunt to carry a portion, generally the major portion, of the line current. If the instrument resistance is known, the shunt resistance can be computed to carry any desired fraction of the line current. In low-range ammeters, especially portables, the shunt may be mounted inside the case. In high-range portables and in most switchboard ammeters, the shunt is mounted separately and connected to the instrument by calibrated leads. The arrangement of instrument and shunt is indicated in Fig. 17-13.

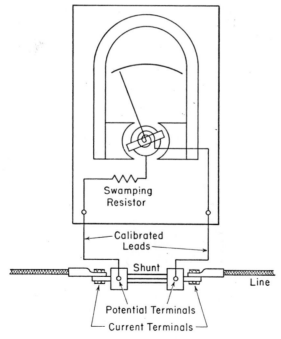

Fig. 17-13. Measurement of current by a millivoltmeter and an external shunt.

Another way of looking at the situation is to regard the PMMC movement as a millivoltmeter that measures current in terms of the potential drop across the shunt. This gives a better idea of the problem in some respects than the parallel-resistor picture and helps in computations.

Instruments and shunts of different ranges can be designed to work interchangeably, if they are designed for the same full-scale voltage drop. Thus, a 50-mv instrument may be used with any shunt designed for a 50-mv drop when carrying rated current. A "10-amp 50-mv shunt" is one that gives a drop of 50 mv when the line current is 10 amp and similarly for other ratings. The resistance of a 100-amp 50-mv shunt is practically 0.050/100 = 0.0005 ohm. For low ranges the computation should take into account the current drawn by the instrument, but since that is a small quantity, probably 5 to 30 ma, it has negligible effect in the high ranges. A drop of 100 mv is used in some applications (e.g., for greater accuracy) for some recording instruments or to permit running the instrument leads to greater distances.

A high-capacity (4000-amp) shunt is pictured in Fig. 17-14. The resistance element consists of sheets of manganin imbedded in end blocks

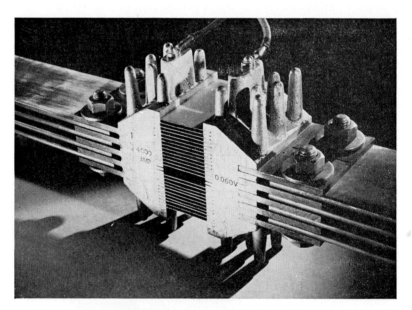

Fig. 17-14. High-capacity external shunt for switchboard service. (*Courtesy of General Electric Company.*)

of copper. Note the separate terminals for the potential leads that go to the instrument. The idea of separate current and potential leads is used on all shunts, even of small current rating, so that the instrument reading is determined only by the IR drop of the shunt and remains unaffected by any contact drop that may exist at the current terminals.

One feature in Fig. 17-13 that should be noted is the resistor in series with the instrument coil. This resistor is made of manganin and is included to reduce the proportional effect of temperature on the resistance of the millivoltmeter circuit. It is frequently called a "swamping resistor" because its main function is to "swamp out" the effect of the resistance change in the copper coil.

A more complete cancellation of temperature effects can be accomplished by the circuit of Fig. 17-15. The resistance of the circuit as a whole

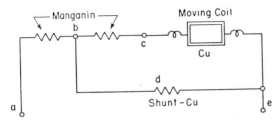

Fig. 17-15. Circuit for temperature compensation of a millivoltmeter.

increases slightly with temperature owing to the presence of copper; therefore the total current is reduced for a fixed value of voltage applied from a to e. However, the branch b-d-e, being all copper, increases more rapidly in resistance than does branch b-c-e, which is part manganin, so a larger fraction of the total current is passed through the moving coil. Correct proportioning of the parts makes possible the cancellation of temperature effects, including those of springs and magnet.

The moving coils for instruments used with shunts are not wound on metal frames, as this would give excessive damping. The moving coil being closed through the shunt gives a low resistance path that provides sufficient damping action.

ELECTRODYNAMOMETER VOLTMETERS AND AMMETERS

17-13. The Electrodynamometer Movement

The electrodynamometer movement is exceedingly important. It is the basic type in the construction of a-c voltmeters and ammeters both in the range of power frequencies and in the lower part of the audio-frequency range as well. Equally important, it serves as a transfer instrument, for it may be calibrated on d-c and then used on a-c, thus giving a connected system of voltage and current magnitudes for measurements with d-c and a-c. The same mechanism is used as wattmeter and varmeter and, with some modification, as power-factor meter and frequency meter.

The idea of this mechanism may be obtained by starting with the PMMC movement and considering how it would behave on a-c. It would have a torque up-scale during one half of the cycle and an equal effect in the opposite direction during the next half cycle. If the frequency were very low, the pointer would swing back and forth around the zero point. For the ordinary meter on power frequencies the inertia is so great that the pointer does not go very far in either direction but merely vibrates slightly around zero. If, however, we were to reverse the direction of the field flux each time that the current through the movable coil reverses, the torque would stay in the same direction, which can be upward along the scale for both halves of the cycle. The flux can be made to reverse at the correct time by obtaining it from a field coil in series with the movable coil. The action of the meters can be likened to that of a d-c motor. If the field flux stays in the same direction, the direction of rotation of a motor reverses if the armature current is reversed in direction. If field and armature are reversed simultaneously, the torque and rotation remain the same. The electrodynamometer mechanism has several points of resemblance to a series-connected motor.

The general arrangement of the parts is indicated in Fig. 17-16, which shows the field coil divided in two sections to give a more uniform field near

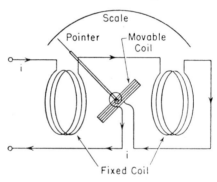

Fig. 17-16. Diagram of an electrodynamometer movement used as a milliammeter.

the center and to allow passage of the instrument shaft. The instrument as shown may be a milliammeter, or may become a voltmeter by the addition of series resistance. Figure 17-17 is a phantom view of an actual instrument. It shows the field coils, movable coil, control springs, counterweights and truss-type pointer. The movement is surrounded by a laminated steel shield to shield it from the effect of stray magnetic fields. The damping is provided by two vanes, shown at the bottom, which move in

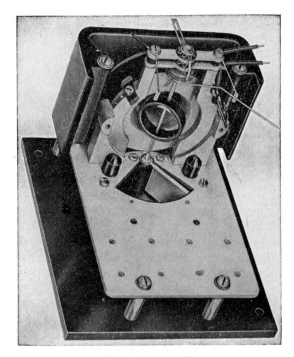

Fig. 17-17. Cutaway of Electrodynamometer movement. (*Courtesy of Sensitive Research Instrument Corporation.*)

sector-shaped chambers. The movable coil is wound either as a self-sustaining coil or else on a nonmetallic form. A metallic form cannot be used, as eddy currents would be induced in it by the alternating field. It can be seen that great care has been taken to make the construction rigid and to avoid any shifting or change of dimension that might affect calibration.

The action of the electrodynamometer mechanism may be analyzed by means of equation (1), above. In this case B depends upon i; therefore, the torque becomes a function of i^2. If this instrument is used on direct current, the most noticeable item is the spreading of the scale by approximately a square law (modified by the effects of angularity between the axes of the fixed and movable coils). If used on alternating current, the torque at any instant is proportional to the square of the current at that instant. As the current varies, the torque varies also (but always in the same direction). The coil has too much inertia to follow the individual torque pulsations but takes up a position in which the average torque is balanced by the torque of the spring. The pointer position is thus a function of the mean-square current, but the scale, for convenience, is calibrated in terms of the square root of this quantity. The instrument reads root-mean-square or effective current by the nature of the fundamental force relationships. If the instrument is calibrated originally on direct current, and a mark is put on the scale at, let us say, 1 amp, and then used to measure an alternating current that causes the pointer to indicate 1 amp, we know that this alternating current has an *rms* value of 1 amp. We thus have a connection between a-c and d-c instruments, as an *rms* current of 1 amp produces the same heating effect in a resistor as 1 amp of d-c.

The electrodynamometer movement is much lower than the PMMC in sensitivity, or, stated in another way, has a much greater power consumption. The reason for this is easily seen, as the PMMC has a field furnished for it by a permanent magnet; hence, power is drawn from the measured circuit only for the movable coil. In the electrodynamometer type, however, power must be drawn to set up the field as well. Electrodynamometer voltmeters and ammeters require much greater power because of the field windings, which must supply a large number of ampere-turns. A higher proportion of copper than in the PMMC type is desirable on this account and is suitable also for temperature compensation. With the springs changing 0.04 per cent per °C, and the resistance of copper 0.4 per cent per °C, the copper should make up about one tenth of the circuit. The manganin-to-copper ratio should, therefore, be approximately nine to one.

Electrodynamometer instruments have a much lower field strength than the PMMC in spite of higher power consumption. Whereas a PMMC movement may have a flux density of 1000 with usual steel and 4000 with special steel magnets, the electrodynamometer may have a density of perhaps 60 gausses. The low value is to a large part the result of the use of a complete air path for the magnetic flux in most electrodynamometer movements. Some have been constructed using special, laminated steel for a portion of the flux path. This reduces the power requirements very greatly, but introduces some trouble in the avoidance of frequency and wave-form effects on the instrument calibration.

17-14. Electrodynamometer Voltmeter

The electrodynamometer movement can be made a voltmeter by the addition of a series resistor. It is the most accurate type of a-c voltmeter and is used where accuracy requirements justify it over the cheaper moving-iron type. The sensitivities are low in comparison with the d-c instruments, as explained above, and range from about 10 to 30 ohms/v, depending upon the accuracy rating and the voltage scale. There is a general tendency for a lower ohms-per-volt sensitivity as the voltage rating is reduced. When an instrument is redesigned for a lower range, the voltage for the copper part of the circuit (fixed plus movable coils) must be reduced to keep a reasonable copper ratio, and with reduced voltage the current must be increased to provide sufficient power. Compensation schemes, as shown in Fig. 17-15, are more needed for temperature compensation in the a-c than d-c instruments because of the higher copper ratio. Another reason for keeping the coil windings a small part of the circuit for alternating current is to minimize inductive effects, so that frequency will not affect the calibration too greatly. Electrodynamometer voltmeters may be used, in general, within their guaranteed accuracy from direct current to about 125 cycles per second. Instruments are available that are compensated to be within the accuracy tolerance for frequencies up to 2500 cycles per second.

17-15. Electrodynamometer Ammeter

The electrodynamometer movement indicated in Fig. 17-16 may be regarded as a milliammeter, but it would be difficult to design it for much more than 100 ma, as the lead-in spirals — given sufficient thickness to carry a larger current without appreciable heating — would become too stiff. It becomes necessary for larger currents to resort to shunting, and this raises the question of where to place the shunt. It would be undesirable to place the shunt across the entire instrument, as that would require a large drop across the shunt and would give large temperature and frequency errors. A better solution is to wind the fixed coils of a wire size that will carry the entire current and then to use a shunt across the movable coil only. The shunt voltage drop is much reduced, so allowance can be made for a swamping resistor in series with the movable coil in order to reduce the temperature error. By this arrangement it is feasible to build ammeters up to a 20-amp range or more. It would be possible to go higher if necessary; however, for large values of alternating current it is generally more desirable to use an instrument current transformer (chap. 15) and a 5-amp ammeter.

MOVING-IRON INSTRUMENTS

17-16. Principles of Moving-iron Instruments

Moving-iron instruments may be divided into two general classifications: the attraction type and the repulsion type. Each of them has several subdivisions.

(a) *Attraction type.*

If a small piece of soft iron is pivoted or otherwise mounted near a coil and current is passed through the coil, the iron vane is attracted and tends to be pulled into the stronger magnetic field inside the coil. The soft iron is attracted for either direction of current in the coil, so the principle may be used as the basis of an a-c instrument. The force is dependent upon the current (actually upon i^2), hence the addition of a control spring and scale make possible a measuring device.

A variation of the idea is to have the vane inside the coil but mounted at an angle on the instrument shaft, which in turn is at an angle from the axis of the coil. Passage of current through the coil tends to rotate the vane so that it lines up with the field direction in the coil. This construction is referred to as the inclined-coil type, as the coil is mounted at an angle with respect to the axis of the instrument shaft. Both voltmeters and ammeters of this design are in use.

(b) *Repulsion type.*

If *two* pieces, or "vanes," of soft iron are mounted close together inside a coil and current is passed through the coil, the iron vanes are magnetized, with north poles at one end and south poles at the other. Repulsion takes place between the two vanes, since like poles are adjacent to one another. The force of repulsion can be used for measuring purposes if one vane is mounted rigidly on the coil support and the other is mounted on the instrument shaft.

Figure 17-18 is a diagram of a repulsion-type instrument of the radial-vane, or "book," form. The diagram indicates the arrangement of the iron vanes inside the coil. Figure 17-19 shows the principle of the concentric-

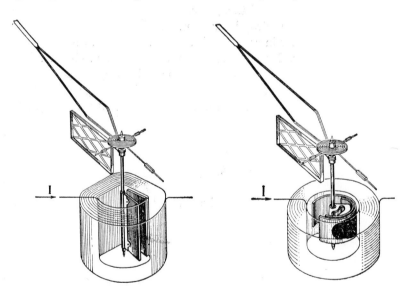

Fig. 17-18. Diagram of a radial-vane form of repulsion-type instrument mechanism. (*Courtesy of Daystrom, Inc., Weston Instruments Division.*)

Fig. 17-19. Diagram of a concentric-vane form of repulsion-type instrument mechanism. (*Courtesy of Daystrom, Inc., Weston Instruments Division.*)

vane form of movement. The fixed vane may be shaped to secure special scale characteristics, opening the scale where desired. Figure 17-20 is a phantom view of an instrument of the concentric-vane type. The picture shows details of the iron vanes, the damping vane, control spring, and pointer.

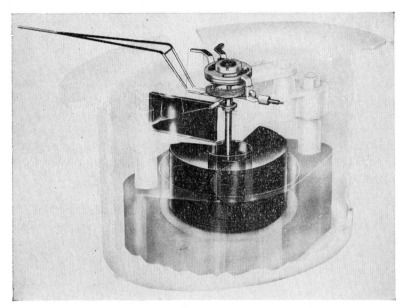

Fig. 17-20. Phantom view of a concentric-vane repulsion-type instrument mechanism. (*Courtesy of Daystrom, Inc., Weston Instruments Division.*)

The moving-iron movement has the advantages of simplicity and ruggedness. There are no current-carrying parts on the moving system, and the one control spring serves as spring only. The moving system is not damaged by even very great overloads. The same movement can be adapted to different ranges merely by change of the stationary field coil. As compared with the electrodynamometer, the moving-iron type of instrument has the advantages of ruggedness and lower cost and the disadvantage of lower accuracy.

Comparison of the two types of repulsion movement shows that the radial-vane type is more sensitive, requiring about 0.1 watt (for the movement alone) as against 0.25 watt for the concentric type, and the scale arrangement is more nearly linear. The concentric type has shorter vanes and is less subject to errors arising from residual magnetism. Both of these movements, it may be noted, are much less sensitive than the PMMC movement for direct current.

The repulsion movement is inherently a root-mean-square device, as may be seen from the fact that the force between magnetic poles, for any given position, depends on the product of the pole strengths, and each of the poles depends on the field strength, which in turn is dependent on the coil current. The instantaneous torque thus depends on the instantaneous i^2; hence the pointer position is a function of the mean-square current, and is independent, basically, of the wave form. There are frequency errors in this movement as a result of the effects of hysteresis and eddy currents in the vanes, and these errors increase with frequency. The high-frequency components of complex wave forms are accordingly not indicated quite correctly, and the instrument is not entirely independent of wave form in actual practice. The moving-iron instruments give readings on direct current, but the readings may be slightly in error because of residual magnetism in the iron vanes. This causes the indication to be high if the instrument is connected in circuit one way and low if the connections are reversed. The error may be eliminated if readings are taken with both polarities and averaged. However, it is preferable to use a PMMC movement for direct current when possible.

Figure 17-21 shows how attraction vanes may be added to the concentric-vane construction. The attraction vanes may be used to modify the scale arrangement or to give longer scale travel. Instruments of the "concentric-scale" design have been constructed with pointer travel of 240°.

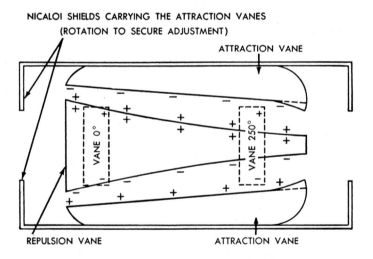

Fig. 17-21. Developed view of a repulsion-attraction system of magnetic vanes. (*Courtesy of General Electric Co.*)

17-17. Moving-iron Voltmeters

Moving-iron voltmeters of the repulsion type may be built with a movement of the sort discussed in the preceding section by using a coil consisting of a large number of turns of fine wire and adding a series resistor. The design must be worked out so that the coil voltage is not too large a fraction of the applied voltage, in order to minimize temperature and frequency errors. For this reason a low-range voltmeter is designed for fewer turns and greater current than a high-range instrument. For example, a 150-v instrument may draw a current at full scale of 25 to 60 ma, depending on the design, and a 30-v instrument from 70 to 200 ma.

A moving-iron voltmeter tends to read high with increasing temperature as the result of weakening of the spring (about 0.04 per cent per °C) and to read low because of the increase of resistance of the copper part of the circuit (0.4 per cent per °C). It is possible to proportion the circuit so that the two effects cancel. These instruments are subject to a considerable amount of "self-heating," that is, heating due to I^2R loss in the instrument circuit. It is important to reduce the temperature error so that inaccuracy is not produced when the voltmeter is left connected to the line for considerable periods of time.

A double-range instrument can be built, but the design must be based on the lower range in order that the copper ratio may be of suitable value. This is illustrated very nicely by the designs in Fig. 17-22, where (a) shows a correct design based on the 75-v range. The 150-v range is secured by adding a series resistance of 6400 ohms. In (b) is shown a single-range design for 150 v, giving about the same copper ratio as on the 75-v range in (a). This is a satisfactory design for 150 v, but if a tap were to be brought out for 75 v as indicated, the 75-v range would have 30 per cent copper, which would result in excessive temperature and frequency errors.

Moving-iron voltmeters have a frequency error as the result of eddy currents and hysteresis in the iron vanes and possibly eddy currents in the frame of the instrument. The eddy currents in the vanes reduce their pole-strength and hence cause the instrument to read low as the frequency is increased. In addition, the inductive reactance of the coil increases with frequency; this increases the impedance of the instrument circuit and reduces the current and indication for a given applied voltage. On account of these frequency effects, a voltmeter should be calibrated at the frequency at which it is to be used. The usual commercial instrument may be used within the accuracy tolerance from 25 to 125 cycles per second. Compensated designs may be secured at the expense of an increase of operating power that are good up to 500, 1000, and, in special designs, even to 2500 cycles. In consequence of the frequency error, distorted voltage waves, which may be regarded as consisting of a fundamental frequency plus

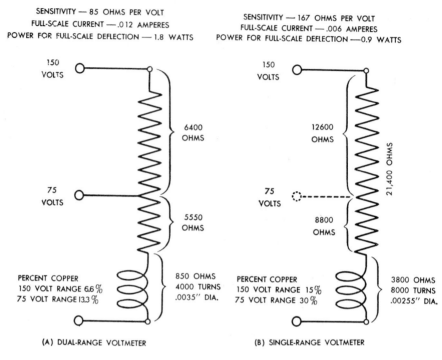

SENSITIVITY — 85 OHMS PER VOLT
FULL-SCALE CURRENT — .012 AMPERES
POWER FOR FULL-SCALE DEFLECTION — 1.8 WATTS

SENSITIVITY — 167 OHMS PER VOLT
FULL-SCALE CURRENT — .006 AMPERES
POWER FOR FULL-SCALE DEFLECTION — 0.9 WATTS

150 VOLTS

150 VOLTS

6400 OHMS

12600 OHMS

21,400 OHMS

75 VOLTS

75 VOLTS

5550 OHMS

8800 OHMS

PERCENT COPPER
150 VOLT RANGE 6.6%
75 VOLT RANGE 13.3%

850 OHMS
4000 TURNS
.0035" DIA.

PERCENT COPPER
150 VOLT RANGE 15%
75 VOLT RANGE 30%

3800 OHMS
8000 TURNS
.00255" DIA.

(A) DUAL-RANGE VOLTMETER

(B) SINGLE-RANGE VOLTMETER

Fig. 17-22. Diagram illustrating design features of the circuit of a moving-iron voltmeter. (*Courtesy of General Electric Co.*)

superposed harmonics, are not indicated correctly, and the error is generally greater than for a pure sine wave of the fundamental frequency only.

17-18. Moving-iron Ammeters

A moving-iron ammeter may be designed for any desired full-scale current, within limits, by winding with wire of adequate current-carrying capacity and putting on enough turns to give the required ampere-turns to operate the mechanism. This means many turns of fine wire for a small current rating and a few turns of heavy wire for a large rating. We meet a limitation at the low-current end in that the impedance of the coil becomes so large that insertion of the ammeter (milliammeter) in the circuit makes an objectionable change of conditions. The impedance of low-range milliammeters becomes prohibitive for many purposes below about 100 ma — a 100 ma instrument of this type has an impedance of about 50 to 100 ohms, and a 15-ma instrument about 2000 to 3000 ohms at 60 cps. The high-current limitation is set by the difficulty of winding a small coil of very heavy wire. It is better generally to use a low-range ammeter (usually 5 amp) and an instrument current transformer for measurement of large currents.

A moving-iron ammeter is subject to temperature error insofar as the control spring is concerned. The change of resistance of the coil does not affect the accuracy, but merely makes a slight change in the voltage drop across the instrument. Similarly, the frequency error of an ammeter is caused by hysteresis and eddy currents, but not by the change of impedance of the windings. Ammeters may be used with accuracy to somewhat higher frequencies than voltmeters, the usual commercial instrument being good from about 25 to 500 cycles per second. Compensated instruments are available that are reasonably accurate from 25 to 2500 cycles.

RECTIFIER INSTRUMENTS

17-19. Principles of Rectifier Instruments

Rectifier instruments are one answer to the search for an a-c voltmeter of higher sensitivity than the electrodynamometer or moving-iron types that have been studied in the foregoing sections. The general idea is to rectify the alternating current that is to be measured and then determine its value by means of a d-c instrument. Rectifier instruments generally use copper-oxide or silicon rectifiers.

The copper-oxide rectifier is based on the discovery that a layer of cuprous oxide formed on one face of a plate or disk of copper gives very different electrical conductivities in the two directions of current flow, and thereby makes possible the "rectification" of an alternating current into a unidirectional current. A perfect rectifier would allow unimpeded current flow in one direction and prevent any current flow in the opposite direction. The copper-oxide rectifier falls far short of this ideal but does have enough dissimilarity of resistance in the two directions of flow to be useful as a rectifier; it is small in size (as constructed for instrument use), needs no auxiliary power supply, and has a long life.

A rectifier for instrument work frequently consists of four rectifying elements formed into a so-called "bridge" circuit. This is illustrated in Fig. 17-23, which shows a series resistance, a bridge rectifier, and a PMMC instrument. The arrow-and-bar symbol is used for the rectifier to represent, by the arrow, the easy direction of flow and, by the bar, the hindrance to flow in the opposite direction. The solid arrows represent the conventional direction of current flow for the half cycle in which the upper terminal is positive. Note that arms b-c and a-d are conducting in the low-resistance direction and the other arms are assumed to allow practically no flow. In the second half-cycle the line polarity is reversed, and current flows in the direction of the dotted arrows, passing this time through arms d-c and a-b.

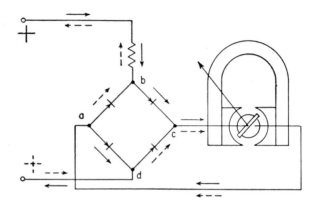

Fig. 17-23. Circuit for a single-range rectifier-type voltmeter.

However, the current flow through the instrument coil is in the same direction as before. The coil thus carries a pulsating unidirectional current which is made up in the ideal case of a series of sine loops. The instrument gives a steady deflection owing to inertia, and since this is a PMMC movement, the reading is proportional to the average value of the current flowing through it.

The last statement indicates one inherent limitation of the rectifier instrument — its indication depends on the *average* value of the current loops and not on their effective (*rms-*) value. These instruments are calibrated on alternating current of sinusoidal form, and their scales are marked in *rms*-values for this form. They should, accordingly, give correct readings as long as the measured quantity is also sinusoidal, but they do not read *rms*-values for distorted waves. There may be a large wave-form error if the measured quantity differs considerably from the sinusoidal shape.

The rectifying action depends to some extent upon frequency because of a capacitive effect at the rectifying layer. The rectifier as a circuit element may be represented by means of a varying resistance shunted by a capacitance. The capacitance causes a reduction of the rectification ratio at the higher frequencies. A rectifier voltmeter may be used up to a frequency of 10,000 to 20,000 cycles per second, or more by special design, but the error may be of the order of 10 per cent as compared with a 60-cycle calibration. Instruments of this type should be calibrated at the frequency at which they are to be used, if accuracy is an important consideration.

The indication of a rectifier instrument is also affected to a considerable extent by a change of temperature. The resistance of a rectifier decreases with increase of temperature for both directions of current flow, and the two effects cancel partially. The remaining effect, however, is of a much greater order of magnitude than is met in instruments of other types. The effect is not too serious (in comparison with other errors) for a change of a few degrees near room temperature, but is a matter of major concern if the instruments are to be used over a wide range of temperature. For accuracy in the latter case the instruments should be enclosed in a temperature-controlled box.

A solid-state rectifier has resistance characteristics of the form shown in Fig. 17-24. As a consequence, the rectification ratio is good for fairly large

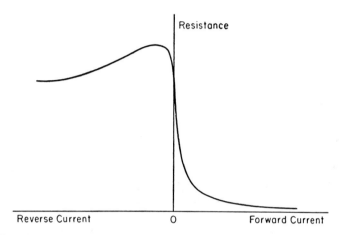

Fig. 17-24. Variation of resistance of a copper-oxide rectifier with direction and amount of current.

values of current. For very small currents, however, only a small portion of the curve is covered, so the resistance ratio and the rectification ratio are poor. An understanding of this fact is important because it explains the crowding met in the lower part of the scale of low-range instruments and also sets a limit to the sensitivity that can be attained in microammeters and voltmeters. Rectifier instruments work out to best advantage in medium- to high-range voltmeters, where the considerable series resistance swamps out the effect of changes of the rectifier resistance. Sensitivities of 1000 and 2000 ohms/volt are regularly available in voltmeters of the rectifier type — a value many times higher than can be attained in other types of a-c voltmeters. Rectifier instruments are used also for microammeters and low-range milliammeters (up to about 10 or 15 ma). For high current ranges the rectifier becomes too bulky if designed for the entire current, and shunting is not practicable on account of the change of resistance of the rectifier under different conditions of operation.

Rectifier instruments are usually guaranteed to ±5 per cent of full scale on wave forms approximating sine shape and at ordinary room temperatures. The repeating accuracy under one set of conditions of frequency, wave form, and temperature should be materially better than this value.

A rectifier undergoes a change early in its life, and this may change the calibration, particularly in low ranges, unless the rectifier is artificially pre-aged.

THERMOCOUPLE INSTRUMENTS

17-20. Principle of Current Measurement by Thermocouple

Thermocouple is the name given to the combination of two wires of dissimilar materials that have the following property: when they are joined to make a complete circuit and one junction is maintained at a higher temperature than the other, current flows in the circuit. This principle is used in temperature measurement; one junction is placed at the point where the temperature is to be measured, and the other junction is placed where the temperature is known or can be maintained at a fixed value. The same principle is used for measurement of current by placing the "hot junction" in thermal contact (but not necessarily electrical contact) with a heater carrying the current. The placement of the cold junction becomes a problem in this case. If the thermocouple wires are run back to the binding posts of the indicating instrument, thus placing the cold junction at that point, an error of measurement is introduced if the temperature there is different from that in the space surrounding the heater. For the measurement of current, the cold junction should be placed near the hot junction in such way that there is no difference of temperature between them unless the heater is carrying current. Copper wires are run from the thermocouple terminals back to the instrument, as illustrated in the diagram of Fig. 17-25. (The thermocouple wires are drawn with different weights of line merely as a reminder that they are of different materials.) The thermocouple wires are frequently spot-welded or hard-soldered to the center of the heater, but sometimes are fastened in place by a drop of insulating material. The insulated construction serves to isolate the indicating instrument electrically from the a-c circuit, but has the disadvantage of making the instrument slow to respond to changes of current in the measured circuit. A thermocouple reads the root-mean-square value of current regardless of wave form, since the action is a matter of the heat (I^2R) liberated in the heater.

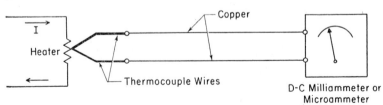

Fig. 17-25. Diagram of a thermocouple ammeter.

17-21. Construction of Thermocouples

For low-current measurements the heater consists of a short piece of small-diameter resistance wire mounted in an evacuated glass tube. A typical vacuum-type thermocouple is shown in Fig. 17-26. The vacuum mounting is used to minimize heat losses so that greater output can be obtained from a small current through the heater. Vacuum-type thermocouples are available in ranges from about 2 to 500 ma. This method of measurement

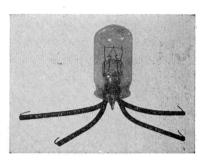

Fig. 17-26. Vacuum thermocouple. (*Courtesy of General Electric Co.*)

places much lower loading on the circuit than electrodynamometer or moving-iron instruments, and the loading is almost purely resistive. The inductance is merely that of the simple loop of wire in and out of the thermocouple mounting. Accordingly, thermocouples can be used for measurement up to high frequencies of the order of 50 to 75 Mc. Thermocouples must be operated with great care, as the fine heater wire may be burned out by relatively small overloads (50 per cent is about the permissible limit).

Another thermocouple arrangement is the bridge mounting shown in Fig. 17-27. No separate heater is used; instead the current to be measured passes directly through the thermocouples and raises their temperature in proportion to the I^2R. The cold junctions are at the pins that are imbedded in the common bakelite base and the hot junctions at splices midway between the pins. The couples are oriented as indicated in (a), so that the resultant thermal voltages give rise to a d-c difference of potential from A to B. The a-c potential between these points is zero because of the balanced resistances in the four arms, so no alternating current flows through the instrument. This construction gives greater output voltage than the single couple used in the vacuum mounting, is more rugged in withstanding overloads, and can be constructed conveniently to measure greater currents. Figure 17-27(b) illustrates the construction of a bridge-type thermo element. Bridge-type elements are made with a-c ratings from about 0.1 to 1 amp.

A compensated type of thermo element is shown in Fig. 17-28. There are two large terminal blocks of copper with the heater element between them. The thermocouple junction is welded or hard-soldered to the midpoint of the heater. The other ends of the thermocouple wires are connected, one to each of two copper strips that bridge the space between the terminal blocks. These strips, though fastened to the blocks mechan-

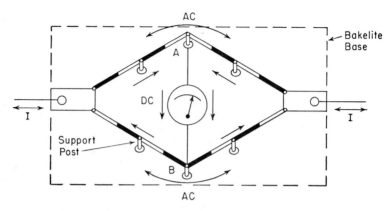

(a) Diagram of connections.

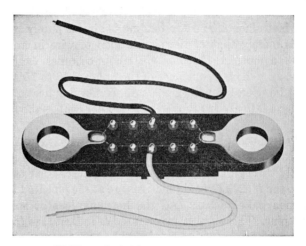

(b) View of a bridge-type thermoelement.

Fig. 17-27. Bridge-type thermoelement. (*Courtesy of Daystrom, Inc., Weston Instruments Division.*)

ically, are separated from them by a thin sheet of mica to give electrical insulation while permitting effective thermal contact. The instrument wires are connected to the two strips. By this means the cold junction is kept effectively at the average temperature of the two end blocks. Elements of this type are made with ratings from 1 to 50 amp. The accuracy of indication depends, of course, on the instrument used with these elements. With a high-grade instrument an accuracy of about 1 per cent is possible up to 50 Mc or somewhat higher. Above the 5-amp range the voltage drop in the heater is 0.15 v for full scale, so the power consumption is 0.15 watt for each ampere of range. Thermocouple ammeters of some types may be

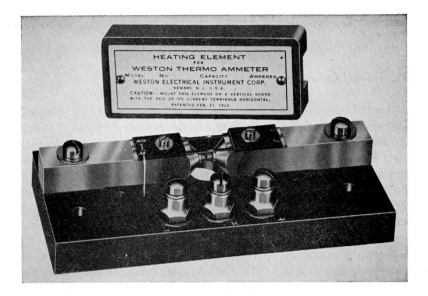

Fig. 17-28. Compensated thermoelement. (*Courtesy of Daystrom, Inc., Weston Instruments Division.*)

used to frequencies of 200 Mc, or even higher, though requiring a correction factor in the upper part of the range.*

Thermocouple instruments are very useful in the high-frequency field of use, since their negligible inductive effect and relatively low power consumption make them usable at frequencies far beyond the range that can be handled by the instruments commonly employed for low-frequency measurements. Their main shortcoming in comparison with the other instruments that we have studied is their limited overload capacity. They must be used with great care to avoid burning out. Nondestructive overload is limited to about 50 per cent, as even this loading more than doubles the heat to be dissipated from the heater.

Voltmeters may be constructed using low-current vacuum thermocouples and series resistors. They are available with sensitivities of 100 to 500 ohms/v in ranges up to 500 v. The usable frequency range for acceptable accuracy is limited more by the characteristics of the series resistor than by the thermocouple. With wire-wound resistors the voltmeters are limited to audio and supersonic frequencies, at least in the higher voltage ranges from 100 to 500 v. Developments in the use of deposited-film resistors in a shielded mounting, some with double shield and coaxial

---

*Reference 7.

transmission line-type design, permit accurate measurements up to 30 to 100 Mc.*

17-22. Thermal Converter

A thermal converter, in general, is a device for converting an alternating current or voltage to a direct current for measurement purposes. The basic component is some form of thermocouple, as discussed in the foregoing section. One important field of use is in high-frequency measurements of current and voltage, made possible by the low inductance and capacitance of the thermal element. However, the thermal principle is used also at power frequencies, both for precision conversion from root-mean-square a-c to d-c in current and voltage determinations, and also in an important commercial way for totalizing power quantities.

A thermal element may be used, as discussed in par. 21, in conjunction with a d-c milliammeter to form a voltmeter or ammeter for high-frequency measurement. (It is not much used in this way at power frequencies owing to lack of overload capacity, since more rugged movements are usable in the low-frequency range.) For the most accurate work the thermal element is used as a *transfer* device, i.e., the output on the unknown a-c is matched by a like output on d-c, and the d-c is measured by some form of potentiometer.† In this way only short-time stability is required of the thermal element, and actual measurement is made by an accurate d-c method. The transfer idea applies also in the low-frequency range for establishment of accurate *rms* calibration of basic standard a-c instruments, and thus performs an essential measuring function for low frequencies.

Figure 17-29 is a diagram of a thermal converter arranged for a single-phase load. The current would ordinarily be fed to the converter from the secondary of a current transformer, but as this makes no difference in the principle the transformer is omitted for simplicity. The current I is constrained by magnetic action to divide equally in the two halves of the transformer winding, giving a current $I/2$ for each set of thermal elements. The symbol r is used to denote the voltage ratio of the potential transformer from the primary to each half of the secondary. With the secondaries in additive connection, as shown, a current $(2rE)/(4R)$ is circulated around the closed loop. This current is marked I_v on the diagram to denote that it is proportional to the load voltage. The two component currents are additive in bridge arms *a-e* and *e-d*, and subtractive in *b-f* and *f-c*. The magnitudes of the resultant currents depend on the phase relationship between them, which is the phase angle θ of the load. The phasor combina-

---

*Reference 6.
†See par. 17-35 and reference 4.

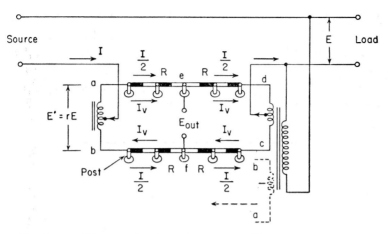

Fig. 17-29. Diagram of a thermal converter. Winding a-b is on same core with c-d, but is shown separately for clearness. R = resistance of thermo-elements in one arm of bridge.

tion of currents and their magnitudes, as found by the law of cosines, are shown in Fig. 17-30. If we assume that the thermal elements give an output voltage proportional to the heat liberated in them ($E = k \times$ watts), and

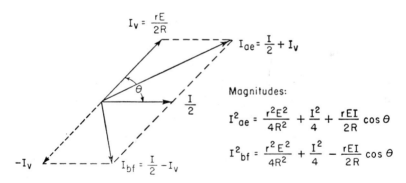

Fig. 17-30. Phasor diagram for the circuit of Fig. 17-29.

connect the element polarities as shown in the figure, we have for the output voltage of the bridge

$$E_{\text{out}} = k(I^2_{ae}R - I^2_{bf}R) \qquad (2)$$

Substitution of values for the currents gives

$$E_{\text{out}} = krEI \cos \theta \qquad (3)$$

That is, the output voltage is directly proportional to the load power.

A three-phase thermal converter can be made by combining two of the single-phase elements, and adding the output voltages in series. The action is equivalent to the two-wattmeter method for three-phase power measurement. The output voltage of a single-phase or three-phase converter can be recorded directly by a self-balancing potentiometer (chap. 8), or combined with the voltages from other loads for totalization.

ELECTROSTATIC VOLTMETERS

17-23. Electrostatic Voltmeters

The action of electrostatic voltmeters depends on the force action between electric charges. If adjacent electrodes are given electric charges, there is force action between them, and for a specified geometry the force action depends on the amounts of charge. This principle is used in several ways in making measurements in electrical systems.

One method is to measure the force between parallel plane disks. This method is capable of very accurate results if the dimensions of the disk and the spacing are measured accurately, if the field around the measured disk is controlled by a guard ring, and if the force on the disk is determined by an accurate beam balance. This arrangement has been used as an absolute method for the measurement of high voltages, as the relationship between force and voltage may be computed directly from basic relationships when the edge-field is controlled by the guard ring. The method is not desirable for low voltages, as the force is so small for practical spacings that the accuracy is poor. For a given spacing the force increases as the square of the voltage, giving improved accuracy. For a high voltage the spacing of the plates must be sufficient, of course, to avoid arc-over or corona formation.

The device described above is a special one for measurement of very high voltages and for the establishment of calibrations at high voltages. Many other forms of electrostatic measuring equipment are used over a wide range of voltage. There are quadrant electrometers of different designs with suspended moving parts for high sensitivity and electrostatic meters of portable type. These other devices, while not having accurately calculable laws of deflection, may be calibrated from other instruments and have characteristics that make them useful in some kinds of work.

Electrostatic voltmeters have the property that, for any particular spacing of the plates, the force is proportional to the square of the voltage. Accordingly, when used on alternating potentials, the instrument reads the root-mean-square value. They may also be used on direct current.

17-24. Characteristics and Uses of Electrostatic Voltmeters

Electrostatic voltmeters have the characteristic mentioned above of being usable on direct current and any wave form of alternating current. When the instrument is first connected to a source of direct current, it draws a momentary charging current which dies away rapidly, the curernt thereafter being determined by the insulation resistance. The insulation resistance, under good conditions, should be in the millions of megohms, and even with considerable humidity should be of the order of thousands of megohms, so the electrostatic voltmeter on direct current can be considered for most purposes a zero-current device. This is one reason for its use in d-c measurements, in which recourse is generally had to the permanent-magnet moving-coil instrument. Some high-voltage, low-current rectified power supplies would be affected considerably by the current drawn by a PMMC instrument, but may be measured with no change by the electrostatic voltmeter.

Figure 17-31 shows the movement of a portable electrostatic voltmeter of 120-v range, which is about the lowest for which the electrostatic type is made. The picture shows the two sets of fixed plates, the moving plates of double-sector form, the pointer, and the control spring. The moving system is carried by two pivot-and-jewel bearings. There is an air vane for damping, mounted in a chamber under the base plate. The number and

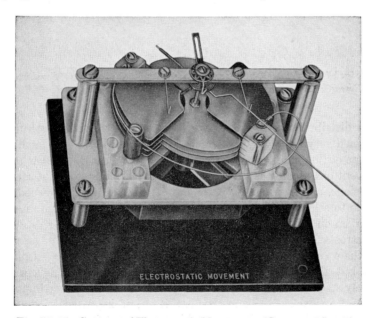

Fig. 17-31. Cutaway of Electrostatic Movement. (*Courtesy of Sensitive Research Instrument Corporation.*)

spacing of the plates depend on the voltage for which the instrument is designed. Instruments of this form are constructed for full-scale ranges from 120 to 5000 v.

Higher voltage instruments must have larger cases and larger insulators. Figure 17-32 shows a 50,000-v instrument which is typical of a line ranging from 5000 to 50,000 v.

Fig. 17-32. A 50,000-volt electrostatic voltmeter. (*Courtesy of Sensitive Research Instrument Corp.*)

The distribution of scale markings of an electrostatic instrument is determined very largely by the square-law relationship, but is modified by the shape of the plates. Inspection of the instruments shows that the first 20 per cent of the range is so compressed that it usually does not carry any subdivisions, and the next 10 or 15 per cent is still quite crowded. It is desirable to select an instrument of such range that readings do not need to be taken on this portion of the scale.

Electrostatic voltmeters read *rms*-values of alternating voltages regardless of the wave form, as mentioned above. They consume practically zero power, but they do draw a small capacitive current. The current is extremely small at power frequencies but may become of some importance at radio frequencies. The instrument itself, if insulated with low-loss insulation, may be used with reasonable accuracy at frequencies as high as 50 Mc. The effect of the instrument capacitance in loading the circuit must be considered at the high frequencies. The instrument of Fig. 17-31, for example, has a capacitance of 225 $\mu\mu$f as built for a 120-v range, and 22 $\mu\mu$f for a 5000-v range.

WATTMETERS AND VARMETERS

17-25. Construction of a Wattmeter

The wattmeter consists essentially of the electrodynamometer movement that was discussed in par. 12. For power measurements the fixed and moving coils are connected in different circuits instead of being in series, as they are in voltmeters and ammeters. The fixed coils, or "field coils," are usually employed as the current coils in series with the load so that they can be wound with the size of wire needed to carry the load current. The moving coil and the control springs, which can carry only a small current, serve very well as the potential element.

The connection of a wattmeter as used for the measurement of the power drawn by a load is shown in Fig. 17-33. The current through the field coils

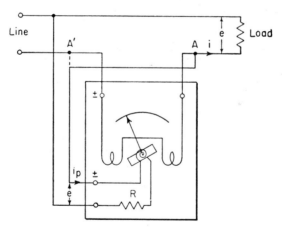

Fig. 17-33. Diagram of a wattmeter connected to read the power to a single-phase load.

is the load current i (assuming for the present that i_p is negligible with respect to i). The current i_p through the potential circuit is practically e/R at any instant, as inductive effects in this circuit are negligibly small. The torque on the moving coil for any particular coil position depends on the product of field-flux density times the current in the moving coil, or more simply for our present purpose, on the product of the currents in the two coils. Then, since $i_p = e/R$, the torque depends on $e \times i$. If e and i vary in time, the torque also varies, and if the variations are rapid, the moving coil cannot follow them but takes up a position so that the control spring torque equals the average value of the torque produced by the electromagnetic action. Since the coil torque at any instant depends on the power ($e \times i$) at that instant, the average torque is a measure of the average power. This holds true regardless of the wave forms of e and i within the frequency limitations of the instrument.

In summary

$$\text{Instantaneous torque} \propto (\text{Inst. } e \times \text{Inst. } i) \tag{4}$$

$$\text{Average torque} \propto \frac{1}{t_2 - t_1} \int_{t_1}^{t_2} ei \, dt \tag{5}$$

For the commonly used sine-wave case, assuming e and i to vary sinusoidally in time with an angle θ of phase difference, or

$$e = \sqrt{2} \, E \sin \omega t \tag{6}$$

$$i = \sqrt{2} \, I \sin (\omega t \pm \theta) \tag{7}$$

the integral (5) over a complete cycle can be shown to yield as the average value of the power

$$W_{\text{ave}} = EI \cos \theta \tag{8}$$

where E and I are the effective (rms-) values of voltage and current, respectively.

In a sense wider than the mere measurement of power in an electrical circuit, the wattmeter may be regarded as a multiplying device. That is, the indication depends, for steady quantities, on the product of the voltage impressed on the potential terminals of the instrument multiplied by the current flowing in the current coil, regardless of the source of these two quantities. If e and i are the voltage and current for the same circuit element, the product represents power. However, the wattmeter may be used in other ways. If a voltage equal to the load voltage in magnitude but at 90° phase angle from it is impressed on the potential circuit, the reading represents the *reactive power*, or "vars" of the load. For an example outside the electrical field, suppose the current to come from a device that gives a current proportional to the torque of an engine of some sort and that the voltage is from a tachometer than gives a voltage proportional to the rotational speed of the engine. In this case, the indication of the wattmeter is proportional to the horsepower of the engine. If the derived e and i quantities vary in time, the wattmeter not only multiplies them together but averages the product within the limits of its inertia as related to the speed of the fluctuations. If the fluctuations are very slow, the pointer follows them closely; if the variations are a little faster and near the period of the moving coil, the pointer may over-indicate or under-indicate, depending on the damping; finally, if the variations are very rapid, as in the cyclic variations at power frequencies, the pointer cannot follow the individual swings but reads the average.

17-26. Wattmeter Connections — Single Phase

Figure 17-33 shows two details of wattmeter connections. The first is the presence of the ± marks on one current terminal and one potential terminal, which should be connected into the circuit as shown. The chief point is that the ± potential terminal be connected to the side of the line in which the current coil is located. As indicated in the figure, this connection brings the current and potential coils to essentially the same electrical potential, as by far the major part of the voltage drop in the potential circuit is across the resistor R. If the potential circuit were to be reversed, there would be a large difference of potential between the two coils, resulting in an electrostatic force between them and producing an error in indication. If the pointer should read backward for any reason, the current connections should be reversed, but not the potential connections. Some portable wattmeters are equipped with a switch that reverses the terminals of the potential coil without changing its position in the circuit; this is a convenient arrangement that obviates the necessity for reversing the heavy current connections.

A wattmeter is rated not only for the number of watts for the full-scale range, but also for the allowable current in the field coils and the allowable voltage to be impressed on the potential circuit. It is important that current and voltage ratings not be exceeded, even if the pointer is below the full-scale point, as may happen in low-power factor circuits. As an extreme example, if the load consisted of static capacitors, we could have a large current flowing in the field coils and yet have very few watts showing on the scale. If we kept adding capacitors, the instrument could become severely overheated before there was much deflection on the scale. Special low-power-factor wattmeters are made that give full-scale deflection with rated volts and amperes if the power factor is 20 per cent. Such wattmeters give better readings in low-power-factor circuits, as in the measurement of the iron loss of a transformer on open-circuit test.

It is customary in measuring loads in which the current is large to use a wattmeter with a 5-amp current coil and provide a multiplying effect by means of an instrument current transformer. If the voltage is above 300 v, an instrument potential transformer may be used, and above 750 v will surely be used. (Between 300 and 750 v an external resistance multiplier is possible.) The potential transformer is built to step down from the line voltage to 115 or 120 v for the wattmeter. Instrument transformers are discussed in chap. 15.

The moving coil of a wattmeter has a small inductance, so the current in the potential circuit is not quite in phase with the voltage applied to it. The phase angle is small (of the order of 1′ or 2′) at the usual power frequencies, hence the effect on the wattmeter reading is negligible for ordinary measurements. The error becomes important for a load with phase angle very close to 90°. Another effect of occasional importance is the voltage induced in the moving coil by the stationary field coils for any position except when the magnetic field directions are at 90°. The induced voltage is small enough to be negligible in most cases, but may cause appreciable error if the circuit impedance is low.

The wattmeter circuits consume some power which may be included in the scale indication along with the load power. The instrument power is small (a matter of a few watts) and frequently is negligible with respect to the large load power. There are times, however, when neglect of the instrument power would cause a serious error. In order to give a correct indication of load power, the wattmeter current coil should carry exactly the load current, and the voltage across the potential circuit should be exactly the load voltage. With the potential winding connected to A in Fig. 17-33, on the *load* side of the wattmeter, the voltage is the correct load value, but the current through the field coil is greater than it should be by the amount of i_p. The wattmeter accordingly reads high by the amount of power drawn by its potential circuit. If the connection is moved to A', the field coil carries the correct load current, but the voltage is high by the drop across the field winding, so the instrument reads high by the amount of the I^2R loss in the field winding. In some cases the error is serious by either connection, and a computed correction must be made from known constants of the instrument. In other cases the error is considerably smaller by one connection than the other, so that the correction either may be neglected, or at least is a smaller fraction of the true power. Connection at A is better in general for a high-current, low-voltage load, and at A' for a low-current, high-voltage load. Another solution is the use of a compensated wattmeter as discussed in the next section.

17-27. Compensated Wattmeter

A wattmeter may be provided with an additional field winding to cancel the effect of the potential-coil current. The arrangement is indicated in Fig. 17-34. The new winding, which may be of small wire since it carries i_p only, should have the same number of turns as the main winding and be so disposed as to have the same magnetizing effect. The effects of i_p are thus cancelled out with respect to setting up any field flux, and the error caused by wattmeter-potential-coil power is avoided.

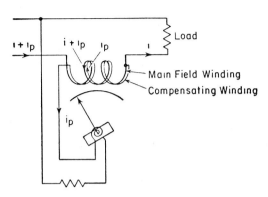

Fig. 17-34. Diagram of a compensated wattmeter.

Generally, measurements of current and voltage must be made in addition to power, and the inclusion of ammeter and voltmeter raise some new questions of the elimination of errors. The problem with the voltmeter is much the same as for the potential coil of the wattmeter. If the voltmeter is connected on the load side of the wattmeter, the voltage is correct but an error is introduced into the wattmeter reading. If the voltmeter is connected on the line side of the wattmeter, the voltmeter reading is slightly in error but the power is correct. A compensating winding could theoretically be added to the wattmeter to take care of the voltmeter, but such instruments are not available. Similar problems arise with respect to placement of an ammeter. If the ammeter is between the load and the voltage connection, the voltage is in error; if on the other side, the ammeter is in error. The use of a compensated wattmeter, it may be noted, corrects for the effect of i_p on the wattmeter, but still leaves i_p to cause error in an ammeter on the line side of the wattmeter.

In summary, there is no combination of voltmeter, ammeter, and wattmeter that is free of error. In many cases the errors are small enough to be neglected. When load quantities are small and instrument power must be considered, the best meter arrangement should be selected for the particular combination of load and instruments and then computations be made to correct the errors.

17-28. Wattmeter Based on the Hall Effect*

The wattmeter based on the Hall effect is presented as an interesting example of a physical phenomenon applied in an ingenious way to provide a new solution of a measuring problem. The Hall effect has been known since 1879, but it is only the recent development of new semiconductor

---

*Reference 1.

materials with greater output that has made an instrument of this sort feasible.

The Hall effect was discussed in par. 16-11. The voltage output was shown to depend on the product of magnetic field density and the current flowing in the element, with directions as shown in Fig. 16-15. The Hall unit was presented there as a means of measuring magnetic field strength, but it was also shown to be primarily a multiplying device. Multiplication is a process needed in many instrumentation situations, so the Hall element appears to have many possibilities.

One use of a multiplying element is in the construction of a wattmeter. The diagram is shown in Fig. 17-35; by this arrangement the magnetic field

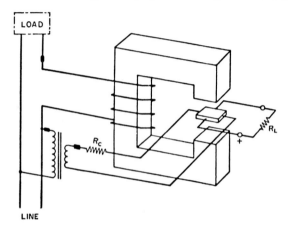

Fig. 17-35. Diagram of a Hall watt transducer. (*Courtesy of Westinghouse Electric Corp.*)

density depends upon the load current, and the element current is proportional to the load voltage. The Hall output voltage is thus proportional at each instant to the load power at that instant. The output device, indicated in the figure as R_L, can be a d-c permanent-magnet movement which averages the instantaneous values of output.

An important consideration in the design of the multiplying device is the selection of the material to be used in it. Development in semiconductor materials has made possible Hall elements of greater output and improved characteristics. Figure 17-36(a) presents a comparison of two materials with respect to the output, and its variation with temperature. Indium arsenide, though it has a smaller Hall coefficient at low temperatures, is affected to a much smaller degree by temperature changes than indium antimonide (a change of 0.02 per cent/°C against 1.3 per cent/°C in the usual working range), which is an important consideration in a measuring device. Figure 17-36(b) shows output as related to the control

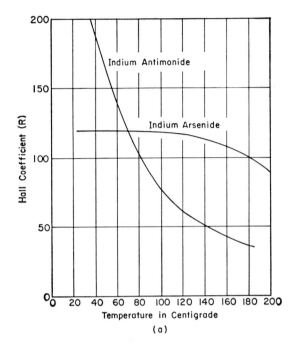

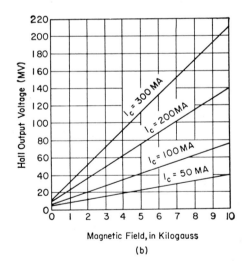

Fig. 17-36. Characteristics of Hall generators.

(a) Variation of the Hall coefficients of InAs and InSb with temperature.

(b) Variation of the output voltage with control current and magnetic field.

(*Courtesy of the Westinghouse Electric Corp.*)

current; the curves are linear, but do not quite pass through zero, but this can be remedied by a small bias voltage. The control current should be kept to a value low enough to minimize heating effects.

The output of several Hall elements can be placed in series if it is desired to totalize the power from several sources. This necessity arises also in measuring polyphase power; as shown in par. 17-29 two elements are needed to measure the total power in a three-wire circuit, and in general, $n - 1$ elements must be used in an n-wire system.

17-29. Three-phase Power Measurement

The power in a three-phase three-wire circuit may be measured by two single-phase wattmeters with current coils connected in two of the line wires and the potential circuit in each instrument connected from its own line wire to the wire that does not contain a wattmeter. This is illustrated in Fig. 17-37. If the neutral wire is carried along with the phase wires,

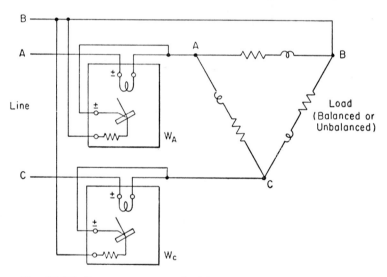

Fig. 17-37. Two-wattmeter method for measurement of power in a three-phase, three-wire circuit.

making a four-wire circuit with some loads from line wires to neutral, then three wattmeters are needed to insure reading the correct total under any condition of unbalance. In fact, there is a theorem for which a generalized proof can be given that the power can be determined in an n-conductor system under any condition of unbalance by $(n - 1)$ wattmeters.

The true total power is given in any of the above cases by taking the algebraic sum of the wattmeter readings. There can be negative readings under some conditions of power factor of the loads, so it is necessary to observe the connections of the instruments and the signs of the readings. It is important to connect the wattmeters symmetrically with respect to the marked terminals in order to establish a reference for determining the signs of the readings. A symmetrical arrangement is illustrated in Fig. 17-37. If one of the wattmeters reads backwards when power is applied to this circuit, the fact should be noted, then the reading brought up-scale by the reversing switch, if one is available, or by reversing the current-coil connections. The reading of the reversed meter is treated as a negative quantity.

17-30. Polyphase Wattmeters

Instruments are constructed containing two electrodynamometer movements on one shaft. The two elements may be used exactly as the two wattmeters in Fig. 17-37. Once the connections are properly made, the instrument carries out the addition of torques and so indicates the total power on one scale. This is a convenience to the operator and also saves switchboard space, as compared with separate instruments. The polyphase wattmeter may also be used in two-phase circuits and in single-phase three-wire circuits.

17-31. Varmeters

Power system operators frequently need information regarding reactive effects as well as the true power in their circuits. One way is to measure the reactive power. The instrument used for this purpose is called a "varmeter" since it measures the volt-amperes-reactive. The device is essentially a wattmeter in conjunction with a phase-shifting device that gives a voltage at a phase angle of 90° from the true load voltage.

A 90° shift can be obtained in a single-phase circuit by a circuit made up of R, L, and C components. This requires careful proportioning, and the resulting circuit is sensitive to changes in frequency. The case of more common interest, however, is three-phase varmetering, which can be accomplished by a simple transformer arrangement. If we have an open-delta transformer connection, as in Fig. 17-38, with the line wires connected to points A, B, and C and have taps at the proper points, we can arrange to have two voltages of the same magnitude as the line voltages, but at 90° from them. If we bring out a tap at D, 57.7 per cent of the winding from B to A, and at point E, 15.4 per cent beyond C, the voltage phasor from D to E is 100 units long just as is A-B, but D-E has a 90° phase angle from A-B. Likewise F-G is 100 units long and 90° from C-B. If we connect instruments as in Fig. 17-35, except that we supply the potential circuit of wattmeter A with the voltage D-E instead of the voltage A-B, and watt-

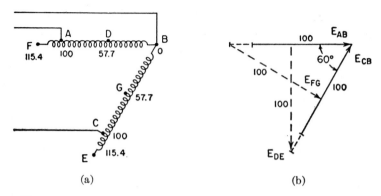

Fig. 17-38. Transformer windings to furnish quadrature voltages for var-metering in a three-phase three-wire circuit. (a) Transformer windings, showing taps. (b) Phasor diagram.

meter C with the voltage F-G instead of C-B, we have accomplished the desired 90° shifts, and the instruments now read the reactive power.

17-32. Power-factor Meters

Power-factor meters are made of several different types. One kind uses a movement of the electrodynamometer construction, except that the moving element has two crossed coils, as shown in Fig. 17-39. For application to a three-phase circuit, the field coil is connected in series in one of the line wires, and the crossed coils are connected from that wire through resistors across two phases of the circuit in such way that the torques are opposed. Since no control springs are used, the balance position of the moving element depends on the ratio of the two torques, which is a function of the power factor of the load. On a single-phase circuit one of the crossed coils is connected in series with a resistor, the other through an inductor across the line wires. The relative torques of the two coils depends on the phase position of the field current and so gives a pointer position that can be calibrated in terms of power factor.

Another type of power-factor meter is the polarized vane instrument as shown in Fig. 17-40. The stator is constructed like a three-phase induction motor and, when connected through resistors to three-phase line wires, sets up a rotating magnetic field as in an induction motor. The central, or "polarizing," coil is connected in series with one of the line wires. The vane takes up the position that the rotating field has at the instant when the polarizing flux is maximum. This instrument may be used on a single-phase circuit if a phase-splitting network is provided to set up a rotating field.

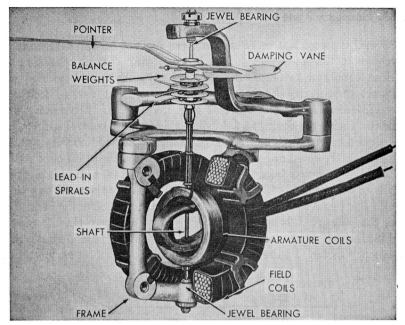

Fig. 17-39. Cut-away view of a crossed-coil power-factor meter. (*Courtesy of General Electric Co.*)

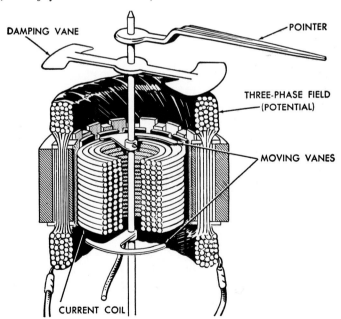

Fig. 17-40. Diagram showing structure of a polarized-vane power-factor meter. (*Courtesy of General Electric Co.*)

17-33. Frequency Meters

Several types of instruments have been devised for the determination of frequency. The moving-iron type has a moving element consisting of a soft-iron vane and two crossed stationary coils that are connected with some sort of frequency-discriminating network, so that one coil is stronger at low frequencies and the other at high frequencies. The moving vane aligns itself with the resultant field since there are no control springs, and thus gives a measure of the frequency. Fluctuations of line voltage affect both coils in the same proportion and hence do not change the indication.

There are several instruments that make use of resonant electrical circuits in the determination of frequency. Generally there are two tuned circuits, one tuned to resonance slightly below the low end of the instrument scale, the other slightly above the high end. These two circuits may be combined with a crossed-coil instrument or an electrodynamometer to make a frequency meter. Figure 17-41 is a schematic diagram of an instrument

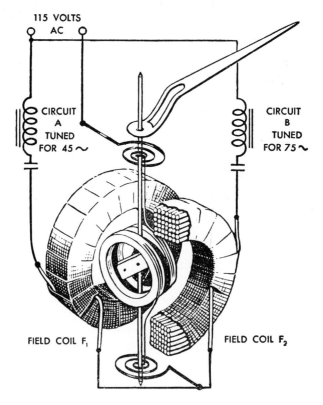

Fig. 17-41. Schematic diagram of a resonant-circuit frequency meter. (*Courtesy of General Electric Co.*)

designed to have 60 cycles at the center of the scale. The small iron vane in the center provides the restoring torque and tends to hold the pointer at the center. The two tuned circuits, tuned for 45 and 75 cycles, respectively, are each in series with half of the field winding, connected in opposition. At frequencies below 60 cycles the field coil F_1 predominates, producing torque in a counterclockwise direction; above 60 cycles the torque is in the opposite direction. A complete study of the action may be made in terms of phasor diagrams, since the deflecting torque is proportional to the product of the resultant field flux, the armature flux (or current), and the cosine of the phase angle between them.

Figure 17-42 shows internal and external views of an instrument that determines frequency in terms of *mechanical* resonance effects. There is a series of reeds fastened to a common base that is flexibly mounted and that carries the armature of an electromagnet whose coil is energized from the a-c line whose frequency is to be measured. The reeds are tuned to have exact natural frequencies of vibration, partly by selection of length and partly by loading the free end with a small amount of solder which can be scraped for exact tuning. The reed that has a natural frequency that is the same as the pull frequency of the electromagnet builds up a large vibration, which makes the small square end of the reed appear to lengthen out to a ribbon. If the line frequency is intermediate between two reeds, both vibrate, and their relative amplitudes indicate which is closer. Accordingly, interpolation can be made between the discrete frequencies of the reeds.

This device has a number of good qualities. It is simple and rugged and maintains calibration well if not abused. (It should not be allowed to operate at excessive amplitude of vibration for a long period.) A tapped series resistance can be provided to adapt it to a range of voltages. The operation is independent of the exact value of the voltage and is not affected by harmonics in the voltage wave, as are some of the electrical instruments. The latter point can be illustrated in an interesting way by impressing two voltages of different frequencies, say 40 cycles and 60 cycles, in series to the instrument terminals. The 40 cycle reed will vibrate and so will the 60 cycle reed, showing the two frequencies separately. Double-range instruments can be made, with frequencies in the ratio of 1 : 2. This is accomplished by providing two operating magnets, one wound on a soft-iron core and the other on a permanent magnet. The soft-iron electromagnet gives two pulls per cycle. In the other electromagnet, if the flux component caused by the a-c coil is less than the flux of the permanent magnet, the total flux is pulsating with one peak per cycle. Thus a reed tuned to vibrate 60 times a second is set into vibration when a frequency of 60 cycles per second is impressed upon the permanent-magnet electromagnet, or when a 30 cycle frequency is impressed upon the soft-iron electromagnet.

Figure 17-43 gives a diagram of a "transducer-type" frequency meter,

(a) Internal view.

(b) External view.

Fig. 17-42. Portable-type "Frahm" tuned-reed frequency meter. (*Courtesy of James G. Biddle Co.*)

so called because the frequency measuring function is entirely separated from the indicating instrument, which in this case is a simple d-c meter.* Two parallel off-resonance circuits are used, one resonant below the instrument range and one above. In an instrument with 60 cps at mid scale one circuit might be resonant at 48 cps and the other at 72 cps. Each

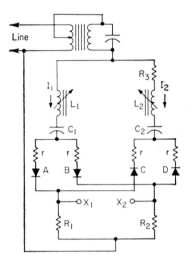

Fig. 17-43. Circuit diagram of a transducer-type frequency meter. (*Courtesy of the Westinghouse Electric Corporation.*)

circuit consists of an inductor, a capacitor, and two rectifiers connected for full-wave rectification. A d-c milliammeter, connected between X_1 and X_2 in the diagram, reads the difference of currents, with the zero point near the middle of the scale. The circuit may be designed for a narrow range, such as 58–62 cps, or 59–61, with corresponding figures for other frequency bands. A voltage-regulating transformer is included to avoid changes of indication caused by fluctuations of the line voltage.

17-34. General Considerations in Calibration of Instruments

Special problems arise in the calibration of each type of instrument, but there are some considerations that apply to many types and so may be discussed in general before attention is turned to specific situations. The general remarks deal in part with the mechanical condition of the instrument and in part with procedures and precautions in making the required measurements. The discussion will deal in the main with the checking of existing instrument scales rather than the making of the original calibration.

---

*Reference 3.

When an instrument is received for calibration it should first be checked carefully for condition to insure that there is no defect that would prevent satisfactory functioning after it has been calibrated. Some of these matters are discussed in the following paragraphs.

(a) *General examination.*

The instrument should be inspected for signs of damage, such as injury to the case or discoloration of the scale by overheating, and for looseness of the glass window or the binding posts. Any trouble of this sort should be corrected before the instrument is calibrated.

(b) *Examination for friction.*

The instrument should be connected in an appropriate circuit, so that the deflection can be controlled smoothly over the entire scale. Increase the reading *slowly* from zero to full scale and back to zero, meanwhile watching for irregular motion of the pointer. Stop at several points on the scale and tap the case *gently* with the finger; friction is indicated by a jump of the pointer upon tapping.

Friction may be the result of a worn or deformed pivot or of a cracked jewel, either of which may be caused by a blow or by rough handling. Sometimes sticking of the pointer is caused by the paper scale coming loose from the metal backing and buckling up because of humidity. Excessive friction may also be caused by a damping vane of an air-damping system becoming slightly deformed and rubbing on the surface of the damping chamber. In d-c instruments, trouble is sometimes experienced by dirt entering the air gap and interfering with the motion of the moving coil. Such dirt is hard to remove if it is magnetic in nature. The troubles caused by dust and dirt give point to an item of advice; if the case of an instrument must be opened, the work should be done in clean surroundings, and great care should be taken to prevent the entry of dirt. The instrument should not be allowed to stand open and uncovered for a long period.

If the test above indicates excessive friction, the cause should be found and the trouble corrected before a calibration is attempted. If the pivot or jewel is defective, it must be replaced, which is an operation calling for an experienced instrument repairman. Some of the other difficulties mentioned above may be cured more easily.

(c) *Examination for balance.*

The pointer of an instrument may have been bent by a heavy overload or reverse kick and then restored to zero by means of the zero-adjuster. After this event, however, the moving system is not in balance as it should be, which may cause considerable error if the instrument is used with its axis horizontal. The calibration may also be affected because of the difference of pointer and moving element position as compared with the original condition. An instrument should be checked for correctness of balance before it is calibrated.

To check the balance, set the pointer on zero, turn the instrument in the horizontal plane until the pointer points directly away from you, lift the instrument to the vertical plane, and observe whether the pointer stays on zero. Return the instrument to the horizontal plane and turn it so that the pointer is directed to your left; then again raise the instrument to the vertical and observe zero. If the pointer stays on zero in both cases, the instrument movement is in balance. If not, examine the movement, straighten the pointer if necessary, rebalance by adjustment of the counter-weights until the test indicates correct balance.

(d) *Precautions in calibration.*

The pointer should be set carefully on zero before calibration is begun. Set the zero, then tap the case lightly to insure a true zero free of any slight frictional effect. Be sure that connections to shunts make good contact and that contact surfaces are clean. Leads used with millivoltmeters during calibration should be the same that will be used in actual service or, if this is impossible, they should be standard leads of the same resistance.

Electrostatic effects of stray origin should be avoided. Do not rub the glass window of the instrument or the rubber case just before a reading is to be taken, as the electrostatic charge thus induced may affect the position of the pointer. (Breathing on the glass helps to dissipate the charge.) Electro-static forces between the current and potential coils of a wattmeter should be avoided by making connections (as indicated by the polarity markings) so that the coils are at the same potential.

In the original calibration of an instrument, the position of the main, or "cardinal," points is found by test measurement, and the intermediate points are determined by a proportional-dividing arrangement. The scales on all but the small, cheap instruments are hand-drawn. Individual instruments may, of course, differ in the quality of draftsmanship, hence it is advisable before calibration to inspect the scale for regularity of the divisions. If the scale appears irregular and if the instrument must be used for precise work, the calibration should be checked at a greater number of points.

(e) *Effect of stray fields.*

Stray fields are particularly likely to be troublesome in testing in industrial plants, where there are many circuits carrying heavy currents. The possibilities of such difficulties should be kept in mind, not only for such testing, but in calibration work as well. The effects of d-c and a-c fields may be considered separately.

A stray d-c field affects the indication of an unshielded permanent-magnet instrument and causes an equal percentage error all along the scale. The presence of a stray field may be detected by turning the instrument to different positions while it is carrying a steady current. An unshielded, or poorly shielded, instrument may have its calibration permanently affected if it is located near a bus bar that is subjected to a heavy short-circuit. An a-c field affects a d-c instrument only if it is great enough to make a change in the strength of the permanent magnet.

An a-c field affects the indication of an a-c instrument, depending on the type of movement and its shielding. An electrodynamometer instrument (unshielded) is affected differently at different parts of the scale, depending on the angle between the moving coil and the stray field. A check may be made by turning the instrument 180° while it is carrying a current of constant strength. A voltmeter or an ammeter of the moving-iron type usually has a coil with axis parallel to the instrument shaft and so is subject to error from fields in this direction. Rotation of the instrument about its axis will not in general disclose the presence of the stray field.

It is important in calibration to avoid the effect of stray fields. Instruments should not be placed too close to conductors that carry heavy currents, and the leads to the instruments should be carefully placed and preferably twisted together if they carry large currents. Instruments should not be placed too close together nor placed on a sheet of iron or a steel-topped bench. On the other hand, instruments that are intended for mounting in a steel panel should be mounted in a similar panel for testing and calibration unless the completeness of internal shielding removes the necessity for such precaution.

17-35. Methods of Calibration of Voltmeters and Ammeters

A d-c voltmeter or ammeter may be calibrated by means of a potentiometer, as discussed in chap. 8. This is the fundamental method and is necessarily used for the basic standard instruments. It is too slow for the general run of production calibration (unless a deflection-type potentiometer is used) and is more precise than is needed. The usual switchboard and portable instruments are calibrated by comparison with a high-grade standard instrument of suitable range. It is probably unnecessary to give circuit diagrams for a calibration setup. In the case of voltmeters, the unknown is connected in parallel with the standard and supplied with power by a potential divider with an additional rheostat for fine adjustment. The voltage is built up to the first point at which a reading is to be made; the setting is usually made to place the unknown exactly on the division mark, so that the reading of the correct voltage can be made on the standard, which presumably has a more finely divided scale than the unknown. Readings are made at the cardinal points until full scale is reached and, again, in decreasing order to provide check points for any effect of spring set. The instrument may be tapped gently before a reading is taken to eliminate any slight frictional effect. (Any serious friction should have been eliminated before the calibration was started, as discussed above.) The calibration of d-c ammeters follows the same pattern as the foregoing, except that the unknown and standard meters are placed in series with one another and with suitable coarse and fine rheostats to a steady low-voltage supply.

The situation is somewhat different for a-c instruments in that calibration is necessarily by comparison with standard instruments. Alternating-current potentiometers have been constructed, but they do not play the same part in measurements that the d-c potentiometers do. The a-c type is more complicated, is not generally available, has wave-form complications, and suffers from the absence of anything comparable with the standard cell for d-c. As a consequence, the general plan of prodecure is to calibrate an electrodynamometer instrument on direct current, and then to use it as a standard for the calibration of a-c instruments. This plan insures, also, that a-c effective values of voltage and current are equivalent to the d-c values, since the electrodynamometer movement depends basically on current-squared values. In calibration of an electrodynamometer on direct current, readings should be taken with both directions of current, in order to eliminate possible effects of stray fields.

The circuits for comparison of a-c ammeters or voltmeters with the standards may be similar to those described above for direct current, except that variable transformers may be used instead of rheostats, if desired, to control instrument settings. The variable transformer gives a desirable combination of smooth control and a minimum of heat dissipation in the equipment.

The usual commercial indicating instrument is calibrated at power frequency, usually 60 cps, and is guaranteed to be within its accuracy tolerance for a rather restricted range of frequency. If an instrument is to be used above the rated frequency range, the calibration should be checked at the working frequency, since several effects enter to make the indication susceptible to frequency variation. The diagram of Fig. 17-44 shows an arrangement that may be used for calibration or for direct measurements of current and voltage over a wide range of frequency.* A thermocouple is employed, but only as a comparison or "transfer" device. The unknown is connected to the thermocouple (with series resistance for voltage, or a shunt for current). A switch-over is made to a d-c source which is adjusted for the same thermocouple output. In the arrangement of Fig. 17-44(a) or (b)

---

*Reference 4.

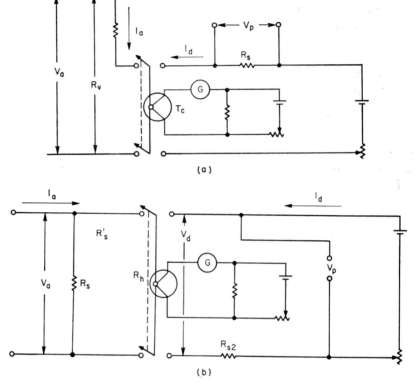

Fig. 17-44. Schematic diagram of the volt-ampere converter, used for a-c measurements over a wide range of frequencies. (a) Circuit for voltage measurement. (b) Circuit for current measurement. (From Ref. 4, by permission)

the thermocouple voltage is bucked by a d-c voltage in the galvanometer circuit to bring the galvanometer to a null when balancing against the unknown. This adjustment is kept constant after the throw-over, and the main d-c supply is adjusted to bring the galvanometer to a null again, so the d-c must produce the same thermocouple voltage (hence, the same heating effect) as the unknown a-c. The d-c is measured by a potentiometer, so the accuracy attained depends upon the potentiometer and upon the resistors of the series or shunt multiplying system. Accuracy within 0.05 per cent can be attained to 20 kc, or somewhat higher, limited mainly by the frequency characteristics of the multiplier resistors. The thermocouple is not employed as a calibrated device, and constancy of its properties is required only for the time needed to make the two balances. No dependence is placed upon deflection readings. Thermal voltage converters with series resistors of special design have been developed, permitting accuracies within a few tenths of a per cent up to about 30 megacycles.* Transfer methods are used for applications requiring high accuracy; an indicating instrument may be used for convenience if lower accuracy is adequate.

17-36. Calibration of Wattmeters

The basic wattmeter calibration is accomplished on direct current, using potentiometers to measure both voltage and current. After an initial reading is taken of E, I, and the wattmeter at a particular scale point, the directions of both current and voltage should be reversed and a second reading taken to eliminate the effect of stray fields. The potentiometer method is slow and is used only for the calibration of a standard wattmeter. A circuit for the calibration of a wattmeter on direct current is shown in Fig. 17-45. The details of the potentiometer circuits are omitted for

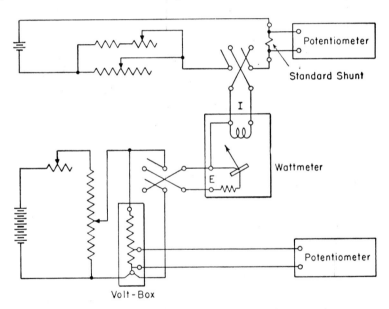

Fig. 17-45. Circuit for calibration of a wattmeter by potentiometers.

simplicity, since they take the standard form discussed in chaps. 7 and 8.

Working instruments may be calibrated by comparison with a standard wattmeter by connecting their current coils in series to a suitably regulated current supply or actual load circuit and connecting their potential circuits in parallel to a voltage supply or the load. Deflections of the instruments may be obtained by many combinations of E and I, but it is general practice

---

*Reference 6.

to set the voltage at a fixed value comparable to the instrument rating, as 115 or 100 v (230 or 200 v if the wattmeter is so rated), and then to control the deflection by variation of the current from zero to full-scale. If the wattmeter has a 500-watt scale and a 5-amp current coil, the voltage might be set at 100 v and readings obtained by varying the current from zero to 5 amp. The rating of the current coil should, of course, not be exceeded.

17-37. Oscilloscopes and Oscillographs

Oscilloscopes and oscillographs are used to obtain information about current or voltage in an electrical circuit either to supplement the information given by indicating instruments or to replace the instruments in cases in which their speed is inadequate. Oscilloscopes and oscillographs permit determination of current and voltage variations that take place very rapidly. These devices are frequently used to obtain qualitative information about a circuit, as, for example, the form of current and voltage waves or time relationships between events in a circuit. Wave form is of frequent concern for electrical generators, in welding equipment, rectifiers, and in many vacuum-tube circuits. Transient happenings may be observed or recorded. Frequency determinations may be made by means of an oscilloscope. The balancing of an a-c impedance bridge may be carried out by observation of the pattern on the screen of an oscilloscope.

Many of the purposes mentioned above are satisfied by qualitative observations of the pattern on the screen of the instrument. In some cases a true measuring function is added to satisfy a need for quantitative information. This is generally true when an oscillograph is used to determine currents and voltages during a fault (short-circuit) on a power system. Oscillographs are also used in a quantitative way for the recording by electrical means of nonelectrical quantities, such as the strain of a loaded structural member, acceleration, or pressure in the cylinder of an automobile engine. The oscillograph picture may be interpreted quantitatively in any of these cases if a calibration is conducted to establish the relationship between the deflection on the screen and the magnitude of the current or voltage that produces it.

Oscillographs are frequently used for direct visual work when a permanent record is not required, and the desired information can be obtained by inspection or measurement on the screen. In other cases a photographic record may be made by some form of camera attachment. This may be desired even in the case of recurring waves, and it becomes essential in capturing a single transient event and recording it for future study.

Oscillographs of several different principles have been constructed. The main division may be made into cathode-ray and mechanical types, each of which may be found in several forms. The cathode-ray oscillograph may be classified by form of cathode (either cold-cathode or thermionic-cathode type) or by the method of maintaining vacuum, into the sealed-off type or the pumped type. The thermionic-cathode, sealed-off type is by far the most commonly used form. The most frequently used mechanical oscillograph is the permanent-magnet moving-coil type, which may be regarded as a modification of the d'Arsonval galvanometer. Another form is the string oscillograph, in which the active member is a single wire or "string" supported between the poles of a magnet. A beam of light is projected laterally through holes in the magnet poles to cast a shadow of the string on the viewing screen or photographic film.

17-38. The Cathode-ray Oscillograph

The cathode-ray oscillograph is an exceedingly useful tool for analysis and measurement in electrical circuits. In the modern form it is rugged, simple, and convenient in operation and is capable of a wide variety of uses. An outstanding feature is the very high speed of response that permits its use to picture events even in the microsecond range. The pattern is "written" on a fluorescent screen or a photographic film by a beam of electrons, thus securing the maximum in mobility and response. Cathode-ray oscillographs are used from the low power frequencies up through the audio range and into the radio frequencies. They are used also to investigate lightning surges either in the natural form or in laboratory surge testing.

Figure 17-46 shows a cut-away view of a cathode-ray tube of the sealed-glass type. The principal components are the glass envelope or bulb, the base, the electron-gun assembly, the deflection-plate assembly and the fluorescent screen. It is the function of the electron gun to furnish a beam of electrons accelerated to a high velocity and focused into a concentrated beam that gives a small spot on the screen. The beam passes between two pairs of plates, one pair oriented in the vertical plane and the other in the horizontal. An electric field between the vertical pair of plates produces a deflection of the beam in the horizontal direction, and a field between the horizontal plates gives a deflection in the vertical direction. (Some oscillographs use deflection by means of magnetic fields transverse to the axis of the tube, produced by current-carrying coils located near the neck of the tube.) Simultaneous control of the beam in the two axes makes possible the production of any desired pattern on the screen of the tube.

Figure 17-47 is a diagram of the electron gun arrangement. Electrons are emitted from the indirectly heated thermionic cathode a, and leave through a hole in the grid, so that the strength of the beam may be controlled by the potential applied to the grid. The electrons are accelerated

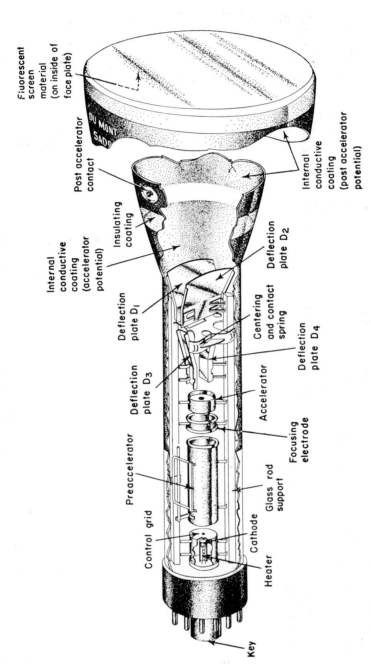

Fig. 17-46. Cut-away view of a cathode-ray tube of the electrostatic deflection type. (*Courtesy of Allen B. DuMont Laboratories, Inc.*)

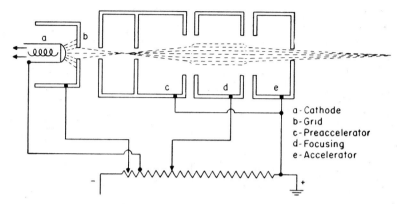

Fig. 17-47. Diagrammatic view of the electron gun of a cathode-ray tube. (*Courtesy of Allen B. DuMont Laboratories, Inc.*)

by the application of a high positive voltage to electrodes c and e, with focusing control by the intermediate electrode d. An increase of luminosity, needed in high-speed writing, may be secured by an increase of the accelerating voltage at the expense of decreased sensitivity to the voltage impressed on the deflecting plates. In some high-voltage tubes, "intensifier bands" are added around the tube between the deflecting plates and the screen and energized at high potential to give additional acceleration *after* the electrons pass the deflecting plates. The electrons can in this way be given high energy without a serious sacrifice of deflection sensitivity.

Figure 17-48 shows the general arrangement of the deflection plates of a cathode-ray tube. There are two pairs of plates, one pair oriented horizon-

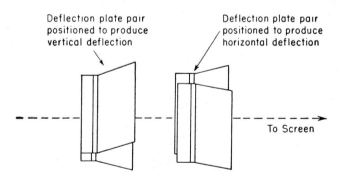

Fig. 17-48. Typical arrangement of deflection plates in a cathode-ray tube. (*Courtesy of Allen B. DuMont Laboratories, Inc.*)

tally, so that the field between them produces a vertical deflection of the electron beam and the other pair oriented vertically to give a horizontal deflection. The plates are usually named in accordance with their effect on the electron beam and not for their orientation in space. Thus, the term "horizontal-deflection plates" (with or without the hyphen) is understood to mean the pair of plates that gives a horizontal, or x-axis, deflection to the electron beam. The deflection plates are sometimes flared as shown in the figure so that the dimension of the plate in the direction of flight of the electrons may be made fairly great to achieve sensitivity and yet not have the electrons strike the plates.

The inside surface of the glass tube is given a coating of colloidal graphite in order to form a conducting layer that prevents the accumulation of charge on the walls of the tube. The coating is electrically connected to the accelerating electrode in order to return any incident electrons to the circuit and to prevent the build-up of charge that would affect the correct action of the electron beam.

The screen material of a cathode-ray tube absorbs the kinetic energy of the incident beam of electrons and gives off luminous energy. Several materials of somewhat different characteristics are available for this purpose, so selection may be made on the basis of the use to which the cathode-ray tube is to be put. The "phosphors" differ in color and "persistence," that is, the length of time that they continue to emit light after the excitation is removed. The Type P1 screen gives a green trace of medium persistence; it is efficient, very good for visual work and reasonably satisfactory photographically. The Type P2 produces a bluish-green trace with a long-persistent yellow phosphorescence that permits observation of the pattern for a period ranging from a fraction of a second to several minutes after removal of excitation, depending upon conditions. The P4 screen is well suited to television tubes because of the white color and a persistence that is sufficient to minimize flicker and yet not cause blurring of moving objects. The P5 and P11 screens give blue traces of short persistence that are good for high-speed photography because of their high actinic value. The P5, though not so efficient as the P11, is used for very high-speed writing because of its shorter persistence.

17-39. Uses of the Cathode-ray Oscillograph

Many applications of the cathode-ray oscillograph are best satisfied by having the horizontal axis represent time to a linear scale. This result may be accomplished by having the voltage between the x-axis plates build up in a linear manner with time, to move the spot at constant speed from left to right across the screen, and then to return to the starting value in a very short period of time. The x-axis voltage, if plotted against time, would present a saw-tooth pattern, with gradual straight-line build-up, and sudden drop to the original value. A voltage of this pattern may be produced by several forms of electron-tube circuits employing vacuum or gas-filled tubes. The "sweep generator" is made adjustable in frequency, so that the pattern on the screen may appear to stand still when a recurring voltage is being viewed. In addition, a synchronizing impulse is introduced into the sweep circuit from the viewed voltage to insure that the sweep voltage is locked to the exact frequency.

Cathode-ray oscillographs are frequently used for inspection of wave forms in a-c circuits, and for this purpose are very useful and convenient. They may also be used to observe a transient event in an electrical circuit if the transient can be made to recur many times a second by means of a rotating contact maker or tube-switching device. In these cases a linear time sweep is used, and the voltage to be viewed is connected to the y-axis deflecting circuit, generally through an amplifier. Commercial cathode-ray oscillographs are usually built with complete self-contained amplifiers for both vertical and horizontal deflections. Many of the amplifiers are designed for a-c operation and do not indicate steady (d-c) components; however, oscillographs are also provided with direct-coupled amplifiers that respond to d-c as well as a-c.

Interesting patterns may be produced by the application of alternating voltages to both sets of deflecting plates. If two alternating voltages of the same frequency, wave shape, and time-phase are applied to x-axis and y-axis plates, the result is a straight, inclined line on the screen — the line is at 45° if the same voltage is applied to the two circuits and the over-all sensitivity of deflection plates and amplifiers is made the same in both directions. However, if the voltages are of the same frequency, sinusoidal, and at 90° time-phase, the pattern is an ellipse (a circle if the deflecting effects are equal in the two axes). An ellipse with *inclined* major axis results from two sinusoidal voltages of the same frequency but with a phase difference between zero and 90°. The phase angle may be estimated by measurements of the pattern on the screen. One method is by measurement of horizontal and vertical intercepts of the ellipse, as measured along the x- and y-axes; this measurement is difficult to achieve with accuracy, however, particularly with narrow inclined figures. Figure 17-49 shows a preferable method based upon the easily measured lengths of the major and minor axes of the inclined ellipse. For convenience, the gains are adjusted to fit the ellipse within a square, as marked by co-

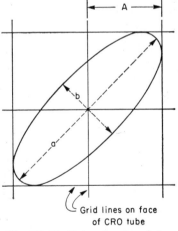

Grid lines on face
of CRO tube

Fig. 17-49. Determination of phase difference of two sinusoidal voltages of same frequency by the pattern on the fact of cathode-ray tube.

ordinate lines on the face of the oscillograph. The dimensions a and b are measured. Then, from a study of the equation of the ellipse it can be shown that the phase angle, ϕ, between the voltages producing the horizontal and vertical deflections is given by,

$$\sin \phi = \frac{ab}{4A^2} \qquad \text{or} \qquad \tan \frac{\phi}{2} = \frac{b}{a}$$

(The amplifiers in an oscillograph produce some phase shift; the phase difference indicated on the face of the tube represents the phase difference of the input voltages only if the phase shifts are the same for horizontal and vertical amplifiers. Equality of phase shifts can be investigated by tying the two inputs to the same source, to see whether the pattern is a single line or an ellipse.)

Two nearly equal frequencies may be compared very precisely by the change of pattern on the screen. With a small difference in frequency, the two voltages appear to drift slowly with respect to the phase difference between them. From a beginning when the voltages are in phase and the pattern is a straight line, the line opens to an elongated ellipse, then to a circle (for equal magnitudes), closes to an ellipse, and then to a straight line with inclination opposite to the original, for a drift of one-half cycle. At the completion of one cycle of difference, the pattern has returned to the starting arrangement. If, for example, two oscillators are being compared, one with a frequency of 1000 cycles per second, and the other 1001, the picture on the screen completes a cycle of change in one second. If the frequency of one oscillator can be adjusted so that several seconds are required for a cycle of change of the pattern, then the two frequencies are known to be equal within a fraction of a cycle per second, which is a small percentage difference in terms of the 1000-cycle frequency.

Voltages of frequencies that are different, but related by a simple integral ratio, give patterns of the type shown in Fig. 17-50. These are

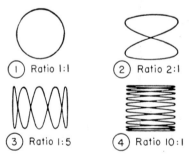

Fig. 17-50. Lissajous figures. (*Courtesy of Allen B. DuMont Laboratories, Inc.*)

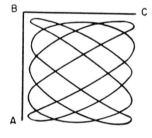

Fig. 17-51. Method of determining the ratio of frequencies in a Lissajous figure. (*Courtesy of Allen B. DuMont Laboratories, Inc.*)

called Lissajous figures. In (2) the spot is swept through two cycles horizontally while going through one vertically. In (3) the ratio is one cycle horizontally for five vertically, and in (4) the ratio of horizontal to vertical frequencies is ten to one. Figure 17-51 shows a complex pattern secured when the ratio is other than $n : 1$ or $1 : n$. The ratio of frequencies may be determined from the observation that five loops touch the vertical line AB, and three the horizontal line BC. The spot is thus being swept back and forth horizontally five cycles for three cycles vertically. The ratio of frequency on the horizontal axis to the frequency on the vertical axis is accordingly $5 : 3$. If one frequency drifts slightly with respect to the other, the pattern appears to rotate, or "barrel" from the integral ratio.

Sometimes it is desired to make a photographic record of a standing pattern on the screen of a cathode-ray tube. This may be done by directing a camera at the screen, and making an exposure in the usual manner. In addition, special camera devices that fit directly on the oscillograph are available for this purpose. The recording of a transient voltage of considerable duration must usually be carried out in a different manner. The transient voltage is applied to one set of plates, and the other pair is not excited. If the transient is applied to the x-axis plates, the spot is moved in a horizontal direction across the screen. The time co-ordinate is given by moving a strip of photographic film past the screen in a vertical direction. The film is contained in a holder provided with suitable drive mechanism and having a horizontal slit to admit the light of the moving spot. Some oscillographs for the recording of fast phenomena (such as lightning surges) have been constructed of metal, with provision for opening so that the film may be placed in the evacuated chamber where the electron beam may impinge directly on it. This construction requires associated pumping equipment to restore the vacuum in the tube after the film is introduced. Usually enough film can be placed in the tube at one time to provide for several exposures, and provision is made for moving successive frames of film into position for exposure by means of a vacuum-tight device that extends through the tube envelope.

One respect in which the cathode-ray oscillograph is at a disadvantage as compared with the mechanical type is in provision for multiple traces. The usual cathode-ray tube is a one-trace device. However, it is often important to inspect or record a number of quantities simultaneously. In the inspection of a simple circuit it is often desired to find the relationship of current and voltage, thus requiring two curves on the screen. In other applications a greater number is desired, as for example, in recording faults in a three-phase power system or in the multiple-point recording from strain gages. One method of showing two traces on the screen is to use an electron-tube switching arrangement so that the two voltages are connected alternately to the y-axis deflecting plates. If the frequency is high enough, the two curves appear stationary because of the persistence of vision of the eye, and to persistence of luminosity of the screen material. Another solution is to build two, three, or even six electron guns into one tube. While the multiple-gun construction is possible, it increases the complexity and cost of the tube very greatly, particularly if the number is made greater than two.

17-40. Mechanical Oscillographs

The most common form of mechanical oscillograph is similar in principle to the d'Arsonval galvanometer, but the construction is modified to give rapid response. As in the galvanometer, the magnetic field is provided by a permanent magnet. (An electromagnet was used in some instruments of older design before the improved permanent-magnet materials were available.) The moving coil in many oscillographs consists of a single loop of wire (the wire being flattened to the form of a ribbon), with the two wires close together. Deflection is observed by means of a beam of light reflected from a small mirror cemented to the moving coil. The construction is indicated in Fig. 17-52. In (a) the oscillograph galvanometer is shown in

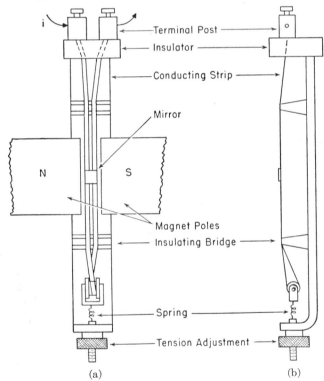

Fig. 17-52. Construction of a permanent-magnet, moving-coil type of oscillograph. (a) Front view. (b) Side view, magnet removed.

place between the poles of the magnet, with the mirror at the center of the active part of the coil. A spring is provided to keep the wires under tension, so that the system is sufficiently rigid to have a high speed of response. If current is passed through the coil as indicated, one wire is pushed toward the observer and the other away from him, thus producing a torque on the coil. The coil conductors are supported by the two insulating bridges, so that the moving part of the coil is only the part between bridges. The mirrors on some oscillograph galvanometers measure 0.015 in. by 0.060 in. (smaller mirrors are sometimes used), which gives an idea of the pains that are taken to keep the mass and inertia of the moving system to a low value. The fact that the two wires forming the sides of the coil are so close together gives a low moment of inertia about the axis of rotation.

Figure 17-53 indicates the optical arrangement used in an oscillograph. A beam of light is thrown upon the mirror of the oscillograph galvanometer,

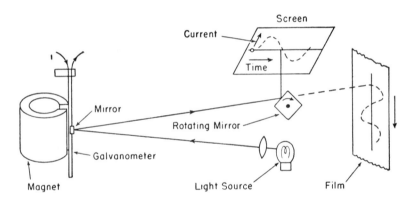

Fig. 17-53. Diagram of the optical system of an oscillograph.

and the return-beam is directed to a screen by means of a rotating mirror for visual observation or to a moving film for photographic recording. Only phenomena that repeat regularly a number of times each second (20 to 60 or more) can be viewed on the screen. If, in such cases, the rotating mirror is driven synchronously with the change of the quantity that is being studied, the light spot traces over the same pattern on the screen many times a second and gives to the eye the impression of a continuous curve whose ordinate is proportional to the current flowing through the galvanometer, and whose abscissa represents time. If the rotating mirror of Fig. 17-53 is removed, the light spot may be allowed to strike a film (or sensitized paper) that is enclosed in a suitable container and driven in a direction at right angles to the motion of the light spot. In this way a record of recurring or transient phenomena may be made.

A very important matter in the design of a recording system is the provision of suitable damping, for otherwise the observed pattern may be distorted to the extent that it is actually misleading. Damping may be obtained by placing the galvanometer in a tight enclosure that is filled with oil. A window is provided for passage of the light beam to the galvanometer mirror and return. The oil must be clear and colorless for optical reasons and must be of the correct viscosity to give the desired amount of damping. The matter of damping will be studied in the next section.

Some oscillographs are provided with pivoted coils and pen arms for direct ink recording, instead of using the light-beam arrangement described above. The direct ink record is convenient, as it permits immediate viewing of the record and does not require photographic processing. The added inertia of the pointer, however, limits the frequency of response, so that oscillographs of this type can be used only for frequencies materially less than 100 cycles per second for accurate recording. The pen-arm type of instrument may be regarded either as a slow oscillograph or as a rapid recording voltmeter or ammeter, as there is no sharp dividing line between the two classes of instruments. The usual d-c recording voltmeter or ammeter, as used in power systems and elsewhere, does not possess satisfactory accuracy for frequencies above approximately one cycle per second. Some oscillographs of the pen-arm type are provided with heat-sensitive paper and a heated stylus, or else with paper sensitive to an electrical discharge from the stylus, as a means of avoiding the use of ink, with its attendant difficulties of blotting and clogging.

17-41. Features of Mechanical Oscillographs

One outstanding feature of the mechanical oscillograph is the ease of providing for multiple-element recording. Even the small portable oscillographs usually have at least two galvanometers, arranged side by side, to project the light spots on the same viewing screen or film. The galvanometers may be provided with adjustable shunts for current measurements or with series resistors for voltage. Larger oscillographs may have three, six, or more galvanometers, forming an instrument that is very useful for the study of currents and voltages in a three-phase power system under fault conditions or for other applications in which it is desired to study the relationships of several different quantities under steady or transient conditions. Some oscillographs designed for strain-gage testing of structures, airplanes, etc., have 12 to 24 galvanometers.

The usual oscillograph galvanometer is of the type described above, with a permanent-magnet field. Special watt elements are available which have a soft-iron frame and a field coil, so that they may be used for rapidly changing voltage and current, just as a wattmeter is used for steady readings. The watt elements, as the name implies, may be used for the direct recording of instantaneous power in power-system studies.

Oscillographs may be adapted for either short- or long-time recording. Short records are made on a strip of film wrapped around a small drum mounted in a container with a slit that matches the back-and-forth motion of the light spots. Provision is usually made for a trip arrangement and shutter that admit light to the film for only one revolution of the drum. A camera for long records consists of a supply roll of film, a drive mechanism, and a take-up reel. In either type of film drive, provision is made for adjustment of the speed of the film to suit the type of record that is being made.

Damping and speed of response are two of the most important matters to consider in the application of an oscillograph to a measurements problem. The equation and discussion of the behavior of a galvanometer, given in chap. 5, may be applied equally well to the oscillograph. Figure 17-54

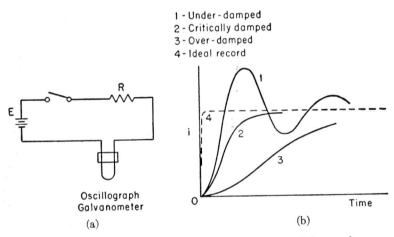

Fig. 17-54. Effect of oscillograph damping on the recorded curve for a simple transient.

illustrates the effect that the characteristics of the oscillograph galvanometer have upon the observed record for a simple circuit change. The circuit, as shown in (a), consists of a battery, a galvanometer element, and a resistor that is as nearly non-inductive as possible. The current in this circuit jumps almost instantly from zero to E/R when the switch is closed. The oscillograph is not able, however, to reproduce the change so rapidly, so that the record, if the time scale is drawn out sufficiently, looks somewhat as in (b). Curve 1 is the record given in case the galvanometer is underdamped, showing the oscillations of the *coil* before it settles to its new position. The overdamped Curve 3 represents unnecessarily sluggish behavior. An approach to the critically damped case, Curve 2, is desirable, as being the closest possible approach to a perfect recording of the true happenings in the circuit. An inexperienced observer may look at an oscillograph record such as Curve 1 and decide that the current in the circuit is oscillating — which is not true for the present circuit. An experienced observer, acquainted with the theory of galvanometer behavior, will study the oscillographic record of a new or unknown event with care, realizing that certain characteristics of the record, such as "wiggles" where a sudden offset of the curve occurs, may be caused by the response of the oscillograph galvanometer and do not represent happenings in the electrical circuit.

Another form of information regarding the trueness of the record given by an oscillograph is disclosed by a study of its response to steady-state alternating currents of different frequencies. This study may be made directly in terms of the mechanical parameters of the galvanometer or it may be made by analogy of the mechanical system to the electrical circuit consisting of R, L, and C in series. The analogy is appropriate, since the differential equation for the R, L, C circuit, written in terms of the charge, q, is identical in form to the galvanometer equation, as written in chap. 5. The analogy associates the following pairs of quantities in the electrical and mechanical equations, respectively: (q,θ), (L,J), (R,D), $(1/C,S)$, (applied voltage, applied torque). The electrical circuit is shown in Fig. 17-55(a),

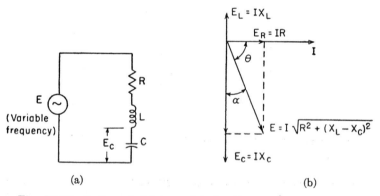

(a)   (b)

Fig. 17-55. R, L, and C in a series circuit with sinusoidal applied voltage. (This circuit is used in analogy to the mechanical behavior of an oscillograph galvanometer.) (a) Circuit. (b) Phasor diagram.

and the phasor diagram in (b) for a low frequency — that is, for a frequency much below the resonant value. The following equations may be derived by ordinary a-c circuit theory:

$$\frac{E_c}{E} = \frac{X_c}{\sqrt{R^2 + (X_L - X_c)^2}} \qquad\qquad (9)$$

and
$$\alpha = 90° + \tan^{-1} \frac{X_L - X_C}{R} \tag{10}$$

The equations are more convenient for the present application if they are "normalized" to express the various quantities in terms of ratios based upon the resonant condition. The letter r will be used as a subscript to refer to the value of the quantity at resonance, and the values of the reactances at resonance will be used as a base. Resonance is defined by the equality of the reactances, thus

$$X_{Lr} = X_{Cr} = X_r \tag{11}$$

Let F be the normalized frequency, that is

$$F = \frac{f}{f_r} = \frac{\text{Operating frequency}}{\text{Resonant frequency}} \tag{12}$$

The reactances for any frequency may now be written in the following way:

$$X_L = FX_r \tag{13}$$

$$X_C = \frac{1}{F} X_r \tag{14}$$

The resistance of the circuit may be expressed in terms of X_r:

$$d = \frac{R}{2X_r} \tag{15}$$

The factor 2 is introduced in (15) to bring the definition of relative damping in the steady-state sinusoidal case into agreement with the critical damping of the transient case, as stated in mechanical units in (24) and (26) of chap. 5. Thus d of (15) is directly comparable with γ used in chap. 5. Equations (9) and (10) may be reduced to the following forms by substitution of (12) to (15), inclusive:

$$\frac{E_C}{E} = \frac{1}{\sqrt{4d^2F^2 + (F^2 - 1)^2}} \tag{16}$$

$$\alpha = 90° + \tan^{-1} \frac{F^2 - 1}{2dF} \tag{17}$$

The ratio of E_C to E is of interest in the mechanical analogy, as it represents the ratio of the spring torque to the applied torque, and the spring torque is proportional to the deflection θ. For the oscillograph galvanometer at very low frequencies, the damping and inertia effects are negligible, and the entire applied torque acts against the spring; similarly $E_C = E$ for the electrical circuit at a very low frequency. If an alternating torque of constant amplitude but varying frequency is developed in the oscillograph galvanometer by the passage of an alternating current of constant amplitude but varying frequency, the amplitude of the oscillograph record will vary with frequency in accord with the same law of variation as E_C/E in the electrical circuit. The curve of E_C/E plotted against frequency may thus be used as a relative-response curve for the oscillograph, with zero frequency representing a very low frequency calibration, or practically, a static calibration. A family of response curves is shown in Fig. 17-56, with the parameter "d" representing the resistance

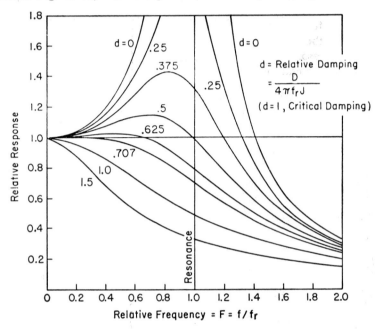

Fig. 17-56. Relative-response curves for steady-state sinusoidal excitation for R, L, C electrical circuit, or D, J, S mechanical system.

(damping) of the system. For zero damping the amplitude builds up as the resonant frequency is approached, going to infinity at resonance, and decreasing toward zero for higher frequencies. At $F = 0.2$, however, the error of overshoot is only 4 per cent [$1/(1 - F^2)$ for $d = 0$]. Note that the curve is nearly flat for a considerable frequency range for $d = 1/\sqrt{2}$, and that this is approximately the best condition for uniformity of a-c response. The parameter $d = 1$ corresponds with the case of critical damping, or $R = 2\sqrt{L/C}$ in the electrical circuit. A damping slightly less than the critical value is generally satisfactory, as it gives fast response and only slight overshooting on step functions such as Fig. 17-54(b). We may therefore conclude that oscillograph damping given by a value of d between $1/\sqrt{2}$ and 1 is satisfactory for transient and steady-state a-c conditions. The damping of an oscillograph may be determined by a test as indicated in Fig. 17-54.

Figure 17-57 shows curves for the angle α in the electrical circuit of Fig. 17-55. This quantity is of interest in the oscillograph case as it represents the angle by which the record on the screen or film lags behind the actual current flowing in the galvanometer. For zero damping there is a sudden change from $\alpha = 0$ to $\alpha = 180°$ for frequencies just below and just above the resonant point. The addition of damping causes α to change much more gradually. The presence of the angle α should be kept in mind in examining oscillographic records, as it represents an error — that is, it is a failure of the observed curve to show correctly the happenings in the system in which measurements are being made.

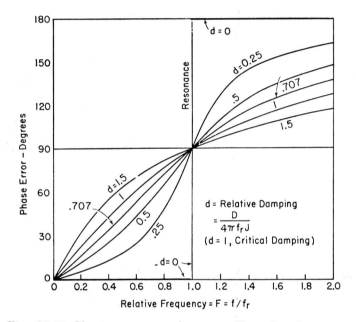

Fig. 17-57. Phase-error curves for an oscillograph galvanometer carrying sinusoidal currents.

The characteristics of oscillograph galvanometers are often given in terms of the resonant frequency. This does not imply that the oscillograph can be used to record currents accurately up to this frequency. The foregoing response study has indicated that an oscillograph galvanometer may be used, depending on the damping, to a frequency of 0.2 to 0.4 of the resonant frequency, or possibly somewhat higher by special attention to the damping, by the use of frequency-correction curves or by the application of special equalizing circuits. Commercial oscillograph galvanometers are available with resonant frequencies up to 4000 to 5000 cycles per second, and somewhat higher by special construction.

There is a definite connection between the resonant frequency of an oscillograph and the current sensitivity (which is usually expressed as the current required per unit deflection on the oscillograph screen). The principal constants that enter into the design of an oscillograph galvanometer may be seen by reference to Fig. 17-52 to be (1) the distance between the bridges, (2) the size and stiffness of the conducting strip, (3) the tension on the stringing, and (4) the strength of the magnetic field. Greater deflection sensitivity can be obtained by placing a greater distance between bridges and using thinner stringing with less tension. These changes work, however, to decrease the resonant frequency, as they give a more compliant (less rigid) system. It is practically axiomatic that greater sensitivity is obtained at a sacrifice in the speed of response. Sensitive, low-speed galvanometers are very useful in some spplications, but for high-speed recording a fast galvanometer must be used, with an attendant increase in the operating power.

REFERENCES

1. Barabutes, T., and W. J. Schmidt, "Principles and Considerations in the Design of a Hall Multiplier," *Trans. A.I.E.E.* Paper No. 59–875, (1959).

2. Thomander, V. S., and R. C. MacIndoe, "Taut Band Suspensions for 250 Degree Instruments," *Trans. A.I.E.E.*, Paper No. 59–159, (1959).

3. Smith, U. L., "A Transducer Type Frequency Meter," *Trans. A.I.E.E.*, p. 497 (Sept. 1959).

4. Hermach, F. L., and E. S. Williams, "A Wide-Range Volt-Ampere Converter for Current and Voltage Measurements," *Trans. A.I.E.E.*, Paper No. 59–161, (1959).

5. Harris, F. K., *Electrical Measurements*, New York: John Wiley and Sons, Inc., 1952.

6. Hermach, F. L., "Thermal Converters for Accurate Voltage Measurements to 30 Megacycles," *Trans. A.I.E.E.*, Paper No. 59–876, (1959).

7. McAninch, O. G., "Thermocouple-Type Ammeters for Use at Very High Frequencies," *Trans. A.I.E.E.*, p. 241 (July 1954).

APPENDIX I

SENSITIVITY OF WHEATSTONE BRIDGE

The usual statement concerning the relative sensitivity of a Wheatstone bridge for the two arrangements of battery and galvanometer is to the effect that the sensitivity is greater if the galvanometer is connected from the junction of the two high-resistance arms to the junction of the two low-resistance arms. This is true for the same voltage applied to the bridge in both cases, but does not take into account the greater heating of one of the bridge members in one case as compared with the other.

The following study shows the relation of sensitivities if comparison is made for the same *maximum power* in the bridge arm subjected to the greatest heating. This seems a more important comparison, as the heating must be considered carefully to avoid inaccuracy or actual damage to the bridge members. The voltage is generally not important apart from considerations concerning heating, as it is usually easy to add another battery.

The notation shown below uses 1 ohm for the lowest resistance arm, and m and n, respectively, for the two adjacent arms. This is no limitation on the generality of the results, as the value R could be taken for the low resistance arm and mR and nR for adjacent arms. The same equations would be obtained as below except for additional R factors throughout, so the relative results would be the same and there would be the additional nuisance of writing the R factors. Conversely, any actual bridge may be reduced to the form shown by dividing the resistance of each member (including the galvanometer) by the resistance of the lowest-valued arm. The solution is obtained by the Thevenin method outlined in chap. 4.

This study is directed toward finding the bridge arrangement that gives the greatest galvanometer deflection. The term "matching" as used here refers to matching the galvanometer to the bridge for the purpose of maximum power transfer. This is distinct from the idea of selection of the components so that the bridge serves to give critical damping to the galvanometer. The importance of the damping criterion depends, as discussed in chap. 5, on whether galvanometer deflections are to be read in order to interpolate beyond the last dial on the bridge; if this should not be so, the damping given by the bridge is of little moment, as the bridge key is closed only momentarily. From the damping standpoint, the bridge gives the choice of the resistances as seen from the two diagonals; these are the R_{01} and R_{02} of the following analysis.

1. Comparison for Equal Maximum Power

Assumptions:

W = maximum allowable watts for any bridge arm

$m > n > 1$

d = small unbalance in 1-ohm arm $(d \ll 1)$

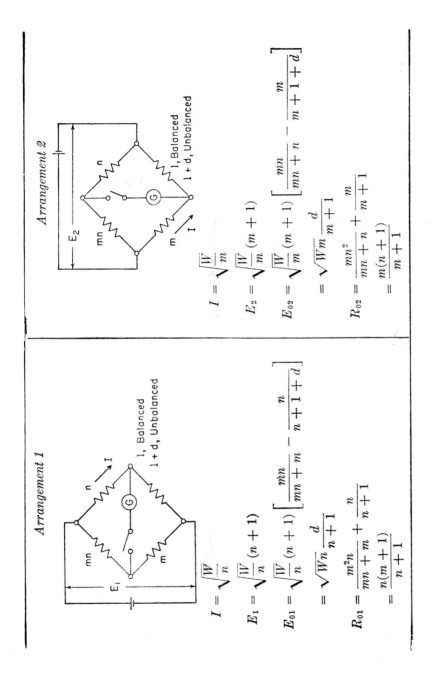

Arrangement 1

1, Balanced
1 + d, Unbalanced

$$I = \sqrt{\frac{W}{n}}$$

$$E_1 = \sqrt{\frac{W}{n}}\,(n+1)$$

$$E_{01} = \sqrt{\frac{W}{n}}\,(n+1)\left[\frac{mn}{mn+m} - \frac{n}{n+1+d}\right]$$

$$= \sqrt{Wn}\,\frac{d}{n+1}$$

$$R_{01} = \frac{m^2 n}{mn+m} + \frac{n}{n+1}$$

$$= \frac{n(m+1)}{n+1}$$

Arrangement 2

1, Balanced
1 + d, Unbalanced

$$I = \sqrt{\frac{W}{m}}$$

$$E_2 = \sqrt{\frac{W}{m}}\,(m+1)$$

$$E_{02} = \sqrt{\frac{W}{m}}\,(m+1)\left[\frac{mn}{mn+n} - \frac{m}{m+1+d}\right]$$

$$= \sqrt{Wm}\,\frac{d}{m+1}$$

$$R_{02} = \frac{mn^2}{mn+n} + \frac{m}{m+1}$$

$$= \frac{m(n+1)}{m+1}$$

CASE (a) $r_o = C$

$$I_{o_1} = \frac{E_{01}}{R_{01}} = \sqrt{\frac{W}{n}} \, \frac{d}{m+1}$$

$$I_{o_2} = \frac{E_{02}}{R_{02}} = \sqrt{\frac{W}{m}} \, \frac{d}{n+1}$$

$$\frac{I_{o_1}}{I_{o_2}} = \sqrt{\frac{m}{n}} \, \frac{n+1}{m+1}$$

$$\therefore I_{o_1} < I_{o_2}, \text{ for } m > n$$

Arrangement 1 | *Arrangement 2*

CASE (b)

$$r_o = \sqrt{R_{01}R_{02}} = \text{Geometric mean of } R_{01} \text{ and } R_{02}$$

$$r_o = \sqrt{mn}$$

Arrangement 1

$$I_{o_1} = \frac{\sqrt{Wn}\,\dfrac{d}{n+1}}{\dfrac{n(m+1)}{n+1} + \sqrt{mn}}$$

$$= \frac{\sqrt{W}\,d}{\sqrt{n}\,(m+1) + \sqrt{m}\,(n+1)}$$

Arrangement 2

$$I_{o_2} = \frac{\sqrt{Wm}\,\dfrac{d}{m+1}}{\dfrac{m(n+1)}{m+1} + \sqrt{mn}}$$

$$= \frac{\sqrt{W}\,d}{\sqrt{m}\,(n+1) + \sqrt{n}\,(m+1)}$$

$$\therefore I_{o_1} = I_{o_2}$$

Matching galvanometer selected for each bridge

$$r_{g_1} = R_{01}$$

$$I_{g_1} = \frac{E_{01}}{2R_{01}} = \frac{1}{2}\sqrt{\frac{W}{n}}\,\frac{d}{m+1}$$

$$I_{g_1}^2 r_{g_1} = \frac{1}{4}\frac{W}{n}\,\frac{d^2}{(m+1)^2}\,\frac{n(m+1)}{n+1}$$

$$= \frac{W}{4}\,\frac{d^2}{(m+1)(n+1)}$$

$$r_{g_2} = R_{02}$$

$$I_{g_2} = \frac{E_{02}}{2R_{02}} = \frac{1}{2}\sqrt{\frac{W}{m}}\,\frac{d}{n+1}$$

$$I_{g_2}^2 r_{g_2} = \frac{1}{4}\frac{W}{m}\,\frac{d^2}{(n+1)^2}\,\frac{m(n+1)}{m+1}$$

$$= \frac{W}{4}\,\frac{d^2}{(m+1)(n+1)}$$

$$\therefore \text{ Same power to each galvanometer}$$

Summary:

(1) The two arrangements are equally good, for the same W, if the galvanometer in each arrangement matches its bridge, and if each galvanometer is equally efficient from a power standpoint. This is Case (c) above.

(2) For the same galvanometer used in either bridge

$$
\begin{aligned}
I_{g_1} < I_{g_2}, \qquad &\text{if } r_g < \sqrt{mn} \qquad &&\text{Case (a)}\\
I_{g_1} = I_{g_2}, \qquad &\text{if } r_g = \sqrt{mn} \qquad &&\text{Case (b)}\\
I_{g_1} > I_{g_2}, \qquad &\text{if } r_g > \sqrt{mn}
\end{aligned}
$$

Note: Arrangement 1 is more sensitive *only* if the galvanometer resistance is greater than $\sqrt{mn}$ times the low-resistance arm.

2. Comparison for Equal Applied Voltages

This comparison is included here, first because it is the one usually met, and second, to show the relationship to Comparison I, above.

$$\text{Let } E_1 = E_2 = E$$

$$E_{01} = E\left(\frac{mn}{mn+m} - \frac{n}{n+1+d}\right)$$

$$= \frac{End}{(n+1)^2}$$

$$I_{g1} = \frac{E_{01}}{R_{01}+r_g}$$

$$= \frac{End}{n(m+1)(n+1)+r_g(n+1)^2}$$

$$E_{02} = E\left(\frac{mn}{mn+n} - \frac{m}{m+1+d}\right)$$

$$= \frac{Emd}{(m+1)^2}$$

$$I_{g2} = \frac{E_{02}}{R_{02}+r_g}$$

$$= \frac{Emd}{m(m+1)(n+1)+r_g(m+1)^2}$$

$$\frac{I_{g1}}{I_{g2}} = \frac{n[m(m+1)(n+1)+r_g(m+1)^2]}{m[n(m+1)(n+1)+r_g(n+1)^2]} = \frac{1+r_g\dfrac{m+1}{m(n+1)}}{1+r_g\dfrac{n+1}{n(m+1)}}$$

Summary:

$$\frac{I_{g1}}{I_{g2}} = 1 \quad \text{if } r_g = 0; \qquad \frac{I_{g1}}{I_{g2}} = 1 \quad \text{if } m = n, \text{ any } r_g$$

$$\frac{I_{g1}}{I_{g2}} > 1 \quad \text{if } r_g > 0 \text{ and } m > n$$

APPENDIX II

WYE-DELTA TRANSFORMATIONS

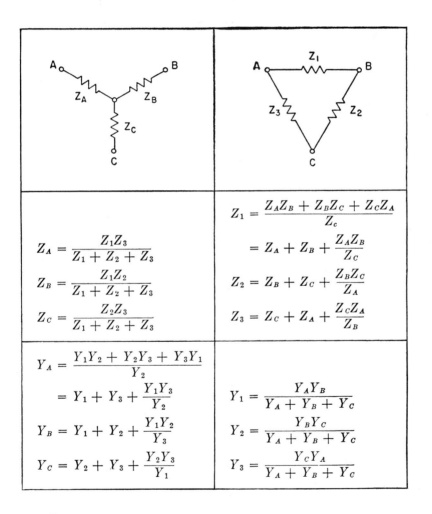

$$Z_1 = \frac{Z_A Z_B + Z_B Z_C + Z_C Z_A}{Z_c}$$

$$= Z_A + Z_B + \frac{Z_A Z_B}{Z_C}$$

$$Z_A = \frac{Z_1 Z_3}{Z_1 + Z_2 + Z_3}$$

$$Z_B = \frac{Z_1 Z_2}{Z_1 + Z_2 + Z_3}$$

$$Z_2 = Z_B + Z_C + \frac{Z_B Z_C}{Z_A}$$

$$Z_C = \frac{Z_2 Z_3}{Z_1 + Z_2 + Z_3}$$

$$Z_3 = Z_C + Z_A + \frac{Z_C Z_A}{Z_B}$$

$$Y_A = \frac{Y_1 Y_2 + Y_2 Y_3 + Y_3 Y_1}{Y_2}$$

$$= Y_1 + Y_3 + \frac{Y_1 Y_3}{Y_2}$$

$$Y_1 = \frac{Y_A Y_B}{Y_A + Y_B + Y_C}$$

$$Y_B = Y_1 + Y_2 + \frac{Y_1 Y_2}{Y_3}$$

$$Y_2 = \frac{Y_B Y_C}{Y_A + Y_B + Y_C}$$

$$Y_C = Y_2 + Y_3 + \frac{Y_2 Y_3}{Y_1}$$

$$Y_3 = \frac{Y_C Y_A}{Y_A + Y_B + Y_C}$$

SUMMARY OF A-C BRIDGE CIRCUITS

	Name	Circuit	Measures (Principal Component)	Balance Equations
1	Comparison (series constants)		L or C	$R_x = R_s \dfrac{R_2}{R_1}$ If inductive, $L_x = L_s \dfrac{R_2}{R_1}$ If capacitive, $C_x = C_s \dfrac{R_1}{R_2}$
2	Comparison (Parallel constants)		C (or L)	$C_x = C_1' - C_1$ $G_x = \dfrac{R_2 - R_2'}{R_2 R_2'}$

3	Maxwell*		L	$L_x = R_2 R_3 C_1$ $R_x = \dfrac{R_2 R_3}{R_1}$
4	Hay*		L	$L_x = \dfrac{R_2 R_3 C_1}{1 + \omega^2 C_1^2 R_1^2}$ $R_x = \dfrac{\omega^2 C_1^2 R_1 R_2 R_3}{1 + \omega^2 C_1^2 R_1^2}$
5	Owen*		L	$L_x = R_2 R_3 C_1$ $R_x = R_2 \dfrac{C_1}{C_3}$

*See chap. 13 for the effect of residuals in this circuit.

	Name	Circuit	Measures (Principal Component)	Balance Equations
6	Inductance of Iron-cored Coils with Direct Current		L	$L_x = R_2 R_3 C_1$ $R_x = \dfrac{R_2 R_3}{R_1}$ (May use Hay Circuit)
7	Anderson		L	$L_x = C R_2 \left(R_3 + R_5 + \dfrac{R_3 R_5}{R_1} \right)$ $R_x = \dfrac{R_2 R_3}{R_1}$

	Stroud and Oates			Same
8	Schering*		C	$C_x = C_s \dfrac{R_3}{R_4}$ $R_x = R_4 \dfrac{C_3}{C_s}$ $D_x = \omega C_3 R_3$
9	Resonance*		L or C (f known) R of coil f (L and C known)	$X_L = X_c$, or $L \cdot C = \dfrac{1}{\omega^2}$ $f = \dfrac{1}{2\pi \sqrt{LC}}$

	Name	Circuit	Measures (Principal Component)	Balance Equations
10	Wien, or R-C Frequency Bridge		f	$f = \dfrac{1}{2\pi \sqrt{R_3 R_4 C_3 C_4}}$ $\dfrac{R_1}{R_2} = \dfrac{R_3}{R_4} + \dfrac{C_4}{C_3}$
11	Carey Foster		C	$C = \dfrac{M}{R_2 R_3}$ $R_1 = \dfrac{R_2(L - M)}{M}$
	Heydweiller		M	$M = R_2 R_3 C$ $L = CR_3(R_1 + R_2)$

| 12 | Heaviside | | M | $$M = \frac{R_2 L_3 - R_1 L_4}{R_1 + R_2}$$ $$R_1 R_4 = R_2 R_3$$ |
| 13 | Campbell | | M | Detector on (1): $$\frac{R_3}{R_4} = \frac{R_1}{R_2} = \frac{L_1}{L_2}$$ Detector on (2): $$\frac{M_1}{M_2} = \frac{R_3}{R_4}$$ |

APPENDIX IV
PROPERTIES OF THE PROBABILITY CURVE*

$$y = \frac{1}{\sqrt{2\pi}} \epsilon^{-t^2/2}$$

t = normalized abscissa = x/σ,

where σ = standard deviation

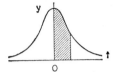

t	y	Area	t	y	Area
0	.3989	0	1.5	.1295	.4332
.05	.3984	.0199	1.55	.1200	.4394
.1	.3970	.0398	1.6	.1109	.4452
.15	.3945	.0596	1.65	.1023	.4505
.2	.3910	.0793	1.7	.0940	.4554
.25	.3867	.0987	1.75	.0863	.4599
.3	.3814	.1179	1.8	.0790	.4641
.35	.3752	.1368	1.85	.0721	.4678
.4	.3683	.1554	1.9	.0656	.4713
.45	.3605	.1736	1.95	.0596	.4744
.5	.3521	.1915	2.0	.0540	.4772
.55	.3429	.2088	2.1	.0440	.4821
.6	.3332	.2257	2.2	.0355	.4861
.65	.3230	.2422	2.3	.0283	.4893
.6745	.3178	.2500	2.4	.0224	.4918
			2.5	.0175	.4938
.7	.3123	.2580	2.6	.0136	.4953
.75	.3011	.2734	2.7	.0104	.4965
.8	.2897	.2881	2.8	.0079	.4974
.85	.2780	.3023	2.9	.0060	.4981
.9	.2661	.3159	3.0	.0044	.4987
.95	.2541	.3289	3.1	.0033	.4990
1.0	.2420	.3413	3.2	.0024	.4993
1.05	.2299	.3531	3.3	.0017	.4995
1.1	.2179	.3643	3.4	.0012	.4997
1.15	.2059	.3749	3.5	.0009	.4998
1.2	.1942	.3849	3.6	.0006	.4998
1.25	.1827	.3944	3.7	.0004	.4999
1.3	.1714	.4032	3.8	.0003	.4999
1.35	.1604	.4115	3.9	.0002	.5000
1.4	.1497	.4192	4.0	.0001	.5000
1.45	.1394	.4265			

*For an extended table, see "Tables of Probability Function, Vol. II," Federal Works Agency, Work Projects Administration for City of New York, 1942.

Index

A

B